£60.00

Molecular and Cell Biology of Yeasts

Molecular and Cell Biology of Yeasts

Edited by
E.F. WALTON
Group Leader, Eukaryotic Molecular Biology
Celltech Ltd

and

G.T. YARRANTON
Head of Molecular Genetics
Celltech Ltd

Blackie
Glasgow and London

Van Nostrand Reinhold
New York

Blackie and Son Ltd
Bishopbriggs, Glasgow G64 2NZ
7 Leicester Place, London WC2H 7BP

Published in the USA and Canada by
Van Nostrand Reinhold
115 Fifth Avenue
New York, New York 10003

Distributed in Canada by
Macmillan of Canada
Division of Canada Publishing Corporation
164 Commander Boulevard
Agincourt, Ontario M1S 3C7

British Library Cataloguing in Publication Data

Molecular and cell biology of yeasts.
1. Yeasts
I. Walton, E.F. II. Yarranton, G.T.
589.2′33

ISBN 0-216-92501-0

For the USA and Canada

ISBN 0-442-20711-5
Library of Congress Catalog Card Number 88-215

Phototypesetting by Thomson Press (India) Ltd, New Delhi
Printed in Great Britain by Bell & Bain (Glasgow) Ltd.

Preface

Our knowledge of the molecular and cell biology of yeast has advanced rapidly in recent years. The accumulation of this knowledge has not only promoted industrial applications but has also aided our fundamental understanding of cellular processes in lower and higher eukaryotes. Advances in the subject have not been reviewed for some time, and so we hope that this book is timely. We have aimed to cover both the fundamentals and the complexities of molecular and cellular function, and have included yeasts other than *Saccharomyces cerevisiae* which have made a considerable impact on our understanding of cellular processes—in particular, *Schizosaccharomyces pombe* and the methylotrophic yeasts, *Pichia pastoris* and *Hansenula polymorpha*.

The study of yeasts has been greatly facilitated by their fast, simple growth, accessible biochemistry and ease of genetic manipulation. Lower eukaryotes, especially *S. cerevisiae*, have long been used as model organisms in studying the cellular biology of higher eukaryotes. This book documents the high degree of conservation in cellular function and process, and demonstrates how yeast facilitates our understanding of both higher and lower eukaryotic biology. We feel, therefore, that the book will be of interest to the non-yeast biologist as well as to those active in yeast research.

There have been extensive studies of gene expression in *Saccharomyces cerevisiae* since the advent of genetic engineering and the development of new techniques of molecular biology. Research into yeast promoter structure and function is important to our understanding of both gene regulation and the means by which heterologous genes can be expressed to high levels for the production of commercially important proteins in the food/drink and pharmaceutical/healthcare industries.

Since gene regulation is of importance to our understanding of cellular differentiation and development in all eukaryotes, we have included a chapter on the negative regulation of gene expression.

After transcription, the fate of mRNA can be quite varied: some mRNA molecules are long-lived, others short-lived. The stability of mRNA is of considerable interest, not only because it reflects a level of control on cellular processes but also because it influences the efficiency with which a particular gene can be expressed to the high levels required for commercial purposes. For this reason, we have included a chapter reviewing the current status of our knowledge of mRNA translation and stability, a topic not normally covered in detail.

Research on the expression and secretion of heterologus proteins has resulted in considerable advances in the techniques of producing proteins and peptides in yeast. The problems and solutions to some of these problems have been included in the book.

A level of control of cellular functions and processes is found in the area of post-translational proteolytic processing and protein degradation. Significant advances have been made here, and so we have included a comprehensive review which will also be of interest to those studying such processes in other organisms, and to those concerned with the impact of these processes in the production of proteins and peptides in yeast.

For over a quarter of a century, the study of killer toxin of the type 1 killer system in *S. cerevisiae* has been pursued with vigour. This area of research has led to the discovery of *ds*RNA genomes, which encode both the killer toxin itself and the information required for the replication and maintenance of *ds*RNA in yeast. The study of this mycovirus has revealed how intracellular protein trafficking and post-translational modification confers immunity on the killer toxin. The recent extensive research is reviewed by the leading scientists in this field.

In higher multi-cellular eukaryotes, individual cells respond to external signals through ligand-receptor interaction. Signal transduction is a complex chain of interactions which is readily studied in yeast. The GTP-binding proteins (G-proteins), key components in this interaction, have been the target for extensive study over the past five years. These proteins have functions identical to those in higher eukaryotes and share homologous structural domains. Their function in yeast is the transmission of signals for the control of cell proliferation in response to external stimuli: thus the study of these proteins is important for understanding the mechanisms of oncogenesis in mammalian cells. Recently, it has been found that G-proteins are involved not only in the transfer of signals across membranes from the external environment to the internal components of the cell but also in the transfer of signals between intracellular membranes—in particular, the membranes of the secretory pathway. Thus the role of G-proteins in cellular functions may be broader than first envisaged.

The regulation of cell proliferation in both *Saccharomyces cerevisiae* and *Schizosaccharomyces pombe* has been investigated for almost two decades, and work on cell division cycle mutants has resulted in the isolation of genes and the discovery of functions that regulate cell proliferation. The isolation of mutants which accelerate progression through the cell cycle as small or '*wee*' mutants in *S. pombe* has added considerably to our knowledge of cell cycle control. We have included a review of these advances, with an emphasis on *S. pombe*, where most of the recent discoveries have been made.

Saccharomyces cerevisiae has been exploited in the brewing and baking industries for centuries. With the aid of genetic engineering, brewing yeast can now be manipulated with greater ease, allowing improvements in beer quality and production costs. We felt it relevant to include a chapter covering this rapid progress with brewing yeast, since it has played an important role in the foundation of one of the earliest biotechnological processes.

Several yeast organisms other than *Saccharomyces cerevisiae* have been investigated as possible host organisms for the expression and secretion of heterologous proteins. Work on the methylotrophic yeasts *Pichia pastoris* and *Hansenula polymorpha* has revealed that these organisms can be extremely efficient at both intracellular production and extracellular secretion of proteins. We have included a chapter reviewing both these organisms in detail.

The 2 μm circle plasmid of *Saccharomyces cerevisiae* is commonly used in part or in whole for the replication of heterologous genes to high copy number. Endogeneous 2 μm remains at steady-state copy number levels and the mechanism of this homeostasis is at present being elucidated. Our understanding of the mechanisms involved in plasmid copy number control and stability is important for the application of 2 μm to high copy number expression vectors in yeast. In the last few years this understanding has approached new levels, and a review of the current status of the subject is included in the book.

We would like to thank our contributors for the speed with which they completed their work to ensure that the volume was published rapidly, and we hope that our readers will find the book a valuable reference to the remarkable advances in yeast molecular and cell biology over the past five years.

E.F.W.
G.T.Y.

Contents

3   Messenger RNA translation and degradation in *Saccharomyces cerevisiae*   70
Alistair J.P. Brown

4   The production of proteins and peptides from *Saccharomyces cerevisiae*   107
D.J. King, E.F. Walton and G.T. Yarranton

5 Yeast (*Saccharomyces cerevisiae*) proteinases: structure, character- istics and function 134

Hans H. Hrisch, Paz Suarez Rendueles and Dieter H. Wolf

9 Genetic manipulation of brewing yeasts   280
Edward Hinchliffe and Dina Vakeria

10 Genetic manipulation of methylotrophic yeasts   304
P.E. Sudbery and M.A.G. Glesson

11 The two-micron circle: model replicon and yeast vector   330
P.A. Meacock, K.W. Brieden and A.M. Cashmore

Contributors

Ken-ichi Arai

Department of Molecular Biology, DNAX Research Institute of Molecular and Cellular Biology, 901 California Avenue, Palo Alto, CA 94301, USA

Keith A. Bostian

Division of Biology and Medicine, Brown University, Providence, RI 02912, USA. *Present address*: Department of Exploratory Microbiology and Genetics, Merck, Sharp and Dohme Research Laboratories, Rahway, NJ 07065-0900, USA

K.W. Brieden

Leicester Biocentre, University of Leicester, University Road, Leicester, LE1 7RH, UK

Alistair J.P. Brown

Biotechnology Unit, Institute of Genetics, University of Glasgow, Church Street, Glasgow G11 5JS, UK

A.M. Cashmore

Department of Genetics, University of Leicester, University Road, Leicester, LE1 7RH, UK

M.A.G. Gleeson

The Salk Institute of Biotechnology Industrial Associates Inc. (SIBIA), 505 Coastal Boulevard, La Jolla CA 92057, USA

Edward Hinchliffe

Delta Biotechnology Ltd, Castle Court, 59 Castle Boulevard, Nottingham, UK

Tatsuya Hirano

Department of Biophysics, Faculty of Science, Kyoto University, Sakyo-ku, Kyoto 606, Japan

Hans H. Hirsch

Biochemisches Institut der Universität Freiburg, Hermann-Herder-Str. 7, D-7800 Freiburg, FRG

Kozo Kaibuchi

Department of Molecular Biology, DNAX Research Institute of Molecular and Cellular Biology, 901 California Avenue, Palo Alto, CA 94304, USA

Yoshito Kaziro

Institute of Medical Science, University of Tokyo, 4-6-1 Shirokanedai, Minatoku, Tokyo 108, Japan

D.J. King — Celltech Ltd, 216 Bath Road, Slough Berkshire SL1 4EN, UK

Kunihiro Matsumoto — Departmant of Molecular Biology, DNAX Research Institute of Molecular and Cellular Biology, 901 California Avenue, Palo Alto, CA 94304, USA

P.A. Meacock — Leicester Biocentre, University of Leicester, University Road, Leicester, LE1 7RH, UK

Jane Mellor — Department of Biochemistry, University of Oxford, South Parks Road, Oxford OX1 3QU, UK

Masato Nakafuku — Institute of Medical Science, University of Tokyo, 4-6-1 Shirokanedai, Minatoku, Tokyo 108, Japan

Stephen L. Sturley — Division of Biology and Medicine, Brown University, Providence, RI 02912, USA. *Present address*: Department of Biochemistry, University of Wisconsin, Madison, WI 53706, USA.

Paz Suarez Rendueles — Biochemisches Institut der Universität Freiburg, Hermann-Herder-Str. 7, D-7800 Freiburg, FRG *Present address*: Departmento de Biologia Funcional, Facultad de Medicina, Universidad de Oviedo, E-33071 Oviedo, Spain

P.E. Sudbery — Department of Genetics, University of Sheffield, Sheffield S10 2TN, UK

Dina Vakeria — Research Department, Bass PLC, 137 High Street, Burton-on-Trent, UK

E. Fintan Walton — Department of Molecular Genetics, Celltech Ltd, 216 Bath Road, Slough, Berkshire SL1 4EN, UK

Dieter H. Wolf — Biochemisches Institut der Universität Freiburg, Hermann-Herder-Str. 7, D-7800 Freiburg, FRG

Mitsuhiro Yanagida — Department of Biophysics, Faculty of Science, Kyoto University, Sakyo-ku, Kyoto 606, Japan

G.T. Yarranton — Celltech Ltd, 216 Bath Road, Slough Berkshire SL1 4EN, UK

1 The activation and initiation of transcription by the promoters of *Saccharomyces cerevisiae*

JANE MELLOR

1.1 Introduction to gene regulation

A coherent explanation for the mechanisms used by bacteria to control the expression of their genetic material in response to environmental conditions was formulated in the early 1960s (Jacob and Monod, 1961; Jacob *et al.*, 1964). A gene can be divided into three distinct functional parts. The structural unit encodes the gene product, the promoter allows the expression of the structural region and the regulatory sequences ensure that expression occurs only in response to the correct environmental stimuli. Gene regulation occurs primarily by controlling the rate of transcription initiation at the promoter which acts in *cis* and is the site recognized by RNA polymerase. The regulatory sequences control the rate of transcription initiation through interaction with specific *trans*-acting regulatory proteins that may act positively to enhance the rate of transcription or negatively to repress transcription. A general outline of the basic principles of gene regulation is shown in Figure 1.1

A prokaryotic promoter consists of two distinct elements located at -10 and -35 relative to the RNA initiation site at $+1$. Both are necessary for interaction with the RNA polymerase holoenzyme and are sufficient for directing basal level constitutive transcription from a discrete initiation site. Different forms of the holoenzyme exist that are distinguished by the sigma subunit. This subunit is essential for promoter recognition. Thus some form of regulation can be imposed by enzymes containing distinct sigma subunits interacting with specific DNA elements in different promoters. Positive control of transcription is achieved by a second protein–DNA interaction between an activator protein and the regulatory site which is usually located just upstream of the promoter elements. Enhancement of the rate of transcription is probably achieved by protein–protein interactions between the RNA polymerase and the activator protein. Negative control of transcription is achieved by preventing RNA polymerase–promoter interaction. The element to which the repressor protein binds is usually located within the region of DNA that normally interacts with RNA polymerase, and this bound repressor prevents functional interactions and transcription is impaired.

Recent developments in recombinant DNA technology have allowed the isolation and characterization of many eukaryotic genes. It is clear that the majority of eukaryotic genes have a far more complex organization than those

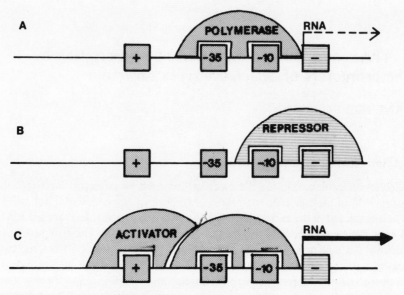

Figure 1.1 Transcriptional regulation in prokaryotes. Promoter (-35 and -10) and regulatory ($-$ and $+$) sequences for a generalized gene are shown. The position of the regulatory elements with respect to the promoter may vary. Panel A represents basal level transcription, with the RNA polymerase interacting with promoter elements. Panel B shows a specific DNA–protein interaction between the repressor protein and the operator site which prevents polymerase binding to the promoter. Panel C describes positive activation of transcription, with an activator protein interacting with both specific DNA sequences and the RNA polymerase.

found in prokaryotes. Structural regions of the gene may be interspersed with non-coding DNA, and the promoter/regulatory sequences are more complex and are often found far upstream or downstream from the structural gene. Despite the many differences that exist between eukaryotic and prokaryotic genes, a major event in the control of gene expression in eukaryotes occurs at the point of transcription initiation, and these events are regulated by positive or negative *trans*-acting proteins. Thus the basic principles of gene regulation may be applied to eukaryotic genes, but it is now clear that the properties of promoter/regulatory elements are quite distinct from prokaryotes and the molecular interactions are different. In this chapter, the organization of, and functional interactions that occur at, promoter and regulatory elements in genes of the baker's yeast, *Saccharomyces cerevisiae*, will be discussed.

1.2 Organization and expression of the yeast genome

Yeast is a simple single-cell eukaryotic organism that shares many characteristics with higher eukaryotic cells (Strathern *et al.*, 1981). The ease with which yeast can be manipulated genetically, coupled with its obvious industrial uses, has made this a favoured system for studying gene regulatory events, and much progress has been made since the advent of recombinant DNA technology and the development of a transformation system for yeast in the late 1970s. The yeast genome contains 16 linear chromosomes. Each chromosome contains multiple origins of replication, two telomeres and a centromere, and is in the form of chromatin in a nucleus. The genome contains about 10^4 kbp of DNA, of which about 50% is transcribed under normal growth conditions (Hereford and Rosbash, 1977; Kaback et al., 1979), and thus contains about 5000 protein coding genes which must be packed quite closely together. The genetic map shows that genes of similar function are generally not grouped together, and so differ from prokaryotic genes which are often clustered into operons. Two well-studied exceptions are the genes involved in galactose utilization (GAL7, GAL10, GAL1) and the genes at the mating type locus (MATα1 and MATα2).

In yeast, most genes are transcribed at similar low levels of approximately one to two molecules of mRNA per cell (St John and Davis, 1979; Struhl and Davis, 1981). Most of the few tens of yeast genes that have been analysed in detail to date were isolated by virtue of their high expression levels or tight regulatory response to given environmental conditions. Thus much of our detailed knowledge of yeast promoters has come from genes that are probably not representative of the majority of yeast genes. Most are likely to be expressed constitutively at the low basal level of one to two molecules per cell under all conditions. In any particular situation where regulation is likely to occur, such as change of carbon source, environmental stress, amino acid starvation, progression through the cell cycle and changes in cell type, transcription rates change for only a small number of genes.

Like all eukaryotes there are three functionally distinct RNA polymerase activities in yeast. RNA polymerase I transcribes only the clustered genes for ribosomal RNA which represents about 70% of the total cellular RNA at steady state; RNA polymerase III transcribes the tRNA genes and the 5 S ribosomal RNA genes which together account for about 30% of the total cellular RNA; RNA polymerase II transcribes all the remaining 5000 or so protein coding genes which together represent only 1% of total cellular RNA at steady state. Much of our current knowledge of promoter structure and function is derived from studies of genes transcribed by RNA polymerase II, and thus this discussion will focus on the promoters of protein coding genes.

The term promoter will be used to describe the region of DNA found upstream of the structural part of a protein coding gene. Three elements within this region are necessary and sufficient for the regulation, efficiency and accuracy of transcription initiation; the upstream regulatory sequences, the TATA element and initiator (I) element. The general organization of the promoter of a typical gene transcribed by RNA polymerase II is shown in Figure 1.2. In any yeast promoter, one of each of these elements can be identified, which together are sufficient for properly regulated transcription, although most yeast promoters have a more complex organization with

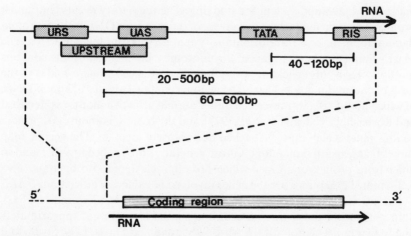

Figure 1.2 A schematic representation of elements and their relative position in a yeast promoter. The position and number of the upstream sequences may vary from that shown.

multiple elements being present. Noticeable features of yeast promoters are their relatively large size and the variability in the distances separating individual elements when compared to their prokaryotic counterparts (Faye *et al.*, 1981; Struhl, 1981; Beier and Young, 1982; Guarente *et al.*, 1982*a*; Struhl, 1982*a*; Dobson *et al.*, 1982*a, b*; Siliciano and Tatchell, 1984; Sarokin and Carlson, 1984; Wright and Zitomer, 1984). The values for distances given in Figure 1.2 are averages, and in some genes essential regulatory functions can be located up to 1.5 kb distal to the structural gene (Miller and Nasmyth, 1984). In other genes, such as *MAT*α, *PET56* and *TRP1*, promoter function lies within a region only 100 bp upstream from the RNA initiation site (Siliciano and Tatchell, 1984; Struhl, 1985*a*; Kim *et al.*, 1986). In the majority of promoters, essential sequences are found dispersed over a region extending about 500 bp upstream from the RNA start site.

1.3 Upstream promoter elements

1.3.1 *General properties of upstream elements*
Sequential deletion of sequences residing 5′ to the structural region of the gene of interest is the method used to determine the limits of the promoter necessary for wild-type levels and correct regulation of transcription. More extensive deletions will alter transcription or its regulation because essential promoter sequences, or upstream elements, are removed and thus their position on the promoter can be estimated. Three functionally distinct upstream elements have been defined in yeast promoters, upstream activator sequences (UASs),

upstream repressor sequences (URSs) and constitutive promoter elements. The upstream activator sequence is defined as the element in a yeast promoter at which positive activation and regulation of transcription occurs (for a brief review see Guarente, 1984, and Struhl, 1987). However, some positive regulatory sites do not correspond with the upstream activation sequence. Thus transcriptional activation and positive control can be functionally distinct properties of yeast promoters. Generally, deletion of the region that defines the UAS results in a significant drop in transcription which is no longer responsive to regulatory stimuli. The activity of the UASs may be modulated by upstream repressor sequences when they are located in the same promoter. The URSs usually have a negative influence on transcription and respond to specific regulatory signals to lower the efficiency of transcription to a greater or lesser extent. These sequences are also known as negative control elements, and when deleted may result in increased transcription or inappropriate regulation. Constitutive promoter elements activate transcription but are not regulatory sites. Deletion of these elements generally reduces transcription to undetectable levels.

Upstream elements generally share common features. They are found, and can function, at long and variable distances upstream from the RNA initiation site. They are functional when inverted with respect to their normal position, and thus in some situations may act bidirectionally. Upstream activating and regulating sequences, both positive and negative, have additional properties not shared with constitutive elements. They are defined by short sequences that confer on a promoter a particular regulatory response. These elements are exchangeable, and if inserted into other promoters they confer the regulatory response characteristic of the promoter in which they are normally found. In this section, the organization and characteristics of elements that mediate positive and negative control of transcription are considered.

1.3.2 Upstream activator sequences

(1) *The upstream activator sequence confers promoter specificity.* Upstream activator sequences (UASs) were first recognized in the early 1980s by 5′ deletion analyses of the *CYC1* and *HIS3* promoters (Guarente and Ptashne, 1981; Struhl, 1981, 1982a, b). Subsequently, similar elements have been found in a number of yeast promoters, most notably *GAL1-10, HIS4, SUC2* and *CYC7* (Guarente et al., 1982a, Donahue et al., 1982; West et al., 1984; Sarokin and Carlson, 1984; Wright and Zitomer, 1984). A number of regulatory genes have been defined genetically as mediators of positive control of transcription. At the time that these loci were defined it was supposed that their products would interact *in trans* at some sequence within the gene they regulate. For example, transcription of the genes involved in galactose utilization at the *GAL1, GAL7* and *GAL10* loci are uninducible in yeast strains bearing recessive mutations in the *GAL4* gene

(Johnson and Hopper, 1982; Laughton and Gesteland, 1982). *GAL4* negative mutations are epistatic to other galactose regulatory mutations, and this lead to the suggestion that the *GAL4* product was a direct activator of transcription. The construction of a *promoter fusion* or *hybrid promoter* containing UAS$_G$ from the *GAL1–10* intergenic region fused on to the *CYC1* mRNA start sites demonstrated that UAS$_G$ is the regulatory region at which the product of the regulatory gene *GAL4* mediates galactose dependent transcription (Guarente *et al.*, 1982a). Direct interaction between the *GAL4* protein and UAS$_G$ has subsequently been demonstrated (Giniger *et al.*, 1985). Other genetic evidence indicated that specific genetic loci were mediators of positive control. Transcriptional induction of the *CYC1* gene at UAS$_C$ by heme, one of the *CYC1* inducers, is diminished by recessive mutations in the *HAP1* locus, suggesting that *HAP1* encodes an activator whose activity or synthesis depends on the cytochrome cofactor heme (Guarente *et al.*, 1983). Similarly, derepression of transcription of genes such as *HIS1*, *HIS3*, *HIS4*, and *TRP5*, all of which are subject to general control, where starvation for one amino acid results in a general increase in transcription of a number of genes involved in amino acid biosynthesis, is prevented by mutations in the *AAS3* locus (now know as *GCN4*) (Hinnebusch and Fink, 1983a, b; Greenberg *et al.*, 1986). *GCN4* mutations are epistatic to other regulatory mutations affecting general control, thus implicating positive control. The proteins encoded by *HAP1* and *GAL4* have subsequently been shown to interact directly with UASs adjacent to the genes they regulate (Selleck and Majors 1987a, b; Pfeifer *et al.*, 1987). The protein encoded by *GCN4* has also been shown to bind directly to the positive control site in *HIS3* and *HIS4*, which in these promoters does not coincide with the upstream activation site (Hope and Struhl, 1985).

(2) *Upstream activator sequences are exchangeable.* The key experiment in defining a role for the UASs in determining promoter specificity was that of Guarente *et al.* (1982a). Here, a hybrid promoter was constructed containing UAS$_G$ linked to the TATA element and initiator element from the *CYC1* promoter. *CYC1* expression is now activated and properly regulated by the *GAL* UAS, but the pattern of expression, that is the use of mRNA start sites, is determined by the *CYC1* sequences. The construction of a hybrid promoter has become a standard test used to identify region(s) of a yeast promoter that contain transcriptional activating sequences. DNA fragments can be inserted into a heterologous promoter lacking its own UAS, and the resulting hybrid analysed for the ability to reactivate transcription. The *CYC1* promoter lacking the UASs and linked to the β-galactosidase coding region is commonly used, but in theory any promoter should function adequately.

(3) *Upstream activator sequences are defined by multiple short sequences.* The upstream elements for a number of different genes have been defined functionally and localized to small regions between 17 and 30 base-pairs. The *CYC1* promoter was the first in which subsites each with discrete UAS

function were defined (Guarente *et al.*, 1984). *CYC1* is subject to complex regulation, responding to several different physiological signals, and this is reflected by the presence of tandem distinctly regulated UASs which respond to heme and carbon source. In the absence of heme, or under anaerobic growth conditions, both UAS1 and UAS2 fail to activate transcription. Under heme-sufficient conditions in glucose medium, UAS1 activity is tenfold greater than that of UAS2. Under derepressed conditions in a gluçoneogenic carbon source such as lactate, UAS1 is derepressed tenfold, whereas UAS2 is derepressed 100-fold. Thus *CYC1* transcription is driven from UAS1 in glucose medium and equally from UAS1 and UAS2 when lactate is the sole carbon source.

Three *trans*-activating genetic loci, *HAP1*, *HAP2* and *HAP3*, have been defined as activators of *CYC1* transcription. *HAP1* activates UAS1 and is also a positive activator of at least two other related genes in the yeast respiratory chain, *CYC7* and cytochrome b_2 (Guarente *et al.*, 1984; Clavilier *et al.*, 1976; Verdiere *et al.*, 1986). *HAP2* and *HAP3* combine to control activation at UAS2 (Pinkham and Guarente, 1985). Mutations in *HAP2* and *HAP3* are pleiotropic, resulting in a reduction in levels of many cytochromes and in an inability to grow on non-fermentable carbon sources, and in combination the two genes may comprise a global activation system for yeast genes involved in respiration with heme as a cofactor.

An analysis of UAS1 by site-directed mutagenesis has defined two subsites, A and B, both of which are required for activity of UAS1 (Lalonde *et al.*, 1986). Two factors from yeast have been identified that bind to the same sequence in site B in a mutually exclusive manner; one is the *HAP1* gene product and the other has been identified as the factor RC2. Both of these proteins are present only in heme-sufficient conditions (Arcangioli and Lescure, 1985). In addition, *HAP1* appears to associate with site A but only when associated with another site A binding factor in the form of a complex. The role of RC2 is unclear, but two possible functions have been suggested; first, that it is a negative regulator that modulates UAS1 activity by competing with HAP1 for binding, and second, that RC2 could be a positive regulator that is involved in the activation of the upstream UAS2. The implications of these ideas are that UAS1 and UAS2 cannot function at the same time, and the function of RC2 may be to regulate which UAS is activating transcription (Pfeifer *et al.*, 1987).

CYC1 is probably one of the best characterized promoters in terms of the organization and regulation at its UAS, but complex organization of upstream activator sequences seems to be a feature of other families of coordinately regulated yeast genes. A notable example is the family of genes encoding ribosomal proteins, where coordinated transcription is required to maintain the protein synthetic capacity of yeast. A computer search of the promoter sequences of 20 of these genes that have been cloned led to the identification of three short sequence motifs present in 15 of them. These consensus sequences are of variable length and complexity, and are generally found at variable distances upstream of the structural region; they are HOMOL1 (consensus AACATCC/TG/ATA/GCA); RPG (consensus ACCCATACATT/CT/A); and a T-rich region. Deletion analysis and hybrid promoter constructions with one ribosomal protein gene *RP39A* showed that all three elements are essential for the expression of this gene, and that the tripartite element has UAS activity and can reactivate the heterologous *CYC1* promoter (Rotenberg and Woolford, 1986). It has since been shown that the TUF factor, which was originally identified as a factor which bound to the HOMOL1- and RPG-related sequences in the *TEF1* and *TEF2* promoters (Huet *et al.*, 1985), binds with differing affinities to the related HOMOL1 and RPG elements in the promoters of a number of ribosomal protein genes (Vignais *et al.*, 1987; Huet and Sentenac, 1987). It appears that HOMOL1 and RPG elements are variants of a consensus UAS in a wide range of genes involved in protein synthesis in yeast. Other yeast promoters contain tandemly duplicated upstream activator elements that share closely related sequences. The *CUP1* gene has two UASs with related sequences 32 and 34 bp long. The effect of the two UASs in this promoter appears to be additive, since deletion of either reduces copper-dependent transcription by half. Duplication of a single copy of one UAS restores wild-type RNA levels (Thiele and Hamer, 1986). Similarly, the *HIS4* promoter contains three copies of the core consensus element (TGACTC) found upstream in one or more copies of 28 or so genes regulated by general amino acid control. At least two copies, together with flanking sequences, are required for wild-type derepressed expression from a hybrid promoter (Hinnebusch *et al.*, 1985). The *SUC2* promoter contains five copies of a core consensus sequence (A/C) (A/G)GAAAT. Multiple insertion of a 32-bp region containing one copy of the 7-bp repeat into a hybrid promoter can activate transcription but no repression is seen indicating a different region of the promoter is required for glucose repression. The significance of the 7-bp repeat came from an analysis of pseudorevertant of a defective *SUC2* promoter. In most cases where significant glucose-repressible invertase activity was seen, changes in an imperfect 7-bp repeat to the consensus sequence had occurred (Sarokin and Carlson, 1986).

Many of the genes (*GAL1*, *GAL10*, *GAL7*, *GAL2*) involved in galactose metabolism share common regulation at UAS_{GAL} by an interplay between the positive (GAL4) and negative (GAL80) regulatory proteins. The *GAL1* and *GAL10* genes are transcribed divergently and share a bidirectional UAS (UAS_G) located midway between the two genes. UAS_G contains four related 17-bp dyad symmetrical sequences at which GAL4 binds to activate transcription. In this case, just one copy of the 17-base-pair sequence is sufficient to give galactose-inducible GAL4 mediated activation of defective *GAL1* or *CYC1* promoters (Giniger *et al.*, 1985).

1.3.3 Negative regulatory elements and mechanisms for the repression of transcription

Although upstream activation at the UAS by *trans*-acting proteins is likely to be a basic mode of gene regulation in yeast, other upstream elements can be identified which may modify the activity of UASs. One such element, the negative element, interacts with transacting proteins to regulate the activity of activator sequences such that transcription is repressed. This may occur at a distance or by antagonizing the binding of transcription factors at the UAS.

(1) *Repression of transcription by negative elements acting at a distance.* Regulation of cell mating type in yeast provides examples of both mechanisms of negative control and has been reviewed by Walton and Yarranton in Chapter 2.

Repression of transcription by negative elements acting at a distance is not limited to the control of the yeast mating-type genes. In other yeast genes described below, the negative elements are also physically separated from the UAS. There are two enolase genes, *ENO1* and *ENO2* in yeast, which are differentially expressed on glycolytic or gluconeogenic carbon sources. *ENO2* mRNA levels are 20-fold higher after growth in a glycolytic carbon source, whereas *ENO1* is constitutively expressed and the mRNA levels are the same regardless of carbon source. Both promoters contain UASs that are functionally similar and mediate glucose-dependent activation of transcription, possibly via the *GCR1* product. However, constitutive expression of *ENO1* in the presence of both glucose and gluconeogenic substrates is achieved by interference of glucose-dependent induction of transcription at the UAS. This promoter contains a negatively acting element located downstream from the UAS that antagonizes UAS function using an as yet unknown mechanism (Cohen *et al.*, 1987).

Another example of differential expression of closely related genes mediated by negative control is emerging for promoters regulated by oxygen. Heme is a known cofactor in the induction of aerobic (oxygen-induced) genes. These include *CYC1* and *CYC7* (Guarente and Mason, 1983; Verdiere *et al.*, 1986) where heme interacts with the *HAP1* gene product to activate transcription, and with a group of aerobic genes (of as yet unknown function) where it interacts with the *ROX1* gene product to activate expression under aerobic conditions. In the anaerobic oxygen-repressed gene, *ANB1*, heme together with the *ROX1* product actually causes inhibition of *ANB1* expression under aerobic conditions. Here the same cofactor and *trans*-acting protein differentially affect the expression of related genes by activation or repression of transcription (Lowry and Lieber, 1986). Negative elements have been found in other yeast promoters functioning to down-regulate but not totally inhibit expression activated by the UAS. In the *CYC7* and *TRP1* genes, deletion of the negative element causes a two- to five-fold increase in transcription (Wright and Zitomer, 1984, 1985; Kim *et al.*, 1986).

(2) *Catabolite repression.* Certain yeast genes are transcribed at reduced levels when the sole carbon source in the medium is glucose. These genes may be involved in the metabolism of other carbon sources such as sucrose,

galactose and ethanol, and include *SUC2*, *GAL1–10* and *ADH2* (Beier and Young, 1982; Sarokin and Carlson, 1984, Guarente *et al.*, 1982*a*). Other glucose-repressible genes include *CYC1* (Guarente and Mason, 1983). This glucose effect is termed catabolite repression, and analysis of promoter fusions indicates that the sequences mediating catabolite repression are located upstream of the TATA box. However, since it is difficult to distinguish the effects of repression from those of activation, both of which may occur in response to the same signal, a mechanism for catabolite repression in yeast has only recently been proposed.

Catabolite repression is a well-known phenomenon in prokaryotes and is mediated by a cAMP-CAP regulatory complex which positively activates transcription by binding to the promoter (deCrombrugghe *et al.*, 1984). In contrast, it appears that, in a least one case, catabolite repression in yeast is mediated by negative control. The cluster of genes (*GAL1*, *GAL10* and *GAL7*) involved in galactose metabolism are highly induced in galactose-containing medium and are subject to glucose repression. By constructing a promoter fusion containing the *GAL1–10* UAS, including the region required for glucose repression and various DNA fragments each of which contained an intact *HIS3* promoter, it was possible to measure the effect of activation and catabolite repression on *HIS3* expression. The *HIS3* promoter is not carbon-source-dependent and is transcribed at a low basal level under most conditions. Activation of UAS_G by galactose caused an increase in basal level expression, whereas repression by glucose caused expression to drop below basal levels, indicating a negative effect on *HIS3* promoter function (Struhl, 1985*b*). In this example, the negatively acting site is located upstream from the *HIS3* regulatory sequences, indicating that the effect can be mediated at a distance. It is interesting that in this example of repression normal promoter functions are maintained. The repression in these two systems is superimposed on normal promoter function; *HIS3* is still induced by amino acid starvation and *CYC1* is subject to catabolite repression.

1.3.4 *Constitutive promoter elements*

Constitutive promoter elements have received little attention compared to regulated promoters which are easier to study. In section 1.2 it was proposed that the majority of yeast genes will be constitutively expressed at low efficiency so that their mRNA will be present at one to two molecules per cell. This will require a constitutive upstream element, such as found in *PET56*, *HIS3*, *DED1* or *TRP1* transcription unit II (Struhl, 1985*a*, Kim *et al.*, 1986; 1988). Under certain physiologically defined conditions, basal mRNA levels may be increased two- to three-fold. It is clear that one or more mechanisms could account for this. One possible way to achieve this state is to relieve a promoter from partial negative repression so that above basal level transcription is driven by the UAS. A different mechanism would be to drive transcription using a constitutive promoter element and effect increases in mRNA levels by specific regulation at an upstream activator sequence. Such a promoter organization is found in the *HIS3* gene (Struhl, 1985).

In the small number of promoters analysed to date the constitutive promoter element is a simple poly(dA-dT) sequence. The length of this tract appears to affect the efficiency of transcription, longer tracts leading to higher levels of transcription. The *DED1* gene has a 34-base-pair region containing 28 thymidine residues, and produces five times more RNA than *HIS3* or *PET56* which share a common 17-base-pair tract containing 15 thymidine residues (Struhl, 1985a). A rare constitutive promoter-up mutation in the *ADR2* gene was shown to be due to the expansion of the normal 20-base-pair dA-dT tract to one 54 to 55-base-pairs long which presumably resulted in the high-level constitutive expression observed (Russell *et al.*, 1983). The *TRP1* promoter has a complex organization; transcription results in two classes of transcript, I and II, that are influenced by different promoter elements. The more distal upstream sequences, composed of a UAS and a negative element, are responsible for class I transcripts, while the proximal upstream element is located within an A/T rich region of dyad symmetry and drives constitutive expression of the class II transcripts. These upstream elements function independently; deletion of the class I elements does not affect class II transcription significantly, and *vice versa*. The poly(dA-dT) tracts are found in a 12-base-pair inverted repeat separated by a 21-base-pair spacer region. Deletion of this region abolishes transcription of the class II RNAs with little effect on class I transcription (Kim *et al.*, 1986, 1988). Poly(dA-dT) tracts have been postulated to occur in a number of other yeast promoters, such as *SUC2* and *URA3*, but await experimental analysis (Struhl, 1986).

There are two possible ways that this relatively simple sequence feature can influence transcription. The tracts may function by association with a poly(dA-dT) specific transcription factor analogous to the regulated upstream elements in yeast promoters. As yet, no (dA-dT) specific transcription factor has been found. A generally favoured hypothesis is that poly(dA-dT) tracts function as constitutive promoter elements by excluding nucleosomes, thus affecting chromatin structure (Kunkel and Martinson, 1981; Prunell, 1982). This is proposed to occur by virtue of the unusual structure adopted by dA-dT tracts in solution; they can form kinks or bends (Marini *et al.*, 1982; Koo *et al.*, 1986; Ryder *et al.*, 1986), and such sequences also have a helix repeat of 10.00 base pairs instead of the normal 10.34 (Peck and Wang, 1981; Rhodes and Klug, 1981). The proposed mechanism for controlling transcription from a promoter such as *HIS3* that contains both constitutive and regulatory upstream sequences is discussed in detail in section 1.5.3.

1.3.5 *Heat shock elements*

Exposure of cells of both prokaryotic and eukaryotic origin to elevated temperatures induces the synthesis of a family of specific proteins called heat shock proteins (HSPs), while the synthesis of several other proteins, especially those involved in ribosome assembly, is dramatically reduced. Little is known about the function of HSPs, although it has been suggested that they are involved in the acquisition of thermotolerance. Heat shock proteins are encoded by a number of heat shock genes, some of which are expressed at a significant basal level before heat shock whereas others are not. For instance, there are three genes in yeast which are heat-inducible and are related to the major heat shock protein in *Drosophila*, *hsp70*. One of these genes, *YG100*, has significant basal level expression before heat shock, whereas *YG106* and *YG107* do not. Other yeast genes not immediately associated with heat shock genes respond by elevated expression. The *PGK* gene encoding the glycolytic enzyme PGK responds to heat shock by increasing mRNA levels at least six-fold when grown on glucose (Piper *et al.*, 1986). No increase is seen on gluconeogenic carbon sources. Perhaps more unusually, the hsp48 protein, which is proposed to have a major role in heat resistance in yeast, has recently been identified as an isoprotein of enolase, a product of the *ENO1* gene (Idia and Yahara, 1985).

A common sequence, the heat shock element (HSE), consensus CnnGAAnnTTCnnG, is found in one or more copies in the promoters of genes responding to heat shock. An analysis of the hsp70 promoter on *YG100* indicates that the heat shock response is very complex (Slater and Craig, 1987). This gene has two regions with good matches to the consensus HSE. One of these (HSE2), together with flanking sequences, functions both as a UAS, maintaining basal level expression, and as a heat shock element allowing a massive increase in expression after stress. It is possible that there are different types of HSEs, some showing good matches to the consensus, that have basal level activity and some without but that are essential for a full heat shock response. Some HSEs may serve as binding sites for constitutive activators as well as heat shock activators, and some may be regulated negatively. Neither the *PGK* or *ENO1* promoter has a perfect copy of the HSE consensus in the upstream DNA. In *PGK*, that match is as good as that found at the functional HSE2 in *YG100*. The HSE in *PGK* has no UAS function; basal level transcription is maintained by the *PGK* UAS located upstream from the HSE. Delection analysis has shown that the HSE sequence is required for elevated expression during heat shock (Piper *et al.*, 1986). In some promoters the HSE can be considered as a conventional upstream activator sequence regulating gene expression in response to a particular physiological state, whereas in other cases the HSE just augments gene expression.

1.4 Transcriptional control at upstream elements

The upstream elements for a number of different genes have been defined in functional terms and localized to small regions. A comparison of the nucleotide sequences within these regions shows that elements for different genes do not generally share common sequences. In this section the effect of the nature and organization of upstream elements on the control and regulation of transcription in different promoters is considered.

1.4.1 *Coregulated genes*

Genes that are subject to similar patterns of regulation may have common short nucleotide sequences, often repeated, upstream from the RNA start site. However, members of a family of coregulated genes may not be induced or repressed to the same extent. The affinity of a binding site for its regulatory protein may allow differential expression levels, and this may be influenced by both the organization of the common short motifs and the nature of the flanking sequences.

Many of the genes involved in amino acid biosynthesis are coregulated and share a hexanucleotide repeat TGACTC, which is the core sequence at which the GCN4 protein binds to enhance transcription under conditions of amino acid starvation (reviewed in Jones and Fink, 1982; see also Hinnebusch and Fink, 1983; Donahue *et al.*, 1982; Hope and Struhl, 1985). Deletion analysis, point mutagenesis and DNase 1 footprinting indicate that the GCN4 protein shows different affinities for the repeated TGACTC core sequence found in some promoters. Flanking sequences, particularly the two bases preceding the core, and a short run of thymidine residues downstream from the core, are important for optimal GCN4 binding to the *HIS3* and *HIS4* promoters, and mutation in any of these bases will significantly reduce GCN4 binding *in vitro* (Hope and Struhl, 1985; Hill *et al.*, 1986; Arndt and Fink, 1986).

A shared sequence with dyad symmetry can be identified in the family of genes (*PHO5* and *PHO7*) encoding acid phosphatases that are regulated by phosphate levels in the medium (Rudolf and Hinnen, 1987). Similarly, all the promoters which are subject to galactose-induced transcription and interact with the GAL4 protein, such as *GAL80*, *GAL2*, *GAL7*, *MEL1* and *GAL1–10*, share a 23-base-pair dyad symmetrical consensus sequence (Bram *et al.*, 1986). The GAL4 protein is bound to the UAS in both induced and uninduced conditions. *GAL80* mediates repression by binding to the a region in carboxy terminus of GAL4 and preventing activation (Johnson *et al.*, 1987). The number of binding sites in the promoters of these various genes ranges from one to four, and the number of binding sites correlates with the extent each gene is repressed by the GAL80 protein. In the cases where there are multiple sites, for instance *GAL1–10*, *GAL2* and *GAL7*, there is one pair with the highest binding affinity located at a dyad–dyad distance of 82 to 87 base-pairs, and these promoters are tightly regulated by GAL80. In contrast, the *GAL80* promoter, which is subject to autogeneous control and contains only a single GAL4 binding site, is repressed only fivefold (Igarashi *et al.*, 1987). Thus, although this family of genes shares a common upstream sequence, the number of copies of this sequence determines subtle regulatory responses with multiple sites, facilitating repression by GAL80 and high induction ratios when galactose is the available carbon source. In contrast with the marked effect on repression, the number of GAL4 binding sites has little effect on the induction response. The single sites in *MEL1* allows transcription at fully induced levels, and insertion of further GAL4 binding sites does not increase RNA levels (Bram *et al.*, 1986).

Multiple copies of a short sequence (consensus PyCACGAAAA) are important for correct cell-cycle-dependent regulation of transcription from the *HO* promoter. Deletion of all 10 copies of this sequence from the wild-type promoter results in transcription both in G_1 and late in the cell cycle. Insertion of a few synthetic copies of this sequence between the upstream activation/regulation region and the TATA box results in proper regulation in G_1, but not later in the cell cycle. As the number of copies of the sequence is increased, the regulation become more authentic, indicating that it is the multiplicity of binding sites that is critical for proper regulation (Nasmyth, 1985*b*).

Sites at which negative regulation is mediated in response to the same regulatory signals show homology over short regions. The site where MATa1/MATα2 binds (see Chapter 2) as a heterodimer to mediate repression of haploid specific genes shares homology to the site where MATα2 binds as a dimer to repress *a*-specific genes. The remainder of the MATa1/MATα2 site is presumed to be the a1 binding site (Miller *et al.*, 1985).

```
BINDING SITE                                                    PROTEIN
(T/C)C(A/G)TGTN-----------6-bp  spacer------TACATCA        MATal/MATα2
        α2                                  a1
        GCATGTA---------19-bp  spacer------TACATGG          MATα2
        α2                                  α2
```

It is likely that many coregulated genes will have upstream sequences that share core regions of homology. However, it is clear that subtle differences in regulation may be mediated by the sequences flanking the core homology. These may, for instance, influence the avidity of binding of a particular transcription factor and thus determine the extent of repression or derepression of transcription (Struhl and Hill, 1987). Alternatively, these flanking sequences may determine the specificity of binding of a transcription factor that may exist in modified forms. It is likely that the spacer sequences separating the core binding sites for MATa1 and MATα2 proteins help differentiate binding of both MATα2, as a dimer resulting in repression of *a*-specific genes, and of MATa1/MATα2, as a heterodimer resulting in repression of haploid specific genes (Miller *et al.*, 1985).

There is one example in the literature of coregulated genes which share a common transcriptional activator that binds to two upstream activation sites which share no obvious sequence similarity (Pfeifer *et al.*, 1987*b*). The HAP1–heme complex binds to UAS1 in the *CYC1* promoter and also to the UAS in the *CYC7* promoter. The sites in *CYC1* and *CYC7* compete for HAP1 binding and have comparable affinities for the protein. The regions protected from DNase 1 digestion in each UAS are shown below:

TGGCCGGGGTTTA**CGGACG**ATGA	*CYC1*
CCCTC**GCTATTATCGCTATTAGC**	*CYC7*

The nucleotides in bold type represent HAP1 contact sites in each UAS. The possible mechanisms whereby HAP1 can bind to two different sequences to mediate the same regulatory response is discussed in section 1.7.4. This example makes it clear that coregulated genes do not necessarily share similar positive control sites, although the same *trans* activator protein may mediate induction.

1.4.2 *Regulation of transcription by upstream elements that function bidirectionally*

Some of the nucleotide sequences of yeast upstream promoter elements do not show dyad symmetry and thus are directional with respect to the structural gene. Others, however, do show clear rotational symmetry. In some yeast promoters the upstream promoter elements coordinately activate two divergently transcribed genes such as *GAL1–GAL10* and *MATα1–MATα2*. Although UAS$_G$ contains core sequences for GAL4 binding that have dyad symmetry, there are four such core sequences arranged asymmetrically between the *GAL1* and *GAL10* structural genes (Giniger *et al.*, 1985). Fine deletion analysis and the inversion of small DNA fragments in the *GAL1–10* promoter tend to support the idea that divergent transcription is controlled by one element functioning with true bidirectionality (West *et al.*, 1984; Giniger *et al.*, 1985).

In several other promoters directing transcription of one gene, the upstream elements or synthetic copies of them have been inverted to examine whether the presence or the lack of symmetry reflects differences in promoter function (Guarente and Hoar, 1984; Struhl, 1984; Hinnebusch *et al.*, 1985; Johnson and Herskowitz, 1985; Sarokin and Carlson, 1986; Ogden *et al.*, 1986). In general, in all derivatives where upstream elements, either positively or negatively acting, have been inverted, the genes are transcribed and regulated with equal efficiency as derivatives containing the element in the normal orientation. In the rare cases where elements do not function normally in the reverse orientation, there is usually an explanation. Often the activity of the element is blocked by sequences that normally reside upstream of the element, and trimming the flanking sequences to yield a minimal fragment is sufficient to demonstrate fully functional bidirectionality. These 'blocking' sequences may normally have an essential function in the wild-type configuration of the gene, ensuring directionality for an upstream element of one gene that is close to the promoter of a second divergently transcribed gene. For instance, a sequence which functions as a transcriptional terminator is found just upstream from the activation sequences in the *URA3* promoter (Yarger *et al.*, 1986). Brent and

Ptashne (1984) have demonstrated that transcriptional terminators can block the transfer of an activation signal from the UAS to the RNA initiation region. The sequence in the *URA3* promoter could function to shield the gene from the influence of neighbouring genes by preventing transfer of an activation signal from an element functioning bidirectionally. Alternatively, the terminator may prevent read-through transcription which may interfere with normal regulation. It is likely that sequences which impose unidirectionality on normally bidirectional elements will be a feature of other yeast promoters and regulatory sequences. However, one other mechanism to explain how a bidirectional element can function in one orientation has been proposed by Struhl (1986). In order to promote transcription, an activation signal must be transferred from the upstream elements to the TATA element and I site. Struhl has proposed that yeast promoters have different classes of TATA elements that can interact with one type of activation signal only. Thus the nature of the TATA element can impose directionality on transcriptional activation. This mechanism is discussed in more detail in section 1.5.3.

1.4.3 *The effect of the position of the upstream element*
A survey of yeast promoters where the positions of the upstream elements has been defined reveals that these elements are located at variable and often long distances from the TATA element and RNA initiation site. Even shared elements that mediate coregulation in different promoters are found at varying positions within the promoter. Evidence that position is relatively unimportant for promoter function comes from hybrid promoter constructions. A number of upstream elements have been tested in a derivative of the *CYC1* promoter in which both UAS1 and UAS2 have been deleted. All are capable of reactivating and specifically regulating *CYC1* directed transcription when located in this fixed position about 250-base-pairs upstream, yet in their own promoter they may be found anywhere between 100-base-pairs to over a kilobase away from the structural gene. In some promoters, the effect of spacing has been systematically studied. The *GAL1* and *CYC1* upstream elements activate transcription only in response to appropriate conditions, and correctly regulated transcription is seen even when these elements are moved over the range of hundreds of base-pairs (Struhl, 1984; Guarente and Hoar, 1984). However, in many cases the question of position effects is only just being considered. In the case of the family of genes induced in the presence of galactose, from one to four copies of the GAL4 binding sites is present, but it has been shown that the number of copies is not related to the levels at which these genes are expressed (Bram *et al.*, 1986); in other words, insertion of additional binding sites does not increase RNA levels. *GAL2*, *GAL1-10* and *GAL7* all contain at least one pair of binding sites, yet are expressed at very different levels. The other two genes, *GAL80* and *MEL1*, both containing just one binding site, are also expressed at markedly different levels. In their paper, Bram and co-authors suggest that the location of the single GAL4 binding site

may reflect expression levels in *GAL80* and *MEL1*. The binding site is found 95 base-pairs upstream of the poorly expressed *GAL80* structural gene, and 230 base-pairs upstream of the efficiently expressed *MEL1* gene. However, this is difficult to reconcile with the three genes that have two optimally spaced GAL4 binding sites located at similar positions in the promoter but which have very different final expression levels.

There are several possible explanations for this discrepancy. The first is that the rate of transcription from these two promoters may be similar, but the rate of degradation of the mRNA is very different, giving markedly different steady-state expression levels. A second explanation is that the upstream sequence does not determine the efficiency of gene expression but this determinant resides within other regions of the promoter, perhaps at the TATA element or I site, and this is determining the different expression levels in, for example, *GAL80* and *MEL1*. This issue is considered in more detail in section 1.4.4. The third possibility, as suggested by Bram *et al.* (1986), is that the position of the upstream sequences in the promoter may affect the expression potential. There is evidence in support of all three explanations. Insertion of UAS_G into various deleted derivatives of the *HIS3* promoter shows a tenfold difference in galactose-inducible expression, depending on the position of insertion of the UAS (Struhl, 1984). Insertion of UAS_{PGK} at various positions within the *TRP1* promoter results in marked position-dependent activation. Here, optimal activation is only seen when UAS_{PGK} is inserted close to the position normally occupied by UAS_{TRP1} in the wild-type promoter. When the converse experiment is done and UAS_{TRP1} is inserted at various positions in the *PGK* promoter lacking UAS_{PGK}, optimal activation is seen over a region spanning 400 base-pairs (Mellor *et al.*, unpublished observations). These experiments indicate that different promoters might use different mechanisms to transfer the activation signal from the UAS to the TATA element and that this is characteristic for each promoter. Thus some hybrid UAS/TATA box combinations might be highly dependent on the position of the UAS in the promoter, whereas other combinations might not be.

A similar situation may also exist in higher eukaryotic genes. Enhancers are defined as functioning in a position and orientation-independent manner, and have similar properties to UASs in yeast promoters. One of the first enhancers described was found in the virus SV40 and was shown to activate many other promoters independent of its position in a promoter. However, when activating transcription of the early genes in SV40 this enhancer is extremely sensitive to its position relative to other elements in the promoter (Takahashi *et al.*, 1986).

1.4.4 *Are upstream sequences the sole determinants of transcriptional efficiency?*

It is generally accepted that upstream sequences determine the efficiency with which a gene is transcribed. The construction of hybrid promoters containing the upstream promoter elements of one gene fused to the RNA initiation region of a second generally results in RNA levels characteristic of a wild-type version of the first gene from which the upstream sequences are taken. However, there are now examples in the literature which support the idea that other regions of the promoter can influence the overall level of transcription. In a set of fusions between the *CUP1* upstream sequences and the *CYC1* TATA and initiation elements, none resulted in the high levels of copper-inducible transcription normally observed for the authentic *CUP1* promoter (Thiele and Hamer, 1986). Reciprocal fusions between the upstream sequences and initiation regions of the *DED1* and *HIS3* promoters also suggested that sequences within or downstream from the initiation region influence transcriptional efficiency. When the *HIS3* initiation region is controlled by the *DED1* upstream sequences, levels of *HIS3* transcripts are similar to wild-type *DED1* levels. Here, the *DED1* upstream sequences are apparently determining transcriptional efficiency. However, high *DED1* RNA levels, similar to the wild-type *DED1* levels, are also seen when the *HIS3* upstream sequences direct transcription from the *DED1* initiation region. As *HIS3* RNA levels are normally fivefold lower than those from the authentic *DED1* promoter, it is likely that the *DED1* sequences are contributing an efficiency determinant (Chen and Struhl, 1986).

In reciprocal fusions between elements in the *TRP1* and *PGK* promoters, the RNA levels produced are independent of the source of the upstream sequences. UAS_{PGK} in the *TRP1* promoter only reactivates transcription to the low levels normally seen for the wild-type promoter, and UAS_{TRP1} reactivates transcription from the *PGK* promoter to the high levels normally seen for the wild-type *PGK* gene (Mellor *et al.*, unpublished observations).

The location of the sequences that influence the overall level of transcription in uncertain; they may reside around or downstream of the RNA initiation region. A region within the coding part of the *PGK* gene has been identified that when deleted results in a six to ten fold drop in rate of transcription initiation from the *PGK* promoter (Mellor *et al.*, 1987). This region, known as the downstream activator sequence, may be one determinant of transcription efficiency in this gene. This may explain in part why insertion of various yeast transcriptional activator sequences into a *PGK* promoter lacking the UAS results in RNA levels similar to those produced by the authentic *PGK* gene (Rathjen *et al.*, 1987, and unpublished observation).

The question as to whether the upstream sequences alone determine the efficiency of transcription and whether position effects are important still remains open. The family of genes induced by galactose have already been mentioned in relation to the question of how differential regulation and expression is achieved. It is likely that thorough analysis of a gene family which share some aspects of regulation but differ in others will shed light on this complex issue.

1.4.5 Complex upstream elements

It is becoming clear that the distinction between upstream activation and repression events is not dependent on the definition of a particular protein–DNA interaction as a repressor or activator, but rather on the overall context within which this interaction occurs in a particular promoter. This is exemplified by the silencer element that functions upstream of the *HMR* locus to repress transcription of mating-type information (see Chapter 2). Individually, two of the three short sequences within the silencer, E and B, activate transcription from heterologous promoter yet when combined together or with the third sequence (A, containing an ARS consensus sequence) they form a very efficient repressor of transcription (Brand *et al.*, 1987). These silencer sequences are also involved in DNA replication and segregation. Two *trans*-acting proteins, RAP-1 and SBF-B, bind to the E and B sequences respectively. The E site is also found at the *HML* locus and is homologous to the upstream activator sequences of ribosomal protein genes (UAS$_{RPG}$), translation elongation factor EF-1α and the mating type genes *MATα1/MATα2*. This suggests that *RAP-1*, which is an essential gene, may be involved in both repressing and activating transcription (Brand *et al.*, 1987; Shore and Nasmyth, 1987). Neither SBF-B nor RAP-1 are products of the *SIR1-4* genes which are also essential for silencer function. It is possible that RAP-1 and SBF-B do not function as activators in the silencer, but rather form a DNA–protein complex that is unique to the silencer and is recognized by the *SIR* gene products. One implication of this observation is that regulation of transcription may be achieved by different combinations of a relatively small number of transcription factors acting at different sites in yeast promoters.

1.4.6 Signals and coeffectors

Another level at which transcription can be regulated at upstream elements is the availability of specific DNA binding proteins or cofactors required for their synthesis or their DNA binding activity. Many transcriptional activators are present at fairly constant amounts under all environmental conditions. Thus some cofactors that act to signal specific changes in the environment are required to modify the activity of the activator protein. For example, galactose or one of its metabolites may be the co-factor required to modify GAL80 activity such that it no longer associates with the carboxy terminus of GAL4. This region is now available for interaction with as yet unknown transcription factors which may result in activation of transcription.

The levels of intracellular heme appear to reflect a global signal and coeffector for a number of different genes. In combination with the *ROX1* gene product, heme acts as both a positive and negative effector at aerobic and anaerobic promoters (Lowry and Lieber, 1986), and when associated with the HAP1 protein, the HAP1–heme complex regulates the expression of the cytochromes. The levels of intracellular heme are a measure of the capacity of the cell to carry out oxidative phosphorylation which reflects whether growth occurs under aerobic and anaerobic conditions and the nature of the available carbon source. Thus complex regulation of many genes can be mediated by a single cofactor or signal molecule in combination with a small number of regulatory proteins. In physiological terms this mechanism is sensible. Rather than using many compounds or their metabolites as signals of carbon source or oxygen availability, one key component is used. It is not known what regulatory mechanisms determine the amount of intracellular heme in a cell.

A different mechanism exists to regulate expression of GCN4, the global activator of amino acid biosynthetic genes. *GCN4* RNA levels are fairly constant even under conditions of amino acid starvation. However, the levels of GCN4 protein are 50 times higher when amino acids are limiting. In this example, transcription of the many genes involved in amino acid biosynthesis is regulated by changes in the levels of the GCN4 protein. The key to the mechanism used to regulate GCN4 levels is the structure of the *GCN4* mRNA (Thireos *et al.*, 1984; Hinnebusch, 1984). The mRNA has an extremely long 5′ untranslated region on which 4 AUGs followed by in-frame termination codons precede the initiating AUG for the structural gene. Generally, eukaryotic mRNAs have short leader sequences, translation is initiated at the 5′ proximal AUG, and reinitiation at more downstream AUGs is extremely inefficient (Kozak, 1980). Thus, according to the rules of eukaryotic translation initiation, no GCN4 protein should be made. This is the case under normal growth conditions; only under conditions of amino acid starvation are the rules bent, resulting in GCN4 expression. Mutations in one of the gene products required *in trans* for GCN4 expression also generally affects the translation of other genes. Thus although the mechanism for GCN4 expression is not fully understood, it is likely that it is closely related to the mechanism for control of translation. This idea also goes a little way to explaining how starvation for any amino acid can produce a global effect on many genes.

The capacity of a regulatory protein to respond to physiological signals and activate transcription may also be regulated by post-transcriptional modifications. Conformation changes may allow a constitutively expressed protein to bind to its cognate sequence or, if bound, to associate with other proteins of the transcriptional machinery. Some genes identified genetically as *trans*-acting regulators of transcription have recently been shown to encode protein kinase activities. In both prokaryotic and higher eukaryotic systems, post-translational modification by phosphorylation is known or proposed as a mechanism for modulating the activity of transcription factors (Zillig *et al.*, 1975). In yeast there is also evidence that protein kinases play a critical function in regulating transcription. The gene *SNF1* encodes a serine-threonine-specific protein kinase that is required for depression of glucose-repressible genes (Celenza and Carlson, 1986). One of the substrates for *SNF1* activity is the *SSN6* gene (Schultz and Carlson, 1987) which is a negative regulator of *SUC2*. *SNF1* activity may directly phosphorylate *SNN6* to inactivate it, or the effect may be mediated through a second gene such as *SNF4*. Like many transcriptional regulators, mutations in *SNN6* also affect the expression of many other genes. Thus protein phosphorylation serves as a signal in the regulatory circuit for catabolite repression and is likely to be a common feature in many yeast regulatory circuits.

The models that have been proposed to explain how an activation signal is transferred from the upstream sequences to the RNA initiation regions are discussed in detail section 1.8.

1.5 The TATA element

A short A/T-rich sequence is a ubiquitous feature of prokaryotic and eukaryotic promoters. In prokaryotic genes it is known as the Pribnow box and is located 10 base-pairs upstream from the RNA initiation site. The Pribnow box together with the *cis*-acting sequence at − 35 form the prokaryotic promoter at which RNA polymerase associates to initiate transcription in response to regulatory signals. When the sequences of many eukaryotic promoters are compared the only common feature is a short A/T-rich region with the canonical sequence TATA(A/T)A. In higher eukaryotes, the TATA box is invariably located 25 to 30 base-pairs away from the 5′ end of the corresponding mRNA (Breathnach and Chambon, 1981). In *S. cerevisiae* no such clearly defined sequence exists, and because yeast promoters are extremely A/T-rich, it is difficult to describe which out of several potential sequences is a TATA box. An experimental approach is usually required to differentiate functional TATA elements from other A/T-rich regions.

The first indication of a second critical promoter element came from the same sequential 5′ deletion analyses that had been used to define upstream promoter elements (Faye *et al.*, 1981; Struhl, 1981). In deletion mutants lacking the upstream element, transcription was reduced considerably compared to the wild-type promoter. Further deletion did not result in a drop in RNA levels until a second boundary was reached. Downstream from this point was a region containing one or more of the canonical TATA sequences and deleting these sequences resulted in a further drop in RNA levels, usually to undetectable levels. Mapping the 3′ boundary of the TATA element is more complicated because the deletion series must begin just upstream from the transcription initiation site. This has been done for the *HIS3* promoter and the results indicate that the whole TATA element resides within a 20 base-pairs region extending from − 32 to − 52 upstream from the RNA initiation site (Struhl, 1982a). More evidence suggested that two separate regions of DNA are necessary to constitute a promoter, neither of which is sufficient alone. Mutations for several genes were obtained that reduced expression approximately tenfold, but still retained intact upstream elements. These mutations were located in the region of the promoter around the TATA element (Struhl, 1982a; Guarente and Mason, 1983; Siliciano and Tatchell, 1984).

1.5.1 *The TATA element is found at variable distances upstream from the RNA initiation site*

The preliminary experimental data outlined above indicated that the function of the TATA element in yeast promoters may be different from that in higher eukaryotic genes. The functional TATA elements do not reside a fixed distance from the RNA initiation site; instead, the distance from the RNA start site usually averages 60 base-pairs, but this distance is much more variable and can be 100 base-pairs or more (Dobson *et al.*, 1982a; Burke *et al.*, 1983; Williamson *et al.*, 1983). Furthermore, many yeast promoters contain more than one sequence with homology to the TATA consensus (Faye *et al.*, 1981; Donahue *et al.*, 1982; Dobson *et al.*, 1982a; Guarente and Mason 1983; Yocum *et al.*, 1984; Kim *et al.*, 1986). The functional significance of these multiple potential TATA elements has only been analysed for a small number of genes (see section 1.5.3).

1.5.2 *The TATA element is directional*

One feature of upstream elements is their ability to function with little or no dependence on their orientation with respect to the structural gene. Two pieces of experimental evidence suggest that the TATA element functions in one orientation only. The *HIS4* promoter contains four potential TATA elements, but only one is required for promoter function. Insertion of synthetic oligonucleotides into a mutant form of the *HIS4* promoter where this TATA element is removed revealed that the sequences TATA or TATAA cannot restore promoter function, although TATAAA works almost as well as the wild-type sequence. Reversing the TATAAA oligonucleotide so that the promoter now contains the sequence TTTATA does not restore promoter function, indicating that the TATA element is orientation-specific (Nagawa and Fink, 1985). In experiments where extreme care was taken not to alter the spatial arrangement between the TATA element and the rest of the *HIS3* promoter, Struhl (1982a, b) demonstrated that a small fragment from coliphage M13 containing a sequence resembling the *HIS3* TATA element can functionally replace the *HIS3* element, but only when inserted in the sense orientation.

1.5.3 *Functionally distinct TATA elements in yeast promoters*

Although the TATA element is a general feature of yeast promoters, it is becoming clear that individual elements, even when located in the same promoter, have distinct functions. The early experiments that defined the TATA element as essential for promoter function in yeast genes indicated that deletion of the element resulted in a substantial drop in RNA levels, usually in the region of tenfold (Struhl, 1982a; Guarente and Mason, 1983; Siliciano and

Tatchell, 1984). In some promoters, such as *PGK*, deletion of the region containing the two potential TATA elements causes a marked reduction in *PGK* protein levels, but RNA levels as judged by Northern blotting are similar to the wild-type levels (Ogden *et al.*, 1986). However, contrary to the results of Ogden *et al.*, the majority of transcripts do not map to the wild-type initiation site, but initiate heterogeneously at a number of sites mainly downstream from the ATG (J. Rathjen and J. Mellor, unpublished results). It is probable that in this situation a cryptic TATA element with perfect homology to the consensus, located at + 30 with respect to the wild-type initiation site, is activated.

One explanation for the presence of multiple TATA elements in some yeast promoters has been proposed by Struhl (1986). The *HIS3* promoter contains one constitutive and one regulated upstream element. Struhl suggests that each upstream element interacts specifically with a TATA element to give the distinct patterns of transcription initiation seen under repressed or dere-pressed conditions. The consensus element (TATAAA) located between − 35 and − 45 is proposed to be the TATA element (T_R) with which the regulatory upstream element interacts to increase *HIS3* transcription under conditions of amino acid starvation. Constitutive transcription is mediated by the poly(dA-dT) tract in conjunction with a different TATA element (T_C) located somewhere between − 45 and − 83. The best match resides at − 49 to − 54 (TATACA). This proposition explains why the *PET56* gene is not induced under conditions of amino-acid starvation. *HIS3* and *PET56* are transcribed divergently and share a poly(dA-dT) tract which serves as a bidirectional constitutive upstream element. Also located between these two genes is the regulated upstream element which allows *HIS3* but not *PET56* to be induced under amino acid starvation, yet this element can function bidirectionally. It is suggested that the TATA element in front of *PET56* can only function in conjunction with the constitutive upstream element and thus transcription is unresponsive to induction. *HIS3* has both constitutive (T_C) and regulatory (T_R) TATA elements, and is inducible.

This organization is unlikely to be limited to the *HIS3* promoter. Transcription unit I of the *TRP1* promoter contains two regions with homology to the TATA element. TATA 1 functions to set the multiple RNA start sites, and in its absence TATA 2 can function to set a different pattern of initiation sites. When both elements are deleted, a heterogeneous pattern of start sites located further down the promoter is seen (Kim *et al.*, 1986). Hybrid promoter fusions have been created where the *PGK* UAS replaces the *TRP1* UAS and one or both TATA elements. An analysis of transcript initiation sites indicates that the *PGK* UAS can interact with *TRP1* TATA 1 to set a wild-type transcription pattern as expected. However, unlike the *TRP1* UAS, the *PGK* UAS cannot interact with TATA 2. Thus when TATA 1 is deleted, activation by the *PGK* UAS only results in the heterogeneous pattern of start sites, typically seen only when both TATAs are deleted. This indicates that in the *TRP1* promoter the two TATA elements interact in a functionally distinct manner with different upstream sequences. However, the significance of this organization in terms of regulating promoter function is not fully understood.

1.6 The transcription initiation site

In higher eukaryotes, the TATA box functions to determine efficient transcription initiation at a site about 30 base-pairs downstream. There is a spatial relationship between the TATA box and the actual site (I) of transcription initiation, but no actual requirement for specific sequences at the I site; its location is determined by the position of the TATA box. Deletion of intervening sequences results in initiation at a site 30 base-pairs downstream from the TATA box (Ghosh et al., 1981; Wasylyk et al., 1983). Yeast appears to have a novel element that, in conjunction with the TATA element, determines the RNA initiation site(s) (I).

1.6.1 Preferred sequences for I sites

Shared sequences around the RNA initiation regions of yeast genes were first noted by Dobson et al. (1982a). A number of highly expressed yeast genes have a CT-rich block about 20 base-pairs long, followed 9 to 12 base-pairs downstream by the sequence CAAG. RNA initiation usually occurs at or near the CAAG sequence. The significance of this sequence organization for expression from efficient yeast promoters has not yet been determined. However, the CAAG sequence is very similar to a consensus I sequence observed by Hahn et al. (1985) during a survey of yeast promoters. 55% of the promoters surveyed had either RRYRR (38%; R = purine, Y = pyrimidine) or TC(G/A)A (17%) at the I site. The RRYRR and TC(G/A)A sequences were particularly prominent when the I site was greater than 50 base-pairs downstream from the TATA box.

The sequence specificity at the I site was demonstrated by replacing some of the normal RNA start sites in the CYC1 promoter with an oligonucleotide containing one of the preferred sequences TCGA; transcription initiated efficiently at this newly introduced sequence (Hahn et al., 1985). Further evidence for the specificity of the I site came from reciprocal hybrid promoter fusions between HIS3 and DED1 that contained the upstream promoter element and the TATA element of one gene fused to the I site and mRNA coding region of the other (Chen and Struhl, 1985). Transcription initiation patterns characteristic of the wild-type gene containing the particular I site were seen. In the HIS3 promoter, both major I sites conform to the RRYRR homology. This experiment also indicates that TATA elements can function with I elements from different genes, and that in this case the I element contains the information for determining where transcription begins. A detailed analysis of the sequences determining mRNA start site selection has also been carried out on the PHO5 promoter. In this promoter a TATA element at −101 is absolutely required for transcription initiation at two major sites containing the RRYRR motif. Sequential deletion into the I region results in the utilization of new I sites, all of which have the RRYRR motif (Rudolf and Hinnen, 1987). In the downstream *TRP1* promoter a seven-base-pair motif with the sequence CACGTGA is required for RNA initiation at sites a few bases downstream. Deletion of the motif, but not the RNA start sites, results in no mRNA initiation from the promoter. A factor from yeast nuclear extracts that binds specifically to this CACGTGA motif has also been found. Cloning and mutagenesis of this gene is in progress.

1.6.2 *Spatial organization of the TATA element and the I sites suggests an initiation window*

For some yeast genes, transcription begins at a unique site (e.g. *HIS4*; Donahue *et al.*, 1982), others may have two to four major initiation sites (e.g. *TRP1* transcript II, Dobson *et al.*, 1982b; *PGK*, Ogden *et al.*, 1986; *PHO5*, Rudolf and Hinnen, 1987) and there are several examples of extreme 5' heterogeneity (e.g. *CYC1*, Faye *et al.*, 1981, Guarente and Mason, 1983; *HIS1*, Hinnebusch and Fink, 1983; *URA3*, Rose and Botstein, 1983; *TRP1* transcript I, Kim *et al.*, 1986). In addition, yeast promoters may have one or more potential TATA elements. Experiments that address the spatial relationships between TATA elements and I sites suggest that the spacing between these elements is not critical, and define the concept of an initiation window (40–120 base-pairs downstream from the TATA) which is the range over which a TATA/I site combination is functional.

The *CYC1* promoter has a complex arrangement of TATA boxes and RNA initiation sites. There are over twenty mRNA initiation sites clustered into six major groups at $+1$, $+10$, $+16$, $+25$, $+34$ and $+43$. Four potential TATA elements lie upstream at positions -154, -106, -52 and -22. Three of these TATA sequences are functional and contribute to transcription initiation; the -106 TATA promotes initiation at $+1$, $+10$ and $+16$, the -52 TATA at $+16$, $+25$ and $+34$ and the -22 TATA at $+34$ and $+43$. Altering the spacing to a limited degree between the TATA elements and the initiation region does not alter the start site usage, although changes in the efficiency of usage are seen (Hahn *et al.*, 1985). Similar conclusions, that each TATA element potentiates a specific subset of mRNA start sites, were reached by McNeil and Smith (1986). Each TATA element is capable of activating transcription only within a window, the initiation window, 55–110 base-pairs downstream of the element. The concept of the initiation window was strengthened by the work of Nagawa and Fink (1985) on the spatial relationships in the *HIS4* promoter. Here there is one I site at -60, and four potential TATA sequences, of which only one (at -123) is required for transcription initiation. In this promoter, the potential I site is recognized only if it is located within a defined distance downstream from the TATA. When the distance between the TATA and the I site is shortened, the I site is not recognized, and transcription begins at a new site about 60 base-pairs downstream from the TATA. If the TATA is moved to new positions upstream from its normal location, the new site of transcription initiation is always at least 60 base-pairs from the TATA. There is a maximum distance from the TATA beyond which a potential I site will not be recognized. At 74 base-pairs, initiation is specific for the I site, but at 110 base-pairs, it is not recognized. In this case, the initiation window in which the TATA element can recognize an I site is a minimum of 60 base-pairs and a maximum of 110 base-pairs.

1.6.3 *Mechanism of mRNA start site selection*

The TATA element functions primarily to determine the frequency but not the position of an initiation event in a yeast promoter. This is in direct contrast to the proposed function for the TATA element in higher eukaryotic promoters; here it determines not only the frequency but also the position of the initiation event. Potential I sites are located within an initiation window 40–120 base-pairs downstream from a TATA element. The initiation window may contain multiple I sites competent for mRNA synthesis, as shown clearly for the *PHO5* promoter during gradual deletion over the initiation window (Rudolf and Hinnen, 1987). However, the actual start sites within this window are probably selected by a mechanism that prefers sites proximal to the TATA element. The efficient use of TATA proximal start sites will prevent the use of potential sites located downstream in the window. If there are no strong I sites within the window, then other weaker sites in the vicinity will be used; this suggests that in the absence of a strong start site a yeast gene might have many initiation sites. However, genes such as *HIS1* and *CYC1* which have multiple initiation sites also have multiple TATA elements. The presence of multiple functional TATA elements and multiple I sites in these two promoters is difficult to reconcile with the results from an analysis of the *HIS4* promoter where a second TATA element is inserted upstream of the normal element. In this case, only the more proximal of the two potential TATA elements is recognized. It would appear, based on the analysis of a limited number of promoters, that different TATA/I site relationships exist in different promoters. The existence of the I site is the major difference between yeast and higher eukaryotic promoters, and may explain some of the observed mechanistic differences in promoter function. The observed sequence specificity (RRYRR, TCGA, CAAG and other motifs) at the I site is likely to be reflected in specific initiator proteins. Supporting evidence for this comes from hybrid promoter fusions. Insertion of the I site from the *PET56* gene between the UAS_G and the *HIS3* promoter and structural gene (containing the *HIS3* TATA element and all sequences downstream) blocked galactose dependent activation of *HIS3* expression. Here the I site appears to be acting as a repressor of transcription similar to a negative element whose effect is usually mediated by association with transacting proteins (Struhl, 1984). The existence of the I site may also help to explain a second major difference between yeast and mammalian promoter and that is that upstream sequences are not functional when placed downstream of the RNA initiation site in yeast genes (Guarente and Hoar, 1984; Struhl, 1984). It is possible that the I site blocks activation at the TATA element located upstream of the I site. Finally, it is likely that the whole process of transcription initiation in yeast promoters is more complex than in higher eukaryotes since both TATA binding and I site binding proteins might be required. Again this may be a possible explanation for the failure to reconstitute yeast transcription *in vitro*.

1.7 Regulatory proteins

Throughout the previous sections where the properties of the three main elements required for promoter function in yeast have been considered, the basic assumption has been made that *trans*-acting proteins are specifically associated with each individual DNA element. This assumption is essential so that a detailed molecular picture of transcription initiation and regulation can be drawn. However, these presumptive *trans*-acting proteins have been shown to bind to specific DNA sequences in only a few cases, and much of our knowledge is inferred from genetic and physiological criteria. Thus many presumptive *trans*-acting proteins are defined only on the basis of mutations that abolish the ability of a particular gene or family of genes to be regulated properly, and some are based on simple prediction and speculation. Many regulatory mutations can cause a particular phenotype in many different and often indirect ways. Without biochemical evidence, such as purified proteins and an *in vitro* transcription system, it is difficult to determine if a mutation directly affects transcription by altering a specific DNA–protein interaction. However, in this section four different *trans*-acting activator proteins that have been shown to bind specifically to unique DNA sequences will be considered. The approaches used are different in each case, and exemplify the methods available to identify the specific protein–DNA interactions.

1.7.1 *Binding by the positive regulatory protein GCN4*

Extensive deletion analysis of a number of genes under general amino acid control such as *HIS3* and *HIS4*, coupled with genetic evidence, suggested that the *GCN4* gene product had a direct role in regulating transcription at the consensus TGACTC core sequence (Hinnebusch and Fink, 1983a; Penn *et al.*, 1983; Donahue *et al.*, 1983; Struhl *et al.*, 1985; Hinnebusch *et al.*, 1985). A novel but generally useful approach was adopted to demonstrate the DNA binding properties of GCN4 (Hope and Struhl, 1985). This method involves cloning the *GCN4* coding region into a vector containing a promoter which is recognized by SP6 RNA polymerase, allowing the template to be transcribed *in vitro*. The pure mRNA can then be translated into a ^{35}S labelled protein by *in vitro* translation. DNA binding activity is demonstrated by incubating the labelled protein with specific DNA fragments and separating the protein–

DNA complexes from free protein on a native polyacrylamide gel. Hope and Struhl showed that the *in vitro* translated product of the *GCN4* gene bound specifically to a 20-base-pair region of the *HIS3* gene that had previously been shown to be critical for transcriptional regulation, and which contains the TGACTC sequence common to other coregulated genes. A different approach which yielded similar results to those of Hope and Struhl was used by Arndt and Fink (1986). Here the GCN4 protein was synthesized in *E. coli* and purified before use in DNase 1 footprint analysis. In both methods, the assumption that the GCN4 protein did not require any post-translational modifications for binding activity was made, and in the case of GCN4 is valid. This may not be so for other proteins, and may preclude the use of either of these techniques. Further evidence that *GCN4* encodes a specific DNA binding protein included the following:

(i) The GCN4 protein binds specifically to the promoter regions of genes subject to general control (e.g. *HIS3*, *HIS4*, *TRP5*, *ARG4*, *ILV1* and *ILV2*), whereas it does not bind to analogous regions of four non-regulated genes (*TRP1*, *DED1*, *URA3* and *GAL1–10*) (Hope and Struhl, 1985; Arndt and Fink, 1986).

(ii) The site on the GCN4 protein that is required for DNA binding *in vitro* was localized by synthesizing a mutant form lacking the 40 carboxy-terminal amino acids which failed to bind DNA *in vitro*. This synthetic GCN4 protein is similar to a *GCN4* mutant gene that is non-functional *in vivo* (Hope and Struhl, 1985).

(iii) The DNA binding and transcriptional activation regions of the protein are separable. The activation region of the protein was localized, using a hybrid gene containing the binding domain of the *E. coli* lexA protein fused to serially deleted regions of the *GCN4* gene and testing for binding activity *in vitro* and stimulation of transcription, either from the authentic *HIS3* promoter or from the lexA operator located in a derivative of the *CYC1* promoter *in vivo*. The region required for transcriptional activation resides in a 19-amino-acid stretch that is located in an acidic region in the centre of the protein, whereas the binding domain is located within the carboxy-terminal 40 amino acids (Hope and Struhl, 1986).

1.7.2 *Binding by the GAL4 activator protein and regulation by GAL80*

Two independent experiments demonstrated that the regulation of the genes involved in galactose metabolism depends on GAL4 binding to the appropriate sequence in the upstream region (Bram and Kornberg, 1985; Giniger *et al.*, 1985). In the first experiment, GAL4 binding was demonstrated using a standard *in vitro* filter binding assay. This allowed identification and partial purification of a binding protein only from strains of yeast engineered to overexpress *GAL4*. This protein was shown to protect the 17-base-pair core consensus sequence from DNase 1 digestion (Bram and Kornberg, 1985). The second approach demonstrated GAL4 binding *in vivo*, using either photofootprinting or methylation protection of bound sequences combined with indirect end labelling to specifically detect the *GAL1–10* intergenic region in the yeast genome. These experiments showed that when compared with purified genomic DNA, specific guanine residues show altered reactivity to methylation or ultraviolet light when yeasts are grown on galactose, but only in a *GAL4*-producing strain (Giniger *et al.*, 1985; Selleck and Majors, 1987*a*, *b*). To show that these changes were due to GAL4 binding and that GAL4 would bind to its cognate sequences in the absence of other yeast proteins, the experiment was repeated in *E. coli*. Two plasmids were transformed into bacteria, one in which *GAL4* is overexpressed by the *tac* promoter and one which contains the *GAL* UAS. The DNA in the bacteria was methylated, and specific modification of guanine residues was seen only when GAL4 was expressed (Giniger *et al.*, 1985). GAL4 is one of the first proteins shown to bind DNA but not to result in direct transcriptional activation.

Separable binding and activation domains were first demonstrated for the GAL4 protein (Brent and Ptashne, 1985). A similar approach was subsequently adopted for demonstrating domains in GCN4, as described in the previous section (Hope and Struhl, 1986). A modified version of a *GAL1–lacZ* fusion was constructed in which UAS$_G$ is replaced by the *E. coli lexA* operator site. No expression of lacZ activity is seen in a yeast strain engineered to overproduce lexA. However, a lexA–GAL4 hybrid protein containing the carboxy-terminal part of GAL4 is able to activate the *GAL* promoter, and results in lacZ expression. The protein binds to the *lexA* operator but stimulates transcription by virtue of the GAL4 domain. This hybrid protein can no longer bind to UAS$_G$, indicating the binding and activation regions are separable. Other experiments by Brent and Ptashne (1985) and Ma and Ptashne (1987*a*) also lead to the conclusion that GAL4 is divided into distinct functional domains. At the amino-terminal end are the regions that direct nuclear localization and binding to UAS$_G$, while functions for activating transcription, oligomerization of GAL4 and interaction with the product of *GAL80* are found towards the carboxy terminus of GAL4 (Silver *et al.*, 1984; Laughon and Gesteland, 1984). The *GAL80* gene product inhibits transcriptional activation by GAL4 which can be relieved by galactose. Catabolite repression (the glucose effect) is independent of *GAL80* and is presumed to mediate repression by preventing GAL4 binding to UAS$_G$ by an unknown mechanism. However, GAL80 has been shown to recognize and bind to the carboxy-terminal 30 amino acids of the GAL4 protein to form a complex that is bound to UAS$_G$ but prevents transcriptional activation. In the presence of galactose, the complex dissociates, thereby activating transcription (Ma and Ptashne, 1987*b*; Johnston *et al.*, 1987). On the basis of these observations for GAL80 function, it is highly likely that GAL80 prevents GAL4 interacting with other proteins to stimulate transcription.

1.7.3 *Binding by the activator protein TUF*

A factor known as TUF has been identified that interacts specifically with the RPG and HOMOL1 motifs that are variants of a consensus upstream activator sequence (UAS$_{RPG}$) found upstream of many genes coding for the yeast translational apparatus (see section 1.3.2). TUF factor was identified and shown to be a specific DNA binding protein, using two independent and unusual approaches (Huet and Sentenac, 1987). Firstly, the protein–DNA complex was isolated directly by pore-limited electrophoresis in polyacrylamide gradient gels, and shown to contain a single 150-kD protein. Secondly, a two-step 'blot and footprint' procedure was used. Here the TUF factor is separated electrophoretically and blotted on to a nitrocellulose filter, then incubated with uniquely end-labelled DNA containing the binding site for sufficient time to allow binding. Excess DNA is then washed off the filter. The bound DNA–protein complex is then incubated briefly with DNase 1, and the resulting DNA fragments separated on a sequencing gel. This second procedure demonstrated that the 150-kD protein blotted on to nitrocellulose exhibited the same specific DNA binding domain as the TUF factor in whole nuclear extracts. The binding domain of TUF was localized to a 50-kD region by limited proteolysis, suggesting distinct function domains on this binding protein.

1.7.4 *The HAP1 protein binds to two different sequences*

The role of the HAP1 protein in controlling the expression of *CYC1* has been discussed at length in sections 1.3 and 1.4. The DNA binding properties of HAP1 have been inferred from the following data. First is the ability of various cloned and frameshifted derivatives of the *HAP1* gene expressed from a high copy number vector to complement a *hap1* mutant; second is the *in vitro* DNA binding activity of protein extracts from yeast containing the *HAP1* derivatives. As in all the previous examples, these data show that binding and transcriptional activation are located in separate domains on the HAP1 protein. The UAS1 binding domain, which also coincides with the heme binding domain, is located at the amino terminus, and the activation domain is at the carboxy terminus (Pfeifer *et al.*, 1987*a*). HAP1 binds to both UAS1 of *CYC1* and UAS of *CYC7*; the binding to both UASs is stimulated by heme, yet the binding sites bear no obvious similarity (Pfeifer *et al.*, 1987*b*). There is one *hap1* mutation that selectively abolishes UAS1 binding in *CYC1*, but has no effect on binding the UAS in *CYC7*, and this suggests that *HAP1* might have two separate binding domains. However, UAS1 and *CYC7* sequences compete equally well with each other for HAP1 binding, suggesting that the HAP1 protein contains only one binding domain. Assuming that HAP1 only contains one binding domain, Pfeifer *et al.* (1987*b*) propose that HAP1 can recognize two different sequences, partly by virtue of the conformation adopted at the binding sites. It was noted that HAP1 has a preponderance of minor groove contacts at both UASs. HAP1 binding is unusual in that it is the only example where there is no consensus sequence at the binding sites. In all the previous examples, the activator or repressor proteins bind to multiple sites that adhere to strict consensus sequences.

1.7.5 Regulatory proteins contain functionally distinct domains

In the four examples discussed previously activation of transcription and DNA binding activity are located in functionally distinct domains on the proteins. Separable functional domains are also found in two well-characterized prokaryotic positive regulatory proteins, Lambda *cI* and phage P22*cII* (Guarente *et al.*, 1982*b*; Hochschild *et al.*, 1983). In all cases, the DNA binding domains are located in regions of the protein where there is a preponderance of basic amino acids, whereas transcriptional activator domains are found in the most acidic regions of the protein. An extensive deletion analysis of the *GAL4* gene which resulted in small regions of the protein being fused to the DNA binding domain showed that GAL4 contains two short acidic regions for activating transcription, presumably by interacting with other proteins (Ma and Ptashne, 1987*a*). Mutant forms of a derivative of GAL4 protein that increase the activation function invariably increase the acidity in activating region 1 (Gill and Ptashne, 1987). Hybrid proteins containing regions of *E. coli* genomic DNA linked to the DNA binding region of GAL4 can activate transcription from the *GAL1* UAS in yeast. Analysis of residues encoded by the *E. coli* DNA shows that all of the activating sequences are acidic, but show no obvious homology with the activating regions of GAL4 or GCN4 (Ma and Ptashne, 1987*b*). The secondary structure at the acidic regions rather than the primary sequence may be important for transcription activation. In some of the mutant forms of GAL4 tested for the ability to activate transcription, up to 80% of the protein between the binding and activator domains could be deleted without significant loss of function. Other activator proteins can also have large portions of the protein deleted without loss of activation function, such as GCN4 (Hope and Struhl, 1986) and ADR1 (Hartshorne *et al.*, 1986). This raises the question as to the function of these regions that are dispensable for transcriptional activation. It is likely that these are the regions through which more complex levels of control are mediated by other proteins or regions that allow cooperative binding to several binding sites within a UAS. Further analysis should resolve these issues.

1.8 Mechanisms for transcriptional activation

Transcription initiation in yeast can be described as a process in which RNA polymerase II initiates mRNA synthesis at discrete sites. The available evidence suggests that the polymerase does not recognize specific sequences on the promoter, but rather a complex of transcription factors assembled on a DNA scaffold. The interactions required for the formation and stabilization of this complex constitute a further level for the control and regulation of transcription. In simplistic terms, this complex can be envisaged as containing activator proteins (such as GAL4 or HAP1), and more general factors such as those proposed to interact at the TATA and I sites. Protein–protein interactions between these factors bound to the DNA, and other general factors required for transcription initiation, may then stabilize the complex. In view of the many different activator proteins required to regulate the expression of the genes transcribed by RNA polymerase II, they must share common features. In the small number of activator proteins sequenced to date, there is no significant homology in the primary amino acid sequence. However, more general structural features can be identified. One feature is the basic regions which correspond to the DNA binding domains. The second structural feature is the short regions composed of acidic residues that are sufficient for transcriptional activation. These acidic activation regions are likely to be interaction sites at which protein–protein contacts are made to form the initiation complex before transcription. In some situations, they may also be target sites which are modified in some way by regulatory proteins that result in the repression of transcription. Regulation at UAS_G by the GAL4 activator protein and the GAL80 repressor exemplify this idea. The site at the carboxy terminus of GAL4 at which GAL80 interacts maps to the same region required for transcriptional activation. GAL4 is bound to regulatory sites in UAS_G in both the induced and uninduced states. Transcriptional activation is determined by competition between GAL80 and other transcription factors for the same region on GAL4. On induction, galactose (or a derivative) interacts with GAL80 and changes the conformation of the protein, so that its affinity for the region on GAL4 is lowered and factors required for transcription can interact. Thus the on/off state of transcription is determined by the relative affinities of the negative regulator, GAL80, and other transcription factors for this region.

A mechanism to explain the observation that activator proteins function at long and variable distances from the TATA element, coupled with the idea that the activator and TATA proteins must interact, directly or indirectly, to form a transcription complex, is required. As many different upstream elements can be functionally associated with different RNA initiation regions, it is likely that any mechanism to explain activation of transcription must be a general one. The ability of DNA–protein complexes situated between the upstream sequences and the TATA element to repress transcription to a greater or lesser extent may suggest that some signal is transferred between these elements. The mechanism used to transfer the activation signal is unknown, but there are speculative models. Ptashne (1986) has proposed a model that involves looping out of the intervening DNA to bring disperse protein–DNA interactions into closer proximity. For instance, interactions between the TATA element and upstream sequences may be mediated in this way.

Struhl (1986) has presented experimental evidence in support of the suggestion that the TATA binding protein is a candidate for interaction with upstream activator proteins, although this does not rule out intermediary factors yet to be identified. The regulatory TATA element T_R is required for properly regulated activation of *HIS3* transcription by either GCN4 or GAL4, whereas activation is not observed in combination with the constitutive TATA element T_C. The implication of this observation is that the protein that binds to T_R can associate with the activation regions on some proteins that bind to upstream regulatory regions whereas transcription initiation using poly(dA–dT) tracts and T_C uses a different molecular mechanism. The protein that binds to the constitutive TATA element may interact more directly with RNA polymerase II.

A gene is not represented by a naked DNA sequence, but exists in the form of chromatin, a DNA–histone complex, as one of many units of genetic information found on a chromosome. The basic unit of chromatin is the nucleosome, and the organization of nucleosomes over the region of a gene may be critical in determining whether or not a gene is expressed. The constitutive promoter elements are proposed to function by excluding nucleosomes from key regions of the promoter. This supports the idea that the chromatin structure may play an important role in some aspect of gene regulation.

Generally genes which are not being expressed exist in a state which is relatively impermeable to nucleases; activation of transcription increases sensitivity of genes to nuclease digestion, particularly in the region of the promoter. Are the changes observed in chromatin structure part of the activation signal? The nuclease-sensitive sites may be a prerequisite for subsequent gene activation by, for example, generating torsional stress required for transcription (Weintraub, 1985). Alternatively, the increased sensitivity to nucleases may be a consequence of the binding of transcription factors which prevent the region of DNA being organized into nucleosomes. The *GAL1–10* promoter displays hypersensitivity to nucleases in both the induced and uninduced state, which may simply reflect GAL4 binding to the four available sites in both states (Giniger *et al.*, 1985; Proffitt, 1985). However, photofootprinting *in vivo* shows that the promoter undergoes changes at the TATA box which are dependent on ongoing transcription (Selleck and Majors, 1987*b*). In the *PHO5* promoter, clear changes in chromatin structure, with the loss of three to four nucleosomes, occur after induction (Almer *et al.*, 1986). In the uninduced state, only one element in the activation site is found in a hypersensitive state; the other three elements are contained within positioned nucleosomes. Although the presence of a nucleosome may not preclude interaction with the regulatory protein, the changes that occur on induction suggest that chromatin organization may be critical for activation in this gene. The single hypersensitive site in the uninduced state may reflect binding of a regulatory protein. On induction, one or more activating proteins may modify the interaction of the regulatory protein with the DNA in such a way as to locally unfold the chromatin, perhaps by enhancing changes in local DNA conformation, transiently exposing the DNA and allowing other regulatory proteins to bind. Cooperative binding of more transcription factors may occur, allowing a propagation of the open chromatin structure in both directions throughout the promoter. This may expose the TATA element and I site and allow the formation of an active transcription complex. A change in the nuclease sensitivity around the TATA element is observed after activation of transcription (Struhl, 1984). Some mechanism, whether the speculative model proposed above or some other, is required to free the RNA initiation region from association with nucleosomes, as it has been demonstrated that nucleosomes inhibit transcription initiation *in vitro* (Lorch *et al.*, 1987).

Using the idea of propagation of an activation signal by altering chromatin structure in both directions from the upstream site, it is also possible to explain how negative elements repress transcription when located at a distance. It may not be necessary for negative repressor proteins to interact directly with activation proteins if repression occurs by simply antagonizing these local changes in chromatin structure. Repression at a distance may also be achieved by repressors mediating interactions at different sites on the nuclear matrix from those required for activation. Association of an active transcription complex with the transcription machinery may occur on membrane surfaces in the nucleus. The *SIR* gene products, which are essential for silencer function at the *HMR* locus, are found associated with the nuclear matrix (Kimmerly and Rine, 1987). Their role in silencer function may be as an attachment site on the nuclear membranes. This interaction may preclude transcription or transcriptional activation by preventing local changes in chromatin structure or access to the transcription machinery.

Thus, binding of one protein to a site on a promoter may displace nucleosomes, revealing other sites at which proteins essential to activation may bind. Providing both upstream and downstream proteins are bound to the DNA and contain available sites required for activation, a functional complex can be formed. One constraint on this interaction that may cause repression of transcription is the nature of the protein–DNA interactions in the intervening sequences. The chromatin structure resulting from binding of one or more factors may potentiate the formation of a complex. Loss of nucleosomes may create stress in the DNA which may be relieved by looping of the intervening DNA and/or the cooperative binding of other intermediary factors, and thus facilitate the formation of the active complex. However, different mechanisms for activation may exist for different upstream/initiation region combination. Looping explains why the upstream sequences in many yeast promoters function in either orientation and in a position-independent manner. In those promoters where activation is clearly position-dependent, there may be constraints on the looping out of intervening DNA. Here the activation signal may be transferred by protein–protein contacts along one side of the DNA helix, such those as seen in the SV40 early promoter (Takahashi *et al.*, 1985). Binding of an activator protein at the upstream site may result in cooperative binding of different factors at several sites, displacing nucleosomes in the process, and potentiating further protein–protein interactions until an active complex is formed. Repression of transcription by protein–DNA interactions may disrupt the formation of a loop, prevent the transfer of the signal through further cooperatively bound proteins, or change the phasing and position of key nucleosomes. It is clear that no one model can explain the mechanisms for transcriptional activation in yeast genes. Different genes are likely to use elements of each factor that can be correlated with transcriptional activation to regulation expression in response to physiological signals.

1.9 Industrial application of yeast promoters

Saccharomyces cerevisiae is an excellent host for the large-scale production of many natural and novel proteins (for recent reviews, see Kingsman *et al.*, 1985, 1987). Generally the expression of any synthetic or foreign gene in yeast requires the use of yeast sequences to direct efficient transcription and translation. Much of the fundamental research described in this chapter is directed towards understanding how transcription directed by yeast promoters is controlled and regulated. This information is now being transferred and applied to industrial needs. It is now recognized that the choice of a promoter is no longer dictated by availability; a promoter can be designed especially to meet the economic and logistic factors involved in the commercial production of one particular product.

Two types of promoter can be used to express a foreign gene in yeast. Constitutive promoters allow a gene product to be synthesized continuously throughout the culture period. These promoters are suitable for the expression of products which are stable and non-toxic at high levels to the host. Regulated promoters offer greater flexibility, and allow the gene product to be expressed only in response to defined conditions.

High-efficiency expression vectors are often used at the present time for producing mammalian or other proteins in yeast. However, expressing genes at high copy number can often present unforeseen problems. The replication origin from the endogenous yeast plasmid, 2-micron, allows vectors to be maintained at between 50 and 200 copies per cell. All other regulatory sequences present on these plasmids will also be present at high copy number and this must be considered with regard to promoter function. It is now clear that some specific *trans*-acting proteins are present at low concentrations and can severely limit the expression potential of a promoter present at high copy number (Baker *et al.*, 1987; Irani *et al.*, 1987). Maximal expression from promoters induced by galactose can be achieved only by increasing the concentration of GAL4 and decreasing the concentration of GAL80 (Baker *et al.*, 1987). Other factors as yet unidentified may also be limiting the expression from some promoters at high copy number, such as those induced by galactose. Other promoters are relatively insensitive to limiting *trans*-acting factors at high copy number. For instance, the *PGK* gene, which encodes the glycolytic enzyme phosphoglycerate kinase is normally expressed at 1–2% of total cell protein. Placing the intact gene on a high copy number (100 per cell) vector yields PGK at about 50–80% of total cell protein (Mellor *et al.*, 1985a). Thus, directing the expression of any heterologous coding region with the *PGK* promoter should theoretically result in substantial yields. Results with *PGK* and many other promoters have been disappointing; generally yields are only 1–5% of total cell protein. Overall yields can be improved if the vector is designed to secrete the protein continuously into the medium. Much effort has been directed towards identifying the defect in heterologous gene expression from the *PGK* promoter which appears to have a reduction in the efficiency of transcription (Mellor *et al.*, 1987). Levels of 5% of total cell protein, although lower than theoretically possible, are still within the range that makes the production of a foreign protein economically viable in yeast. As more yeast genes are isolated and their promoters characterized, the scope for designing efficient strategies for foreign gene expression increases. Rather than express genes from regulated promoters at high copy number, one or two hybrid genes could be integrated into the host genome. Switching on gene expression from these integrated genes towards the end of a fermentation after a high biomass is reached may result in more economical yields than those obtained when a high-copy-number vector is used. Promoters from highly expressed genes that are switched on only when yeasts reach stationary phase could be used as well as promoters that can exploit by-products of other industrial processes. Yeast can be grown to a high biomass on a 'waste' carbon source such as whey. A promoter can be designed such that a gene is switched on only after such a carbon source is exhausted. Thus biomass can be achieved before the product is made. Heat can also be utilized in this way by driving expression from a heat shock promoter. Many yeast promoters are 'leaky', and small amounts of a product may be produced even before a promoter is induced. If a product is particularly toxic, this may result in problems in achieving sufficient biomass. The silencer of transcription at the *HMR* locus could be exploited together with heat to design a tightly regulated expression system. The silencer could be regulated by a temperature-sensitive mutation in one or more of the *SIR* genes, and expression of the foreign gene induced using a heat shock promoter at the same time the silencer is inactivated. As more is understood about the mechanisms used to regulate gene expression from yeast promoters, greater sophistication in the design of economic strategies for foreign gene expression in yeast will result.

References

Almer, A., Rudolph. H., Hinnen, A. and Horz, W. (1986) Removal of positioned nucleosomes from the yeast *PHO5* promoter upon *PHO5* induction releases additional upstream activating DNA elements. *EMBO J.* **5**: 2689.

Arcangioli, B. and Lescure, B. (1985) Identification of proteins involved in the regulation of yeast iso-1-cytochrome c expression by oxygen. *EMBO J.* **4**: 2627.

Arndt, K. and Fink, G. (1986) GCN4 protein, a positive transcription factor in yeast, binds general control promoters at all 5'TGACTC3' sequences. *Proc. Natl. Acad. Sci. USA* **83**: 8516.

Baker, S.M., Johnston, S.A., Hopper, J.E. and Jaehring, J.A. (1987) Transcription of multiple copies of the yeast *GAL7* gene is limited by specific factors in addition to *GAL4*. *Mol. Gen. Genet.* **208**: 127.

Beier, D.R. and Young, E.T. (1982) Characterisation of a regulatory region upstream of the *ADR2* locus of *Saccharomyces cerevisiae*. *Nature* **300**: 724.

Berg, P.E., Popovic, Z. and Anderson, W.F. (1984) Promoter dependence of enhancer activity. *Mol. Cell. Biol.* **4**: 1664.

Bram, R.J. and Kornberg, R.D. (1985) Specific protein binding to far upstream activating sequences in polymerase II promoters. *Proc. Natl. Acad. Sci. USA* **82**: 43.

Bram, R.J. and Kornberg, R.D. (1987) Isolation of a *Saccharomyces cerevisiae* centromere DNA-binding protein, its human homologue, and its possible role as a transcription factor. *Mol. Cell. Biol.* **7**: 403.

Bram, R.J., Lue, N.F. and Kornberg, R.D. (1986) A *GAL* family of upstream activating sequences in yeast: roles in both induction and repression of transcription. *EMBO J.* **5**: 603.

Brand, A.H., Breeden, L., Abraham, J., Sternglanz, R. and Nasmyth, K. (1985) Characterisation of a silencer in yeast: a DNA sequence with properties opposite to those of a transcriptional activator. *Cell* **41**: 41.

Brand, A.H., Micklem, G. and Nasmyth, K. (1987) A yeast silencer contains sequences that can promote autonomous replication and transcriptional activation. *Cell* **51**: 709.

Breathnach, A. and Chambon, P. (1981) Organization and expression of eukaryotic split genes coding for proteins. *Ann. Rev. Biochem.* **50**: 349.

Brent, R. and Ptashne, M. (1984) A bacterial repressor protein or a yeast transcriptional terminator can block upstream activation of a yeast gene. *Nature* **312**: 612.

Brent, R. and Ptashne, M. (1985) A eukaryotic transcriptional activator bearing the DNA specificity of a prokaryotic repressor. *Cell* **43**: 729.

Broach, J.R. (1983) Construction of high copy number yeast vectors using 2 μm circle sequences. In *Methods in Enzymology*, eds. R. Wu, L. Grossman and K. Moldave, Vol. 101, Academic Press, New York, 307.

Burke, R.L., Tekamp–Olsen, P. and Najarian, R. (1983) The isolation, characterisation and sequence of the pyruvate kinase gene of *Saccharomyces cerevisiae*. *J. Biol. Chem.* **258**: 2193.

Celenza, J.L. and Carlson, M. (1986) A yeast gene that is essential for release from glucose repression encodes a protein kinase. *Science* **233**: 1175.

Chen, W. and Struhl, K. (1985) Yeast mRNA initiation sites are determined primarily by specific sequences, not by the distance from the TATA element. *EMBO J.* **4**: 3273.

Clavilier, L., Pere Aubert, G., Somlo, M. and Slonimski, P. (1976) Réseau d'interactions entre des gènes non liés: regulation synégique ou antagoniste de la synthèse de l'iso-1-cytochrome c, de l'iso-2-cytochrome c et du cytochrome b2. *Biochemie* **58**: 155.

Cohen, R., Yokoi, T., Holland, J. P., Pepper, A.E. and Holland, M. J. (1987) Transcription of the constitutively expressed yeast enolase gene *ENO1* is mediated by positive and negative *cis*-acting regulatory sequences. *Mol. Cell. Biol.* **7**: 2753.

deCrombrugghe, B., Busby, S. and Buc, H. (1984) In *Biological Regulation and Development*. Vol. IIIB, ed. Yamamoto, K., Plenum, New York, 129.

Dobson, M.J., Tuite, M.F., Roberts, N.A., Kingsman, A.J., Kingsman, S.M., Perkins, R.E., Conroy, S.C., Dunbar, B. and Fothergill, L.A. (1982a) Conservation of high efficiency promoter sequences in *Saccharomyces cerevisiae*. *Nucleic Acids Res.* **10**: 2625.

Dobson, M.J., Tuite, M.F., Mellor, J., Roberts, N.A., King, R.M., Burke, D.C., Kingsman, A.J. and Kingsman, S.M. (1982b) Expression in *Saccharomyces cerevisiae* of human interferon-alpha directed by the *TRP1* 5' region. *Nucleic Acids Res.* **11**: 2287.

Donahue, T.F., Farabaugh, P.J. and Fink, G.R. (1982) The nucleotide sequence of the *HIS4* region of yeast. *Gene* **18**: 47.

Errede, B., Company, M., Ferchak, J.D., Hutchinson, C.A. and Yarnell, W.S. (1985) Activation regions in a yeast transposon have homology to mating type control sequences and to mammalian enhancers. *Proc. Natl. Acad. Sci. USA.* **82**: 5423.

Faye, G., Leung, D., Tatchell, K., Hall, B.D. and Smith, M. (1981) Deletion mapping of sequences essential for *in vivo* transcription of the iso-1-cytochrome c gene. *Proc. Natl. Acad. Sci. USA.* **78**: 2258.

Ghosh, P.K., Lebowitz, P., Frisque, R.J. and Glutzman, Y. (1981) Identification of a promoter component involved in positioning the 5′ terminus of simian virus 40 mRNAs. *Proc. Natl. Acad. Sci. USA.* **77**: 100.

Gill, G. and Ptashne, M. (1987) Mutants of GAL4 protein altered in an activation function. *Cell* **51**: 121.

Giniger, E., Varnum, S. and Ptashne, M. (1985) Specific DNA-binding of GAL4, a positive regulatory protein of yeast. *Cell* **40**: 767.

Greenberg, M.L., Myers, P.L., Skvirsky, R.C. and Greer, H. (1986) New positive and negative regulators for general control of amino acid biosynthesis in *Saccharomyces cerevisiae*. *Mol. Cell. Biol.* **6**: 1820.

Guarente, L. (1984) Yeast promoters: positive and negative elements. *Cell* **36**: 799.

Guarente, L. and Hoar, E. (1984) Upstream sites of the *CYC1* gene of *Saccharomyces cerevisiae* are active when inverted but not when placed downstream of the TATA box. *Proc. Natl. Acad. Sci. USA.* **81**: 7860.

Guarente, L. and Mason, T. (1983) Heme regulates transcription of the *CYC1* gene of *Saccharomyces cerevisiae*. *Cell* **32**: 1279.

Guarente, L. and Ptashne, M. (1981) Fusion of *Escherichia coli lacZ* to the cytochrome c gene of *Saccharomyces cerevisiae*. *Proc. Natl. Acad. Sci. USA* **78**: 2199.

Guarente, L., Yocum, R.R. and Gifford, P. (1982a) A *GAL10-CYC1* hybrid yeast promoter identifies the *GAL4* regulator as an upstream site. *Proc. Natl. Acad. Sci. USA* **79**: 7410.

Guarente, L., Nye, J.S., Hochschild, A. and Ptashne, M. (1982b) Mutant Lambda phage repressor with a specific defect in its positive control function. *Proc. Natl. Acad. Sci. USA* **79**: 2236.

Guarente, L., Lalonde, B., Gifford, P. and Alami, E. (1984). Distinctly regulated tandem upstream activation sites mediate catabolite repression of the *CYC1* gene of *Saccharomyces cerevisiae*. *Cell* **36**: 503.

Hahn, S., Hoar, E.T. and Guarente, L. (1985) Each of three "TATA elements" specifies a subset of the transcription initiation sites at the *CYC1* promoter of *Saccharomyces cerevisiae*. *Proc. Natl. Acad. Sci. USA* **82**: 8562.

Hartshawn, T.A., Blumberg, H. and Young, E.T. (1986) Sequence homology of the yeast regulatory protein *ADR1* with Xenopus transcription factor TFIIIA. *Nature* **320**: 283.

Hereford, L.L. and Rosbash, M. (1977) Number and distribution of polyadenylated RNA sequences in yeast. *Cell* **10**: 453.

Hill, D.E., Hope, I.A., Macke, J.R. and Struhl, K. (1986) Saturation mutagenesis of the yeast *HIS3* regulatory site: requirements for transcriptional induction and for binding by GCN4 activator protein. *Science* **234**: 451.

Hinnebusch, A.G. (1984) Evidence for translational regulation of the activator of general amino acid control in yeast. *Proc. Natl. Acad. Sci. USA* **81**: 6442.

Hinnebusch, A.G. and Fink, G.R. (1983a) Positive regulation in the general amino acid control of *Saccharomyces cerevisiae*. *Proc. Natl. Acad. Sci. USA* **80**: 5374.

Hinnebusch, A.G. and Fink, G.R. (1983b) Repeated DNA sequences upstream from *HIS1* also occur at several other co-regulated genes in *Saccharomyces cerevisiae*. *J. Biol. Chem.* **258**: 5238.

Hinnebusch, A.G., Lucchini, G. and Fink, G.R. (1985) A synthetic *HIS4* regulatory element confers general amino acid control on the cytochrome c gene (*CYC1*) of yeast. *Proc. Natl. Acad. Sci. USA* **82**: 498.

Hochschild, A., Irwin, N. and Ptashne, M. (1983) Repressor structure and the mechanism of positive control. *Cell* **32**: 319.

Hope, I.A. and Struhl, K. (1985) GCN4 protein, synthesized *in vitro*, binds *HIS3* regulatory sequences: implications for general control of amino acid biosynthetic genes in yeast. *Cell* **43**: 177.

Hope, I.A. and Struhl, K. (1986) Functional dissection of a eukaryotic transcriptional activator protein, GCN4 of yeast. *Cell* **46**: 885.

Huet, J. and Sentenac, A. (1987) TUF, the yeast DNA-binding factor specific for UAS_{rpg} upstream activation sequences: Identification of the protein and its DNA-binding domain. *Proc. Natl. Acad. Sci. USA* **84**: 3648.

Huet, J., Cottrelle, P., Cool, M., Vignais, M.-L., Thiele, D., Marck, C., Buhler, J.-M., Senetac, A. and Fromageot, P. (1985) A general upstream binding factor for the genes of the yeast translational apparatus. *EMBO J.* **4**: 3539.

Igarashi, M., Segawa, T., Nogi, Y., Suzuki, Y. and Fukasama, T. (1987) Autogeneous regulation of the *Saccharomyces cerevisiae* regulatory gene *GAL80*. *Mol. Gen. Genet.* **207**: 273.

Iida, H. and Yahora, I. (1985) Yeast heat-shock protein of Mr 48 000 is an isoprotein of enolase. *Nature* **315**: 688.

Irani, M., Taylor, W.E. and Young, E.T. (1987) Transcription of the *ADH2* gene in *Saccharomyces cerevisiae* is limited by positive factors that bind competitively to its intact promoter region on multicopy plasmids. *Mol. Cell. Biol.* **7**: 1233.

Jacob, F. and Monod, J. (1961) Gene regulatory mechanisms in the synthesis of proteins. *J. Mol. Biol.* **3**: 318.

Jacob, F., Ullman, A. and Monod, J. (1964) *C.R. Hebd. Séanc. Acad. Sci., Ser.* X, **258**: 3125.

Johnson, A.D. and Herskowitz, I. (1985) A repressor protein (MATα2 product) and its operator control expression of a set of cell type specific genes is yeast. *Cell* **42**: 237.

Johnson, S. and Hopper, J.E. (1982). Isolation of the yeast regulatory gene *GAL4* and analysis of its dosage effects on the galactose–melibiose region. *Proc. Natl. Acad. Sci. USA*, **79**: 6971.

Johnson, S.A., Salmeron, J.M. and Dincher, S.S. (1987) Interaction of positive and negative regulatory proteins in the galactose regulon of yeast. *Cell*. **50**: 143.

Jones, E.W. and Fink, G.R., (1982) Regulation of amino acid biosynthesis and nucleotide biosynthesis in yeast. In *The Molecular Biology of the Yeast Saccharomyces: Metabolism and Gene Expression*, eds. J.N. Strathern, E.W. Jones and J.R. Broach, Cold Spring Harbor, New York, 181.

Kaback, D.B., Angerer, L.M. and Davidson, N. (1979) Improved methods for the formation and stabilisation of R loops. *Nucleic Acids Res.* **6**: 2499.

Kim, S., Mellor, J., Kingsman, A.J. and Kingsman, S.M. (1988) An AT rich region of dyad symmetry is a promoter element in the yeast *TRP1* gene. *Mol. Gen. Genet.* **211**: 472.

Kim, S. (1986) D. Phil. Thesis, University of Oxford, UK.

Kim, S., Mellor, J., Kingsman, A.J. and Kingsman, S.M. (1986) Multiple control elements in the *TRP1* promoter of *Saccharomyces cerevisiae*. *Mol. Cell. Biol.* **6**: 4251.

Kimmerly, W. and Rine, J. (1987) The replication and segregation of plasmids containing *cis*-acting regulatory sites of silent mating type genes in *Saccharomyces cerevisiae* is regulated by *SIR*. *Mol. Cell. Biol.* **7**: 4225.

Kingsman, S.M., Kingsman, A.J. and Mellor, J. (1987) The production of mammalian proteins in *Saccharomyces cerevisiae*. *Trends in Biotechnol.* **5**: 53.

Kingsman, S.M., Kingsman, A.J., Dobson, M.J., Mellor, J. and Roberts, N.A. (1985) Heterologous gene expression in *Saccharomyces cerevisiae*. *Biotechnol. Genet. Eng. Rev.* **3**: 377.

Koo, H.-S., Wu, H.-M. and Crother, D.M. (1986) DNA bending at adenine–thymine tracts. *Nature* **320**: 501.

Kozak, M. (1980) Binding of wheat germ ribosomes to bisulphite modified reovirus messenger RNA: Evidence for a scanning mechanism. *J. Mol. Biol.* **144**: 291.

Kunkel, G.R. and Martinson, H.G. (1981) Nucleosomes will not form on double-stranded RNA or over poly(dA)-poly(dT) tracts in recombinant DNA. *Nucleic Acids Res.* **9**: 6869.

LaLonde, B., Arcangioli, B. and Guarente, L. (1986) A single *Saccharomyces cerevisiae* upstream activation site (UAS1) has two distinct regions essential for its activity. *Mol. Cell. Biol.* **6**: 4690.

Laughon, A. and Gesteland, R. (1982) Isolation and preliminary characterisation of the GAL4 gene, a positive regulator of transcription in yeast. *Proc. Natl. Akad. Sci. USA*, **79**: 6827.

Lorch, Y., LaPointe, J. W. and Kornberg, R. D. (1987) Nucleosomes inhibit the initiation of transcription but allow chain elongation with displacement of histones. *Cell* **49**: 203.

Lowry, C.V. and Lieber, R.H. (1986) Negative regulation of the *Saccharomyces cerevisiae ANB1* gene by heme, as mediated by the *ROX1* gene product. *Mol. Cell. Biol.* **6**: 4145.

Ma, J. and Ptashne, M. (1987*a*) Deletion analysis of GAL4 defines two transcriptional activating segments. *Cell*. **48**: 847.

Ma, J. and Ptashne, M. (1987b) The carboxy-terminal 30 amino acids of GAL4 are recognized by GAL80. *Cell* **50**: 137.

Marini, J.C., Levene, S.D., Crothers, D.M. and Englund, P.T. (1982) A bent helix in kinetoplast DNA. *Cold Spring Harb. Symp. Quant. Biol* **47**: 279.

McNeil, J.B. and Smith, M. (1985) *Saccharomyces cerevisiae CYC1* mRNA 5' end positioning: analysis by *in vitro* mutagenesis, using synthetic duplexes with random mismatch base pairs. *Mol. Cell. Biol.* **5**: 3545.

McNeil, J.B. and Smith, M. (1986) Transcription initiation of the *Saccharomyces cerevisiae* iso-1-cytochrome c gene: multiple independent TATA sequences. *J. Mol. Biol.* **187**: 363.

Mellor, J., Dobson, M.J., Roberts, N.A., Tuite, M.F., Emtage, J.S., White, S., Lowe, P.A., Patel, T., Kingsman, A.J. and Kingsman, S.M. (1983) Efficient synthesis of enzymatically active calf chymosin in *Saccharomyces cerevisiae*, *Gene* **24**: 1.

Mellor, J., Dobson, M.J., Roberts, N.A., Kingsman, A.J. and Kingsman, S.M. (1985a) Factors affecting heterologous gene expression in *Saccharomyces cerevisiae*. *Gene* **33**: 215.

Mellor, J., Fulton, A.M., Dobson, M.J., Wilson, W., Kingsman, S.M. and Kingsman, A.J. (1985b) A retrovirus-like strategy for expression of a fusion protein encoded by the yeast transposon Ty1. *Nature* **313**: 243.

Mellor, J., Dobson, M.J., Kingsman, A.J. and Kingsman, S.M. (1987) A transcriptional activator is located in the coding region of the yeast *PGK* gene. *Nucleic Acids Res.* **15**: 6243.

Miller, A.M. and Nasmyth, K.A. (1984) Role of DNA replication in the repression of silent mating type loci in yeast. *Nature* **312**: 247.

Miller, J.H., MacKay, V.L. and Nasmyth, K.A. (1985) Identification and comparison of two sequence elements that confer cell type transcription in yeast. *Nature* **314**: 598.

Nagawa, A. and Fink, G.R. (1985) The relationship between the "TATA" sequences and transcription initiation sites at the *HIS4* gene of *Saccharomyces cerevisiae*. *Proc. Natl. Acad. Sci. USA* **82**: 8557.

Nasmyth, K.A. (1982) Molecular genetics of yeast mating type. *Ann. Rev. Genet.* **16**: 439.

Nasmyth, K.A. (1983) Molecular analysis of a cell lineage. *Nature* **302**: 670.

Nasmyth, K.A. (1985a) At least 1400 base pairs of 5'-flanking DNA is required for the correct expression of *HO* gene in yeast. *Cell* **42**: 312.

Nasmyth, K.A. (1985b) A repetitive DNA sequence that confers cell-cycle START (*CDC28*)-dependent transcription of the *HO* gene in yeast. *Cell* **42**: 225.

Nasmyth, K.A. (1987) The determination of mother cell specific mating type switching in yeast by a specific regulator of *HO* transcription. *EMBO J.* **6**: 243.

Nasmyth, K.A., Stillman, D.J. and Kipling, D. (1987) Both positive and negative regulators of *HO* transcription are required for mother cell-specific mating-type switching in yeast. *Cell* **48**: 579.

Oettinger, M.A. and Struhl, K. (1985) Suppressors of *Saccharomyces cerevisiae his3* promoter mutations lacking the upstream element. *Mol. Cell. Biol.* **5**: 1910.

Ogden, J.E., Stanway, C., Kim, S., Mellor, J., Kingman, A.J. and Kingsman, S.M. (1986) Efficient expression of the *Saccharomyces cerevisiae PGK* gene depends on an upstream activation sequence but does not require TATA sequences. *Mol. Cell. Biol.* **6**: 4335.

Peck, L.J. and Wang, J.C. (1981) Sequence dependence of the helical repeat of DNA in solution. *Nature* **292**: 375.

Petes, T.D., Hereford, L.M. and Konstanin, K.G. (1978) Characterisation of two types of yeast ribosomal DNA genes. *J. Bacteriol.* **134**: 295.

Pfeifer, K., Arcangioli, B. and Guarente, L. (1987a) Yeast *HAP1* activator competes with the factor RC2 for binding to the upstream activation site UAS1 of the *CYC1* gene. *Cell* **49**: 9.

Pfeifer, K., Prezent, T. and Guarente, L. (1987b) Yeast *HAP1* binds to two upstream activation sites of different sequences. *Cell* **49**: 19.

Pinkham, J.L. and Guarente, L. (1985) Cloning and molecular analysis of the *HAP2* locus: a global regulator of respiration gene in *Saccharomyces cerevisiae*. *Mol. Cell. Biol.* **5**: 3410.

Piper, P.W., Curran, B., Davis, M.W., Lockheart, A. and Reid, G. (1986) Transcription of the phosphoglycerate kinase gene of *Saccharomyces cerevisiae* increases when fermentative cultures are stressed by heat-shock. *Eur. J. Biochem.* **161**: 525.

Proffitt, J.H. (1985) DNase1-hypersensitive sites in the galactose gene cluster of *Saccharomyces cerevisiae*. *Mol. Cell. Biol.* **5**: 1522.

Prunell, A. (1982) Nucleosome reconstitution on plasmid-inserted poly(dA)-poly(dT). *EMBO J.* **1**: 173.

Ptashne, M. (1986) Gene regulation by proteins acting nearby and at a distance. *Nature* **322**: 697.

Rathjen, P.D., Kingsman, A.J. and Kingsman, S.M. (1987) The yeast ROAM mutation: identification of the sequences mediating host gene activation and cell-type control in the yeast retrotransposon Ty. *Nucleic Acids Res.* **15**: 7309.

Rhodes, D. and Klug, A. (1981) Sequence–dependent helical periodicity in DNA. *Nature* **292**: 378.

Rose, M. and Botstein, D. (1983) Construction and use of gene fusions to *lacZ* (β-galactosidase) that are expressed in yeast. *Meth. Enzymol.* **101**: 167.

Rotenberg, M.O. and Woolford, J.L. (1986) Tripartite upstream promoter element essential for expression of *Saccharomyces cerevisiae* ribosomal protein genes. *Mol. Cell. Biol.* **6**: 674.

Rudolf and Hinnen, A. (1987) The yeast *PHO5* promoter: phosphate-controlled elements and sequences mediating mRNA start-site selection. *Proc. Natl. Acad. Sci. USA* **84**: 1340.

Russell, D.W., Smith, M., Cox, D., Williamson, V.M. and Young, E.T. (1983) DNA sequences of two yeast promoter-up mutants. *Nature* **304**: 652.

Ryder, K., Silver, S., DeLucia, A.L., Fanning, E. and Tegmeyer, P. (1986) An altered conformation in origin region I is a determinant for the binding of SV40 large T antigen. *Cell* **44**: 719.

Sarokin, L. and Carlson, M. (1984) Upstream region required for regulated expression of the glucose-repressible *SUC2* gene of *Saccharomyces cerevisiae*. *Mol. Cell. Biol.* **4**: 2750.

Sarokin, L. and Carlson, M. (1986) Short repeated elements in the upstream regulatory region of the *SUC2* gene of *Saccharomyces cerevisiae*. *Mol. Cell. Biol.* **6**: 2324.

Scherer, S. and Davis, R.W. (1979) Replacement of chromosmal segments with altered DNA sequences constructed *in vitro*. *Proc. Natl. Acad. Sci. USA* **76**: 4951.

Schultz, J. and Carlson, M. (1987) Molecular analysis of *SSN6*, a gene functionally related to the *SNF1* protein kinase of *Saccharomyces cerevisiae*. *Mol. Cell. Biol.* **7**: 3637.

Selleck, S.B. and Majors, J. (1987a) *In vivo* DNA-binding properties of a yeast transcription activator protein. *Mol. Cell. Biol.* **7**: 3260.

Selleck, S.B. and Majors, J. (1987b) Photofootprinting *in vivo* detects transcription-dependent changes in yeast TATA boxes. *Nature* **325**: 173.

Shore, D. and Nasmyth, K.A. (1987) Purification and cloning of a DNA binding protein from yeast that binds to both silencer and activator elements. *Cell* **51**: 721.

Siliciano, P.G. and Tatchell, K. (1984) Transcription and regulatory signals at the mating type locus in yeast *Cell* **37**: 969.

Siliciano, P.G. and Tatchell, K. (1986). Identification of the DNA sequences controlling the expression of the *MATα* locus of yeast. *Proc. Natl. Acad. Sci. USA* **83**: 2320.

Silver, P.A., Keegen, L. and Ptashne, M. (1984) Amino terminus of the yeast *GAL4* gene product is sufficient for nuclear localisation. *Proc. Natl. Acad. Sci. USA* **81**: 5951.

Slater and Craig. (1987) Transcriptional regulation of an *hsp70* heat shock gene in the yeast *Saccharomyces cerevisiae*. *Mol. Cell. Biol.* **7**: 1906.

Sternberg, P.W., Stern, M.J. Clark, I. and Herskowitz, I. (1987) Activation of the yeast *HO* gene by release from multiple negative controls. *Cell* **48**: 567.

St John, T.P. and Davis, R.W. (1979) Isolation of galactose sequences from *Saccharomyces cerevisiae* by differential plaque filter hybridization. *Cell* **16**: 443.

Strathern, J.N., Jones, E.W. and Broach, J.R. (eds.) (1982) *The Molecular Biology of the Yeast Saccharomyces*, Vols 1 and 2. Cold Spring Harbor Laboratory, New York.

Struhl, K. (1981) Deletion mapping a eukaryotic promoter. *Proc. Natl. Acad. Sci. USA* **78**: 4451.

Struhl, K. (1982a) Regulatory sites for *his3* expression in yeast. *Nature* **300**: 284.

Struhl, K. (1982b) Promoter elements, regulatory elements, and chromatin structure of the yeast *HIS3* gene. *Cold Spring Harb. Symp. Quant. Biol.* **47**: 901.

Struhl, K. (1982c) The yeast *HIS3* promoter contains at least two distinct elements. *Proc. Natl. Acad. Sci. USA* **79**: 7385.

Struhl, K. (1983a) The new yeast genetics. *Nature* **305**: 391.

Struhl, K. (1984) Genetic properties and chromatin structure of the yeast *gal* regulatory element: an enhancer-like sequence. *Proc. Natl. Acad. Sci. USA* **81**: 7865.

Struhl, K. (1985a) Naturally occurring poly(dA-dT) sequences are upstream promoter elements for constitutive transcription in yeast. *Proc. Natl. Acad. Sci. USa* **82**: 8419.

Struhl, K. (1985b) Negative control at a distance mediates catabolite repression in yeast. *Nature* **317**: 822.

Struhl, K. (1986) Constitutive and inducible *Saccharomyces cerevisiae* promoters: Evidence for two distinct molecular mechanisms. *Mol. Cell. Biol.* **6**: 3847.

Struhl, K. (1987) Promoters, activator proteins and the mechanism of transcriptional activation in yeast. *Cell* **49**: 295.

Struhl, K. and Davis, R.W. (1981) Transcription of the *his3* gene region in *Saccharomyces cerevisiae*. *J. Mol. Biol.* **152**: 535.

Struhl, K. and Hill, D.E. (1987) Two related regulatory sequences are required for maximal induction of *Saccharomyces cerevisiae his3* transcription. *Mol. Cell. Biol.* **7**: 104.

Struhl, K., Chen, W., Hill D.E., Hope, I.A. and Oettinger, M.A. (1985) Constitutive and coordinately regulated transcription of yeast genes: Promoter elements, positive and negative regulatory sites and DNA binding proteins. *Cold Spring Harb. Symp. Quant. Biol.* **50**: 489.

Takahashi, K., Vigneron, M., Matthes, H., Wilderman, A., Zenke, M. and Chambon, P. (1986) Requirement for stereospecific alignments for initiation from the SV40 early promoter. *Nature* **319**: 121.

Thiele, D.J. and Hamer, D.H. (1986) Tandemly duplicated upstream control sequences mediate copper-induced transcription of the *Saccharomyces cerevisiae* copper-metallothionin gene. *Mol. Cell. Biol.* **6**: 1158.

Thireos, G., Penn, M.D. and Greer, H. (1984) 5′ untranslated sequences are required for the translational control of a yeast regulatory gene. *Proc. Natl. Acad. Sci. USA* **81**: 5096.

Tschumper, G. and Carbon, J. (1980) Sequence of a yeast DNA fragment containing a chromosomal replicator and the *TRP1* gene. *Gene.* **10**: 157.

Verdiere, J., Greusot, F., Guarente, L. and Slonimski, P. (1986) The overproducing *CYP1* and the underproducing *hap1* mutations are alleles of the same gene which regulates in *trans* the expression of the structural genes encoding the iso-1-cytochrome c. *Curr. Genet.* **10**: 339.

Vignais, M.-L., Woudt, L.P., Wasswnaar, G.M., Mager, W.H. Sentenac, A. and Planta, R.J. (1987) Specific binding of TUF factor to upstream activation sites of yeast ribosomal protein genes. *EMBO J.* **6**: 1451.

Wasylyk, B., Wasylyk, C., Augereau, P. and Chambon, P. (1983) The SV40 72bp repeat preferentially potentiates transcription starting from proximal natural or substitute promoter elements. *Cell* **32**: 503.

Weintraub, H. (1985) Assembly and propagation of repressed and derepressed chromosomal states. *Cell* **42**: 705.

West, R.W., Yocum, R.R. and Ptashne, M. (1984) *Saccharomyces cerevisiae GAL1–GAL10* divergent promoter region: location and function of the upstream activating sequence UAS_G. *Mol. Cell. Biol.* **4**: 2467.

Williamson, V.M., Cox, D., Yound, E.T., Russel, D.W. and Smith, M. (1983). Characterisation of transposable element-associated mutations that alter yeast alcohol dehydrogenase II expression. *Mol. Cell. Biol.* **3**: 20.

Wilson, W., Malim, M.H., Mellor, J., Kingsman, S.M. and Kingsman, A.J. (1986) Expression strategies of the yeast transposon Ty: a short sequence directs ribosomal frameshifting. *Nucleic Acids Res.* **14**: 7001.

Wright, C.F. and Zitomer, R.S. (1984). A positive regulatory site and a negative regulatory site control the expression of the *Saccharomyces cerevisiae CYC7* gene. *Mol. Cell. Biol.* **4**: 2023.

Wright, C.F. and Zitomer, R.S. (1985) Point mutations implicate repeated sequence as essential elements of the *CYC7* negative upstream site in *Saccharomyces cerevisiae*. *Mol. Cell. Biol.* **5**: 2951.

Yarger, J.G., Armilei, G. and Gorman, M.C. (1986) Transcription terminator-like element within a *Saccharomyces cerevisiae* promoter region. *Mol. Cell. Biol.* **6**: 1095.

Yocum, R.R., Hanley, S., West, R. and Ptashne, M. (1984) Use of *lacZ* fusions to delimit regulatory elements of the inducible divergent *GAL1–GAL10* promoter in *Saccharomyces cerevisiae*. *Mol. Cell. Biol.* **4**: 1985.

Zillig, W., Fujiki, H., Blum, W., Janekowic, D., Schwiger, M., Rahmsdorf, H.J., Ponta, H. and Hirsch-Kauffman, M. (1975) *In vivo* and *in vitro* phosphorylation of DNA-dependent RNA polymerase of *E. coli* by bacteriophage-T-induced protein kinase. *Proc. Natl. Acad. Sci. USA* **72**: 2506.

2 Negative regulation of gene expression by mating type

E. FINTAN WALTON and GEOFFREY T. YARRANTON

2.1 Introduction

The regulation of gene expression in *Saccharomyces cerevisiae*, like that in other prokaryotic and eukaryotic organisms, is controlled in part by the presence of both positive and negative elements found in the 5' untranslated promoter region of genes. These elements can lie up to hundreds of base-pairs upstream from the site of transcriptional initiation. The regulation of gene expression plays an important role in determining both the fate of a particular cell within its life-cycle and its response to specific physiological and environmental conditions. Both negative and positive elements are *cis*-acting and confer their effect by the presence or absence of *trans*-acting regulatory proteins encoded by genes independent of the regulated gene. Each regulatory protein binds to a specific sequence within the positive or negative elements, and can regulate a number of different genes which contain these specific sequences. The most common and prominent positive elements are referred to as Upstream Activator Sequences (UASs) because of their position upstream from the site of mRNA transcription, and are analogous, in some respects, to the enhancer sequences found in mammalian promoters. Activator proteins bind to these elements and activate the transcription of the gene. Conversely, repressor proteins bind to negative elements (operator sequences) normally located between the UAS and the site of transcription initiation to repress transcription. Transcriptional activation of a particular gene or set of genes normally requires the presence of specific activator proteins *and* the absence of specific repressor proteins, whereas repression normally requires the presence of these repressor proteins *and/or* the absence of the activator proteins.

A comprehensive review of the promoters of yeast has been presented by J. Mellor in Chapter 1 of this book; in this chapter, we will review in detail how cell-type is regulated in yeast, with a specific look at the regulation of mating-type genes by the MATa1 and MATα2 repressor proteins. We will, in particular, explore our current views on (i) the structure of these repressor proteins and their homology to other eukaryotic and prokaryotic repressors; (ii) how these repressor proteins recognize and bind to specific operator sequences; and (iii) how they function and act to confer repression of transcription. Moreover, we will describe how these repressors can be used for the effective control of relatively strong glycolytic promoters and for the regulation of heterologous gene expression in yeast. We will also review how these *MAT* genes are themselves regulated when located at the silent *HML* and *HMR* loci.

2.2 Mating-type regulation

2.2.1 *Mating-type and the life-cycle of Saccharomyces cerevisiae*
During its life-cycle the yeast *Saccharomyces cerevisiae* can exist as one of three cell-types, α haploid cells, a haploid cells or a/α diploid cells. Each of these cell-types is quite distinct with respect to its role in the life-cycle of yeast and to the other cell-types. For example, an α haploid cell is, under specific conditions, capable of mating with an a haploid cell to produce an a/α diploid; however, α haploid cells will not mate with α haploid cells, a haploid cells will not mate with a haploid cells and neither will mate with a/α diploid cells (for a review see Thorner, 1981). Mating between a haploid cells and α haploid cells is made possible through the production of mating pheromones secreted by each of these cell types; α haploid cells secrete a tridecapeptide called α-factor while a haploid cells secrete an undecapeptide called a-factor. Each binds to a specific receptor produced by the cell of the opposite mating-type which results in both the arrest of the cell at the beginning of the cell-cycle in G1-phase (Bucking-Throm *et al.*, 1973; Wilkinson and Pringle, 1974) and a change in the morphology of the cell (Hartwell, 1973; Sakai and Yanagishima, 1972; Fehrenbacher *et al.*, 1978). These changes mediate the mating process between the two cells and result in the formation of a zygote. This zygote will then produce a/α diploid cells.

Diploid cells behave quite differently from haploid cells under specific physiological conditions. For example, under certain starvation conditions, poor carbon and nitrogen source, these cells undergo meiosis and sporulate to produce asci containing four haploid spores of either a or α mating-type (for a review see Esposito and Klapholz, 1981). Each of the mating-types has distinct characteristics which are mediated through tight regulation of a specific set of genes which determine the particular mating-type of these cell-types.

2.2.2 *Information at the MAT locus determines mating-type*
Mating-type is determined by genetic information residing in a single locus, *MAT*, located on chromosome III (Mortimer and Hawthorne, 1969). The *MAT* locus can contain either of two alleles, *MATa* or *MATα*, and in haploid cells this results in the a and α mating-type, respectively. In diploid cells both alleles are present in a hemizygous form. Each of these alleles encodes two genes which are divergently transcribed; the *MATa* locus contains the *MATa1* and *MATa2* genes while the *MATα* locus contains the *MATα1* and *MATα2* genes (Klar *et al.*, 1981; Nasmyth *et al.*, 1981; Siliciano and Tatchell, 1984; Tatchell *et al.*, 1981). Although *MATa2* is transcribed in both a haploid cells and a/α

diploid cells (Klar *et al.*, 1981), the function of its gene product is unknown, since mutations in this gene appear not to alter the phenotype of either *a* or *a/α* cells (Tatchell *et al.*, 1981; Nasmyth *et al.*, 1981a). The function of the *MATa1*, *MATα1*, and *MATα2* gene products is more clearly understood.

The identification of the function of the *MAT* genes has been aided by the analysis of *STE* genes and others which are required for mating and sporulation. The *STE* genes were originally identified by the isolation of mutants defective in mating (MacKay and Manney, 1974a; MacKay and Manney, 1974b; Hartwell, 1980; Sprague *et al.*, 1981; Leibowitz and Wickner, 1976; Wickner, 1974; Rine *et al.*, 1979; Blair *et al.*, 1979). These genes are unlinked to the *MAT* locus, and some are required for mating in both *MATa* and *MATα* cells whereas others are specific to one cell type or the other (*a*-specific or α-specific *STE* genes). Genetic analysis of these genes and the *MAT* genes has led to a model (Figure 2.1) for the regulation of cell-type by the *MAT* genes (Strathern *et al.*, 1981).

The *MATα1* gene product is a positive activator of α-specific genes, e.g. *STE3*, *MFα1* and *MFα2* genes (Sprague *et al.*, 1983; Brake *et al.*, 1983; Emr *et al.*, 1983; Bender and Sprague, 1987).

The *MATα2* gene product is a negative regulator of *a*-specific genes, e.g. *STE2*, *STE6*, *MFa1*, *MFa2*, and *BAR1* in *MATα* cells (Hartig *et al.*, 1986; Wilson and Herskowitz, 1986; Chan *et al.*, 1983; Brake *et al.*, 1985; Miller *et al.*, 1985; Johnston and Herskowitz, 1985; Kronstadt *et al.*, 1987).

There is no apparent function for either *MATa1* or *MATa2* in *MATa* haploid cells, since disruption of these genes in the *MAT* locus confers a *MATa* phenotype (Nasmyth *et al.*, 1981a; Strathern *et al.*, 1980; Tatchell *et al.*, 1981). Thus, the phenotype of *MATα* cells appears to be conferred through the function of both the *MATα1* and *MATα2* genes, while the phenotype of *MATa* haploid cells is conferred by the absence of these genes.

However, in *MATa/MATα* diploid cells, the *MATa1* gene product functions by combining with the *MATα2* gene product to form a MATa1/MATα2 heterodimer which represses the expression of haploid-specific genes, e.g. *STE5*, *HO* and *RME* (Jensen *et al.*, 1983; Mitchell and Herskowitz, 1986). Moreover, MATa1/MATα2 represses the expression of *MATα1* in *MATa/MATα* diploid cells so that α-specific genes can no longer be activated (Klar *et al.*, 1981; Nasmyth *et al.*, 1981; Tatchell *et al.*, 1981). The net result of this is that neither *a*-specific genes, α-specific genes nor haploid-specific genes are expressed in diploid *MATa/MATα* cells, resulting in a specific diploid cell-type. The function of *MATa2* product is unknown in *MATa/MATα* diploid cells since mutations in the gene result in no apparent mutant phenotype (Nasmyth *et al.*, 1981a; Strathern *et al.*, 1980; Tatchell *et al.*, 1981). Thus, in yeast there are three sets of genes which determine cell-type: *a*-specific, α-specific, and haploid-specific, each of which is differentially expressed depending on the presence of either *MATα1* and *MATα2*, or *MATa1* and *MATα2* gene products.

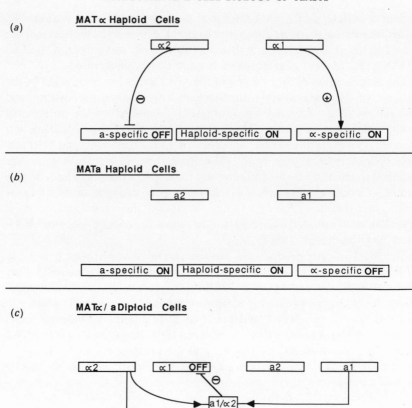

Figure 2.1 Schematic diagram of mating-type regulation of target genes. (a) Haploid MATα cells: MATα1 activates α-specific genes, MATα2 represses the constitutive expression of a-specific cells, while haploid-specific genes are constitutively expressed. (b) Haploid MATa cells: a-specific genes and haploid-specific genes are constitutively expressed while α-specific genes are not expressed in the absence of MATα1. (c) Diploid MATα/MATa cells: MATα2 represses a-specific cells, MATa1/MATα2 represses both haploid-specific and MATα1 expression, while in the absence of MATα1 α-specific genes are not expressed.

Several of the other non-specific STE genes e.g. STE4, STE5, STE7, STE11, and STE12, may also, directly or indirectly, control transcription of a- and α-specific genes (MacKay and Manney, 1974; Hartwell, 1980; Jenness et al., 1983; Fields and Herskowitz, 1985). Both STE2 and STE3 are thought to encode the α-factor and a-factor receptors, respectively, and appear to be activated by their respective mating pheromones (MacKay and Manney, 1974; Hartwell, 1980; Jenness et al., 1983; Burkholder et al., 1985; Nakayama et al., 1985; Hagen et al., 1986; Nakayama et al., 1987).

Repression in prokaryotes is mediated through the binding of a repressor protein to an operator sequence located in the promoter of the regulated gene (for review see Pabo and Sauer, 1984). Could either MATα2 or MATa1 and MATα2 function by binding to an operator sequence located in the 5′ upstream region of target genes? Evidence for the existence of operator sequences was first provided by Miller et al. (1985), after a deletion analysis of the promoter of the haploid specific gene, HO, which is repressed by MATa1/MATα2 in diploid cells. This analysis revealed a region that appeared to be essential for the repression of transcription of HO which comprised a sequence that is repeated 10 times in the promoter region. Further computer searches of other promoters of haploid-specific genes revealed homology to the HO sequences already identified. They found that the consensus sequence for this set of MATa1/MATα2 operator sequences was **(T/C)C(A/G)TGTNN(A/T)NANNTACATCA**. When two examples of these HO sequences were inserted into the CYC1 promoter between the UAS and the TATA box, expression came under MATa1/MATα2 control.

In the same study, Miller et al. (1985) performed a computer search of the promoter regions of a-specific genes regulated by MATα2 to look for homologous sequences within the promoter regions. The search revealed a common 30-bp sequence **GCATGTAATTACCCA AAAAGGAAATTTACATGG** in the promoter regions of MFa1, BAR1 and STE2 genes. When a 97-bp sequence from the STE2 promoter, containing the MATα2 operator, was inserted in either orientation into a site between the UAS and TATA of the CYC1 promoter, repression of transcription from the promoter was observed in both MATα haploid cells and MATa/MATα diploid cells, each of which produces MATα2 repressor.

In a similar experiment, Johnston and Herskowitz (1985) placed the MATα2 operator sequence from STE6 into the promoter of CYC1 and found that expression was repressed in MATα cells. They were able to show, by in-vitro DNAse I protection studies, that the STE6 MATα2 operator sequence also agreed with the consensus sequence identified by Miller et al. (1985).

Both MATa1/MATα2 and MATα2 operator sequences have similar short inverted sequences, i.e. **ATGT.....ACAT**, at the 5′ and 3′ ends of their sequences. Miller et al. (1985) have proposed that the MATα2 operator sequence may consist of two half-sites which are spaced so that they can allow binding of a MATα2 homodimer. In contrast, they propose that the MATa1/MATα2 operator sequence consists of a MATa1 half-site and a MATα2 half-site each of which is capable of binding its respective repressor molecule cooperatively (see below) as a heterodimer. In summary, the repressible target genes which determine mating-type in yeast have specific operator sequences which appear to confer repression of transcription through the presence of their respective repressor proteins. This would suggest that both MATα2 and MATa1 and MATα2 are capable of binding to these operator sequences.

Initial evidence to suggest that the *MATα2* gene product functions directly at the DNA level came from studies on nuclear localization. Hall *et al.* (1984), made various deletions of *MATα2* which were then fused to the *β*-galactosidase (*LACZ*) gene of *E. coli* to provide a marker for localization. By investigating the ability of the various hybrid MATα2-LACZ proteins to be transported to the nucleus in a set of experiments using subcellular fraction-ation and indirect immunofluorescence techniques, they established that the first 13 amino-terminal residues of the MATα2 protein are sufficient for targeting *β*-galactosidase to the nucleus, but these may not be sufficient for transport into the nucleus.

Some indirect evidence for the DNA binding properties of either MATa1 or MATα2 has been suggested by their homology to other eukaryotic and prokaryotic repressor proteins.

2.2.3 *Homology of MAT repressor proteins to homeo-domains of higher eukaryotic regulatory proteins*

The function of the MATa1 and MATα2 repressor proteins bears similarities to the gene products of homeotic genes which determine developmental fate of specific cells in *Drosophila melanogaster* and *Xenopus laevis* (Gehring, 1985). This similarity in function appears to extend to homology in structure. Shepherd *et al.* (1984) observed some homology of the homeo-domains of the *Ultrabithorax*, *Antennapedia* and *fushi tarazu* genes of *Drosophila*, and the *AC1* gene of *Xenopus laevis* with the carboxyl ends of MATa1 and MATα2.

Homology of the homeo-domain regions in MATα2 extends also to the DNA binding regions of the prokaryotic repressors, cI, cro and *E. coli* catabolite-activating protein (CAP) (Laughon and Scott, 1984). Crystallo-graphic studies of these proteins have produced a model in which a common α-helix-turn-α-helix structure is in contact with the major groove of the DNA of the operator (Pabo and Sauer, 1984). The second helix lies within the groove of DNA in contact with specific nucleotides, while the first helix holds the second helix in the correct position (Pabo and Sauer, 1984; Wharton and Ptashne, 1985; Wharton and Ptashne, 1986). Although the tertiary structure of the MATα2 protein is unknown, genetic analysis of these homologous sequences has suggested that they are responsible for binding to the operator DNA.

2.2.4 *MATα1 and MATα2 are DNA-binding proteins*

To investigate whether the 210-amino-acid MATα2 repressor protein is a DNA-binding protein, Porter and Smith (1986) introduced random missense mutations into the putative DNA-binding region of *MATα2* by site-directed mutagenesis using a mixed synthesis oligonucleotide. When the highly conserved Ile_{176} amino acid, which is the point of contact between the two helices in prokaryotic repressors, is changed to Asn, both the *a*-specific and haploid-specific functions of the MATα2 repressor are lost (Figure 2.2). Similarly, when the conserved hydrophobic Trp_{179} residue is changed to Arg

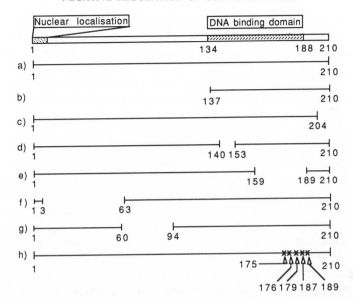

Figure 2.2 Mutant forms of MATα2. (*a*)–(*e*), deletions in MATα2 constructed by Hall and Johnston, (1987); (*f*) deletion constructed by Walton and Yarranton (unpublished results); (*g*) location of missense mutations constructed by Porter and Smith (1986). See text for details of mutant phenotype.

or the Gln$_{175}$ is changed to His, total loss of function is observed, suggesting that this region is involved in DNA binding. These amino acids are the most highly conserved sequences found in the homeo-domains of *Drosophila*, *Xenopus*, mouse and man. Mutations C-terminal to the putative binding domain, i.e. Lys$_{187}$ to Asn and Lys$_{189}$ to Ile, result in loss of the *a*-specific function but not the haploid-specific function of the MATα2 repressor. This suggests that this region and these basic residues in particular play a different role in the function of MATα2 when it interacts with MATa1 (see below).

More direct evidence for the role of the homeo-domain of MATα2 in DNA binding comes from a study by Hall and Johnston (1987), who constructed fifteen in-frame deletions within its coding region. They analysed both the phenotypes of these mutants *in vivo* and the ability of the gene products to bind to the operator sequence *in vitro*. To perform their *in vitro* analysis, they made in-frame fusions of the *MATα2* deletion mutants to the *E. coli* β-galactosidase gene, expressed these in yeast, made crude cell extracts and analysed the ability of these mutant proteins to bind to the MATα2 operator sequence *in vitro* by precipitating the proteins with anti-β-galactosidase monoclonal antibody coupled to *Staphylococcus aureus* cells.

Mutants missing up to the first 136 amino-terminal residues of MATα2 protein retained the ability to bind to the operator sequence, as did mutants missing the last 6 C-terminal residues (Figure 2.2). Mutants lacking

C

either residues 141–152 or 160–188 are incapable of binding to the operator sequence. These *in vitro* studies suggest that the DNA-binding sequences lie between residues 137 and 204 and that this region is capable of folding independently of the amino-terminal domain to form a functional DNA-binding protein. The results are also consistent with the observation that residues 134–188 bear homology to the homeo-domain of homeotic proteins.

2.2.5 *The interaction of MATa1 and MATα2 repressor proteins*

Earlier (section 2.2.2), we described how, in the presence of MATa1 and MATα2, haploid-specific genes are repressed. The mechanism of how MATa1 and MATα2 proteins interact to confer their own specific repression on haploid-specific genes has been investigated by Goutte and Johnston (1988). In a series of *in vitro* experiments, similar to those of Hall and Johnston (1987) described above, they found that increasing ratios of MATa1 to MATα2 shifted the specificity of MATα2 binding to the MATa1/MATα2 operator. In a second set of DNA competition experiments, where the concentration of both types of operator sequence was varied, they found that . DNA-binding specificity did not alter. These results suggest that it is the influence of the MATa1 protein, rather than just the presence of the MATa1/MATα2 operator sequence, which determines the DNA-binding specificity of MATα2. The observation (Porter and Smith, 1986) that mutations in two basic residues C-terminal to the putative DNA-binding domain results in loss of MATa1-mediated repression but not MATa1/MATα2 repression is consistent with the model that the DNA-binding specificity of MATα2 can be altered. However, these results may also suggest that these residues are important for the interaction and stabilization of MATα2 homodimers.

By using deletion mutants of *MATα2*, as described previously by Hall and Johnston (1987), Goutte and Johnston (1988) were able to determine which portion of the protein was important for its MATa1/MATα2 binding activity. They found that a deletion of residues 4–62 (see Figure 2.2) abolished MATa1/MATα2 binding activity but not MATα2 binding activity, suggesting that these residues are important for the interaction of the two proteins and for conferring the altered ability of MATα2 to bind to MATa1/MATα2 operator sequences. Alternatively, it is possible that these deletions alter the tertiary structure of MATα2 so that it can no longer interact with MATa1.

The function and interaction of proteins can be investigated by expressing mutant forms of proteins in the presence of the wild-type form (Herskowitz, 1987). The observation of inhibition or inactivation of wild-type function (dominant negative phenotype) by these mutant proteins can provide clues to their function and interaction. This is especially useful for studying proteins which form homo- or hetero-multimers such as regulatory DNA-binding proteins. The expression of some *MATα2* deletion alleles in the presence of *MATa1* gene product and wild-type *MATα2* gene product show that they are capable of inactivating the function of MATa1/MATα2 in a *trans*-dominant manner. Previous results by Hall and Johnston (1987) showed that, in the presence of wild-type MATα2 protein, deletions 4–62, 2–135, and 204–210 *trans*-inactivate only MATα2-mediated repression. In contrast, deletions 141–152 and 160–188 (both of which lack the DNA-binding homeo-domain) *trans*-inactivate only MATa1/MATα2 repression (Figure 2.2).

These results suggest that interaction between MATa1 and MATα2 occurs between residues 4 and 62 and that a deletion of the homeo-domain prevents DNA-binding. This results in a decrease in the ratio of MATa1 to wild-type MATα2, causing a derepression of MATa1/MATα2 target genes (*trans*-inactivation). Moreover, this also suggests that MATa1 and MATα2 interaction can occur in the absence of DNA-binding.

We have made in-frame deletions of residues 61–93 of MATα2 and have found *trans*-inactivation of the MATa1/MATα2 repression of *HO* expression in the presence of wild-type MATα2 repressor (Figure 2.2; Walton and Yarranton, unpublished). If MATα2 interacts with MATa1 prior to binding to the operator sequence, as suggested by Goutte and Johnston (1988), then this result suggests that residues 61–93 may be important for DNA-binding but not MATa1 interaction, since the mutant MATα2 is capable of inactivating (i.e. competing with) the wild-type molecule. This hypothesis would also be consistent with the model that residues 4–62 (i.e. the amino-terminal domain of MATα2) are required for MATa1 interaction (Goutte and Johnston, 1988) since that particular domain is still present.

The results of Goutte and Johnston (1988) suggest that MATa1 is capable of changing the DNA-binding specificity of MATα2; thus there must be a marked decrease in the concentration of free MATα2 in diploids to repress *a*-specific genes if one takes into consideration that there is also a fivefold reduction in *MATα2* transcripts by the action of MATa1/MATα2 (Nasmyth *et al.*, 1981*b*). Since both MATa1/MATα2 and MATα2 repressor activities are effective in diploid cells, there must be sufficient levels of MATα2 present to carry out both functions.

Recent studies by Sauer *et al.* (1988) show that the MATα2 repressor forms homodimers by interaction through its N-terminal domain. Thus MATa1 may interact with MATα2 repressor protein by competing with MATα2–MATα2 interaction sites. Further, they suggest that MATα2, as a dimer binds to the ends of its operator sequence. The question of whether MATα2 exists as a dimer or a multimer has been addressed by Sauer *et al.* (as cited in Goutte and Johnston, 1988) who have shown that when purified MATα2 protein is in an oxidized form, it exists as a dimer, but when it is reduced it exists as a monomer, suggesting that it can naturally form dimer molecules.

Miller *et al.* (1985) proposed that MATa1 and MATα2 bind as a hetero-dimer to two different half-sites on the MATa1/MATα2 operator – a model similar to that proposed for the cooperative binding of prokaryotic repressors.

The results of Goutte and Johnston (1988) extend this model to suggest that the interaction of MATa1 protein with MATα2 protein alters the DNA-binding specificity of MATα2. The probability that only one MATα2 molecule interacts with MATa1 is demonstrated by the observation of Sauer et al. (as cited in Goutte and Johnston, 1988) that purified MATα2 in an oxidized form (as a dimer) has five- to ten fold less MATa1/MATα2 DNA-binding activity than when it is in a reduced form (as a monomer).

In-vitro studies suggest that the homeo-domain is the DNA-binding region of MATα2, while the answer to the question of which region confers ability to repress can be provided by the analysis of these mutant alleles in vivo. Expression of the MATα2 deletion mutants showed that all those which retained the ability to bind to the a-specific operator in vitro lost their ability to repress either a-specific expression or haploid-specific expression (Hall and Johnston, 1987). Therefore the DNA-binding region is not sufficient for repression in vivo. However, the answer is not complete, for although the mutant forms of MATα2 are transported to the nucleus, as determined by indirect immunofluorescence, evidence for efficient transport across the nuclear membrane needs to be attained. Nevertheless, the possibility that these mutants have lost their repressor function while retaining their capacity of binding to the operator site cannot be ruled out.

2.2.6 How does MATα2 repress transcription?

The investigation of the role and mechanism of MATα2 in DNA-binding and repression is beginning to reveal some intriguing aspects about interrelationship of activation and repression. Kronstadt et al. (1987) have found that the MATα2 operator sequence of the BAR1 promoter overlaps with its major Upstream Activator Sequence (UAS), which is indirectly stimulated by the presence of α-factor, and that deletion of the operator results in a drastic reduction in both constitutive and α-factor stimulated expression. The close proximity of the operator in relation to the UAS suggests that MATα2 may act in repressing transcription by displacing the α-factor stimulated activator. Intriguingly, the STE2 MATα2 operator sequence was found to be capable of behaving as an Upstream Activator Sequence when placed in either orientation within the BAR1 promoter lacking its own endogenous operator sequence. Moreover, they found that when the operator sequence was inserted as two tandem copies, the activation obtained was less than that for a single copy, but was more than that obtained for two copies inserted as inverted repeats. These studies suggest that there may be other factors which bind within or proximal to the MATα2 operator sequence which are capable of activating transcription. Recent work indicates that this may be the case.

Bender and Sprague (1987) have studied a general transcription factor called the Pheromone and Receptor Transcription Factor (PRTF) which appears to be required for the activation of the α-specific genes; MFα1, MFα2 (the two α-factor genes) and STE3 (the a-factor receptor gene). Activation is probably achieved by both MATα1 and PRTF binding synergistically to adjacent sequences within the UASs of these genes as suggested by a series of *in-vitro* studies using gel retardation, DNAse I protection experiments and DNA competition assays. However, the activation of α-specific genes requires other genes such as *STE4, STE5, STE7, STE11*, and *STE12*, none of which appears to be PRTF. Fields and Herskowitz (1985) have suggested that the *STE12* gene product is required for the activation of both *a*-specific and α-specific genes. If this is the case then, in addition to PRTF and MATα1, another protein is required for activation of α-specific genes. This would also suggest combinatorial control of both *a*-specific and α-specific genes.

Bender and Sprague (1987) revealed that PRTF is present in all three cell-types and thus may be a general activator of several genes. A PRTF binding site is also found in the promoter region of the α-factor receptor (*STE2* gene) which overlaps a MATα2 operator sequence, thus the MATα2 repressor may function by displacing PRTF and subsequently preventing transcriptional activation. If PRTF is an activator of *a*-specific genes then maybe the role of MATα2 is to displace PRTF and to prevent it from activating transcription. The mechanism of how MATα2 protein displaces PRTF may be resolved by studying the interaction of PRTF with mutants of *MATα2*.

The binding of PRTF and MATα1 in an interdependent manner to DNA may be similar to the binding of the HAP2 and HAP3 activator proteins to UAS2 of the yeast *CYC1* promoter (Olesen *et al.*, 1987), where neither can bind or function in the absence of the other. The possibility that different repressor and activator molecules confer their effect through displacement and competition for DNA-binding sites is also supported by studies in *Drosophila* which show that different homeotic proteins may bind with different affinities to the same DNA-binding site (Hoey and Levine, 1988). Thus, the transcriptional activity of a promoter may largely depend on the presence of the protein (either activator or repressor) with the greatest binding affinity that can successfully compete for the same binding site. Alternatively, the interaction of one molecule with another may alter its structure so that it may no longer bind to the binding site.

Displacement of activators by repressors may be a common method of transcriptional inactivation. It is intriguing that the *HO* promoter has ten MATa1/MATα2 operator sequences. Since this gene is regulated by a multiple of specific activators binding to various sites within the 1400-bp promoter (see Nasmyth and Shore, 1987, for a review) there is the possibility that multiple MATa1/MATα2 operator sites are required to displace each type of activator complex to exert continuous transcriptional inactivation in diploid cells.

2.2.7 *Operator position influences the degree of repression*
There is evidence to suggest that the MATα2 repressor may exert its inhibitory influence on activators relatively distal to its position. When Miller *et al.* (1985) placed the MATα2 operator of *STE2* within the promoter of *CYC1*

between the UAS and the TATA box, normal regulation of the UAS was observed in $MAT\alpha1^-$, $MAT\alpha2^-$ cells suggesting that the $CYC1$ UAS was the prominent activator sequence. However, in $MAT\alpha1^-$, $MAT\alpha2^+$ cells activation by the $CYC1$ UAS was lost. Similarly, as described in section 2.3, we have found that the relatively strong phosphoglycerate kinase (PGK) promoter can be totally repressed when the MAT $\alpha2$ operator sequence of $MFa1$ is placed as multiple copies between the UAS and TATA box (Yarranton and Walton, unpublished; see section 2.3). Further, Johnston and Herskowitz (1985) demonstrated that by inserting the $MAT\alpha2$ operator sequence of $STE6$ upstream from the UAS of the $CYC1$ promoter, partial but not complete repression is possible. However, the degree of repression was an order of magnitude greater when the operator was placed between the UAS and TATA box, suggesting that the position of the operator in relation to the UAS is important. Similar results have been obtained for insertion of the $MAT\alpha2$ operator sequence upstream of the UAS of the $ADH2$ promoter (Sledziewski et al., 1988).

2.2.8 Application of mating-type regulation

Saccharomyces cerevisiae is widely used as a host organism for the expression of heterologous genes (see Chapter 4 for a review). Although much is understood about promoter function and construction of high-copy-number plasmid expression vectors, reliable high-level protein production during process scale-up is often not achieved. Several factors can influence cell productivity during large scale culture, but the most common ones are: (i) segregational instability; (ii) copy number fluctuation, and (iii) gene rearrangement. The selective advantage associated with a reduction in heterologous protein production can lead to a rapid overgrowth of non-producing cells within the population. This problem is particularly acute with S. cerevisiae when the heterologous protein is engineered for secretion from the cell, with non-producing variants arising at high frequency (Yarranton and Walton, unpublished results). One solution to this problem is to use a negatively regulated gene expression system, such that the growth and production phases of a process can be separated, thereby reducing the selective advantage of non-producing cells.

Mating-type regulation is one of the few naturally occurring negative regulation systems in yeast and should be readily adaptable for use in heterologous gene expression systems. The insertion of $MAT\alpha2$ and $MATa1/MAT\alpha2$ operator sequences into the $CYC1$ promoter demonstrates the ability of these to regulate a heterologous promoter (Miller et al., 1985). However, whether such a system would work for the regulation of strong glycolytic promoters needs to be tested.

The mating-type control system shares several features of the bacteriophage λ lysogeny control system (Pabo and Sauer, 1984). The latter has been employed extensively in prokaryotic expression vector construction and heterologous gene expression (Remaut, 1981). One important feature of the λ system is the requirement for tight negative regulation of gene transcription, and this is achieved through the use of a repressor protein and multiple operator binding sites (Ptashne *et al.*, 1976). This tight control is retained when the promoter-operator is used with the repressor in multicopy expression plasmids. The mating-type system also requires tight negative regulation of 'sets' of target genes for distinct cell-types to be obtained, and in some cases control involves multiple operator sequences, such as the *HO* promoter (Miller *et al.*, 1985). We demonstrate here the use of the mating-type control system for the tight regulation of the phosphoglycerate kinase (*PGK*) gene promoter, on a multicopy plasmid in *S. cerevisiae*.

2.3 Regulation of the PGK gene promoter

2.3.1 Construction of negatively regulatable PGK gene promoter
The *PGK* gene promoter of *S. cerevisiae* has been widely used for the expression of heterologous genes (see Chapters 1 and 4 for reviews and also Kingsman *et al.*, 1985). The structural elements within the promoter are defined partly by sequence homology to known eukaryotic elements (Dobson *et al.*, 1982) and partly by function (Kingsman *et al.*, 1985). TATA box sequences are located at − 111 and − 143, a CT block at − 69 to − 51 and a Upstream Activator Sequence (UAS) at − 403 to − 480 bp from the ATG (Figure 2.3). In order to establish mating-type regulation of the *PGK* gene promoter, operator sequences were inserted at a PvuII restriction site between the putative UAS and TATA elements of the promoter located at − 290 bp from the ATG. This position was chosen as Miller *et al.* (1985) had demonstrated mating-type control of the *CYC1* promoter by insertion of operator sequences between the UAS and TATA box elements.

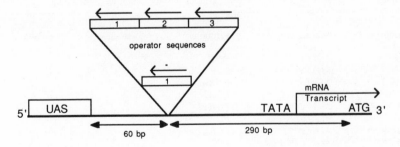

MATα2 or a1/α2 regulated PGK promoter

Figure 2.3 Location of MATα2 and MATa1/MATα2 operator sequences within the *PGK* gene promoter. Arrows show the orientation of operator sequences.

The operator sequences chosen for this study were representative of both *MATa1/MATα2* and *MATα2* regulation; they were the *MFa1/MATα2* operator, **GTGTGTAATTACCCAAAAAGGAAATTTACATGT** and the −411 *HO MATa1/MATα2* operator, **TCATGTTATTATTTACATCA**. These were inserted either as single or multiple copies and screened for their ability to effectively regulate expression of the gene product. The gene chosen as a marker for this study was one encoding the dextrinase protein, (*DEX1*), which is a secreted protein produced in *Saccharomyces diastaticus*, a variant of *S. cerevisiae* (Meaden *et al.*, 1985). This protein is secreted from *S. cerevisiae* when introduced on a expression vector and can be readily detected by growth of cells on agar plates containing soluble starch. The expression of the dextrinase gene from a variety of modified *PGK* gene promoters carried on the same vectors was analysed following transformation of strains of different mating-type. The results in Table 2.1 demonstrate that (i) single copies of *MATα2* or *MATa1/MATα2* operator sequences are insufficient for tight regulation of the *PGK* gene promoter on a multicopy plasmid; (ii) multiple copies of the *MATα2* operator sequence can regulate the *PGK* gene promoter; and (iii) the *MATα2* operator sequences are functional in both diploid and haploid cells.

Regulation of the *PGK* gene promoter comprising the *MATα2* operator sequences (*PGKα2*) was also demonstrated using a dextrinase-*β*-galactosidase translational fusion and measuring *β*-galactosidase activity in cell extracts. From these analyses the reduction in *β*-galactosidase activity measured, for constructs with and without the operators, was between 50- and 300-fold. These results suggest a repression of transcription of at least two orders of magnitude for the multiple *MATα2* operator system.

Table 2.1 Regulation of a modified *PGK* gene promoter by mating-type.

Strain	Mating-type	Promoter	Operator sequences	Dextrinase secretion
AH22	*MATa*	*PGK*	−	+
AH22	*MATa*	*PGKα2*	1–3	+
AH22	*MATa*	*PGKa1/α2*	1	+
MD40/4C	*MATα*	*PGK*	−	+
MD40/4C	*MATα*	*PGKα2*	1	+
MD40/4C	*MATα*	*PGKα2*	3	−
MD40/4C	*MATα*	*PGKa1/α2*	−	+
MD50	*MATa/MATα*	*PGK*	−	+
MD50	*MATa/MATα*	*PGKα2*	1	+
MD50	*MATa/MATα*	*PGKα2*	3	−
MD50	*MATa/MATα*	*PGKa1/α2*	1	+

Dextrinase secretion from yeast strains of different mating-type. The dextrinase gene was expressed from *PGK* promoter comprising either *MATα2* or *MATa1/MATα2* operator sequences and secretion was scored by staining starch plates with iodine.

2.3.2 *Controllable expression of modified PGK gene promoter*

In order to obtain inducible gene expression and to investigate the induction kinetics from the *PGKα2* promoter, a conditional supply of MATα2 repressor protein was necessary. To obtain this, the temperature-sensitive *sir3^{ts}* mutation (Ivy *et al.*, 1986) was used in conjunction with a haploid strain Y49 (*HMLα, MAT::LEU2, hmr::TRP1, leu2, trp1, ade2, ura3, his3, sir3^{ts}*). Hence, at 23°C the strain is functionally an *a* strain, whereas, at 34°C it is an α strain due to derepression of *HMLα* (see section 2.5). The *PGKα2* promoter should be repressed at 34°C and derepressed at 23°C, with derepression following a temperature switch, depending upon the kinetics of MATα2 repressor decay and *HMLα* activation.

Conditional control of *PGKα2* transcription in Y49 was initially tested by growing cells at either 23°C or 34°C and analysing transcripts by Northern blotting and dot blotting. To avoid cross-hybridization with yeast chromosomal genes, the transcription of a heterologous gene encoding human gastric

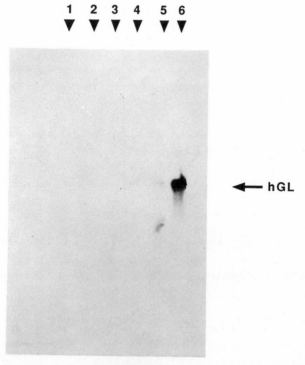

Figure 2.4 Northern blot of total RNA prepared from Y49 transformed with *PGKα2-hGL* and Y48, the parent strain of Y49 that is *MATα*, hence this strain is an α at both temperatures. The filter was probed with nick-translated hGL-specific DNA restriction fragment. Lanes 1 and 2, RNA prepared from Y48 *PGKα2-hGL* grown at 34°C and 23°C. Lanes 3–6, RNA prepared from Y49 *PGKα2-hGL* at 34°C, 30°C and 23°C. Lanes 5 and 6 have identical RNA samples loaded except that lane 5 has tenfold less total RNA.

lipase (hGL, Bodmer *et al.*, 1987) was studied. Probing RNA preparations made from Y49 transformed with *PGKα2-hGL* demonstrated that conditional regulation of the *PGKα2* promoter was obtained (Figure 2.4). Repression of the promoter was evident at both 34°C and 30°C, and clear derepression was evident at 23°C. The differences in RNA levels were quantified by RNA dot blotting, using the plasmid-derived *URA3* mRNA as an internal standard. At least a hundredfold difference between the levels of hGL mRNA at 23°C and 34°C was obtained.

The hGL protein accumulates to 1–5% total cell protein when expressed from the *PGK* gene promoter. Expression from the *PGKα2* promoter in Y49 led to similar levels of accumulation at 23°C, but no hGL protein was detected in extracts grown at 34°C (Figure 2.5). Western blot analysis of cell extracts made from Y49 *PGK 2-hGL* grown at 34°C also demonstrated that protein levels were at the limit of detection by this sensitive technique (data not shown). These results demonstrate that heterologous protein synthesis is significantly reduced as a result of the transcriptional regulation of the *PGKα2* promoter.

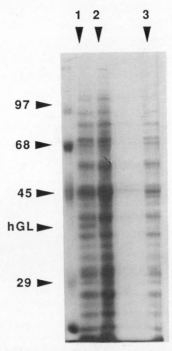

Figure 2.5 Polyacrylamide-SDS gel of total cell extracts prepared from Y49 *PGKα2-hGL* grown at different temperatures (see text). The 10% gel was stained with Coomassie blue dye. Lanes 1 and 2, Y49 *PGKα2-hGL* grown at 23°C and 34°C, respectively; lane 3, Y48 *PGKα2-hGL* grown at 23°C.

2.3.3 Induction kinetics for PGKα2

Since the *PGKα2* promoter is controlled in Y49 by using the *sir3^{ts}* mutation, it is the supply of the MATα2 protein rather than the direct inactivation of the repressor that determines induction. Hence induction kinetics will depend upon the stablity of the MATα2 repressor and the efficiency of establishing *HMLα* repression. To determine induction kinetics a plasmid construction, which carried the dextrinase gene fused to the hGL gene giving an in-frame translation fusion, was used to monitor product formation. Expression of the fusion protein can be detected by Western blotting as soon as fifteen minutes after temperature shift from 34°C to 23°C (Figure 2.6), and full induction of the gene is achieved within one generation time. This result suggests that the MATα2 repressor has a short half-life, and that controlling MATα2 repressor supply by the *sir3^{ts}* system is a viable route to rapid induction of gene expression.

2.3.4 Control of intact PGK gene and promoter

The *PGK* gene product comprises up to 5% of the total cell protein when expressed as a single chromosomal copy, and this rises to between 30–50%

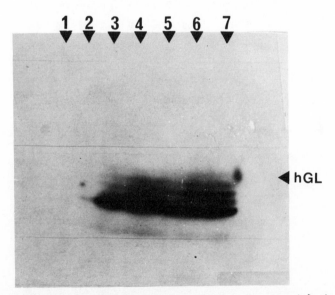

Figure 2.6 Western blot of *DEX*-hGL expression following temperature induction of Y49 *PGKα2-DEX-hGL*. Cells growing exponentially at 34°C were shifted to 23°C and samples removed at 0–9 hours after induction. Cell extracts were electrophoresed on a 10% polyacrylamide-SDS gel, transferred to nitrocellulose paper and probed with rabbit polyclonal antisera to hGL. The Western blot was developed with I^{125}-ProteinA. Lane 1, 0 h; lane 2, 0.25 h; lane 3, 4.25 h; lane 4, 5.75 h; lane 5, 7.25 h; lane 6, 9 h; lane 7 hGL control. The multiple banding pattern represents different glycosylation variants of the fusion protein (Yarranton, unpublished results).

when the gene is cloned on a multicopy plasmid (Mellor *et al.*, 1985). Considerable interest has been shown in the fact that heterologous gene expression does not approach that of the native *PGK* gene, even though the 5′ non-coding sequences may be identical. The steady-state mRNA levels for heterologous genes are lower than for the *PGK* gene, and two possible explanations for this have been suggested. One explanation is that the heterologous mRNAs are less stable (Chen *et al.*, 1984), whereas the other explanation is that the *PGK* gene has a Downstream Activator Sequence (DAS) within the coding sequence (Mellor *et al.*, 1987).

Since the *PGK* gene promoter may require activation from within the gene coding sequence, the *PGKα2* promoter was used to direct the expression of the intact *PGK* gene. This experiment tests the ability of the *MATα2* system to regulate a 'full-strength' *PGK* transcription unit on a multicopy plasmid. Promoter regulation was assessed by SDS-PAGE of total cell extracts made from cells grown at either 23°C (expressing) or 34°C (repressed) (Figure 2.7). The results show that the *MATα2* control system works efficiently on the full-strength *PGK* promoter, and that the promoter's activity is not greatly reduced by insertion of *MATα2* operator sequences.

The negative regulation of mating-type specific genes involves the use of operator sequences (DNA-binding sites) and repressor proteins. This regul-

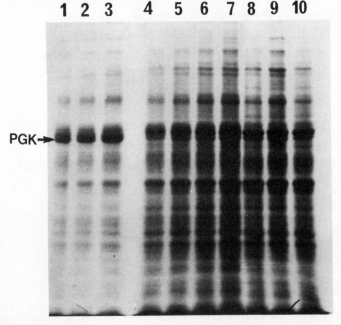

Figure 2.7 Polyacrylamide-SDS gel of total cell extracts prepared from Y49 *PGKα2-PGK*, and Y48 *PGKα2-PGK*. Lanes 1, 2 and 3, extracts prepared from Y49 *PGKα2-PGK* grown at 23°C; lanes 4, 5, 6, 7, and 10, at 34°C; lanes 8 and 9, Y48 *PGKα2-PGK* grown at 23°C and 34°C, respectively.

ation can be applied to other yeast promoters giving efficient regulation of very strong promoters cloned on to multicopy plasmids. Negative regulation of yeast promoters by mating-type sequences can be mimicked using bacterial repressors and operator sequences, although the positioning of the operator sequences suggest that mechanistically the two systems may function differently.

2.4 The behaviour of prokaryotic repressors in yeast

The behaviour of repressor molecules in yeast can be further understood by the study of the ability of heterologous repressors to function in the repression of yeast transcription. Brent and Ptashne (1984) placed the *E. coli* LexA operator between the UAS and TATA box of the *GAL1* promoter and expressed the *LexA* gene. They found that LexA was capable of repressing *GAL1* transcription to varying strengths depending on the position of the LexA operator. When the operator was placed upstream from the UAS they found that no repression occurred. They also observed that two copies of the LexA operator sequence gave consistently greater repression than one copy. This repression can be explained, either by an ability to displace the various transcriptional activators (including RNA polymerase II if the action of LexA is similiar to its normal function in *E. coli*), or by directly blocking the interaction of transcriptional activators with RNA polymerase II. The inability of an operator site upstream from the UAS to confer repression suggests that the LexA protein may block the binding of transcriptional activators at or near the TATA box. The observation that two copies of operator sequence confer a greater degree of repression can be explained by cooperative binding where one bound dimer facilitates the binding of another.

In a separate set of experiments, Brent and Ptashne (1985) fused the DNA-binding portion of LexA to GAL4, the activator protein of the *GAL1* promoter. When the fusion protein was expressed in *E. coli*, they found that it bound to the LexA operator and was capable of repressing transcription. Upon expression of the fusion protein in yeast they found that it was capable of activating transcription from LexA operator sequences located up-stream from the transcription start sites of either *CYC1* or *GAL1*. Absence of the LexA operator or expression of wild-type LexA resulted in no transcriptional activity. Activation of transcription was less when the operator site was positioned 590 nucleotides, as opposed to 178 nucleotides, upstream from the transcriptional start site. Activation was also observed if the LexA operator was placed downstream from the transcription start site. Thus, there appear to be molecular and biochemical aspects to the GAL4 protein which confer its ability to activate transcription. This is supported by evidence which shows that the presence of charged basic residues on activators confer activation of transcription (Ma and Ptashne, 1987; Gill and Ptashne, 1987), possibly by direct interaction with RNA polymerase II. These results demonstrate that the component parts of a protein binding to the promoter region, rather than its physical presence, can influence whether activation or repression will occur.

So far we have described the influence of repressor proteins on transcription when they bind to operator sequences within or proximal to the promoter region of genes. However, in yeast there is another type of complex negative regulation exerted on copies of the mating-type regulatory genes residing at a considerable distance from the *MAT* locus. This repression is exerted by sequences over 2.5 kb away, and the mechanisms of this effect are just being elucidated.

2.5 Mating-type switching and repression at the silent *HML* and *HMR* loci

Mating-type switching is a natural phenomenon found in most 'wild' strains and variants of *Saccharomyces cerevisiae* and such strains are referred to as homothallic. Most laboratory strains of yeast are heterothallic and consequently do not undergo mating-type switching. A homothallic haploid strain of yeast undergoes a change in mating-type from *MATa* to *MATα* and from *MATα* to *MATa* by a change in the mating-type information residing at the *MAT* locus. This switching effectively reduces the period during which yeast cells remain in the haploid phase of the life cycle, because mother and daughter cells can now mate with each other at the expense of a reduced ability to mate with other strains. However, this mechanism may have evolved as a means for wild yeast to have improved survival properties as diploid cells, which are capable of sporulating under conditions of nutritional stress.

Mating-type switching is mediated through an endonuclease encoded by the *HO* gene which facilitates the transfer of the mating-type information from either of the two loci located on chromosome III, *HML* or *HMR*, to *MAT* such that the information previously residing at the *MAT* locus is destroyed and is replaced with information of the opposite mating-type (Herskowitz and Oshima, 1981; Nasmyth *et al.*, 1980; Nasmyth *et al.*, 1981; Nasmyth, 1982; Klar *et al.*, 1984). The *HML* and *HMR* loci usually contain α and *a* information, respectively, which remains transcriptionally inactive or silent when they are in this location. This mechanism of repression appears to be quite different from that regulating mating-type described above, but some features may be similar.

Heterothallic yeast have both *HML* and *HMR* loci but lack a functional *HO* gene and therefore do not switch. It is the transcriptional regulation of the *HO* gene in homothallic yeast which determines the specific behaviour of switching. First, switching occurs in mother cells but not in daughter cells; second, switching occurs only at the beginning of the cell-cycle prior to DNA replication; and finally, diploid cells do not switch. This latter behaviour is mediated through the repression of *HO* transcription by MATa1/MATα2.

The other characteristics of switching are mediated through the *SWI* genes (for a review see Nasmyth and Shore, 1987).

The repression of α and *a* information at *HML* and *HMR* is thought to be conferred both in *cis* at a distance from each locus (Abraham *et al.*, 1984; Feldman *et al.*, 1984; Klar *et al.*, 1981*a*) and in *trans* by four *trans*-acting genes, *SIR1*, *SIR2*, *SIR3* and *SIR4* (Klar *et al.*, 1984; Shore *et al.*, 1984; Ivy *et al.*, 1986; Rine and Herskowitz, 1987). Mutations in any one of these genes results in derepression of the silent loci. The temperature-senstive mutation *sir3*[ts] has been successfully used for regulating the expression of relatively strong promoters containing the MATα2 operator as described in section 2.3. There is also evidence to show that there is a silencer sequence located to the left of each silent locus (Brand *et al.*, 1985). This silencer can repress transcription up to 2.5 kb away, and can function in both orientations and on different sides of the loci (Brand *et al.*, 1985). The silencer regions have Autonomously Replicating Sequences (ARSs), and thus could be origins of chromosomal replication. Study of the *HMR* silencer region has revealed three elements, A, E and B, contained within a sequence of 120 bp (Brand *et al.*, 1987). The A element is homologous to the ARS consensus sequence and has been shown to have ARS activity.

The E element can bind a protein called RAP1 (Repressor/Activator binding Protein) which is also capable of binding to the silencer at *HML* (Shore *et al.*, 1987; Shore and Nasmyth, 1987). Deletion of the *RAP1* gene is lethal to the cell, so its function may be essential (Shore and Nasmyth, 1987). This observation may be explained by the fact that RAP1 binding sites are homologous to the UASs of *MATα1* and, more particularly, several ribosomal genes. Introduction of the RAP1 binding site upstream from the TATA box of the *CYC1* gene confers activation of transcription (Brand *et al.*, 1987). Thus RAP1 is capable of mediating both activator and repressor activity depending on the position of the RAP1 binding site. At present there is no direct evidence that RAP1 itself acts directly as a repressor. It is not encoded by any of the *SIR* genes, and the presence of a RAP1 binding site within the silencer region may not imply a direct function in conferring repression. It is possible that the binding sites are recognized by either the *SIR* proteins or other repressor proteins which compete for binding with RAP1. Alternatively, the RAP1 protein may mediate the binding of repressor proteins in a cooperative manner to other sites within the region so that they can exert their influence in repression.

The B region of the silencer sequence contains a site for the binding of another factor called SBF-B (Silencer Binding Factor-B) which is also not encoded by any of the *SIR* genes. However, SBF-B does not bind to the silencer at *HML* but does bind to other ARS sequences. Thus SBF-B may play a role in the integrity of the structure or function of at least some ARS sequences. When the B region is placed upstream from the TATA box of *CYC1* promoter it is partially capable of activating transcription (Brand *et al.*, 1987).

Another feature of the silencer sequences is that they exhibit properties of centromeres when used as origins of replication on vectors in yeast. They remain completely stable during cell division in the absence of any chromosomal centromeres. However, they lose this stability in absence of any functional *SIR* gene products (Kimmerly and Rine, 1987).

Thus, the silencer regions of *HML* and *HMR* are complex regions which have the features common to DNA replication and chromosome partitioning. The three main elements recognized so far can be each deleted in turn without greatly affecting silencer function, whereas deletion of any two does affect silencer activity. Therefore these elements appear to be functionally redundant. Both RAP1 and SBF-B appear to have activator activity, suggesting that their function at the silencer is to promote transcription; however, there is no evidence to suggest either that this occurs or that it is responsible for conferring silencer function. The various components of the silencer may not individually be responsible for silencer function, but rather it is the combination of these components which exert the repression of transcription at large distances. Since the only proteins identified so far which are known to directly affect silencer function are the SIR proteins, it is likely that these interact with proteins already bound to the silencer, such as RAP1 and SBF-B, to exert their effect. Alternatively, both RAP1 and SBF-B may promote directly or indirectly the binding of the SIR proteins to other proteins bound to the silencer. In this respect the function of these two activators may not be to activate transcription at the silencer but rather to facilitate the binding of the SIR proteins. Whether the DNA replication and centromere properties of the silencer are directly related to its repressor activity has yet to be determined.

2.6 Conclusions

The picture of how MATa1 and MATα2 function as repressors of target genes is becoming more fully understood. It is clear that MATα2 and MATa1 and MATα2 must form either homodimers and heterodimers, respectively, in order to bind to their respective operator sequences. The putative sites for protein–protein binding of MATa1 with MATα2 are likely to be in the N-terminal region of the MATα2 protein. The interaction of both these molecules changes the binding specificity of MATα2 so that it can bind to the MATa1/MATα2 operator sequence. The sites of interaction between MATα2 proteins also appear to be in the N-terminal region.

The structure of MATa1 and MATα2 is similar to other eukaryotic and prokaryotic repressor proteins with a defined DNA-binding region probably consisting of an α-helix-turn-α-helix motif.

The evidence for the function of MATα2 and MATa1/MATα2 suggests that they exert their repressor activity either (i) by displacing transcriptional activators from their DNA-binding sites proximal or within the operator sequences, or (ii) by preventing the interaction between transcriptional activator complexes and RNA polymerase II. Displacement of activation may be conferred through simple competitive binding kinetics for DNA-binding sites, or directly through protein–protein interaction sites within the transcriptional activator complexes when they are bound to DNA. The inhibition of interaction between transcriptional activator complexes and RNA polymerase II may be either by direct competition for the interaction between proteins or by the disturbance of any looped DNA (Ptashne, 1986), between the two interacting complexes (at the UAS and TATA elements), to which the repressors bind.

The fact that operator sequences within relatively strong promoters can tightly regulate expression shows the effectiveness of mating-type regulation in heterologous gene expression. The repression of the *PGK* promoter is probably through the inhibition of interaction between transcriptional activator and RNA polymerase II since the location of the operator sequences is distal both to the putative UAS and to TATA elements. Our observation that multiple operator sequences confer the greatest degree of repression of *PGK* suggests that MATα2 may bind cooperatively to exert this effect. Alternatively, the presence of multiple operator sites may simply result in a greater possibility that the repressor may bind within the promoter to exert its effect. Indeed, there is no direct evidence that the MAT repressors bind cooperatively, and the absence of precisely spaced operator sequences within yeast genes suggests this does not occur in yeast. The only example of naturally occurring multiple copies of operator sequence within a promoter is the MATa1/MATα2 operator sequences in *STE5* and *HO* promoters. It is possible that their function is not for cooperative binding but rather for the displacement of different activator complexes which can bind within or close to the operator sequences.

Studies with a prokaryotic repressor (LexA) function in yeast suggest that repression in this case is mediated by a different mechanism from MATa1 and MATα2. It is apparent that proximity to the TATA element gives the greatest degree of repression, suggesting that LexA may function by blocking the binding of either transcriptional activators, RNA polymerase or both in an analogous manner to its function in *E. coli*.

The function of the silencer sequence in repressing transcription at *HML* and *HMR* loci appears to be more complex than mating-type regulation. The presence of both DNA replication and mitotic segregation functions within the silencer sequence is provocative. It is still not clear whether these features are directly involved in silencer function or whether the silencer has two functions. Just as provocative is the observation that activator proteins bind to this sequence. The studies of MATα2 repression demonstrate that repressors may exert their effect by displacing activators; this may also be the case for silencer function. Alternatively, the RAP1 and SBF-B activator proteins may mediate the interaction of SIR proteins either with themselves or with other proteins bound proximal to or within the silencer sequence. The SIR proteins are the only proteins that a show direct effect on repression. Their isolation as mutations show that they are not essential proteins, unlike RAP1, suggesting that their sole function is to mediate silencer function. The fact that there are four SIR proteins, each of which is essential for silencer function, suggests that they may form a protein complex at the silencer region. Moreover, their association with the nuclear matrix (Kimmerly and Rine, 1987) suggests that the silencer may be part of a major chromatin structure.

It is evident that negative regulation, and in particular that associated with mating-type in yeast, is becoming more clearly understood. The similarities between this system and others in both eukaryotes and prokaryotes suggest that at least some functions and structures are conserved. This suggests that our knowledge of mating-type regulation will greatly assist explorations of other types of negative regulation. Moreover, the application of mating-type regulation to effectively control heterologous expression in an eukaryotic host suggests that similar successful applications could be used in higher eukaryotes. No doubt these will emerge in a very short time.

Acknowledgements

We would like to thank Alan Bennett and Cath Catterall for critical reading of the manuscript and help with the proofs. We would also like to thank Marisa Dunn for providing data in Figure 2.6.

References

Abraham, J., Nasmyth, K.A., Strathern, J.N. and Klar, A.J.S. (1984) Regulation of mating type information in yeast. *J. Mol. Biol.* **176**: 307.
Bender, A. and Sprague, G.F. (1987) MATα1 protein, a yeast transcription activator, binds synergistically with a second protein to a set of cell-type-specific genes. *Cell* **50**: 681.
Blair, L.C., Kushner, P.J. and Herskowitz, I. (1979) Mutations of *HMa* and *HMα* loci and their bearing on the cassette model of mating type interconversion in yeast. *ICN-UCLA Symp. Mol. Cell. Biol.* **14**: 13.
Bodmer, M.W., Angal, S., Yarranton, G.T., Harris, T.J.R., Lyons, A., King, D.J., Pieroni, G., Riviere, C., Verger, R. and Lowe, P.A. (1987) Molecular cloning of a human gastric lipase and expression of the enzyme in yeast. *Biochem. Biophys. Acta* **909**: 237.
Brake, A.J., Julius, D.J. and Thorner, J. (1983) A functional prepro-α-factor gene in *Saccharomyces* yeasts can contain three, four, or five repeats of the mature pheromone sequence. *Mol. Cell.* **3**: 1440.
Brake, A.J., Brenner, C., Najarian, R., Laybourn, P. and Thorner, J. (1985) Structure of genes encoding precursors of the yeast peptide mating pheromone a-factor. In *Protein Transport and Secretion*, ed. M.-J. Gething, Cold Spring Harbor Laboratory, Cold Spring Harbor, 103.
Brand, A.H., Breeden, L., Abraham, J., Sternglanz, R. and Nasmyth, K. (1985) Characterization of a "silencer" in yeast: a DNA sequence with properties opposite to those of a transcriptional enhancer. *Cell* **41**: 41.
Brand, A.H., Micklem, G. and Nasmyth, K.A. (1987) A silencer contains sequences that can promote autonomous plasmid replication and transcriptional activation. *Cell* **51**: 709.
Brent, R. and Ptashne, M. (1984) A bacterial repressor protein or a yeast transcriptional terminator can block upstream activation of a yeast gene. *Nature* **312**: 612.
Brent, R. and Ptashne, M. (1985) A eukaryotic transcriptional activator bearing the DNA specificity of a prokaryotic repressor. *Cell* **43**: 729.

Bucking-Throm, E., Duntze, W., Hartwell, L.H. and Manney, T.R. (1973) Reversible arrest of haploid yeast cells at the initiation of DNA synthesis by a diffusible sex factor. *Exp. Cell. Res.* **76**: 99.

Burkholder, A.C. and Hartwell, L.H. (1985) The yeast α-factor receptor: structural properties deduced from the sequence of the *STE2* gene. *Nucleic Acids Res.* **13**: 8463.

Chan, R.K., Melnick, L.M., Blair, L.C. and Thorner, J. (1983) Extracellular suppression allows mating by pheromone-deficient sterile mutants of *Saccharomyces cerevisiae*. *J. Bacteriol.* **155**: 903.

Chen, C.Y., Oppermann, H. and Hitzeman, R.A. (1984) Homologous versus heterologous gene expression in the yeast *Saccharomyces cerevisiae*. *Nucl. Acids Res.* **12**: 8951.

Dobson, M.J., Tuite, M.F., Roberts, N.A., Kingsman, A.J., Kingsman, S.M., Perkins, R.E., Conroy, S.C., Dunbar, B. and Fothergill, L.A. (1982) Conservation of high efficiency promoter sequences in *Saccharomyces cerevisiae*. *Nucl. Acids Res.* **10**: 2625.

Emr, S.D., Schekman, R., Flessel, M.C. and Thorner, J. (1983) An *MFα1-SUC2* (α-factor-invertase) gene fusion for the study of protein localization and gene expression in yeast. *Proc. Natl. Acad. Sci. USA.* **80**: 7080.

Esposito, R.E. and Klapholz, S. (1981) Meiosis and ascospore development. In *The Molecular Biology of the Yeast Saccharomyces*, eds. J.N. Strathern, E.W. Jones and J.R. Broach, Cold Spring Harbor Laboratory, Cold Spring Harbor, 211.

Fehrenbacher, G., Perry, K. and Thorner, J. (1978) Cell–cell recognition in *Saccharomyces cerevisiae*: regulation of mating-specific adhesion. *J. Bacteriol.* **134**: 893.

Feldman, J., Hicks, J.B. and Broach, J.R. (1984) Identification of sites required for repression of a silent mating type locus in yeast. *J. Mol. Biol.* **178**: 815.

Gehring, W.J. (1985) The homeo box: a key to the understanding of development? *Cell* **40**: 3.

Gill, G. and Ptashne, M. (1987) Mutants of GAL4 protein altered in an activation function. *Cell* **51**: 121.

Goutte, C. and Johnston, A.D. (1988) *a1* protein alters the binding specificity of α2 repressor. *Cell* **52**: 875.

Hagen, D.C., McCaffrey, G., Sprague, G.F. (1986) Evidence the *STE3* gene encodes a receptor for the peptide pheromone *a*-factor: gene sequence and implications for the structure of the presumed receptor. *Proc. Natl. Acad. Sci. USA.* **83**: 1418.

Hall, M.N., Hereford, L. and Herskowitz, I. (1984) Targeting of *E. coli* β-galactosidase to the nucleus in yeast. *Cell.* **36**: 1057.

Hall, M.N. and Johnston, A.D. (1987) Homeo domain of the yeast repressor α2 is a sequence-specific DNA-binding domain but is not sufficient for repression. *Science* **237**: 1007.

Hartig, A., Holly, J., Saari, G. and MacKay, V.L. (1986) Multiple regulation of *STE2*, a mating-type-specific gene of *Saccharomyces cerevisiae*. *Mol. Cell. Biol.* **6**: 2106.

Hartwell, L.H. (1973) Synchronization of haploid yeast cell cycles, a prelude to conjugation. *Exp. Cell. Res.* **76**: 111.

Hartwell, L.H. (1980) Mutants of *S. cerevisiae* unresponsive to cell division control by polypeptide mating hormones. *J. Cell. Biol.* **85**: 811.

Herskowitz, I. (1987) Functional inactivation of genes by dominant negative mutations. *Nature* **329**: 219.

Herskowitz, I. and Oshima, Y. (1981) Control of cell type in *Saccharomyces cerevisiae*: Mating type and mating-type interconversion. In *The Molecular Biology of the Yeast Saccharomyces*, eds. J.N. Strathern, E.W. Jones and J.R. Broach, Cold Spring Harbor Laboratory, Cold Spring Harbor, 181.

Hoey, T. and Levine, M. (1988) Divergent homeo box proteins recognise similiar DNA sequences in *Drosophila*. *Nature* **332**: 858.

Ivy, J.M., Klar, A.J.S. and Hicks, J.B. (1986) Cloning and characterization of four *SIR* genes from *Saccharomyces cerevisiae*. *Mol. Cell. Biol.* **6**: 688.

Jenness, D.D., Burkholder, A.C. and Hartwell, L.H. (1983) Binding of α-factor pheromone to yeast *a* cells: chemical and genetic evidence for an α-factor receptor. *Cell* **35**: 521.

Jensen, R., Sprague, G.F. and Herskowitz, I. (1983) Regulation of yeast mating-type interconversion: feedback control of *HO* gene expression by the mating-type locus. *Proc. Natl. Acad. Sci. USA.* **80**: 3035.

Johnston, A.D. and Herskowitz, I. (1985) A repressor (*MATα2* product) and its operator control expression of a set of cell type specific genes in yeast. *Cell* **42**: 237.

Kimmerly, W. and Rine, J. (1987) The replication and segregation of plasmids containing cis-acting regulatory sites of silent mating-type genes in Saccharomyces cerevisiae is regulated by SIR. Mol. Cell. Biol. 7: 4225.

Kingsman, S.M., Kingsman, A.J., Dobson, M.J., Mellor, J. and Roberts, N.A. (1985) Heterologous gene expression in Saccharomyces cerevisiae. Biotechnol. Genet. Engg Revs, 3: 377.

Klar, A.J.S., Strathern, J.N. and Hicks J.B. (1981a) A position-effect control for gene transposition: state of expression of yeast mating-type genes affects their ability to switch. Cell 25: 517.

Klar, A.J.S., Strathern, J.N., Broach, J.R. and Hicks, J.B. (1981b) Regulation of transcription in expressed and unexpressed mating-type of yeast. Nature 289: 239.

Klar, A.J.S., Strathern, J.N. and Hicks, J.B. (1984) Developmental pathways in yeast. In Microbial Development, eds. R. Losick and L. Shapiro, Cold Spring Harbor Laboratory, Cold Spring Harbor, 151.

Kronstadt, J.W., Holly, J.A. and Mackay, V.L. (1987) A yeast operator overlaps an upstream site. Cell 50: 369.

Laughon, A. and Scott, M.P. (1984) Sequence of a Drosophila segmentation gene: protein structure homology with DNA-binding proteins. Nature 310: 25.

Leibowitz, M. and Wickner, R.B. (1976) A chromosomal gene required for killer plasmid expression, mating, and sporulation in S. cerevisiae. Proc. Natl. Acad. Sci. USA. 73: 2061.

Ma, J. and Ptashne, M. (1987) A new class of yeast transcriptional activators. Cell 51: 113.

MacKay, V.L. and Manney, T.R. (1974a) Mutations affecting sexual conjugation and related processes in Saccharomyces cerevisiae. I. Isolation and phenotypic characterization of non-mating mutaints. Genetics 76: 255.

MacKay, V.L. and Manney, T.R. (1974b) Mutations affecting sexual conjugation and related processes in Saccharomyces cerevisiae. II. Genetic analysis of non-mating mutants. Genetics 76: 273.

Meaden, P., Ogden, P., Bussey, H. and Tubb, R.S. (1985) A DEX gene conferring production of extracellular amyloglucosidase on yeast. Gene 34: 325.

Mellor, J., Dobson, M.J., Kingsman, A.J. and Kingsman, S.M. (1987) A transcriptional activator is located in the coding region of the yeast PGK gene. Nucleic Acids Res. 15: 6243.

Mellor, J., Dobson, M.J., Roberts, N.A., Kingsman, A.J. and Kingsman, S.M. (1985) Factors affecting heterologous gene expression in Saccharomyces cerevisiae. Gene 33: 215.

Miller, A.M., MacKay, V.L. and Nasmyth, K.A. (1985) Identification and comparison of two sequence elements that confer cell-type specific transcription in yeast. Nature 314: 598.

Mitchell, A.P. and Herskowitz, I. (1986) Activation of meiosis and sporulation by repression of the RME1 product in yeast. Nature 319: 738.

Mortimer, R.K. and Schild, D. (1980) The genetic map of Saccharomyces cerevisiae. Microbiol. Rev. 44: 519.

Nakayama, N., Miyajima, A. and Arai, K. (1985) Nucleotide sequences of STE2 and STE3, cell type-specific sterile genes from Saccharomyces cerevisiae. EMBO J. 4: 2643.

Nakayama, N., Miyajima, A. and Arai, K. (1987) Common signal transduction system shared by STE2 and STE3 in haploid cells of Saccharomyces cerevisiae: autocrine cell-cycle arrest results from forced expression of STE2. EMBO J. 6: 249.

Nasmyth, K.A. (1982) The molecular genetics of yeast mating type. Ann. Rev. Genet. 16: 439.

Nasmyth, K. and Shore, D. (1987) Transcription regulation in the yeast life cycle. Science 237: 1162.

Nasmyth, K.A., Tatchell, K., Hall, B.D., Astell, C. and Smith, M. (1980) Physical analysis of mating-type loci in Saccharomyces cerevisiae. Cold Spring Harbor Symp. Quant. Biol. 45: 961.

Nasmyth, K.A., Tatchell, K., Hall, B.D., Astell, C. and Smith, M. (1981) A position effect in the control of transcription at yeast mating type loci. Nature 289: 244.

Olesen, J., Hahn, S. and Guarente, L. (1987) Activators both bind to the CYC1 upstream activation site, UAS2 in an interdependent manner. Cell 51: 953.

Pabo, C.O. and Sauer, R.T. (1984) Protein–DNA recognition. Ann. Rev. Biochem. 53: 293.

Ptashne, M. (1986) Gene regulation by proteins acting nearby and at a distance. Nature 322: 697.

Ptashne, M., Buckman, K., Humayun, M.Z., Jeffrey, A., Maurer, R., Meyer, B. and Sauer, R.T. (1976) Autoregulation and function of a repressor in bacterophage lambda. Science 194: 156.

Porter, S.D. and Smith, M. (1986) Homeo-domain homology in yeast MATα2 is essential for repressor activity. Nature 320: 766.

Remaut, E., Stanssens, P. and Fiers, W. (1981) Plasmid vectors for high-efficiency expression controlled by the P_L promoter of coliphage lambda. *Gene* **15**: 81.

Rine, J., Strathern, J.N., Hicks, J.B. and Herskowitz, I. (1979) A suppressor of mating type mutations in *Saccharomyces cerevisiae*: evidence for and identification of cryptic mating type loci. *Genetics* **93**: 877.

Rine, J. and Herskowitz, I. (1987) Four genes responsible for a position effect on expression from HML and HMR in *Saccharomyces cerevisiae*. *Genetics* **116**: 9.

Sakai, K. and Yanagishima, N. (1972) Mating regulation in *S. cerevisiae*. II. Hormonal regulation of agglutinability of *a* type cells. *Arch. Microbiol.* **84**: 191.

Sauer, R.T., Smith, D.L. and Johnston, A.D. (1988) Flexibility of the yeast α2 repressor enables it to occupy the ends of its operator, leaving the center free. *Genes and Dev.* **2**: 807.

Shepherd, J.C.W., McGinnis, W., Carrasco, A.E., De Robertis, E.M. and Gehring, W.J. (1984) Fly and frog homeo domains show homologies with yeast mating type regulatory proteins. *Nature* **310**: 70.

Shore, D., Squire, M. and Nasmyth, K.A. (1984) Characterization of two genes required for the position-effect control of yeast mating-type genes. *EMBO J.* **6**: 461.

Shore, D. and Nasmyth, K. (1987) Purification and cloning of a DNA-binding protein from yeast that binds to both silencer and activator elements. *Cell* **51**: 721.

Sledziewski, A.Z., Bell, A., Kelsay, K. and MacKay, V.L. (1988) Construction of temperature-regulated yeast promoters using the MATα2 repression system. *Bio/technology* **6**: 411.

Sprague, G.F., Rine, J. and Herskowitz, I. (1981a) Homology and non-homology at the yeast mating type locus. *Nature* **289**: 250.

Sprague, G.F., Rine, J. and Herskowitz, I. (1981b) Control of yeast cell type by the mating type locus: II Genetic interactions between *MATα2* and unlinked α-*specific STE* genes. *J. Mol. Biol.* **153**: 323.

Sprague, G.F., Jensen, R. and Herskowitz, I. (1983) Control of yeast cell type by the mating type locus: Positive regulation of the α-specific *STE3* gene by the *MATα1* product, *Cell* **32**: 409.

Strathern, J.N., Hicks, J.B. and Herskowitz, I. (1980) Control of cell type by mating type locus: The α1-α2 hypothesis. *J. Mol. Biol.* **147**: 357.

Strathern, J.N., Spatola, E., McGill, C. and Hicks, J.B. (1981) The structure and organization of the transposable mating type cassettes in *Saccharomyces*. *Proc. Natl. Acad. Sci.* **77**: 2829.

Tatchell, K., Nasymth, K.A., Hall, B.D., Astell, C. and Smith, M. (1981) *In vitro* mutation analysis of mating-type locus in yeast. *Cell.* **27**: 25.

Thorner, J. (1981) Pheromonal regulation of development in *Saccharomyces cerevisiae*. In *The Molecular Biology of the Yeasts Saccharomyces*, eds. J.N. Strathern, E.W. Jones and J.R. Broach, Cold Spring Harbor Laboratory, Cold Spring Harbor, 143.

Wharton, R.P. and Ptashne, M. (1985) Changing the binding specificity of a repressor by redesigning an α-helix. *Nature* **319**: 601.

Wharton, R.P. and Ptashne, M. (1986) An α-helix determines the DNA-binding specificity of a repressor. *TIBS* **11**: 71.

Wickner, R.B. (1974) Mutants of *Saccharomyces cerevisiae* that incorporate deoxythymidine-5'-monophosphate into deoxyribonucleic acid *in vivo*. *J. Bacteriol.* **117**: 252.

Wilkinson, L.E. and Pringle, J.R. (1974) Transient G1 arrest of *S. cerevisiae* of mating type α by a factor produced by cells of mating type *a*. *Exp. Cell. Res.* **89**: 175.

Wilson, K.L. and Herskowitz, I. (1984) Negative regulation of *STE6* gene expression by the α2 product of *Saccharomyces cerevisiae*. *Mol. Cell. Biol.* **4**: 2420.

Wilson, K.L. and Herskowitz, I. (1986) Sequences upstream of *STE6* gene required for its expression and regulation by the mating type locus in *Saccharomyces*. *Proc. Natl. Acad. Sci. USA* **83**: 2536.

3 Messenger RNA translation and degradation in *Saccharomyces cerevisiae*

ALISTAIR J.P. BROWN

3.1 Introduction

The rate of expression of any gene is dependent upon the abundance of its mRNA and upon the efficiency with which the mRNA is translated. For most yeast genes the most influential level of control is at transcription. However, rates of mRNA translation and degradation vary greatly for different mRNA sequences. Both translation and degradation therefore exert powerful influences upon the overall pattern of protein synthesis in yeast.

Analysis of heterologous gene constructs has provided a valuable alternative strategy for investigations of transcription in yeast (Kingsman *et al.*, 1985). However, this strategy has proved less useful in studies on mRNA translation and degradation because, with a few notable exceptions, attempts are seldom made to measure *directly* the effects of specific constructs upon these processes. Most frequently, conclusions are drawn from indirect measurements, and this has led to some confusion in the literature. In this chapter I will therefore attempt to clear up some of these ambiguities while reviewing our current understanding of mRNA translation and degradation in the yeast, *Saccharomyces cerevisiae*. I will also highlight some important questions that remain unanswered.

3.2 Translation

Protein synthesis is dependent upon a complex translational apparatus which comprises the ribosomes and a large number of protein factors and cofactors that promote various phases of the process. The ribosome is made up of a large and small subunit; in yeast the small subunit (40 S) contains one RNA molecule (18 S) and between 25 and 35 proteins, whereas the large subunit (60 S) is made up of three RNA molecules (25 S, 5.8 S and 5 S) and between 35 and 45 proteins (Warner, 1982). All these constituents, with a few exceptions, seem to be required in equimolar amounts. Rapid progress has been made in our understanding of ribosome structure, including the sites of interaction with mRNA and tRNA molecules (Lake, 1985).

The induction of ribosome synthesis is dependent upon the coordinate expression of the ribosomal RNAs and proteins. Many of the yeast genes that encode ribosomal proteins have now been cloned and sequenced. This has revealed two consensus sequences (Teem *et al.*, 1984; Leer *et al.*, 1985) which appear to be essential for transcription of at least some ribosomal protein genes (Wouldt *et al.*, 1986; Rotenberg *et al.*, 1986). On top of this, regulatory mechanisms exist to ensure that the appropriate proportions of ribosomal proteins are synthesized. Interestingly, the synthesis of excess ribosomal protein is prevented in a number of ways including translational regulation and degradation (Warner *et al.*, 1985). This will be discussed further in section 3.2.4.

For some time, biochemical investigations of yeast protein synthesis lagged behind the analogous studies in higher eukaryotes. This was due to the relatively late development of yeast cell-free translation systems (Gasior *et al.*, 1979; Tuite *et al.*, 1980; Hofbauer *et al.*, 1982; Tuite and Plesset, 1986). Procedures for *in vitro* translation using rabbit reticulocyte lysates were published much earlier (Housman *et al.*, 1970). However, the gap is rapidly closing because of the ease and power of molecular genetics in yeast (Struhl, 1983). For example, the recent cloning and sequencing of some genes encoding yeast translation factors has opened the way for detailed functional analyses (Nagashima *et al.*, 1986; Altmann *et al.*, 1987; Ernst *et al.*, 1987; Qin *et al.*, 1987).

3.2.1 *Initiation*

In *E. coli*, the codon used for initiation of translation is usually determined by the close proximity of a ribosome binding site on the mRNA. The ribosome binding site is recognized by the small ribosomal subunit through sequence complementarity with the 3′ end of the 16 S rRNA (Shine and Dalgarno, 1974; Shine and Dalgarno, 1975). It has been suggested that a similar mechanism might operate in eukaryotes since partial sequence homology can be detected between the 5′-untranslated regions (5′-UTRs) of some eukaryotic mRNAs and the highly conserved 3′ end of the 18 S rRNA (Ziff and Evans, 1978, Sargan *et al.*, 1982). However, the distance between the initiation codon and the putative ribosome binding site on eukaryotic mRNAs is variable (Sargan *et al.*, 1982). Also, the existence of sequences corresponding to the putative ribosome binding sites has been attributed to the clustering of purines and pyrimidines, an effect which is common to noncoding sequences (De Wachter, 1979). Support for this model has therefore waned.

The model that is now widely accepted for initiation of translation in eukaryotes is based upon the 'scanning' mechanism originally proposed by Kozak (Kozak, 1978; Kozak, 1980*a*). According to this model, a complex containing the 40 S subunit does not bind directly at a ribosome binding site, but binds to the 5′ end of the mRNA and scans along the mRNA to the first AUG (Figure 3.1). At this point the complex stops, the 60 S subunit and associated factors bind to the complex and elongation begins. Initiation requires various protein factors, binding of a methionyl tRNA to the 40 S subunit, and both GTP and ATP (Figure 3.2). The beauty of this model is its

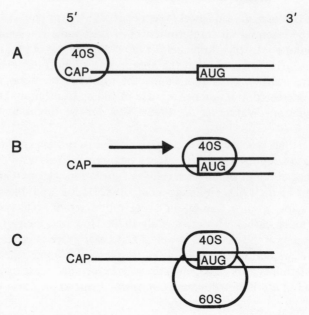

Figure 3.1 Scanning model for initiation of eukaryotic protein synthesis. According to this model (Kozak, 1978), a complex containing the 40 S ribosomal subunit binds at the 5′-end of the mRNA (*A*), and scans in a 3′-direction towards the initiation codon, where it stops (*B*), allowing association of the 60 S subunit (*C*) to complete initiation.

ability to account for the following significant features of eukaryotic initiation:

  (i) an absolute requirement for a 5′ end on the mRNA (Kozak, 1979)

 (ii) the great variability in the length of the 5′-UTR on eukaryotic mRNAs (Kozak, 1983*a*)

(iii) the initiation by most eukaryotic mRNAs using the AUG closest to their 5′ ends (Kozak, 1978).

The ability of the 40 S subunit to migrate along mRNA sequences has been confirmed (Kozak and Shatkin, 1978).

    More recently, the scanning model has been modified to account for the small proportion of mRNAs that do not initiate at the 5′-proximal AUG (Kozak, 1983*b*). It was suggested that the sequence environment around an AUG influences its ability to be recognized by the 40 S subunit, and hence the efficiency with which the sequence acts as a 'stop signal' for scanning. Analyses both *in vivo* and *in vitro* have confirmed the importance of sequences around the initiation codon for efficient translation (Kozak, 1981; Kozak, 1984; Morle *et al.*, 1986). The optimal sequences as defined by Kozak (Kozak, 1983*b*) are ANNAUGGN and GNNAUGR, whereas the least effective sequences are GNNAUGY, YNNAUGR and YNNAUGY (where N = any base, R = purine, Y = pyrimidine). Most 'functional' AUGs have an optimal sequence

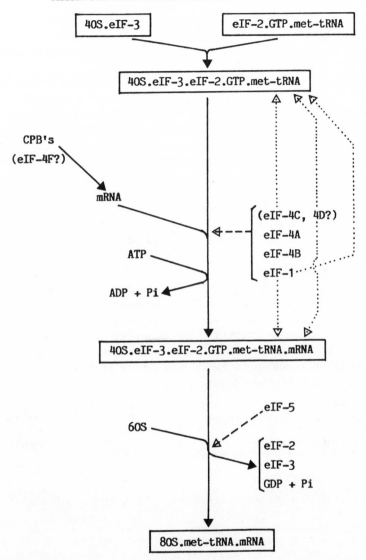

Figure 3.2 The involvement of factors and cofactors in eukaryotic translational initiation. The boxes show major intermediates formed during the initiation process. Eukaryotic initiation factor (eIF); cap binding proteins (CPBs); methionyl-tRNA (met-tRNA). Reactions(→); catalysing effect (— →); stabilizing effect (······→). Information taken from Moldave (1985).

environment, and most 'nonfunctional' AUGs located within the 5′-UTR occur in suboptimal sequence environments (Kozak, 1983*b*). Presumably, a proportion of 40 S subunits is able to bypass an upstream AUG if it has a suboptimal environment, and continue scanning to the functional AUG located downstream.

Detailed comparisons of sequences around the initiation codons of yeast mRNAs have been performed (Dobson *et al.*, 1982; Hamilton *et al.*, 1987). Clear similarities in these sequences are apparent for efficiently expressed mRNAs, suggesting that the environment around the initiation codon influences translational efficiency in yeast. The sequences of the 'Kozak consensus' derived from eukaryotic mRNAs in general, and of the 'yeast consensus' are compared in Table 3.1. It has been demonstrated that in yeast there is no absolute requirement for a particular sequence around the AUG (Sherman *et al.*, 1980; Sherman and Stewart, 1982). In an elegant set of experiments, Sherman and coworkers isolated revertants of a mutation which had disrupted the normal initiation codon in the gene encoding iso-1-cytochrome c (*CYC1*). Many of the revertants had new 5'-proximal, in-frame AUGs that acted as new initiation codons, and most contained roughly normal amounts of iso-1-cytochrome c. These results emphasized the importance of the 5'-proximal AUG, but reduced the importance attributed to the sequence around the AUG. However, no direct measurements of translation were performed (Sherman *et al.*, 1980).

Further analysis of the *CYC1* mRNA by Sherman's group provided the first direct evidence for the influence of the sequence environment around the initiation codon upon translation in yeast (Baim *et al.*, 1985a). This analysis, in which oligonucleotide-directed mutagenesis was used to insert new sequences into the *CYC1* gene, showed that translational efficiency can be affected two- to threefold by altering the sequence environment. More recent data suggests that the presence of certain sequences immediately 5' to the initiation codon

Table 3.1 Comparison of ATG sequence contexts. The sequences were taken from the following references: (*a*) Losson *et al.* (1983); (*b*) Hinnebusch, (1984); (*c*) Werner *et al.* (1985); (*d*) Strick and Fox (1987); (*e*) Hamilton *et al.* (1987); (*f*) Kozak (1983b). Only the sequence contexts of ATGs which are in upstream ORFs are included in the Table (i.e not those which are in-frame with the main coding region and have no in-frame stop codons; see section 3.2.4).

Gene	Upstream AUGs	'Functional' AUG
PPR1[a]	(1), (2) ACGAAG*ATG*A*TG*ATTAAA	TAAATC*ATG*AAGCAG
GCN4[b]	(1) TGAAAA*ATG*GCTTGC	AATAAA*ATG*TCCGAA
	(2) AGAATT*ATG*TGTTAA	
	(3) GCTATC*ATG*TACCCG	
	(4) TTCAAG*ATG*TTTCCG	
CPA1[c]	(1) CATTAT*ATG*TTTAGC	TTTCAA*ATG*TCCTCC
PET3[d]	(1) GAGACA*ATG*AAAAAA	GAACTT*ATG*TTACAA
	(2) ACTGAA*ATG*AACAAG	
	(3) ATTTTT*ATG*TTGATC	
	(4) CTTTTC*ATG*ATTCTA	
Yeast consensus[e]	AAAAAA*ATG*TCT	
	T C C C	
Kozak optimal consensus[f]	ANN*ATG*GGN or GNN*ATG*R	
Kozak suboptimal consensus[f]	GNN*ATG*GY or YNN*ATG*R or YNN*ATG*Y	

can significantly inhibit translation of a heterologous mRNA in yeast (Kinskern *et al.*, 1986). When 10 contiguous G residues are deleted from the construct, expression of the hepatitis B virus core antigen in yeast increases dramatically (Hamilton *et al.*, 1987).

It has been suggested that the sequence environment around the initiation codon 'plays a role in the selection of the AUG for initiation' in yeast (Baim *et al.*, 1985a). This is consistent with the observations of Kozak (1983b), and is particularly relevant to those yeast mRNAs which have short open reading frames (ORFs) upstream from the main coding region. In some cases, these upstream ORFs have been shown to play a role in regulating translation of the main reading frame (see section 3.2.4). (There is good evidence that these ORFs are translated. However, for the purposes of this discussion, sequences 5' to the main coding region will still be referred to as the 5'-untranslated region.) Most upstream AUGs have relatively poor sequence environments compared with the 'functional' start codon (Table 3.1). The *GCN4* mRNA provides a good example of this (Hinnebusch, 1985). However, this 'rule' is not applicable to the *PPR1* or *PET3* mRNAs in which the 'functional' start codons are located in non-ideal sequence environments (Losson *et al.*, 1983; Strick and Fox, 1987). This raises a question about the means by which the translation apparatus recognizes the 'functional' start codon in these mRNAs.

Under certain circumstances, codons other than AUG are able to initiate translation in yeast (Zitomer *et al.*, 1984; Baim *et al.*, 1985a). These codons are between ten- and 1000-fold less efficient than AUG for initiation. The codon UUG is possibly the best of these surrogate initiation codons. Depending upon its sequence environment, UUG appears to be capable of up to 6.9% of the activity of the wild-type start codon (Zitomer *et al.*, 1984).

Most eukaryotic mRNAs have a methylated guanosine residue (m^7G) linked to the 5' end by an unusual 5'–5' phosphotriester linkage (the 5'-cap). In higher eukaryotes the m^7G can be linked to any base, and often this 5' base is itself methylated (Adams and Cory, 1975). However, only two types of 5'-cap are found on yeast mRNAs; m^7GpppA and m^7GpppG (Sripati *et al.*, 1976; De Kloet and Andrean, 1976). The 5'-cap is added post-transcriptionally by the nuclear enzyme, mRNA guanylyltransferase (Itoh *et al.*, 1987). It is now clear that the 5'-cap is required to promote translation of most eukaryotic mRNAs (Banerjee, 1980). Translation of yeast mRNAs also seems to be dependent upon the 5'-cap, since the addition of cap analogues inhibits translation in cell-free extracts (Gasior *et al.*, 1979; Tuite *et al.*, 1980).

eIF-4E appears to be one of a number of factors which binds the 5'-cap in yeast (Altmann *et al.*, 1987). The single yeast gene encoding eIF-4E is essential for growth (Altmann *et al.*, 1987), suggesting that this cap-binding protein is important for initiation. Experiments using mammalian systems have led to suggestions that the recognition and binding of specific factors to the 5'-cap is one of the early stages in the initiation process (Shatkin, 1985). This may lead to binding of the other components of the 'preinitiation complex' (e.g. the 40 S ribosomal subunit, the met-tRNA:GTP:eIF-2 tertiary complex, and other initiation factors) to the mRNA in an ATP-dependent reaction. Another role of 5'-cap binding proteins might be to promote melting of secondary structures in the 5'-UTR that might inhibit scanning (Sonenberg et al., 1981; Shatkin, 1985). However, secondary structure formation close to the 5' end of the mRNA inhibits the formation of a stable complex between the mRNA and 5'-cap-binding proteins (Pelletier and Sonenberg, 1985a; Lawson et al., 1986).

The early observations of Sherman and coworkers indicated that there are no *absolute* requirements for a particular sequence in the 5'-UTR for translation of a yeast mRNA to proceed (Sherman et al., 1980). However, sequences around the start codon, and secondary structure formation within the 5'-UTR can influence translational initiation. What influence does the remainder of the 5'-UTR have upon initiation?

The length of yeast 5'-UTRs can range from less than 30 to over 500 nucleotides, and no obvious regions of strong sequence homology are apparent even within many groups of functionally related mRNAs. However, there is evidence that sequences within the 5'-UTR can subtly influence the efficiency with which an mRNA is translated. For example, mRNAs which encode abundant proteins in yeast, and therefore are presumed to be translated efficiently, have few G residues in their 5'-UTRs (Hamilton et al., 1987). Also, the introduction of G-rich sequences close to the initiation codon decreases the translatability of the *CYC1* mRNA (Baim et al., 1985a). Hence G residues clearly have a deleterious effect on translation, but the basis for this is unknown. G residues may directly inhibit scanning, or they may do so indirectly by allowing the formation of higher-order structures in the 5'-UTR. Certainly the formation of strong secondary structures in the 5'-UTR can disrupt translational initiation in both yeast (Baim et al., 1985b) and in mammalian cells (Pelletier and Sonenberg, 1985b). This seems to be due to the inhibition of scanning by such structures (Kozak, 1980b). Computer-aided analyses using the programme FOLD (Zuker and Steigler, 1981) suggest that few stable secondary structures should form in the 5'-UTRs of some yeast glycolytic mRNAs. This is significant because these mRNAs are thought to be translated efficiently in yeast. Those mRNAs which have strong secondary structures in the 5'-UTR may rely heavily upon the putative RNA unwinding activity of cap-binding proteins (Shatkin, 1985), or upon other unwinding activities analogous to those found in higher eukaryotes (Bass and Weintraub, 1987; Rebagliati and Melton, 1987).

To summarize, the length of the 5'-UTR does not seem to influence translational efficiency (Baim et al., 1985a). However, the presence of strong secondary structures or a high abundance of G residues within the 5'-UTR appear to inhibit initiation on yeast mRNAs. The negative effect of G residues is more marked if they are located around the initiation codon. Initiation is efficient if the start codon is provided by a 5'-proximal AUG that is surrounded by the optimal sequence environment (Table 3.1). No doubt these features have evolved to modulate the relative rates of translational initiation for specific yeast mRNAs.

3.2.2 *Elongation*

The elongation phase of translation can begin once the 60 S ribosomal subunit binds to complete the initiation complex which is located over the start codon. This phase is divisible into four main stages (for reviews see Moldave, 1985; Tuite, 1987). The first stage involves the charging of tRNAs by aminoacyl-tRNA synthetases. This reaction, which requires the hydrolysis of an ATP molecule, clearly must be extremely accurate to place the appropriate amino acid upon a particular tRNA. In the second stage, an aminoacyl-tRNA is bound to the A-site on the ribosome with the help of EF-1. The aminoacyl-tRNA is specified by the mRNA sequence via the codon/anticodon interaction. A peptide bond is then formed in the third stage, which results in the transfer of the peptide chain from the tRNA in the P-site, to the tRNA in the A-site. In the final stage, the elongating complex translocates, such that it moves along one codon on the mRNA and the new peptidyl-tRNA is moved from the A-site to the P-site. This stage requires EF-2. The cycle is then repeated until the complex reaches the termination codon (see Section 3.2.3).

Unlike translation in mammalian cells, the yeast apparatus has a third elongation factor (EF-3). The exact function of this factor is unknown, although it is required for each cycle in the elongation phase (Hutchison *et al.*, 1984). It seems to fractionate with polysomes, but is not associated with ribosomal subunits (Hutchison *et al.*, 1984). While EF-3 has ribosome-dependent GTPase and ATPase activities, it is not involved in binding the aminoacyl-tRNA to the ribosome, nor in peptide bond formation, nor in translocation (Skogerson and Engelhardt, 1977; Skogerson, 1979; Dasmahapatra and Chakraburtty, 1981). Recently, the gene encoding EF-3 has been isolated, and therefore its function can now be investigated using 'reverse genetics' (Qin *et al.*, 1987).

The normal elongation cycle, in which the three-base codon is translated using the three-base tRNA anticodon, is sometimes disrupted in yeast by a specific event which causes the reading frame to be shifted (Sherman, 1982; Craigen and Caskey, 1987). Furthermore, this frame-shifting must take place during translational elongation for the correct synthesis of some proteins to take place in yeast. Frame-shifting occurs within a small region of the *Ty* mRNA to achieve synthesis of one of the retrotransposon proteins, p3 (Mellor *et al.*, 1985; Wilson *et al.*, 1986). Therefore, the *Ty* element appears to take advantage of a low level of natural, sequence-specific frame-shifting in yeast. However, it is not known how this frame-shifting is controlled (Craigen and Caskey, 1987).

There are several features within the coding regions of mRNAs which suggest that the rate of cycling during elongation might be dependent upon the mRNA sequence. In many yeast mRNAs there is a strong bias towards the use

Table 3.2 Codon bias in yeast. The genetic code is tabulated in a standard format with the subset of preferred codons used in highly expressed yeast genes underlined (Bennetzen and Hall, 1982).

First position		Second position			Third position
U	U	C	A	G	
U	Phe	Ser	Tyr	Cys	U
	Phe	Ser	Tyr	Cys	C
	Leu	Ser	Term	Term	A
	Leu	Ser	Term	Trp	G
C	Leu	Pro	His	Arg	U
	Leu	Pro	His	Arg	C
	Leu	Pro	Gln	Arg	A
	Leu	Pro	Gln	Arg	G
A	Ile	Thr	Asn	Ser	U
	Ile	Thr	Asn	Ser	C
	Ile	Thr	Lys	Arg	A
	Met	Thr	Lys	Arg	G
G	Val	Ala	Asp	Gly	U
	Val	Ala	Asp	Gly	C
	Val	Ala	Glu	Gly	A
	Val	Ala	Glu	Gly	G

of 25 out of the 61 possible codons for the 20 amino acids (Bennetzen and Hall, 1982). For example, the *PYK1* mRNA which encodes the glycolytic enzyme pyruvate kinase contains 36 leucine codons. Of these, 32 are UUG, while UUA, CUU, CUC, CUA and CUG are only used four times in total (Burke *et al.*, 1983). The subset of preferred codons for yeast is outlined in Table 3.2. Interestingly, while codon bias exists in most organisms, the preferred codons differ (Bennetzen and Hall, 1982; Ikemura, 1982). However, the subsets of preferred codons in yeast and *Escherichia coli* correlate strongly with the relative abundances of the isoaccepting tRNAs in each organism (Ikemura, 1981*a*; Ikemura, 1981*b*; Ikemura, 1982, Bennetzen and Hall, 1982). For example, the major leucine isoacceptor tRNA in yeast has the anticodon CAA which pairs with UUG, the preferred codon for leucine (Bennetzen and Hall, 1982). Furthermore, the degree of bias towards the utilization of preferred codons in a particular gene clearly correlates with the level at which that gene is expressed. While most highly expressed genes have a strong codon bias, those genes that are expressed only at low levels have little or no bias (Bennetzen and Hall, 1982; Sharp *et al.*, 1986). A comprehensive reivew of the codon usage for individual genes has been published by Murayama and coworkers (Murayama *et al.*, 1986).

Why does this codon bias exist? Several theoretical analyses have shed light on this question. One common feature in these analyses is the hypothesis that the non-preferred codons which use low-abundance tRNAs are translated more slowly than the preferred codons (Bennetzen and Hall, 1982). This presumably is due to the greater time taken to select a low-abundance aminoacyl-tRNA compared with a highly abundant isoaccepting aminoacyl-tRNA. Evolutionary pressure may therefore be exerted towards codons which utilize abundant tRNAs to prevent excessive drainage of tRNA pools during elongation. However, there may also be pressure exerted by codon usage upon tRNA abundance. According to a recent mathematical model (Bulmer, 1987), coevolution of tRNA abundance and codon usage in response to the pressures described above is capable of generating the observed division of genes into highly expressed genes with strong codon bias and genes with random codon usage that are expressed at low levels.

Holm has argued that codon usage has evolved to maximize rates of translation by reducing the drainage on tRNA pools (Holm, 1986). Non-preferred codons are thought to limit gene expression only if their utilization exceeds normal levels (i.e. if mRNAs with poor codon bias are expressed at abnormally high levels), and therefore, differential rates of gene expression are thought not to be a consequence of variable codon bias (Holm, 1986). This model seems to be consistent with that of Varenne and Lazdunski (1986) which addresses the translational problems encountered when attempting to achieve high levels of expression for heterologous sequences with poor codon bias. They predict that unfavourable distributions of non-preferred codons within a gene can set an upper limit on the maximum rate of synthesis of its protein. In particular, a contiguous string of the same four non-preferred codons seems capable of severely reducing expression levels (Varenne and Lazdunski, 1986). This is consistent with experimental data on the expression of a modified chloramphenicol acetyltransferase (CAT) gene in *E. coli* (Robinson *et al.*, 1984).

Some have expressed the view that codon bias has little effect on the levels of expression of heterologous sequences in yeast (see for example, Kingsman *et al.*, 1985; Kinskern *et al.*, 1986). This conclusion is based upon the observation that reasonable amounts of protein can be produced from heterologous constructs with poor codon bias. For example, the hepatitis B virus core antigen sequence has been expressed using the yeast glyceraldehyde phosphate dehydrogenase (*GPD1*) gene promoter (Kinskern *et al.*, 1986). Although the heterologous sequence has extremely poor codon bias in yeast (codon bias index = 0.075; Kinskern *et al.*, 1986), whereas the wild-type *GPD1* gene has an extremely strong bias (codon bias index = 0.99; Bennetzen and Hall, 1982), relatively high levels of the foreign protein are made in yeast.

Why is there a discrepancy between these observations and those expected from the theoretical analyses discussed above? This apparent discrepancy is probably due to a combination of factors; firstly, the indirect experimental observations described above may be misleading, as possible effects of codon bias may be masked by other variables, and secondly, these rates of heterologous gene expression may not be high enough to see dramatic effects of codon bias. For example, no direct measurements of hepatitis B virus core antigen mRNA translation were made (Kinskern *et al.*, 1986). Furthermore, the hepatitis B virus core antigen itself is stable in yeast (Kinskern *et al.*, 1986), and therefore, the high level of expression may have been due to the accumulation of this protein rather than extremely high rates of synthesis.

Evidence for the influence of codon usage upon translational elongation has been obtained in both prokaryotic and eukaryotic systems. Distinct pauses during polypeptide elongation have been correlated with codon usage in *E. coli* (Morlon *et al.*, 1983) and in the silk moth (Lizardi *et al.*, 1979). Furthermore, translation of the yeast *PGK1* mRNA appears to be affected by substituting a large proportion of preferred codons with non-preferred codons at the start of the coding region (Hoekema *et al.*, 1987). Unfortunately, this analysis of *PGK1* must be treated with caution, as no direct measurements of translation were performed, and the codon replacement disrupted the region of the gene that is thought to contain the Downstream Activation Site (DAS; Mellor *et al.*, 1987). Nevertheless, although *PGK1* transcription appears to have been affected by this mutagenesis, some of the reduction in expression may have been due to reduced rates of translational elongation (Hoekema *et al.*, 1987). Direct evidence for codon bias effects in yeast has been obtained by comparing the translation of the *PYK1* mRNA with a *PYK1/LacZ* mRNA (Purvis *et al.*, 1987c). The *PYK1/LacZ* mRNA was derived by replacing the coding region of the *PYK* gene with that of the *E. coli LacZ* gene. The codon bias of the *PYK1/LacZ* mRNA in yeast is − 0.05, whereas that of the *PYK1* mRNA is 0.95 (Purvis *et al.*, 1987c). When the *PYK1/LacZ* mRNA was present at low levels, no significant effects of poor codon bias were observed upon the distribution of this mRNA across sucrose density gradients of yeast polysomes. However, the translation of this mRNA appeared to be decreased when its levels were increased. Other yeast mRNAs with good codon bias (*PYK1*, Actin and Rp1) were unaffected by the elevated levels of *PYK1/LacZ* mRNA, but the translation of an mRNA with poor codon bias (*TRP2*) was inhibited (Purvis *et al.*, 1987c). These results are consistent with the translation of elevated levels of the *PYK1/LacZ* mRNA causing a depletion in the pools of non-preferred tRNAs. This presumably leads to reduced rates of elongation for other cellular mRNAs with poor codon bias. Therefore, the experimental and theoretical evidence on the effects of codon bias upon translation are not inconsistent.

The formation of strong secondary structures within the coding region of an mRNA also seems to inhibit translational elongation. Temporary, but specific pauses during polypeptide elongation have been attributed to secondary structure formation on the mRNA encoding the bacteriophage MS2 coat protein (Chaney and Morris, 1979). Direct evidence has been obtained for the influence of secondary structure formation upon the translation of a yeast mRNA, *CYC1* (Baim *et al.*, 1985*b*). In this case, the introduction of a reasonably stable hairpin ($\Delta G = -19.6$ kcal/mol) near the N-terminus of the coding region caused the synthesis of iso-1-cytochrome c to be reduced to about 20% of the wild-type level. Direct analysis of the distribution of the mutant *CYC1* mRNA on polysome gradients confirmed that its translation was affected (Baim *et al.*, 1985*b*).

Hence, translational elongation for some mRNAs is discontinuous (Chaney and Morris, 1979; Lizardi *et al.*, 1979; Morlon *et al.*, 1983). Why should specific pauses exist during elongation? Clearly many sequences may not have evolved to achieve maximum rates of translation throughout the coding region. However, it has been suggested that in some cases, translational pauses may have evolved to improve the efficiency with which some proteins fold *in vivo* (Purvis *et al.*, 1987*b*). The existence of specific translational pauses may allow portions of a polypeptide to start folding correctly before the remainder is synthesized. This may be particularly important for the folding of some complex, multidomain proteins. However, as yet there is no direct experimental evidence to support this idea; only indirect evidence is available. Polypeptide chains do start to fold before their synthesis is complete (Bergman and Kuehl, 1979*a*; Bergman and Kuehl, 1979*b*; Peters and Davidson, 1982), and hence specific translational pauses are likely to affect how they fold. Also, in some yeast genes, non-random usage of non-preferred codons seems to correlate with the structural organization of their proteins (Purvis *et al.*, 1987*b*). For example, the long strings of non-preferred codons close to the interdomain regions of the *ARO1* gene (Figure 3.3) may help the individual domains to fold relatively independently. Attempts to test this hypothesis are in progress using site-directed mutagenesis of the long string of non-preferred codons in *TRP3* (Figure 3.3).

This hypothesis is relevant to heterologous gene expression, since the codon bias in yeast is frequently different from that found in the organism from which the heterologous gene is derived. Inappropriate translational pauses may, in some cases, reduce the proportion of a protein synthesized in its *active* conformation.

To summarize, strong secondary structures and poor codon bias can have deleterious effects upon translational elongation. However, codon bias effects may only become significant when an mRNA with poor codon bias is at abnormally high levels. The folding of some proteins may be influenced by controlled rates of translational elongation.

3.2.3 Termination

There are no known, normal tRNAs which correspond to the three termination codons UAA, UAG and UGA (Moldave, 1985). However, yeast mutants are isolatable in which translational termination is suppressed using an altered tRNA molecule that is capable of translating a termination codon

D

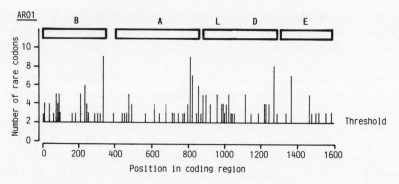

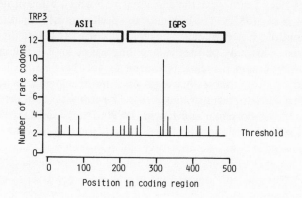

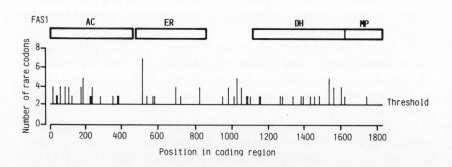

Figure 3.3 Potential translational pauses in yeast mRNAs: a link with protein folding? The length of strings of contiguous 'non-preferred' codons are plotted against the position of the strings within the coding region of three yeast mRNAs; (A) *ARO1*; (B) *TRP3*; (C) *FAS1*. Each encodes a multifunctional enzyme, and the domain organization of each enzyme is illustrated using the rectangles above the graphs. (Domains *L* and *D* within *ARO1* appear to overlap.) Note the correlation between the domain organization and the location of long strings of 'non-preferred' codons in *ARO1* and possibly *FAS1*. (A) and (B) are adapted from Purvis *et al.* (1987b). The information on the gene sequence and domain organization of *FAS1* is taken from Schweizer *et al.* (1986).

(Gesteland *et al.*, 1976; Sherman, 1982). Under normal circumstances, when a termination codon moves into the A-site on the ribosome, a single protein factor, the Release Factor (RF), is thought to bind to the ribosome. The ribosomal peptidyltransferase then catalyses the release of the polypeptide chain from the last tRNA, the RF dissociates from the ribosome, and GTP is hydrolysed to GDP and Pi. The ribosome then dissociates into subunits which become available to initiate a new round of translation (Moldave, 1985; Tuite, 1987).

A low level of natural suppression of termination has been observed for the UGA codon in yeast cell-free translation extracts (Tuite and McLaughlin, 1982). Interestingly, UGA is only used to terminate translation in about 20% of yeast mRNAs, whereas UAA and UAG are used approximately 50% and 30% of the time, respectively (Murayama *et al.*, 1986). Also, a second in-frame termination codon immediately follows the first in many yeast mRNAs (Tuite, 1987). These observations may reflect evolutionary pressure against innaccurate termination. Certainly, efficient nonsense suppressor mutations have deleterious effects on growth (Cox, 1971; Singh, 1977).

Sequences in the 3'-untranslated region (3'-UTR) can influence the translation of some yeast mRNAs. Translation of the *CYC1* mRNA appears to be affected in some *CYC1* mutants in which the 3'-UTR is extended, but no effects are observed in others (Zaret and Sherman, 1984). Similarly, while some mutations in the 3'-UTR reduce the translatability of the *PYK1* mRNA, other related mutations have no effect (Purvis *et al.*, 1987a). In this case, well-defined deletions and insertions were made within the 3'-UTR, and the effects on translation were analysed directly by determining the distribution of the modified mRNA on sucrose density gradients of yeast polysomes. It is not clear how these effects are mediated. It is possible that changes in the 3'-UTR can alter secondary structure formation on the mRNA, which in turn, could affect translation (Zaret and Sherman, 1984). However, there is no obvious homology between the 3'-UTR of the *PYK-1* mRNA and sequences in the 5'-UTR or coding region.

3.2.4 *Regulation of translation*

Rates of translational elongation vary amongst mRNA species, and frameshifting is required for the synthesis of at least one yeast protein (see section 3.2.2). Also, the biosynthesis of the translational apparatus (including the ribosomal subunits and possibly some translational factors) is coordinately regulated in response to the growth conditions (Huet *et al.*, 1985; Nagashima *et al.*, 1986; Huet and Sentenac, 1987). Hence, it is possible that the overall pattern of protein synthesis in the yeast cell is partly controlled at the levels of translational elongation and even termination. However, translation is thought to be regulated primarily at the level of initiation, and there is now a considerable body of evidence to support this supposition. Therefore, only regulatory mechanisms that act at the level of initiation will be discussed in this section. First general, then specific, mechanisms of translational regulation will be discussed.

In yeast, the overall pattern of protein synthesis may be regulated by the intracellular concentration of a specific protein factor. Lodish predicted that the translation of mRNAs which have a relatively low affinity for a rate-limiting initiation factor will be inhibited to a greater extent than the translation of mRNAs with a high affinity for the factor (Lodish, 1974). It would then be feasible for the overall pattern of protein synthesis to be controlled by regulating the intracellular concentration of this rate-limiting initiation factor. However, a weakness in this and some more recent mathematical analyses (Lodish, 1974; Bergmann and Lodish, 1979; Chu and Rhoads, 1980) is the unlikely assumption that everything except the 'discriminatory factor' is in saturating concentrations (e.g. ribosomal subunits and other initiation and elongation factors). Nevertheless, attempts to correct this weakness (Godefroy-Colburn and Thach, 1981) have yielded essentially the same conclusions.

These theoretical analyses are worth serious consideration because they are consistent with the available experimental evidence. For example, artificial manipulation of the overall rate of translational initiation affects the relative amounts of alpha- and beta-globin synthesized in rabbit reticulocyte lysates (Lodish, 1974), and the relative amounts of reoviral and host proteins synthesized in SC-1 cells (Walden et al., 1981). Furthermore, there is some evidence that yeast mRNA populations isolated from glucose repressed or derepressed cells, differ in their abilities to compete for initiation factors (Parets Soler et al., 1987), but it should be noted that this study was conducted in a heterologous translation system. Nevertheless, it seems likely that eukaryotic protein synthesis can be partly controlled via a specific, rate-limiting initiation factor (in its active form), and that this phenomenon is mediated through the differing affinities of specific mRNAs for this factor.

The probable candidate for the rate-limiting eukaryotic initiation factor as eIF-2. The importance of this factor for the repression of translation in reticulocyte lysates in response to haemin deprivation has been demonstrated (for review see Revel and Groner, 1978). Further evidence for the significance of eIF-2 has been obtained by Rosen and coworkers, who demonstrated that the addition of eIF-2 relieved the translational competition by specific mRNAs in reticulocyte lysates (Rosen et al., 1982). It is possible that eIF-2 is also the key initiation factor for global control of translation in yeast. However, the alpha subunit of yeast eIF-2 appears to be phosphorylated throughout exponential growth (Romero and Dahlberg, 1986). This is in direct contrast to rabbit reticulocytes in which phosphorylation converts eIF-2 into its *inactive* form (Revel and Groner, 1978). Also, the analysis of yeast RNA populations in a heterologous cell-free translation system led Parets Soler and coworkers to conclude that this role was fulfilled by cap-binding factors (Parets Soler et al., 1987). Now that the genes encoding yeast initiation factors are being cloned (Altmann *et al.*, 1987; Ernst *et al.*, 1987; for review see Tuite, 1988), it is likely that this ambiguity will be resolved in the near future. An important aspect of this mechanism of translation control is the means by which the affinity of a particular mRNA sequence for the rate-limiting factor is determined. Direct analysis of such interactions will only become possible when the identity of the determinant factor is established.

Under certain circumstances, specific sets of mRNAs in higher eukaryotes are translationally repressed. In both the sea urchin (Jenkins *et al.*, 1978) and the surf clam (Rosenthal *et al.*, 1980), different mRNA populations are utilized in a stage-specific manner, the nonfunctional population being present in a translationally repressed form. In both systems, binding of an undefined protein factor(s) appears to be responsible for blocking translation. The best example of this mechanism of translational regulation is provided by the ferritin mRNAs in mammalian cells. In the absence of ferrous ions, translation of ferritin mRNA, appears to be repressed through the binding of a specific protein factor to a conserved sequence in their 5′-UTRs (Munro *et al.*, 1985). On addition of ferrous iron, derepression occurs, the protein factor is released, and the ferritin mRNAs move from translationally inactive mRNP into polysomal fractions (Aziz and Munro, 1986).

No analogous situation has yet been described in yeast. However, the yeast *CPA1* mRNA is subject to a different form of translational repression (Werner *et al.*, 1987). This gene encodes a subunit of the carbamoyl-phosphate synthetase involved in arginine biosynthesis. (A second carbamoyl-phosphate synthetase is used in the synthesis of pyrimidines.) Expression of *CPA1* is regulated at the transcriptional level and is also subject to translational repression in the presence of arginine (Messenguy *et al.*, 1983). The mechanisms of translational regulation are mediated through an unusual 5′-UTR of about 250 bases (Werner *et al.*, 1987), which contains an open reading frame (ORF) of 25 codons and several potential hairpin structures (Werner *et al.*, 1985). Disruption of this upstream ORF by site-directed mutagenesis causes translation of the *CPA1* mRNA to be derepressed in the presence of arginine. Also, seven *cis*-dominant mutants of *CPA*1, that were selected on the basis of their translationally derepressed phenotype, all map within the ORF (Werner *et al.*, 1987). In these mutants the ORF is disrupted either by the loss of the start codon, the introduction of a nonsense codon, or a change in the amino acid sequence. This strongly suggests that the leader peptide itself is crucial for the mechanism of translational repression.

Werner and coworkers have proposed a model for the arginine-specific repression of *CPA1* mRNA translation (Figure 3.4) that accounts for the inability of the wild-type leader peptide to complement a mutant peptide in *trans*, and the involvement of a regulatory gene (*CPAR*) in repression of *CPA1* expression (Thuriaux *et al.*, 1972). They have suggested (Werner *et al.*, 1987) that in the presence of arginine, the product of *CPAR* interacts with the leader peptide of *CPA1* on the ribosome. Formation of this complex would prevent

5'   3'

upstream ORF   major ORF

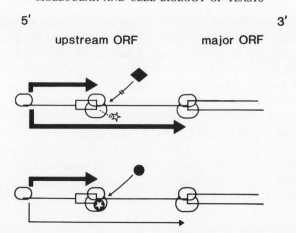

Figure 3.4 Model for translational regulation of *CPA1*. Werner *et al.* (1987) propose that the 40 S complex may scan to the upstream ORF (where translation of the leader peptide will take place) or to the major ORF (where synthesis of carbamoyl-phosphate synthetase will take place). In the presence of arginine, the product of *CPA1* (black circle) recognizes the leader peptide (*), forming a complex on the ribosome that inhibits 40 S scanning to the major ORF. In the absence of arginine, the CPA1 product (black diamond) cannot recognize the leader peptide, and 40 S scanning to the major ORF is not inhibited by the ribosome: leader peptide: *CPA1* product complex.

other 40 S ribosomal subunits from scanning further downstream to the start codon of the 'functional' coding region. When arginine is absent, the *CPAR* gene product would be unable to recognize the leader peptide, the ribosome: *CPAR* product: leader peptide complex would not form, 40 S scanning could proceed to the next start codon, and translation of the carbamoyl-phosphate synthetase subunit would occur. Inherent in this model is the assumption that 40 S ribosomal subunits are capable of scanning beyond the upstream ORF to the major ORF. The initiation codons in the upstream and major ORFs do not appear to lie in optimal sequence environments (Table 3.1).

Expression of the *GCN4* gene is also subject to translational regulation via ORFs within the 5'-UTR. *GCN4* encodes a positive transcriptional activator for a family of genes involved in amino acid biosynthesis (Hinnebusch and Fink, 1983; Driscoll Penn *et al.*, 1984). In the absence of certain amino acids, expression of the *GCN4* gene is induced, yet the abundance of its mRNA does not change significantly (Thireos *et al.*, 1984). This translational induction is dependent upon a hierarchy of *trans*-acting factors encoded by *GCD1*, *GCN1*, *GCN2* and *GCN3*. The *GCN1*, *GCN2*, and *GCN3* proteins are thought to induce *GCN4* mRNA translation by inhibiting the repression mediated by *GCD1* (Hinnebusch, 1985).

The *GCN4* mRNA has an unusually long 5'-UTR of about 590 bases which contains four AUGs (Mueller and Hinnebusch, 1986). Deletion of 280 bases containing all the upstream AUGs leads to derepression of *GCN4* mRNA translation (Thireos *et al.*, 1984). The fact that each of these AUGs is followed rapidly by an in-frame termination codon suggests that they are capable of acting as translational start codons. In fact each one of the upstream AUGs seems to be required to differing extents for the regulatory mechanism to work normally (Mueller and Hinnebusch, 1986; Tzamarias *et al.*, 1986). The *trans*-acting factors may operate by influencing the efficiency of recognition of the upstream AUGs by the 40 S subunit, or the frequency of reinitiation by ribosomes after translation of the ORFs. Interestingly, the sequence environment of the start codon for the major reading frame in *GCN4* is closer to the optimum than those for the upstream AUGs (section 3.2.1; Table 3.1).

The *PET3* mRNA, which encodes a protein required for the translation of a mitochondrial mRNA (Poutre and Fox, 1987), also contains four ORFs within its 5'-UTR (Strick and Fox, 1987). The 4 ORFs within the 5'-UTR of about 470 bases range in length from 6 to 31 codons. The ORFs overlap not only with each other, but also with the main coding region. This complex organization within the 5'-UTR would suggest that the *PET3* mRNA is subject to translational control. However, as yet there is no evidence to support this.

A similar situation exists for the *PPR1* gene which encodes a protein required for positive regulation of the unlinked genes *URA1* and *URA3* (Losson *et al.*, 1985). The *PPR1* mRNA contains 2 contiguous AUGs upstream from the major coding region, and these are followed by an in-frame termination codon located within the coding region, but out of frame with the coding region (Losson *et al.*, 1983). Hence, two short overlapping ORFs of four and five codons, respectively, start within a 5'-UTR of only 20–50 bases. (The length of the 5'-UTR of the *PPR1* mRNA depends upon which of the four transcriptional start points are used; Losson *et al.*, 1983.) Once again, such an arrangement would seem to suggest that the *PPR1* mRNA is subject to translational control, but there is no evidence to support this. In fact, the extremely short half-life of this mRNA (about 1 minute; Losson *et al.*, 1983) might indicate that expression of the *PPR1* gene is regulated at the level of mRNA abundance.

Several other yeast genes have multiple AUGs that could potentially act as translational start codons. They include the *SUC2, HTS1, LEU4, TRM1* and *MOD5* genes (Carlson and Botstein, 1982; Natsoulis *et al.*, 1986; Bletzer *et al.*, 1986; Ellis *et al.*, 1986; Najarian *et al.*, 1987). In these cases the 'upstream AUG' is in-frame with the major reading frame, and no termination codon exists to stop translation from the 'upstream AUG' into the major reading frame. It is now clear in the case of *SUC2* that synthesis of the cytoplasmic enzyme is performed using the downstream start codon, and the 'upstream AUG' is used for translation of an extra leader sequence to target the polypeptide for secretion (Carlson and Botstein, 1982). Similarly, *HTS1* codes for the cytoplasmic and mitochondrial forms of histidine tRNA synthetase (Natsoulis *et al.*, 1986). Therefore, in these cases differential targeting of a particular enzyme is achieved using a single gene. However, differential targeting, at least in the case of *SUC2*, does not seem to be controlled at the level of translation as might initially appear, but at the transcriptional level (Carlson and Botstein, 1982). Transcription of *SUC2* initiates at different points so that the upstream AUG (from which the leader sequence is translated) is either included or omitted from the mRNA.

A form of translational regulation, which does not involve upstream ORFs, affects the expression of the *TCM1* gene (Pearson *et al.*, 1982). This gene encodes ribosomal protein L3, the synthesis of which is tightly coordinated with other ribosomal proteins (Warner *et al.*, 1985). When extra copies of this gene are transformed into yeast, dosage compensation mechanisms operate to maintain a normal level of synthesis of the L3 protein. The dosage compensation occurs at several levels including translation. In cells containing 3.5 times the normal level of L3 mRNA, only 1.2 times the normal amount of L3 protein is synthesized (Pearson *et al.*, 1982).

In summary, translation in yeast is probably regulated at a global level by mRNA competition for a specific initiation factor(s). This would appear to be mediated by the differential affinities of individual mRNA sequences for the factor(s). On top of this, the translation of some specific mRNAs is individually regulated, often through the interaction of upstream ORFs with *trans*-acting factors.

3.3 mRNA degradation

mRNA half-lives range from about 1 minute to over 100 minutes in *S. cerevisiae* (Koch and Friesen, 1979; Chia and McLaughlin, 1979; Losson *et al.*, 1983; Santiago *et al.*, 1986). Therefore, the differential stability of various mRNA sequences must fundamentally effect the overall profile of protein synthesis in yeast. A similar situation exists in other eukaryotes, yet very little is known about the general pathway(s) of mRNA degradation in any eukaryote. This is a direct result of the lack of attention that has been paid to this field in comparison with other aspects of gene expression. However, the observations that the expression of some genes is partly regulated at the level of mRNA turnover (Guyette *et al.*, 1979; Brock and Shapiro, 1983; Mangiarotti *et al.*, 1985; Gay *et al.*, 1987; Lycan *et al.*, 1987; Paek and Axel, 1987; for review see Shapiro *et al.*, 1987) and that the stabilization of mRNAs from some oncogenes can contribute to cellular transformation (Eick *et al.*, 1985; Piechaczyk *et al.*, 1985; Rabbitts *et al.*, 1985; Fort *et al.*, 1987), have led to increased interest in the mechanisms and regulation of eukaryotic mRNA degradation. Some aspects of mRNA degradation in yeast have been reviewed recently (Brown and Lithgow, 1987).

3.3.1 *mRNA degradation mutants*
To my knowledge, no major genetic investigations of mRNA degradation pathways have been performed in yeast, or in any other eukaryote. A number of the conditional-lethal yeast mutants originally isolated by Hartwell

(Hartwell, 1967) have lesions in RNA metabolism. In most of these strains, for example in *rna2* strains, RNA splicing is disrupted (Lee *et al.*, 1986). However, *rna1* does not appear to be involved in splicing. It has been suggested that the transport of mRNA from nucleus to cytoplasm is inhibited in this mutant (Hutchison *et al.*, 1969). The *RNA1* gene has now been cloned (Atkinson *et al.*, 1985), and therefore detailed functional analysis of the *RNA1* gene product is now possible.

Greater progress has been made in *E. coli*. However, the isolation of *E. coli* polynucleotide phosphorylase (PNPase) and RNase II mutations was achieved using tedious biochemical screening procedures (Reiner, 1969; Nikolaev *et al.*, 1976), and it was at least 10 years before their importance in mRNA degradation was confirmed (Donovan and Kushner, 1986). This was due to the wide range of nucleases potentially involved in mRNA degradation, and the functional complementation of the PNPase and RNase II activities (Donovan and Kushner, 1986). An analogous situation is likely to exist in yeast, and therefore direct selection for mRNA degradation mutants in yeast will not be trivial. However, some progress has been made in the biochemical characterization of yeast ribonucleases.

3.3.2 Ribonucleases

Yeast RNases have been divided into three major groups (Van Ryk and von Tigerstrom, 1985):

(i) Mg^{2+}-dependent endoribonucleases
(ii) Mg^{2+}-independent endoribonucleases
(iii) Mg^{2}-dependent exoribonucleases

The biological significance of these enzymes is not known, although possible roles for some have been inferred from their biochemical properties.

Over 50% of the nonspecific nuclease activity in yeast exists in the mitochondrion, and most of this is encoded by a single nuclear gene (*NUC1*) which has now been cloned (Hofmann *et al.*, 1986). The enzyme, which belongs to the group of Mg^{2+}-dependent endoribonucleases, also has DNase activity (von Tigerstrom, 1982). A 500-fold difference has been observed in the rates at which poly(A) and poly(C) are digested by this enzyme (von Tigerstrom, 1982). Three observations would seem to suggest that this RNase is not involved in the degradation pathway(s) for cytoplasmic mRNAs. Firstly, the enzyme is membrane-bound in the mitochondrion (Van Ryk and von Tigerstrom, 1985; Hofmann *et al.*, 1986). Secondly, overexpression of the *NUC1* gene seems to have little effect upon cell viability (Hofmann *et al.*, 1986). Thirdly, the Mg^{2+}-dependent RNase activity decreases during the transition from exponential growth to stationary phase, when mRNA turnover is probably increased (Van Ryk and von Tigerstom, 1985). Other Mg^{2+}-dependent RNases are present in yeast, but little is known about their properties (Lee *et al.*, 1968; Nakao *et al.*, 1968).

A number of studies have revealed the presence of Mg^{2+}-independent RNases in yeast (Nakao *et al.*, 1968; Shetty *et al.*, 1980). Several RNases which belong to this group have been purified and characterized. One of these cleaves RNA to mononucleotides with little base specificity (Shetty *et al.*, 1980). This enzyme has a pH optimum of 6.0, and has no DNase activity. A second Mg^{2+}-independent RNase seems to have a high specificity for double-stranded RNA (Mead and Oliver, 1983). The activity of this enzyme decreases in late exponential growth phase and increases in the early stages of sporulation. The role of this RNase could be to degrade regions of RNA secondary structure that are relatively resistant to other activities. Alternatively, it may be functionally analogous to the double-stranded endonuclease RNase III, which is involved in the post-transcriptional processing of some RNAs in *E. coli* (Mead and Oliver, 1983).

A pyrimidine-specific, Mg^{2+}-independent endonuclease which degrades single-stranded RNA has been isolated and characterized by Stevens (Stevens, 1985). Almost 90% of the cleavage points generated by this enzyme are between a pyrimidine and an A residue. In particular, poly(A, U) is specifically and rapidly cleaved at U—A bonds, whereas homopolymers are degraded relatively slowly (Stevens, 1985). This enzyme is particularly interesting with respect to mRNA degradation because AU sequences have been implicated in mRNA degradation both in *E. coli* and in mammalian cells. Specific points of clevage occur at U—A bonds during the degradation of the *Lac* mRNA in *E. coli* (Cannistraro *et al.*, 1986). Also, a conserved, AU-rich sequence acts as a destabilizing element for a specific set of mammalian mRNAs (Shaw and Kamen, 1986). It is therefore tempting to speculate that the pyrimidine-specific RNase may be important for the degradation of some mRNAs in yeast (Stevens, 1985).

Mg^{2+}-dependent exoribonucleases, the third group of yeast RNases, may also be functionally important in mRNA degradation. An Mg^{2+}-dependent RNase, with a molecular weight of about 160 000 and a broad pH optimum around 8.0, has been shown to have 5′-to-3′ exonuclease activity (Stevens, 1980; Stevens, 1987). Interestingly, it was originally purified from preparations of yeast ribosomes (Stevens, 1980). Furthermore, polyribonucleotides with a capped 5′ end are relatively resistant to this enzyme when compared with uncapped molecules containing a 5′-phosphate (Stevens, 1978; Stevens, 1987). These observations led Stevens to suggest that this RNase is involved in mRNA degradation (Stevens, 1980), and an increase in Mg^{2+}-dependent 5′-exonuclease activity during stationary phase would seem to be consistent with this (Van Ryk and von Tigerstrom, 1985). However, this enzyme also has RNase H activity, suggesting that it may have an alternative role in the hydrolysis of the RNA primers used in DNA replication (Stevens, 1987).

A new approach to the characterization of RNases involved in mRNA degradation has been initiated (Ross and Kobs, 1986). The design of a cell-free mRNA degradation system derived from mammalian cells has now led to the description of a 3'-to-5' exonuclease that is involved in the degradation of a histone mRNA (Ross *et al.*, 1987; Peltz *et al.*, 1987). An analogous 3'-to-5' activity has been found in the yeast cytoplasm (Stevens, 1987). However, as yet little is known about this enzyme.

Therefore, biochemical investigations have revealed the existence of a range of RNase activities in yeast. These studies must now be combined with genetic and molecular approaches to elucidate the functional significance of these enzymes.

3.3.3 *mRNA structure and degradation*
Investigations of the relationship between mRNA structure and degradation have provided the greatest contribution to our understanding of the general mechanisms of mRNA turnover in yeast.

As described in section 3.2.1, two types of 5'-cap are observed on yeast mRNAs. However, the stability of an mRNA does not appear to be influenced by the type of 5'-cap it carries (Piper *et al.*, 1987). Nevertheless, both cap structures presumably fulfil the stabilizing role seen in other eukaryotic systems. The 5'-cap seems to protect an mRNA from 5'-to-3' exonucleolytic attack in *Xenopus* oocytes, in wheat-germ extracts and in a cell-free system derived from mouse L cells (Furiuchi *et al.*, 1977; Drummond *et al.*, 1985).

In a similar fashion, the 3'-poly(A) tail is thought to protect eukaryotic mRNAs against 3'-to-5' exonucleolytic digestion (Marbaix *et al.*, 1975; Littauer and Soreq, 1982; Drummond *et al.*, 1985). Although the poly(A) tail on yeast mRNAs is only about 50 nucleotides in length, compared with the length of approximately 200 residues seen in mammalian mRNAs (McLaughlin *et al.*, 1973), it is likely to serve the same stabilizing function. Some mRNAs, including histone mRNAs, are not polyadenylated in mammalian cells. Interestingly, the histone mRNAs are degraded by 3'-to-5' exonucleases (Peltz *et al.*, 1987; Ross *et al.*, 1987). In yeast, most mRNAs appear to have a poly(A) tail including histone mRNAs and aberrant transcripts (Fahrner *et al.*, 1980; Zaret and Sherman, 1982; Sutton and Broach, 1985). Therefore, it has been argued that a poly(A) tail is added post-transcriptionally to all yeast transcripts (Brown and Lithgow, 1987).

Gradual shortening of the poly(A) tail as an mRNA ages is thought to be due to slow digestion from the 3'-end (Sheiness *et al.*, 1975; Phillips *et al.*, 1979; Brown and Hardman, 1981). However, the length of the poly(A) tail is not thought to have a significant influence upon the stability of an mRNA (Nudel *et al.*, 1976; Krowczynska *et al.*, 1985; Santiago *et al.*, 1987), the possible exception being when the poly(A) tail shortens below a certain critical length. At this point the mRNA appears to be degraded as rapidly as deadenylated mRNA (Nudel *et al.*, 1976).

Since most yeast mRNAs have a 5'-cap and 3'-poly(A) tail, these structures are unable to provide the basis for the great variation in stability observed for different mRNAs (Chia and McLaughlin, 1979; Koch and Friesen, 1979; Santiago *et al.*, 1986). This variation must be mediated by internal features of individual mRNAs that lie between the cap and poly(A) tail.

Yeast mRNas are clearly divisible into two populations on the basis of their half-lives and length. The stability of mRNAs of similar length, but which lie in the different populations can differ by about tenfold (Santiago *et al.*, 1986). Also, within each population, there is a crude inverse relationship between mRNA length and half-life (Santiago *et al.*, 1986). Hence, the length of an mRNA partly determines its stability in yeast. Although a number of variables might affect the stability of an mRNA to a minor extent, a single factor probably accounts for the significant division between the relatively stable and unstable populations. Therefore, the *relative* stability of an mRNA in yeast is determined to a major extent by its length and a second internal feature. This observation, which was originally made with 15 mRNAs, holds firm with the analysis of additional mRNAs (Purvis *et al.*, 1987c), and is reasonably consistent with data generated on other mRNAs using alternative techniques for measurement of mRNA half-lives (Santiago *et al.*, 1986). Furthermore, it has now been demonstrated that an heterologous mRNA conforms to this 'stability rule' (Purvis *et al.*, 1987c). However, there is no correlation between the length and abundance of aberrant *CYC1* mRNAs. These mRNAs have extended 3'-regions caused by mutations in the sequences responsible for transcriptional termination of the *CYC1* gene (Zaret and Sherman, 1984).

It is possible that differences in translation, mediated by internal variations amongst mRNAs, might account for the division between the stable and unstable mRNA populations. Ribosomes may passively protect an mRNA from degradation during translation in yeast. There is now strong evidence for a link between mRNA stability and translation in *E. coli*. For example, *E. coli*, mRNAs are destabilized by premature stop codons (Morse and Yanofsky, 1969; Hiraga and Yanofsky, 1972), and drugs which inhibit translation also influence mRNA stability (Mangiarotti *et al.*, 1971; Gupta and Schlessinger, 1976; Schneider *et al.*, 1978). Similarly, premature stop codons appear to destabilize some human mRNAs (Chang and Kan, 1979; Gorski *et al.*, 1982). Furthermore, the yeast *URA1* and *URA3* mRNAs are dramatically affected by nonsense mutations (Losson and Lacroute, 1979; Pelsey and Lacroute, 1984). However, the stability of two other mRNAs in yeast does not appear to be significantly affected by the introduction of a stop codon at the 5'-end of their coding regions (Purvis *et al.*, 1987a).

The question of whether differences in ribosomal protection can account for the dramatic difference between the stable and unstable mRNA populations in yeast has been addressed directly (Santiago *et al.*, 1987). In this study, the distribution of 10 mRNAs of known half-life across sucrose density gradients of yeast polysomers was analysed. There was no obvious relationship between the number of ribosomes loaded on an mRNA and its stability.

Suggestive evidence for a direct relationship between mRNA degradation and translation in yeast has been published recently (Hoekema *et al.*, 1987). In this study, mutagenesis of the *PGK1* gene led to decreased translation rates and reduced mRNA levels. No direct measurements of mRNA translation or degradation were performed. Nevertheless, since only the coding region was subjected to mutagenesis, the low levels of *PGK1* mRNA were presumed to be due to increased rates of mRNA turnover (Hoekema *et al.*, 1987). However, sequences within the coding region influence *PGK1* transcription (Kingsman *et al.*, 1985; Mellor *et al.*, 1987), and therefore, the reduced levels of *PGK1* mRNA may have been due to the disruption of the transcriptional Downstream Activation Site. Hence it is not possible to infer a relationship between mRNA translation and stability from this data (Hoekema *et al.*, 1987).

There is therefore no *simple* relationship between mRNA translation and degradation in yeast. Ribosomal protection is not a major factor in determining the stablity of all wild-type mRNAs; premature stop codons will destabilize some mRNAs but leave others relatively unaffected. What then is the internal feature of an mRNA which has such a bearing on its half-life?

Evidence is now emerging which indicates that specific internal sequences or structures can strongly influence mRNA stability in a wide range of organisms (Brawerman, 1987). In *E. coli*, a highly conserved hairpin sequence (REP) stabilizes some mRNAs against 3'-to-5' exonucleolytic attack (Higgins *et al.*, 1982; Newbury *et al.*, 1987). A heterologous hairpin structure with a sequence unrelated to REP, is also capable of stabilizing mRNAs in *E. coli* (Wong and Chang, 1986). Also, some prokaryotic or bacteriophage mRNAs are stabilized by sequences at their 5'-ends (Gorski *et al.*, 1985; Belasco *et al.*, 1986). The stabilization of the bacteriophage T4 gene 32 mRNA appears to be dependent upon the interaction of sequences at the 5'-end of the mRNA with phage-derived, *trans*-acting factors (Gorski *et al.*, 1985).

There are several examples of eukaryotic mRNAs that are stabilized when portions of their sequence are deleted. During recovery from heat shock, the *Drosophila* hsp70 mRNA is abnormally stable if sequences at the 3'-end of the gene are delted (Simcox *et al.*, 1985). Similarly, the rate of degradation of the human c-*myc* mRNA is decreased if the first 5'-exon is absent (Eick *et al.*, 1985; Rabbitts *et al.*, 1985). These results imply the existence of destabilizing elements within the wild-type mRNAs, but as yet, the sequences responsible for this effect have not been identified. An element which destabilizes a family of mammalian mRNAs has been characterized in some detail (Shaw and Kamen, 1986). A conserved AU sequence lies in the 3'-UTRs of a number of mRNAs which encode inflammatory mediators (Caput *et al.*, 1986; Shaw and Kamen, 1986). The insertion of this AU sequence into the 3'-UTR causes the rate of degradation of the rabbit beta-globin mRNA to be increased about 30-fold. Therefore, this *cis*-acting element imposes a short half-life on specific mRNA sequences. Interestingly, the degradation of mRNAs which carry this destabilizing element may be controlled via a novel regulatory mechanism involving a cytoplasmic RNA with sequence complementarity to the element (Clemens, 1987).

Evidence has now been obtained for the existence of a destabilizing element in a member of the unstable mRNA population in yeast. Many major modifications of the *PYK1* mRNA do not transfer this relatively stable mRNA into the unstable population (Purvis *et al.*, 1987*a*; Purvis *et al.*, 1987*c*). However, the introduction of sequences from the unstable *URA3* mRNA into the 3'-UTR of the *PYK1* mRNA reduces its half-life from 59.8 ± 7.8 to 5.15 ± 0.23 minutes (Brown, A.J.P. and Moore, J., unpublished results). This change in half-life is not only consistent with the introduction of a destabilizing element into the *PYK1* mRNA, but also with the increased length of the mRNA (from about 1.6 kb to 2.4 kb) reducing its stability still further. The existence of such elements in yeast may provide the basis for the difference between the stable and unstable populations in this organism (Santiago *et al.*, 1986).

Based upon the observations described above, it is possible to design a model that outlines the general mechanisms of mRNA degradation in yeast (Figure 3.5; Brown, A.J.P. and Purvis, I.J., unpublished). This model may be applicable to eukaryotes in general. The initial, rate-limiting event in the degradation pathway is an endonucleolytic cleavage. The frequency with which this occurs determines the stability of an mRNA, and is dependent upon the size of the target (i.e. the length of the mRNA) and upon the presence or absence of a destabilizing element. Once the endonucleolytic cleavage has taken place, the mRNA fragments are rapidly degraded by 5'-to-3' and 3'-to-5' exonucleases. The intact mRNA is protected against these enzymes by the 5'-cap and 3'-poly(A) tail. According to this model, the 5'-cap and 3'-poly(A) tail

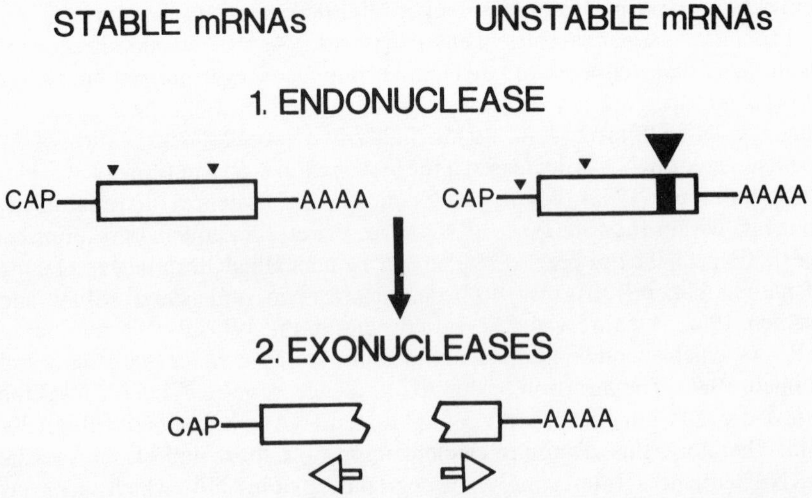

Figure 3.5 Model for the general pathway of mRNA degradation in yeast. See text for details.

stabilize mRNAs, but the *relative* stabilities of individual mRNAs are primarily determined by their length and destabilizing elements. All of the necessary RNase activities have been identified in yeast (section 3.3.2).

This model is capable of accounting for the inconsistent effects of nonsense mutations upon the stabilities of different yeast mRNAs (Losson and Lacroute, 1979; Pelsey and Lacroute, 1984; Purvis *et al.*, 1987a). Presumably, ribosomes will temporarily mask a destabilizing element during the translation of an unstable mRNA. (A ribosome protects about 50 nucleotides of an mRNA against RNase; Kang and Cantor, 1985.) Therefore, if the translation of an unstable mRNA is disrupted, the destabilizing element will become more exposed to endonucleolytic attack, and the stability of the mRNA will be significantly affected. In contrast, the half-life of a stable mRNA would not be dramatically influenced by nonsense codons, because a destabilizing element would not be exposed by the removal of ribosomes. Interestingly, the degree of destabilization caused by a premature stop codon depends upon its location in an *E. coli* mRNA (Nilsson *et al.*, 1987) and in the unstable yeast mRNAs, *URA1* and *URA3* (Losson and Lacroute, 1979; Pelsey and Lacroute, 1984).

It is important to stress that this model possibly describes the *general* pathway used by both homologous and heterologous mRNAs in yeast. The degradation of some specific mRNAs may involve additional, or alternative, steps.

3.3.4 *Regulation of mRNA stability*

The existence of destabilizing elements within some yeast mRNAs can be viewed as a general mechanism for differentially regulating the stability of specific sequences. Specific mRNAs may have evolved to have short half-lives in response to pressure for a reduction in the rate of expression of their genes. However, short mRNA half-lives may also have evolved to increase the rate of response of some genes to transcriptional regulation (Shapiro *et al.*, 1987). mRNAs with long half-lives will remain functional for a significant period after transcription of their genes has ceased.

In addition to this general control mechanism, the half-lives of some specific mRNAs change, relative to other mRNAs in the population, in response to specific stimuli. For example, some mammalian mRNAs are specifically stabilized in the presence of a particular hormone (Guyette *et al.*, 1979; Brock and Shapiro, 1983; Paek and Axel, 1987; for review see Shapiro *et al.*, 1987), while others are affected by viral infection (Strom and Frenkel, 1987). In some of these cases mRNA stability is altered by over tenfold.

There are few well-defined examples of regulation at the level of mRNA degradation in yeast. The best examples are the histone 2A and 2B mRNAs (H2A and H2B, respectively), their abundance being tightly regulated during the cell cycle (Hereford *et al.*, 1981; Hereford *et al.*, 1982). Histone mRNA levels are closely coupled with DNA synthesis; they start to accumulate during early S-phase and decrease late in S-phase. Both transcriptional and post-transcriptional mechanisms are involved in regulating H2A and H2B mRNA levels (Hereford et al., 1981). The 20-fold decrease in mRNA levels late in S-phase is achieved mostly through reduced transcription, but also by a threefold increase in the rate of histone mRNA degradation (Lycan et al., 1987).

This is similar to the situation in mammalian cells where the stability of histone mRNAs is also regulated during the cell cycle (DeLisle et al., 1983). In both systems, sequences at the 3'-end of the histone mRNA are required for post-transcriptional regulation (Graves et al., 1987; Lycan et al., 1987). It has been demonstrated in mammalian cells that this regulatory mechanism is dependent upon translation of the histone mRNA to within 300 nucleotides of the 3'-end (Graves et al., 1987). Mammalian histone mRNAs are not polyadenylated, and their controlled degradation is also dependent on a short hairpin structure at the 3'-end of the mRNA (Graves et al., 1987). In contrast, yeast histone mRNAs are polyadenylated (Fahrner et al., 1980). The mechanisms of regulating histone mRNA stability may, therefore, be similar but not identical in yeast and mammalian cells.

Histone mRNA stability is also regulated to control the ratio of the different histone proteins synthesized in yeast. When the copy number of the H2A and H2B genes is increased by transformation, their transcription is elevated yet the abundance of their mRNAs remains normal (Osley and Hereford, 1981). Direct measurements of mRNA stability in this study demonstrated that the half-life of the H2B mRNA was reduced from about 15 to 7 minutes in the strain carrying the extra copy of the gene.

3.4 Conclusions

Sufficient information is now available to make certain basic predictions about the ways in which the rate of expression of specific genes might be manipulated at the levels of mRNA translation and degradation. For example, to achieve efficient translational initiation, it would seem prudent to ensure that sequences theoretically capable of forming strong secondary structures are deleted from the 5'-regions of chimeric mRNAs, that no AUGs lie 5' to the start codon, and that the start codon is surrounded by the appropriate sequence environment (section 3.2.1). This approach is not only important for increasing the efficiency with which foreign proteins of commercial value can be synthesized in yeast, but also in furthering our understanding of eukaryotic gene expression.

Clearly, many major questions relating to mRNA translation and degradation remain to be answered. Firstly, we do not yet know which translational factor mediates general mRNA competition at the level of initiation. This must be established before it is possible to elucidate the molecular mechanisms by which the competition occurs.

The validity of the 'optimum' sequence environment for the translational start codon still needs to be tested rigorously, using a number of genes in yeast.

While on the surface this might seem a relatively simple problem, such analyses are complicated by the need to consider the effects of mutating the sequence around the start codon upon mRNA secondary structure formation.

Are the upstream ORFs present in some mRNAs translated, or do ribosomal subunits somehow scan over these ORFs to initiate translation at the 'functional' start codon? While analysis of the *CPA1* mRNA has suggested that an upstream ORF can be translated (section 3.2.4), this remains to be proved. This question has fundamental implications for the mechanisms of eukaryotic translational initiation (and possibly termination).

Does codon utilization influence the maximum rate of expression of a gene in yeast? As described in section 3.2.2, preliminary evidence suggesting this is the case has now been published, but additional data are required to confirm this theoretical prediction.

Do controlled rates of translational elongation help some polypeptides fold *in vivo*? To some, this is a radical question which infers that the amino acid sequence itself does not contain *all* of the information required for the folding of some proteins!

How are eukaryotic mRNAs degraded? A considerable amount of work must be done to test the predictions made in this chapter, and to characterize the destabilizing elements that are thought to be present in some yeast mRNAs.

The last question that will be raised here concerns the structure of mRNA: protein complexes *in vivo*. mRNAs do not exist in a naked state *in vivo*: they are complexed with specific proteins in messenger ribonucleoprotein (mRNP) particles (Greenberg, 1981; Swanson *et al.*, 1987). For example, a poly(A)-binding protein, which is encoded by a single gene in yeast (Sachs *et al.*, 1986), has been found in a number of eukaryotes. The poly(A)-binding protein and other RNA-binding proteins seem to have highly conserved domains near their amino-termini that are thought to mediate their interaction with RNA (Grange *et al.*, 1987; Swanson *et al.*, 1987). It is possible that the distribution of these RNA-binding proteins varies depending upon the sequence and secondary structure formation on an mRNA. Also, the interaction of these proteins could influence the translation and degradation of eukaryotic mRNAs. However, these questions remain to be answered.

Research using the model eukaryote *S. cerevisiae* should make an increasingly important contribution to our understanding of mRNA translation and degradation in the future. This is because it affords such efficient use of biochemical, genetic and molecular approaches. In combination, these approaches have the power to answer the major questions in this field.

Acknowledgement

I wish to thank the Science and Engineering Research Council and the Wellcome Foundation for supporting the work on yeast mRNA structure and function in my laboratory. I also thank Dr Ian Purvis for critically reviewing this MS, and Dr Bernard Cohen for his computing expertise.

References

Adams, J.M. and Cory, S. (1975) Modified nucleosides and bizarre 5'-termini in mouse myeloma mRNA. *Nature* **255**: 28.

Altmann, M., Handschin, C. and Trachsel, H. (1987) mRNA cap-binding protein: cloning of the gene encoding protein synthesis initiation factor eIF-4E from *S. cerevisiae. Mol. Cell. Biol.* **7**: 998.

Atkinson, N.S., Dunst, R.W. and Hopper, A.K. (1985) Characterization of an essential *S. cerevisiae* gene related to RNA processing: cloning of *RNA1* and generation of a new allele with a novel phenotype. *Mol. Cell. Biol.* **5**: 907.

Aziz, N. and Munro, H.N. (1986) Both subunits of rat liver ferritin are regulated at a translational level by iron induction. *Nucleic Acids Res.* **14**: 915.

Baim, S.B., Goodhue, C.T., Pietras, D.F., Eustice, D.C., Labhard, M., Friedman, L.R., Hampsey, D.M., Stiles, J.I. and Sherman, F. (1985a) Rules of translation: studies with altered forms of the yeast *CYC1* gene. In *'Sequence Specificity in Transcription and Translation*, eds. Calander, R. and Gold, L., Alan R. Liss, New York, 351.

Baim, S.B., Pietras, D.F., Eustice, D.C. and Sherman, F. (1985b) A mutation allowing an mRNA secondary structure diminishes translation of *S. cerevisiae* iso-1-cytochrome c. *Mol. Cell. Biol.* **5**: 1839.

Banarjee, A.K. (1980) 5'-terminal cap structure in eukaryotic mRNAs. *Microbiol Rev.* **44**: 175.

Bass, B.L. and Weintraub, H. (1987) A developmentally regulated activity that unwinds RNA duplexes. *Cell* **48**: 607.

Belasco, J.G., Nilsson, G., von Gabian, A. and Cohen, S.N. (1986) The stability of *E. coli* gene transcripts is dependent on determinants localized to specific mRNA segments. *Cell* **46**: 245.

Beltzer, J.P., Chang, L.-F.L., Hinkkanen, A.E. and Kohlhaw, G.B. (1986) Structure of yeast *LEU4*: the 5' flanking region contains features that predict two modes of control and two productive translation starts. *J. Biol. Chem.* **261**: 5160.

Bennetzen, J.L. and Hall, B.D. (1982) Codon selection in yeast. *J. Biol. Chem.* **257**: 3026.

Bergman, J.E. and Lodish, H.F. (1979) A kinetic model of protein synthesis: application to haemoglobin synthesis and translational control. *J. Biol. Chem.* **254**: 11927.

Bergman, L.W. and Kuehl, W.M. (1979a) Formation of intermolecular disulphide bonds on nascent immunoglobulin polypeptides. *J. Biol. Chem.* **254**: 5690.

Bergman, L.W. and Kuehl, W.M. (1979b) Formation of an intrachain disulphide bond on nascent immunoglobulin light chains. *J. Biol. Chem.* **254**: 8869.

Brawerman, G. (1987) Determinants of mRNA stability. *Cell* **48**: 5.

Brock, M.L. and Shapiro, D.J. (1983). Oestrogen stabilizes vitellogenin mRNA against cytoplasmic degradation. *Cell* **34**: 207.

Brown, A.J.P. and Hardman, N. (1981) The effect of age on the properties of poly(A)-containing mRNA in *P. polycephalum. J. Genet. Microbiol.* **122**: 143.

Brown, A.J.P. and Lithgow, G.J. (1987) The structure and expression of nuclear genes in *S. cerevisiae.* In *Gene Structure in Eukaryotic Microbes, Spec. Publ.* **22**, ed. Kinghorn, J.R., Society of General Microbiology, London, 1.

Bulmer, M. (1987) Coevolution of codon usage and tRNA aundance. *Nature* **325**: 728.

Burke, R.L., Tekamp-Olson, P. and Najarian, R. (1983) The isolation, characterization and sequence of the pyruvate kinase gene of *S. cerevisiae. J. Biol. Chem.* **258**: 2193.

Cannistraro, V.J., Subbarao, M.N. and Kennell, D. (1986) Specific endonucleolytic cleavage sites for decay of *E. coli* mRNA. *J. Mol. Biol.* **192**: 257.

Caput, D., Beutler, B., Hartog, K., Thayer, R., Brown-Shimer, S. and Cerami, A. (1986) Identification of a common nucleotide sequence in the 3'-untranslated region of mRNA molecules specifying inflammatory mediators. *Proc. Natl. Acad. Sci. USA* **83**: 1670.

Carlson, M. and Botstein, D. (1982) Two differentially regulated mRNAs with different 5' ends encode secreted and intracellular forms of yeast invertase. *Cell* **28**: 145.

Chaney, W.G. and Morris, A.J. (1979) Nonuniform size distribution of nascent polypeptides: the effect of mRNA structure upon the rate of translation. *Arch. Biochem. Biophys.* **194**: 283.

Chang, J.C. and Khan, Y.W. (1979) Beta-thalassemia, a nonsense mutation in man. *Proc. Natl. Acad. Sci. USA* **76**: 2886.

Chia, L.-L. and McLaughlin, C. (1979) The half-life of mRNA in *S. cerevisiae. Molec. Gen. Genet.* **170**: 137.

Chu, L.-Y. and Rhoads, R.E. (1980) Inhibition of cell-free mRNA translation by 7-methyl-guanosine 5-triphosphate: effect of mRNA concentration. *Biochemistry* **19**: 184.

Clemens, M.J. (1987) A potential role for RNA transcribed from B2 repeats in the regulation of mRNA stability. *Cell* **49**: 157.

Cox, B.S. (1971) A recessive-lethal super-suppressor mutation in yeast and other ψ phenomena. *Heredity* **26**: 211.

Craigen, W.J. and Caskey, C.T. (1987) Translational frameshifting: where will it stop? *Cell* **50**: 1.

Dasmahapatra, B. and Chakraburtty, K. (1981) Protein synthesis in yeast: 1 purification and properties of elongation factor 3 from *S. cerevisiae. J. Biol. Chem.* **256**: 9999.

De Kloet, S.R. and Andrean, B.A.G. (1976) Methylated nucleosides in poly(A)-containing yeast mRNA. *Biochim. Biophys. Acta* **425**: 401.

De Lisle, A.J., Graves, R.A., Marzluff, W.F. and Johnson, L.F. (1983) Regulation of histone mRNA production and stability in serum stimulated mouse 3T6 fibroblasts. *Mol. Cell. Biol.* **3**: 1920.

De Wachter, R. (1979) Do eukaryotic mRNA 5 noncoding sequences base-pair with the 18 S rRNA 3′ terminus? *Nucleic Acids Res.* **7**: 2045.

Dobson, M.J., Tuite, M.F., Roberts, N.A., Kingsman, A.J., Kingsman, S.M., Perkins, R.E., Conroy, S.C., Dunbar, B. and Fothergill, L.A. (1982) Conservation of high efficiency promoter sequences in *S. cerevisiae. Nucleic Acids Res.* **10**: 2625.

Donovan, W.P. and Kushner, S.R. (1986) Polynucleotide phosphorylase and ribonuclease II are required for cell viability and mRNA turnover in *E. coli* K12. *Proc. Natl. Acad. Sci. USA* **83**: 120.

Driscoll Penn, M., Thireos, G. and Greer, H. (1984) Temporal analysis of general control of amino acid biosynthesis in *S. cerevisiae*: role of positive regulatory genes in initiation and maintenance of mRNA derepression. *Mol. Cell. Biol.* **4**: 520.

Drummond, D.R., Armstrong, J. and Colman, A. (1985) The effect of capping and polyadenylation on the stability, movement and translation of synthetic mRNAs in *Xenopus* oocytes. *Nucleic Acids Res.* **13**: 7375.

Eick, D., Piechaczyk, M., Henglein, B., Blanchard, J.-M., Traub, B., Kofler, E., Wiest, S., Lenoir, G.M. and Bornkamm, G.W. (1985) Aberrant c-*myc* RNAs of Burkitt's lymphoma cells have longer half-lives. *EMBO J.* **4**: 3717.

Ellis, S.R., Li, J.-M., Hopper, A.K. and Martin, N.C. (1986) Characterization of the *TRM1* locus of *S. cerevisiae*: a gene essential for the modification of both cytoplasmic and mitochondrial tRNA. *Yeast* **2** (Spec. issue): S102.

Ernst, H., Duncan, R.F. and Hershey, J.W.B. (1987) Cloning and sequencing of cDNAs encoding the alpha-subunit of translation initiation factor eIF-2. *J. Biol. Chem.* **262**: 1206.

Fahrner, K., Yarger, J. and Hereford, L.M. (1980) Yeast histone mRNA is polyadenylated. *Nucleic Acids Res.* **8**: 5725.

Fort, P., Rech, J., Vie, A. Piechaczyk, M., Bonnieu, A., Jeanteur, P. and Blanchard, J.-M. (1987) Regulation of c-*fos* gene expression in hamster fibroblasts: initiation and elongation of transcription and mRNA degradation. *Nucleic Acids Res.* **15**: 5657.

Furiuchi, Y., LaFiandra, A. and Shatkin, A.J. (1977) 5′-terminal cap structure and mRNA stability. *Nature* **266**: 235.

Gasior, E., Herrera, F., Sadnik, I., McLaughlin, C.S. and Moldave, K. (1979) The preparation and characterization of a cell-free system from *S. cerevisiae* that translates natural mRNA. *J. Biol. Chem.* **254**: 3965.

Gay, D.A., Yen, T.J., Lau, J.T.Y. and Cleveland, D.W. (1987) Sequences that confer beta-tubulin autoregulation through modulated mRNA stability reside within exon 1 of a beta-tubulin mRNA. *Cell* **50**: 671.

Gesteland, R.F., Wolfner, N., Grisafi, P., Fink, G., Botstein, D. and Roth, J.R. (1976) Yeast suppressors of UAA and UAG nonsense codons work efficiently *in vitro* via tRNA. *Cell* **7**: 381.

Godefroy-Colburn, T. and Thach, R. (1981) The role of mRNA competition in regulating translation: kinetic model. *J. Biol. Chem.* **256**: 11762.

Gorski, J., Fiori, M. and Mach, B. (1982) A new nonsense mutation as the molecular basis for beta-thalassemia. *J. Mol. Biol.* **154**: 537.

Gorski, K., Roch, J.-M., Prentki, P. and Krisch, H.M. (1985) The stability of bacteriophage T4 gene 32 mRNA: a 5′ leader sequence that can stabilize mRNA transcripts. *Cell* **43**: 461.

Grange, T., Martins de Sa, C., Oddos, J. and Pictet, R. (1987) Human mRNA poly(A)-binding protein: evolutionary conservation of a nucleic acid binding motif. *Nucleic Acids Res.* **15**: 4771.

Graves, R.A., Pandey, N.B., Chodchoy, N. and Marzluff, W.F. (1987) Translation is required for regulation of histone mRNA degradation. *Cell* **48**: 615.

Greenberg, J.R. (1981) The polyribosomal mRNA-protein complex is a dynamic structure. *Proc. Natl. Acad. Sci. USA* **78**: 2923.

Gupta, R.S. and Schlessinger, D. (1976) Coupling rates of transcription, translation, and mRNA degradation in streptomycin-dependent mutants of *E. coli. J. Bacteriol.* **125**: 84.

Guyette, W.A., Matusik, R.J. and Rosen, J.M. (1979) Prolactin-mediated transcriptional and post-transcriptional control of casein gene expression. *Cell* **17**: 1013.

Hamilton, R., Wanatabe, C.K. and de Boer, H.A. (1987) Compilation and comparison of the sequence context around the AUG start codons in *S. cerevisiae* mRNAs. *Nucleic Acids Res.* **15**: 3581.

Hartwell, L.H. (1967) Macromolecular synthesis in temperature-sensitive mutants of yeast. *J. Bacteriol.* **93**: 1662.

Hereford, L., Bromley, S. and Osley, M.A. (1982) Periodic transcription of yeast histone genes. *Cell* **30**: 305.

Hereford, L., Osley, M.A., Ludwig, J.R. and McLaughlin, C.S. (1981) Cell-cycle regulation of yeast histone mRNA. *Cell* **24**: 367.

Higgins, C.F., Ames, G.F.L., Barnes, W.M., Clement, J.M. and Hofnung, M. (1982) A novel intercistronic regulatory element of prokaryotic operons. *Nature* **298**: 760.

Hinnebusch, A.G. (1985) A hierarchy of *trans*-acting factors modulates translation of an activator of amino acid biosynthetic genes in *S. cerevisiae. Mol. Cell. Biol.* **5**: 2349.

Hinnebusch, A.G. and Fink, G.R. (1983) Positive regulation in the general amino acid control of *S. cerevisiae. Proc. Natl. Acad. Sci. USA* **80**: 5374.

Hiraga, S. and Yanofsky, C. (1972) Hyper-labile mRNA in polar mutants of the tryptophan operon of *E. coli. J. Mol. Biol.* **72**: 103.

Hoekema, A., Kastelein, R.A., Vasser, M. and de Boer, H.A. (1987) Codon replacement in the *PGK* gene of *S. cerevisiae*: experimental approach to study the role of biased codon usage in gene expression. *Mol. Cell. Biol.* **7**: 2914.

Hofbauer, R., Fessl, F., Hamilton, B. and Ruis, H. (1982) Preparation of a mRNA-dependent cell-free translation system from whole cells of *S. cerevisiae. Eur. J. Biochem.* **122**: 199.

Hofmann, T.J., Uthayashanker, R., Srinivasan and Zassenhaus, H.P. (1986) Mitochondrial nuclease: purification, characterization and gene cloning. *Yeast* **2**: S460.

Holm, L. (1986) Codon usage and gene expression. *Nucleic Acids Res.* **14**: 3075.

Housman, D., Jacobs-Lorena, M., RajBhadary, U.L. and Lodish, H.F. (1970) Initiation of haemoglobin synthesis by methionyl-tRNA. *Nature* **227**: 913.

Huet, J. and Sentenac, A. (1987) TUF, the yeast DNA-binding factor specific for UAS$_{rpg}$ upstream activating sequences: identification of the protein and its DNA-binding domain. *Proc. Natl. Acad. Sci. USA* **84**: 3648.

Huet, J., Cottrelle, P., Cool, M., Vignais, M.-L., Thiele, D., Marck, C., Buhler, J.-M., Sentanac, A. and Fromageot, P. (1985) A general upstream binding factor for genes of the yeast translational apparatus. *EMBO J.* **4**: 3539.

Hutchison, H.T., Hartwell, L.H. and McLaughlin, C.S. (1969) Temperature-sensitive yeast mutant defective in RNA production. *J. Bacteriol.* **99**: 807.

Hutchison, J., Feinberg, B., Rothwell, T.C. and Moldave, K. (1984) Monoclonal antibody specific for yeast elongation factor 3 *Biochemistry* **23**: 3055.

Ikemura, T. (1981a) Correlation between the abundance of *E. coli* tRNAs and the occurrence of the respective codons in its protein genes. *J. Mol. Biol.* **146**: 1.

Ikemura, T. (1981b) Correlation between the abundance of *E. coli* tRNAs and the occurrence of the respective codons in its protein genes: a proposal for a synonymous codon choice that is optimal for the *E. coli* translational system. *J. Mol. Biol.* **151**: 389.

Ikemura, T. (1982) Correlation between the abundance of yeast tRNAs and the occurrence of the respective codons in protein genes: differences in synonymous codon choice patterns of yeast and *E. coli* with reference to the abundance of isoaccepting tRNAs. *J. Mol. Biol.* **158**: 573.

Itoh, N., Yamada, H., Kaziro, Y. and Mizumoto, K. (1987) mRNA Guanylyltransferase from *S. cerevisiae*: large scale purification, subunit functions and subcellular localization. *J. Biol. Chem.* **262**: 1989.

Jenkins, N.A., Kaumeyer, J.F., Young, E.M. and Raff, R.A. (1978) A test for masked message: the template activity of mRNP particles isolated from sea urchin eggs. *Devel. Biol.* **63**: 279.

Kang, C. and Cantor, C.R. (1985) Structure of ribosome-bound mRNA as revealed by enzymatic accessibility studies. *J. Mol. Biol.* **181**: 241.

Kingsman, S.M., Kingsman, A.J., Dobson, M.J., Mellor, J. and Roberts, N.A. (1985) Heterologous gene expression in *S. cerevisiae. Biotechnol. Genet. Engg. Rev.* **3**: 377.

Kniskern, P.J., Hagopian, A., Montgomery, D.L., Burke, P., Dunn, N.R., Hoffman, K.J., Miller, W.J. and Ellis, R.W. (1986) Unusually high-level expression of a foreign gene (hepatitis B virus core antigen) in *S. cerevisiae. Gene* **46**: 135.

Koch, H. and Friesen, J.D. (1979) Individual mRNA half-lives in *S. cerevisiae. Mol. Gen. Genet.* **170**: 129.

Kozak, M. (1978) How do eukaryotic ribosomes select initiation regions in mRNA? *Cell* **15**: 1109.

Kozak, M. (1979) Inability of circular mRNA to attach to eukaryotic ribosomes. *Nature* **280**: 82.

Kozak, M. (1980*a*) Evaluation of the scanning model for initiation of protein synthesis in eukaryotes. *Cell* **22**: 7.

Kozak, M. (1980*b*) Influence of mRNA secondary structure on binding and migration of 40 S ribosomal subunits. *Cell* **19**: 79.

Kozak, M. (1981) Possible role of flanking nucleotides in recognition of the AUG initiator codon by eukaryotic ribosomes. *Nucleic Acids Res.* **9**: 5233.

Kozak, M. (1983*a*) Comparison of initiation of protein synthesis in prokaryotes, eukaryotes and organelles. *Microbiol Rev.* **47**: 1.

Kozak, M. (1983*b*) How do eukaryotic ribosomes recognize the unique AUG initiator codon in mRNA? *Biochem. Soc. Symp.* **47**: 113.

Kozak, M. (1984) Point mutations close to the AUG initiator codon affect the efficiency of translation of rat preproinsulin *in vivo. Nature* **308**: 241.

Kozak, M. and Shatkin, A.J. (1978) Migration of 40 S ribosomal subunits on mRNA in the presence of edeine. *J. Biol. Chem.* **253**:6568.

Krowczynska, A., Yenofsky, R. and Brawerman, G. (1985) Regulation of mRNA stability in mouse erythroleukaemia cells. *J. Mol. Biol.* **181**: 231.

Lake, J.A. (1985) Evolving ribosome structure: domains in archaebacteria, eubacteria, eocytes and eukaryotes. *Ann. Rev. Biochem.* **54**: 507.

Lawson, T.G., Ray, B.K., Dodds, J.T., Grifo, J.A., Abramson, R.D., Merrick, W.C., Betsch, D.F., Weith, H.L. and Thach, R.E. (1986) Influence of 5' proximal secondary structure on the translational efficiency of eukaryotic mRNAs and on their interaction with initiation factors. *J. Biol. Chem.* **261**: 13979.

Lee, M.G., Lane, D.P., and Beggs, J.D. (1986) Identification of the *RNA2* protein of *S. cerevisiae. Yeast* **2**: 59.

Lee, S.Y., Nakao, Y. and Bock, R.M. (1968) The nucleases of yeast: purification, properties and specificity of an endonuclease from yeast. *Biochem. Biophys. Acta* **151**: 126.

Leer, R.J., van Raamsdonsk-Duin, M.M.C., Mager, W.H. and Planta, R. (1985) Conserved sequences upstream of yeast ribosomal protein genes. *Curr. Genet.* **9**: 273.

Littauer, U.Z. and Soreq, H. (1982) The regulatory function of poly(A) and adjacent 3' sequences in translated RNA. *Nucleic. Acid. Res.* **27**: 53.

Lizardi, P.M., Mahdavi, V., Shields, D. and Candelas, G. (1979) Discontinuous translation of silk fibroin in a reticulocyte cell-free system and in intact silk-gland cells. *Proc. Natl. Acad. Sci. USA* **76**: 6211.

Lodish, H.F. (1974) Model for the regulation of mRNA translation applied to haemoglobin synthesis. *Nature* **251**: 385.

Losson, R. and Lacroute, F. (1979) Interference of nonsense mutations with eukaryotic mRNA stability. *Proc. Natl. Acad. Sci. USA* **76**: 5134.

Losson, R., Fuchs, R.P.P. and Lacroute, F. (1983) *In vivo* transcription of a eukaryotic regulatory gene. *EMBO J.* **2**: 2179.

Losson, R., Fuchs, R.P.P. and Lacroute, F. (1985) Yeast promoters *URA1* and *URA3*: examples of positive control. *J. Mol. Biol.* **185**: 65.

Lycan, D.E. Osley, M.A. and Hereford, L.M. (1987) Role of transcriptional and post-transcriptional regulation in expression of histone genes in *S. cerevisiae. Mol. Cell. Biol.* **7**: 614.

Mangiarotti, G., Giorda, R., Ceccarelli, A. and Perlo, C. (1985) mRNA stabilization controls the expression of a class of developmentally regulated genes in *D. discoideum. Proc. Natl. Acad. Sci. USA* **82**: 5786.

Mangiarotti, G., Schlessinger, D. and Kuwano, M. (1971) Initiation of ribosome-dependent breakdown of T4-specific mRNA. *J. Mol. Biol.* **60**: 441.

Marbaix, G., Huez, G., Burney, A., Cleuter, Y., Hubert, E., Leclercq, M. Chantrenne, H., Soreq, H., Nudel, U. and Littauer, U.Z. (1975) Absence of poly(A) segment in globin mRNA accelerates its degradation in *Xenopus* oocytes. *Proc. Natl. Acad. Sci. USA* **72**: 3065.

McLaughlin, C.S., Warner, J.R., Edmonds, M., Nakazato, H. and Vaughan, M.H. (1973) Polyadenylic acid sequences in yeast mRNA. *J. Biol. Chem.* **248**: 1466.

Mead, D.J. and Oliver, S.G. (1983) Purification and properties of a double-stranded ribonuclease from the yeast *S. cerevisiae*. *Eur. J. Biochem.* **137**: 501.

Mellor, J., Fulton, S.M., Dobson, M., Wilson, W., Kingsman, S.M. and Kingsman, A.J. (1985) A retrovirus-like strategy for expression of a fusion protein encoded by yeast transposon Ty1. *Nature* **313**: 243.

Mellor, J., Dobson, M.J., Kingsman, A.J. and Kingsman, S.M. (1987) A transcriptional activator is located in the coding region of the yeast *PGK* gene. *Nucleic Acids Res.* **15**: 6243.

Messenguy, F., Feller, A., Crabeel, M. and Pierard, A. (1983) Control mechanisms acting at the transcriptional and post-transcriptional levels are involved in the synthesis of the arginine pathway carbamoylphosphate synthase of yeast. *EMBO J.* **2**: 1249.

Moldave, K. (1985) Eukaryotic protein synthesis. *Ann. Rev. Biochem.* **54**: 1109.

Morle, F., Starck, J. and Godet, J. (1986) Alpha-thalassemia due to the deletion of nucleotides − 2 and − 3 preceding the AUG intiation codon affects translation efficiency both *in vitro* and *in vivo*. *Nucleic Acids Res.* **14**: 3279.

Morlon, J., Lloubes, R., Varenne, S., Chartier, M. and Lazdunski, C. (1983) Complete nucleotide sequence of the structural gene for colicin A, a gene translated at nonuniform rate. *J. Mol. Biol.* **170**: 271.

Morse, D.E. and Yanofsky, C. (1969) Polarity and the degradation of mRNA. *Nature* **224**: 329.

Mueller, P.P. and Hinnebusch, A.G. (1986) Multiple upstream AUG codons mediate translational control of *GCN4*. *Cell* **45**: 201.

Nagashima, K., Kasai, M., Nagata, S. and Kaziro, Y. (1986) Structure of the two genes for polypeptide chain elongation factor EF–1 alpha from *S. cerevisiae*. *Gene* **45**: 265.

Munro, H.N., Leibold, E.A., Vass, J.K., Aziz, N., Rogers, J., Murray, M. and White, K. (1985) Ferritin gene structure, expression and regulation. In *Proteins of Iron Storage and Transport*, eds. Spik, G., Monkeiul, J., Crichton, R.R. and Mazurier, J., Elsevier, Amsterdam, 331.

Murayama, T., Gojobori, T., Aota, S. and Ikemura, T. (1986) Codon usage tabulated from the GenBank genetic sequence data. *Nucleic Acids Res.* **14** (suppl.): r151.

Nagashima, K., Kasai, M., Nagata, S. and Kaziro, Y. (1986) Structure of the two genes coding for polypeptide chain elongation factor EF–1alpha from *S. cerevisiae*. *Gene* **45**: 265.

Najarian, D., Dihanich, M.E., Martin, N.C. and Hopper, A.K. (1987) DNA sequence and transcript mapping of *MOD5*: features of the 5′ region which suggest two translational starts. *Mol. Cell. Biol.* **7**: 185.

Nakao, Y., Lee, S.Y., Halvorson, H.O. and Bock, R.M. (1968) The nucleases of yeast: properties and variability of ribonucleases. *Biochim. Biophys. Acta* **151**: 114.

Natsoulis, G., Hilger, F. and Fink, G.R. (1986) The *HTS1* gene encodes both the cytoplasmic and mitochondrial histidine tRNA synthetases of *S. cerevisiae*. *Cell* **46**: 235.

Newbury, S.F., Smith, N.H., Robinson, E.C., Hiles, I.D. and Higgins, C.F. (1987) Stabilization of translationally active mRNA by prokaryotic REP sequences. *Cell* **48**: 297.

Nikolaev, N., Folsom, V. and Schlessinger, D. (1976) *E. coli* mutants defficient in exoribonucleases. *Biochem. Biophys. Res. Commun.* **70**: 920.

Nilsson, G., Belasco, J.G., Cohen, S.N. and von Gabain, A. (1987) Effect of premature termination of translation on mRNA stability depends on the site of ribosome release. *Proc. Natl. Acad. Sci. USA* **84**: 4890.

Nudel, U., Soreq, H., Littauer, U.Z., Marbaix, G., Huez, G., Leclercq, M., Hubert, E. and Chantrenne, H. (1976) Globin mRNA species containing poly(A) segments of different lengths: their functional stability in *Xenopus* oocytes. *Eur. J. Biochem.* **64**: 115.

Osley, M.A. and Hereford, L.M. (1981) Yeast histone genes show dosage compensation. *Cell* **24**: 377.

Paek, I. and Axel, R. (1987) Glucocorticoids enhance the stability of human growth hormone mRNA. *Molec. Cell. Biol* **7**: 1496.

Parets Soler, A., Casanova, M., Gozalbo, D. and Sentandreu, R. (1987) Differential translational efficiency of the mRNAs isolated from derepressed and glucose repressed *S. cerevisiae*. *J. Gen. Microbiol.* **133**: 1471.

Pearson, N.J., Fried, H.M. and Warner, J.R. (1982) Yeast use translational control to compensate for extra copies of a ribosomal protein gene. *Cell* **29**: 347.

Pelletier, J. and Sonenberg, N. (1985a) Photochemical crosslinking of cap binding proteins to eukaryotic mRNAs: effect of mRNA 5' secondary structure. *Mol. Cell. Biol.* **5**: 3222.

Pelletier, J. and Sonenberg, N. (1985b) Insertion mutagenesis to increase secondary structure within the 5' noncoding region of a eukaryotic mRNA reduces the translational efficiency. *Cell* **40**: 515.

Pelsey, F. and Lacroute, F. (1984) Effect of ochre mutations on yeast *URA1* mRNA stability. *Curr. Genet.* **8**: 277.

Peltz, S.W., Brewer, G., Kobs, G. and Ross, J. (1987) Substrate specificity of the exonuclease activity that degrades H4 histone mRNA. *J. Biol. Chem.* **262**: 9382.

Peters, T. and Davidson, L.K. (1982) The biosynthesis of rat serum albumin: *in vivo* studies on the formation of the disulphide bonds. *J. Mol. Biol.* **257**: 8847.

Phillips, S.L., Tse, C., Serventi, I. and Hynes, N. (1979) Structure of polyadenylic acid in the RNA of *S. cerevisiae*. *J. Bacteriol.* **138**: 542.

Piechaczyk, M., Yang, J.-Q., Blanchard, J.-M., Jeanteur, P. and Marcu, K.B. (1985) Post-transcriptional mechanisms are responsible for accumulation of truncated c-*myc* RNAs in murine plasma cell lines. *Cell* **42**: 589.

Piper, P.W., Curran, B., Davies, W., Hirst, K. and Seward, K. (1987) *S. cerevisiae* mRNA populations of different intrinsic stability in unstressed and heat-shocked cells display almost constant m7GpppA:m7GpppG 5'-cap ratios. *FEBS Letts.* **220**: 177.

Poutre, C.G, and Fox, T.D. (1987) *PET3*, a *S. cerevisiae* nuclear gene required for translation of the mitochondrial mRNA encoding cytochrome *c* oxidase subunit II. *Genetics* **115**: 637.

Purvis, I.J., Bettany, A.J.E., Loughlin, L. and Brown, A.J.P. (1987a) the effects of alterations within the 3'-untranslated region of the pyruvate kinase mRNA upon its stability and translation in *S. cerevisiae*. *Nucleic Acids Res.* **15**: 7951.

Purvis, I.J., Bettany, A.J.E., Santiago, T.C., Coggins, J.P., Duncan, K., Eason, R. and Brown, A.J.P. (1987b) The efficiency of folding of some proteins is increased by controlled rates of translation *in vivo*: a hypothesis. *J. Mol. Biol.* **193**: 413.

Purvis, I.J., Loughlin, L., Bettany, A.J.E., and Brown, A.J.P. (1987c) Translation and stability of an *E. coli* beta-galactosidase mRNA expressed under the control of pyruvate kinase sequences in *S. cerevisiae*. *Nucleic Acids Res.* **15**: 7963.

Qin, S., Moldave, K. and McLaughlin, C.S. (1987) Isolation of the yeast gene encoding eIF-3 for protein synthesis. *J. Biol. Chem.* **262**: 7802.

Rabbitts, P.H., Forster, A., Stinson, M.A. and Rabbitts, T.H. (1985) Truncation of exon 1 from the c-*myc* gene results in prolonged c-*myc* mRNA stability. *EMBO J.* **4**: 3727.

Rebagliati, M.R. and Melton, D.A. (1987) Antisense RNA injections in fertilized frog eggs reveal an RNA duplex unwinding activity. *Cell* **48**: 599.

Reiner, A.M. (1969) Isolation and mapping of polynucleotide phosphorylase mutants of *E. coli*. *J. Bacteriol.* **97**: 1431.

Revel, M. and Groner, Y. (1978) Post-transcriptional and translational controls of gene expression in eukaryotes. *Ann. Rev. Biochem.* **47**: 1079.

Robinson, M., Lilley, R., Little, S., Emtage, J.S., Yarranton, G., Stephens, P., Millican, A., Eaton, M. and Humphreys, G. (1984) Codon usage can effect efficiency of translation of genes in *E. coli*. *Nucleic Acids Res.* **12**: 6663.

Romero, D.P. and Dahlberg, A.E. (1986) The alpha subunit of eIF-2 is phosphorylated *in vivo* in the yeast *S. cerevisiae*. *Mol. Cell. Biol.* **6**: 1044.

Rosen, H., Di Segni, G. and Kaempfer, R. (1982) Translational control by mRNA competition for eIF-2. *J. Biol. Chem.* **257**: 946.

Rosenthal, E.T., Hunt, T. and Ruderman, J.V. (1980) Selective translation of mRNA controls the pattern of protein synthesis during early development of the surf clam *S. solidissima*. *Cell* **20**: 487.

Ross, J. and Kobs, G. (1986) H4 histone mRNA decay in cell-free extracts initiates at or near the 3' terminus and proceeds 3'-to-5'. *J. Mol. Biol.* **188**: 579.

Ross, J., Kobs, G., Brewer, G. and Peltz, S.W. (1987) Properties of the exonuclease activity that degrades H4 histone mRNA. *J. Biol. Chem.* **262**: 9374.

Rotenberg, M.O. and Woolford, J.L. (1986) Tripartite upstream promoter element essential for expression of *S. cerevisiae* ribosomal protein genes. *Mol. Cell. Biol.* **6**, 674.

Sachs, A.B., Bond, W.M. and Kornberg, R.D. (1986). A single gene from yeast for both nuclear and cytoplasmic poly(A) binding proteins: domain structure and expression. *Cell* **45**: 827.

Santiago, T.C., Purvis, I.J., Bettany, A.J.E. and Brown, A.J.P. (1986) The relationship between mRNA stability and length in *S. cerevisiae. Nucleic Acids Res.* **14**: 8347.

Santiago, T.C., Bettany, A.J.E., Purvis, I.J. and Brown, A.J.P. (1987) mRNA stability in *S. cerevisiae*: the influence of translation and poly(A) tail length. *Nucleic Acids Res.* **15**: 2417.

Sargan, D.R., Gregory, S.P. and Butterworth, P.H.W. (1982) A possible novel interaction between the 3' end of 18 S rRNA and the 5' leader sequence of many eukaryotic mRNAs. *FEBS Lett.* **147**: 133.

Schneider, E., Blundell, M. and Kennell, D. (1978) Translation and mRNA decay. *Mol. Gen. Genet.* **160**: 121.

Schweizer, M., Roberts, L.M., Hottke, H.-J., Takabayashi, K., Hollerer, E., Hoffmann, B., Muller, G., Kottig, H. and Schweizer, E. (1986) The pentafunctional *FAS1* gene of yeast: its nucleotide sequence and order of the catalytic domains. *Mol. Gen. Genet.* **203**: 479.

Shapiro, D.J., Blume, J.E. and Nielsen, D.A. (1987) Regulation of mRNA stability in eukaryotic cells. *Bioessays* **6**: 221.

Sharp, P.M., Tuohy, T.M.F. and Mosurski, K.R. (1986) Codon usage in yeast: cluster analysis clearly differentiates highly and lowly expressed genes. *Nucleic Acids Res.* **14**: 5125.

Shatkin, A.J. (1985) mRNA cap binding proteins: essential factors for initiating translation. *Cell* **40**: 223.

Shaw, G. and Kamen, R. (1986) A conserved AU sequence from the 3' untranslated region of GM–CSF mRNA mediates selective mRNA degradation. *Cell* **46**: 659.

Sheiness, D., Puckett, L. and Darnell, J.E. (1975) Possible relationship of poly(A) shortening to mRNA turnover. *Proc. Natl. Acad. Sci. USA* **72**: 1077.

Sherman, F. (1982) Suppression in the yeast *S. cerevisiae*. In *The Molecular Biology of the Yeast Saccharomyces: Metabolism and Gene Expression*, eds. Strathern, J.N., Jones, E.W. and Broach, J.R., Cold Spring Harbor Laboratory, Cold Spring Harbor, 463.

Sherman, F. and Stewart, J.W. (1982) Mutations altering initiation of translation of yeast iso-1-cytochrome c; contrasts between the eukaryotic and prokaryotic initiation process. In *The Molecular Biology of the Yeast Saccharomyces: Metabolism and Gene Expression.*, eds. Strathern, J.N., Jones, E.W. and Broach, J.R., Cold Spring Harbor Laboratory, Cold Spring Harbor, 301.

Sherman, F., Stewart, J.W. and Schweingruber, A.M. (1980) Mutants of yeast initiating translation of iso-1-cytochrome c within a region spanning 37 nucleotides. *Cell* **20**: 215.

Shetty, J.K., Weaver, R.C. and Kinsella, J.E. (1980) A rapid method for the isolation of ribonucleases from yeast (*S. carlsbergensis*). *Biochem. J.* **189**: 363.

Shine, J. and Dalgarno, L. (1974) The 3' terminal sequence of *E. coli* 16 S rRNA: complimentarity to nonsense triplets and ribosome binding sites. *Proc. Natl. Acad. Sci. USA* **71**: 1342.

Shine, J. and Dalgarno, L. (1975) Determinant of cistron specificity in bacterial ribosomes. *Nature* **254**: 34.

Simcox, A.A., Cheney, C.M., Hoffman, E.P. and Shearn, A. (1985) A deletion of the 3' end of the *D. melanogaster* hsp70 gene increases stablity of mutant mRNA during recovery from heat shock. *Mol. Cell. Biol.* **5**: 3397.

Singh, A. (1977) Nonsense suppressors of yeast cause osmotic sensitive growth. *Proc. Natl. Acad. Sci. USA* **74**: 305.

Skogerson, L. (1979) Separation and characterization of yeast elongation factors. *Meth. Enzymol.* **60**: 676.

Skogerson, L. and Engelhardt, D. (1977) Dissimilarity in protein chain elongation factor requirements between yeast and rat liver ribosomes. *J. Biol. Chem.* **252**: 1471.

Sonenberg, N., Guertin, D. and Cleveland, D. (1981) Probing the function of the eukaryotic 5' cap structure by using a monoclonal antibody directed against cap-binding proteins. *Cell* **27**: 563.

Sripati, C.E., Groner, Y. and Warner, J.R. (1976) Methylated blocked 5' termini of yeast mRNA. *J. Biol. Chem.* **251**: 2898.

Stevens, A. (1978) An exoribonuclease from *S. cerevisiae*: effect of modifications of 5'-end groups on the hydrolysis of substrates to 5'-mononucleotides. *Biochem. Biophys. Res. Commun.* **81**: 656.

Stevens, A. (1980) Purification and characterization of a *S. cerevisiae* exoribonuclease which yields 5'-mononucleotides by a 5'-to-3' mode of hydrolysis. *J. Biol. Chem.* **255**: 3080.

Stevens, A. (1985) Pyrimidine-specific cleavage by an endoribonuclease of *S. cerevisiae. J. Bacteriol.* **164**: 57.

Stevens, A. and Maupin, M.K. (1987) A 5'-to-3' exoribonuclease of *S. cerevisiae*: size and novel substrate specificity. *Arch. Biochem. Biophys.* **252**: 339.

Strick, C.A. and Fox, T.D. (1987) *S. cerevisiae* positive regulatory gene *PET3* encodes a mitochondrial protein that is translated from an mRNA with a long 5'-leader. *Mol. Cell. Biol.* **7**: 2728.

Strom, T. and Frenkel, N. (1987) Effects of herpes simplex virus on mRNA stability. *J. Virol.* **61**: 2198.

Struhl, K. (1983) The new yeast genetics. *Nature* **305**: 391.

Sutton, A. and Broach, J.R. (1985) Signals for transcription initiation and termination in the *S. cerevisiae* plasmid 2-μm circle. *Mol. Cell. Biol.* **5**: 2770.

Swanson, M.S., Nakagawa, T.Y., Levan, K. and Dreyfuss, G. (1987) Primary structure of human nuclear RNP particle C proteins: conservation of sequence and domain structures in hnRNA, mRNA and pre-rRNA binding proteins. *Mol. Cell. Biol.* **7**: 1731.

Teem, J.L., Abovich, N., Kaufer, N.F., Schwindinger, W.F., Warner, J.R., Levy, A., Woolford, J., Leer, R.J., van Raamsdonk-Duin, M.M.C., Mager, W.H., Planta, R.J., Schultz, L., Freisen, J.D., Freid, H. and Rosbash, M. (1984) A comparison of yeast ribosomal protein gene DNA sequences. *Nucleic Acids Res.* **12**: 8295.

Thireos, G., Driscoll Penn, M. and Greer, H. (1984) 5' untranslated sequences are required for the translational control of a yeast regulatory gene. *Proc. Natl. Acad. Sci. USA* **81**: 5096.

Thuriaux, P., Ramos, F., Pierard, A., Grenson, M. and Waime, J.M. (1972) Regulation of the carbamoylphosphate synthetase belonging to the arginine biosynthesis pathway of *S. cerevisiae. J. Mol. Biol.* **67**: 277.

Tuite, M.F. (1988) The genetics and biochemistry of yeast protein synthesis. In *The Yeasts*, eds. Rose, A.H. and Harrison, J.S., Vol. **4**: in press.

Tuite, M.F. and McLaughlin, C.S. (1982) Endogenous readthrough of a UGA termination codon in a *S. cerevisiae* cell-free system: evidence for involvement of both a mitochondrial and a nuclear tRNA. *Mol. Cell. Biol.* **2**: 490.

Tuite, M.F. and Plesset, J. (1986) mRNA-dependent yeast cell-free translation systems: theory and practice. *Yeast* **2**: 35.

Tuite, M.F., Plesset, J., Moldave, K. and McLaughlin, C.S. (1980) Faithful and efficient translation of homologous and heterologous mRNAs in an mRNA-dependent cell-free system from *S. cerevisiae. J. Biol. Chem.* **255**: 8761.

Tzamarias, D., Alexandraki, D. and Thireos, G. (1986) Multiple *cis*-acting elements modulate the translational efficiency of *GCN4* mRNA in yeast. *Proc. Natl. Acad. Sci. USA* **83**: 4849.

Van Ryk, D.I. and von Tigerstrom, R.G. (1985) Activity changes of three nucleolytic enzymes during the life cycle of *S. cerevisiae. Can. J. Microbiol.* **31**: 1095.

Varenne, S. and Lazdunski, C. (1986) Effect of distribution of unfavourable codons on the maximum rate of gene expression by an heterologous organism. *J. Theoret. Biol.* **120**: 99.

von Tigerstrom, R.G. (1982) Purification and characteristics of a mitochondrial endonuclease from the yeast *S. cerevisiae. Biochemistry* **21**: 6397.

Walden, W.E., Godefroy-Colburn, T. and Thach, R.E. (1981) The role of mRNA competition in regulating translation: demonstration of competition *in vivo. J. Biol. Chem.* **256**: 11739.

Warner, J.R. (1982) The yeast ribosome: structure, function and synthesis. In *The Molecular Biology of the Yeast Saccharomyces: Metabolism and Gene Expression*, eds. Strathern, J.N., Jones, E.W. and Broach, J.R., Cold Spring Harbor Laboratory, Cold Spring Harbor, 529.

Warner, J.R., Mitra, G., Schwindinger, W.F., Studeny, M. and Fried, H.M. (1985) *S. cerevisiae* coordinate accumulation of yeast ribosomal proteins by modulating mRNA splicing, translational initiation, and protein turnover. *Mol. Cell. Biol.* **5**: 1512.

Werner, M., Feller, A. and Pierard, A. (1985) Nucleotide sequence of the yeast gene *CPA1* encoding the small subunit of arginine-pathway carbamoylphosphate synthetase: homology of the deduced amino acid sequence to other glutamine aminotransferases. *Eur. J. Biochem.* **146**: 371.

Werner, M., Feller, A., Messenguy, F. and Pierard, A. (1987) The leader peptide of yeast gene *CPA1* is essential for the translational repression of its expression. *Cell* **49**: 805.

Wilson, W., Malim, M.H., Mellor, J., Kingsman, A.J. and Kingsman, S.M. (1986) Expression strategies of the yeast retrotransposon Ty: a short sequence directs ribosomal frameshifting. *Nucleic Acids Res.* **14**: 7001.

Wong, H.C. and Chang, S. (1986) Identification of a positive retroregulator that stabilizes mRNAs in bacteria. *Proc. Natl. Acad. Sci. USA* **83**: 3233.

Wouldt, L.P., Smit, A.B., Mager, W.H. and Planta, R. (1986) Conserved sequence elements upstream of the gene encoding yeast ribosomal protein L25 are involved in transcription activation. *EMBO J.* **5**: 1037.

Zaret, K.S. and Sherman, F. (1982) DNA sequence required for efficient transcription termination in yeast. *Cell* **28**: 563.

Zaret, K.S. and Sherman, F. (1984) Mutationally altered 3′ ends of yeast *CYC1* mRNA affect transcript stability and translational efficiency. *J. Mol. Biol.* **176**: 107.

Ziff, E.B. and Evans, R.M. (1978) Coincidence of the promoter and capped 5′-terminus of RNA from the adenovirus 2 major late transcription unit. *Cell* **15**: 1463.

Zitomer, R.S., Walthall, D.A., Rymond, B.C. and Hollenberg, C.P. (1984). *S. cerevisiae* ribosomes recognize non-AUG initiation codons. *Mol. Cell. Biol.* **4**: 1191.

Zuker, M. and Stiegler, P. (1981) Optimal computer folding of large RNA sequences using thermodynamics and auxiliary information. *Nucleic Acids Res.* **9**: 133.

4 The production of proteins and peptides from *Saccharomyces cerevisiae*

D.J. KING, E.F. WALTON and G.T. YARRANTON

4.1 Introduction

Since the advent of recombinant DNA technology the desire to produce recombinant proteins, mainly for use in the pharmaceutical industry, has driven considerable effort into understanding the expression and production of proteins in a variety of hosts. The first host organism to be used was *Escherichia coli*. This was because the techniques of gene cloning and expression were first developed using *E. coli* and more was understood about *E. coli* molecular biology than about that of any other organism. However, attention soon turned to other potential host systems, due in part to some of the problems encoutered with *E. coli*, and the development of transformation techniques for *Saccharomyces cerevisiae* (Hinnen *et al.*, 1978; Beggs, 1978) opened up the possibility of using *S. cerevisiae* as an alternative host organism.

The use of *S. cerevisiae* as a host for protein production offers several advantages, and for some proteins yeast may be the organism of choice for production. With *E. coli*, recombinant proteins can be expressed at very high level, commonly 10–20% of cell protein. However, several problems have been encountered with intracellular expression in *E. coli*, including: (i) many recombinant proteins are produced in an insoluble and inactive form, and (ii) small polypeptides are often degraded rapidly and must be produced as fusion proteins if accumulation to high levels is to be achieved (Harris, 1983). This means that sophisticated and expensive techniques are required for the recovery of active protein products. These usually include the use of denaturation agents for solubilization of the protein followed by complicated controlled refolding (Marston, 1986). With fusion proteins, proteolytic cleavage and purification of active peptides must be achieved (e.g. Nagai *et al.*, 1985). Also, the mechanisms of transcription, translation and post-translational modification are different in *E. coli* from those in eukaryotes such that proteins produced in *E. coli* may not be identical to the authentic mammalian protein.

The production of many pharmaceutical proteins is achieved using mammalian cells in culture as a host system. Mammalian cells are capable of specific post-translational modifications (e.g. glycosylation) similar to those found on some natural proteins. Also, active proteins are secreted into the culture medium (Bebbington and Hentschel, 1985). As such mammalian cells are attractive hosts for protein production and are widely used on a commercial scale. However, mammalian cells are more difficult to manipulate genetically and are still relatively expensive to grow on a large scale. Thus, although this is the system of choice at present for many proteins, a high-value product is required to enable the use of mammalian cells on a commercial scale.

The yeast *Saccharomyces cerevisiae* is a unicellular eukaryote and has many features in common with mammalian cells. It offers the ease of genetic manipulation of a microorganism together with a well-defined secretory system to enable the location of products outside the cell in a soluble, active form. Intracellular expression can also be used to accumulate proteins to high levels, commonly about 5% of total cell protein and in exceptional cases much higher levels. For example, 40% of total cell protein in the case of hepatitis B core antigen (Kniskern *et al.*, 1986) and 30–70% of total cell protein for superoxide dismutase (Hallewell *et al.*, 1987). *S. cerevisiae* has a long history of use in the brewing and baking industries and is non-pathogenic. This may facilitate its use to produce proteins and peptides free from toxic contaminants for use in the pharmaceutical and food industries. Also, *S. cerevisiae* is relatively simple and inexpensive to grow on a large scale, and this, coupled with the use of a secretory system which can export proteins and peptides into a culture medium with few other proteins present, can lead to the development of economic production processes.

4.2 Expression vectors

Several types of DNA vector have been employed for the expression of heterologous genes. These include 2μm-based vectors, *ARS-* and *ARS CEN-*based vectors and integrating vectors. Most of these have been reviewed sufficiently in detail elsewhere (Broach *et al.*, 1983; Gunge, 1983; Parent *et al.*, 1985) and they are also referred to in the individual papers on heterologous expression, so we will only briefly describe their use in heterologous expression.

4.2.1 Common features of yeast vectors
The majority of yeast vectors require sequences which allow selection and replication in *S. cerevisiae*. In addition, they usually carry equivalent sequences for replication in *E. coli* since this organism is relatively easier to handle for genetic manipulations. The selective sequences for yeast are usually prototrophic markers, such as *LEU2, URA3, TRP1*, and *HIS3*, (see below) which can complement recessive auxotrophic mutations in the host strain. Dominant selectable markers can also be used especially for transforming industrial polyploid yeast, e.g. *CUP1* (Henderson *et al.*, 1985); the dihydrofolate reductase gene (dhfr) (Zhu *et al.*, 1985; Miyajima *et al.*, 1984); thymidine kinase gene (McNeil and Friesen, 1981); G418 resistance gene (Jimenez and Davies, 1980; Webster and Dickson, 1983); hygromycin B resistance gene

(Gritz and Davies, 1983) and chloroamphenicol resistance gene (Hadfield et al., 1986).

4.2.2 2-μm-based vectors

An important aspect of heterologous expression is the ability to stably maintain the heterologous gene within the host (both during storage as production cell banks and during production processes) and, in some but not all cases, at high copy number. There are several approaches to achieving both these aims. One is to place the heterologous expression unit on to a 2μm-based vector (see Chapter 11), which can achieve relatively high (20–200) copies in the host cell (Clarke-Walker and Miklos, 1984; Gerbaud and Guerineau, 1980). 2-μm plasmid is a naturally-occurring autonomously replicating circular 6.3-kb DNA molecule of S. cerevisiae (Hartley and Donelson, 1980). It was one of the first plasmids to be utilized for expression of both homologous and heterologous genes in yeast. Vectors based on 2-μm usually contain its origin of replication and the STB locus, in which the latter confers stability through the interaction of the REP1 and REP2 proteins encoded by endogenous 2-μm plasmids, e.g. pJDB207, (Beggs, 1981), YEp13, (Broach et al., 1979) and YEp24, (Botstein et al., 1979). Hence, these types of vectors require endogenous 2-μm plasmid unless they are modified. During heterologous expression it can be difficult in certain cases to maintain these vectors at high copy number and stability. We have found that expression of peptides by fusion to the prepro-α-factor sequence does not cause problems but the expression of larger proteins, especially those with toxic effects, can result in instability and low copy number (10–20 copies per cell). This can be largely circumvented by using tightly regulated expression (see Chapter 2). However, during the production phase when expression is switched on the problem can reappear especially when the vector uses URA3 as a selection marker (YEp24-based vectors).

Vectors based on or derived from pJDB219 (Beggs, 1978), contain the LEU2 gene inserted into the PstI site of 2-μm. This gene, designated leu2-d, contains a truncated promoter which appears to result in relatively higher copy numbers for these vectors than other LEU2 2-μm-based vectors (Erhart and Hollenberg, 1983).

Most readily available 2-μm-based vectors, such as pJDB207, pJDB219 (Beggs, 1978), and YEp24, require endogenous 2-μm for maintaining high copy number because they do not contain a functional FLP recombinase or two inverted repeats (with the exception of pJDB219). The FLP recombinase is responsible for 'flipping' normal 2-μm plasmid from an A form to a B form and vice versa, by promoting recombination between two 599-bp inverted repeats (Broach and Hicks, 1980; Broach et al., 1982; see Cashmore and Meacock, Chapter 11). The function of 'flipping' appears to be for the amplification of 2-μm plasmid during DNA synthesis when greater than the normal two copies can be produced (Futcher, 1986; Som et al., 1988). Hence, the absence of FLP recombinase leads to an inability to correct low copy number which may occur through disproportional segregation of the vector between mother and daughter cells. Thus, vectors which contain functional REP1 and REP2 (e.g. pJDB219) are still dependent on endogenous 2-μm for amplification. One possible way to ensure high copy number independent of endogenous 2-μm is to construct vectors which produce functional REP1, REP2 and FLP (either on the expression vector or independently) as well as containing the STB locus, origin of replication (ORI) and two inverted repeats. However, if one wants to amplify copy number in a controlled way then one could place FLP under the control of a regulatable promoter (Som et al., 1988).

4.2.3 ARS- and ARS CEN-based vectors

ARS-based vectors contain the Autonomously Replicating Sequences (ARS) of yeast chromosomal DNA. ARS sequences are thought to be the origins of replication of chromosomal DNA (Campbell, 1983) and are capable of replicating to high copy number. Early vectors used the TRP1 gene, for example YRp7 (Struhl et al., 1979), for selection in auxotrophic hosts. However, these vectors are extremely unstable and are lost at high frequency (Murray and Szostak, 1983). Although these vectors were used in early experiments for the expression of heterologous proteins, they are not suitable for production purposes, and to our knowledge they are not directly used for this purpose.

The introduction of chromosomal centromere (CEN) sequences into these vectors stabilizes them considerably, however; their copy number is reduced to about 1–2 copies (Clarke and Carbon, 1980). Selection is usually maintained by either the TRP1 gene, e.g. pYe (CEN3)11, (Clarke and Carbon, 1980) or the URA3 gene, e.g. YCp50 (Johnston and Davis, 1984). To our knowledge these vectors are rarely used for the direct expression of heterologous genes in production processes.

4.2.4 Integrating vectors

There are two types of integrating vector which can be used for the expression of heterologous proteins in yeast. Both are capable of inserting into a specific site within the chromosome. The first type contains both the expression unit and a selectable yeast gene, the mutant allele of which will be the site for integration into the chromosome. Integrating vectors can be selected by a number of genes such as HIS3, e.g. YIp1, or URA3, e.g. YIp5 (Struhl et al., 1979). The introduction of a single break in the duplex DNA of the selectable gene will stimulate recombination and hence integration at the target site. Upon integration a single copy of the expression unit will be flanked by a copy of the mutant target gene on one side and the wild-type gene on the other. Multiple copies of integrants at one site are possible; however, single copies occur at the highest frequency. Even for single integrations the tandem duplication of the target gene can cause instability and rearrangements, so this type of integration is not always suitable for production purposes.

The second type of site-specific integration involves a process known as gene replacement which does not involve tandem duplication of the target gene and as such is more stable (Rothstein, 1983; Struhl, 1983). The approach is to place on either side of the expression unit sequences of the target gene in the order and orientation of the native chromosomally located gene. This vector is then cut at distal ends of the two target sequences so that a fragment, containing the target gene with the expression unit inserted inside, is transformed into yeast. Integration at the target wild-type gene requires two recombination events to take place, each within the target sequences flanking either side of the expression unit. Integration can be monitored by the loss of the target gene phenotype. Although integration by gene replacement is extremely stable, the number of expression unit copies per site is limited.

4.2.5 Chromosomal amplification of genes

Chromosomal amplification of genes in *S. cerevisiae* has not been reported widely. Attempts to amplify genes in the chromosome have been made using the yeast transposable element Ty (Boeke *et al.*, 1988). Haploid yeast can contain 30–40 copies of Ty and each element contains two open reading frames flanked by two δ sequences. These elements can transpose when the element is transcriptionally active. When this occurs the Ty encoded reverse transcriptase can make complementary DNA which can transpose to other sites in the genome. Boeke *et al.* (1988) placed several marker genes into a Ty element under the transcriptional regulation of the *GAL1* promoter. Upon induction they found that they were able to get up to 10 copies of the hybrid Ty molecule into the yeast genome. They also found that the site at which the marker gene is place within Ty influences the ability to transpose. Moreover, they found that some integrated hybrid elements had lower expression levels for the marker gene than expected. Although this technique requires further development and optimization it will be interesting to see whether it will be applicable to the expression of heterologous proteins.

4.3 Intracellular expression

A large number of heterologous proteins have now been expressed in yeast. For example α-interferon (Hitzemann *et al.*, 1981; Tuite *et al.*, 1982), α 1-antitrypsin (Cabezon *et al.*, 1984), human serum albumin (Etcheverry *et al.*, 1986) and hepatitis B surface antigen (Valenzuela *et al.*, 1982). The strategy for expression utilizes a strong yeast promoter coupled to the gene of interest with a transcriptional terminator to ensure efficient termination of transcription. Promoters from genes encoding glycolytic enzymes have been widely used as these are thought to be particularly strong resulting in high-level production of the recombinant protein. For example, phosphoglycerate kinase (PGK) can be produced at 80% of total cell protein when expressed on a multicopy plasmid from its own promoter (Mellor et al., 1985). However, yields of heterologous proteins are not always as spectacular, the same PGK promoter resulting in the production of prochymosin at 5% of cell protein (Mellor et al., 1983). Promoters for high-level expression of heterologous genes can also be isolated from yeast by selection (Goodey et al., 1986). Yeast promoters are reviewed in detail elsewhere in this volume, and need not be discussed further here.

A large number of factors other than choice of promoter can affect expression levels of a heterologous gene. For example, plasmid copy number, plasmid stability, integration into the chromosome, presence of a secretion signal, yeast growth medium and conditions and toxicity of product to the cell. Regulated expression has now been widely used to increase expression levels by allowing cells to grow to high biomass before turning on transcription and production of the heterologous product. This may improve plasmid stability and avoids problems due to the production of proteins toxic to the cell. Several different systems for regulation have been used, as discussed in detail in Chapter 2.

Hitzemann et al. (1981) were the first to express a mammalian protein in yeast, human leukocyte interferon D. An estimation of the level of interferon produced was 0·4% of total cell protein, although there was a high degree of plasmid loss, and correcting for this, some cells could synthesize interferon at 1–2% of total cell protein. Since this early work, levels of expression routinely found are 1–5% of total cell protein, and in some cases much higher levels are found. Hepatitis B core antigen has been produced to 40% of total cell protein (Kniskern et al., 1986), and human superoxide dismutase has been reported to be expressed at levels between 30–70% of total cell protein (Hallewell et al., 1987).

In both these cases the very high level expression observed may be related to the unusually high stability of these proteins in the cell. Hepatitis B core antigen is made as particles which are resistant to proteolytic attack and superoxide dismutase is an unusually stable protein, more thermostable than any characterized globular protein including those from thermophilic bacteria, resistant to urea and resistant to a variety of proteases (Forman and Fridovich, 1973).

Several groups have used strains of yeast carrying the pep4-3 mutation to increase levels of production (Bitter et al., 1984; Cabezon et al., 1984; Rosenberg et al., 1984). This mutation results in a decrease in the level of activity of several vacuolar proteinases (Zubenko et al., 1982). The PEP4 gene encodes yeast proteinase A which is responsible for the activation of vacuolar proteinases from inactive zymogens to active forms (Ammerer et al., 1986; Woolford et al., 1986). Thus deficiency in this gene knocks out a number of different proteinase activities and can improve production levels of heterologous proteins. When considering protein production one must consider not only the biosynthetic system but also product stability.

With the discovery that heterologous proteins produced in *E. coli* are often produced as insoluble inactive protein much interest has focused on the solubility of proteins expressed in yeast. In many cases the expressed protein is found as insoluble material in yeast. Prochymosin was found as insoluble material when expressed in yeast and a proportion may be membrane- or cell-wall-bound. Unlike expression in *E. coli* a proportion of the prochymosin produced (about 30%) was readily isolated in a soluble form which could be activated to enzymically active chymosin (Mellor *et al.*, 1983).

Human gastric lipase has been expressed in yeast with over 90% of the protein produced found as insoluble inactive material (Bodmer *et al.*, 1987). The remainder of the lipase was produced as active soluble enzyme. Activity was retained although the enzyme was produced in yeast in an unglycosylated form. The nature of insoluble heterologous proteins produced in yeast may be analogous to the inclusion bodies seen in *E. coli*. Many amorphous inclusion bodies have been seen by electron microscopy in yeast cells producing heterologous proteins such as proinsulin-superoxide dismutase fusions (Cousens *et al.*, 1987) and tissue inhibitor of metalloproteinases (Kaczorek *et al.*, 1987). These inclusion bodies are not membrane-bound and appear to contain misfolded heterologous protein. In bacteria, inclusion bodies have been used as an aid to purification by collecting the inclusion bodies by centrifugation after cell lysis (Marston 1986). With yeast lysates however, the purification obtained in this way is not as good, due to the presence of cellular organelles and large cell wall debris which co-sediments with 'inclusion bodies'. Active protein can be recovered from yeast inclusion bodies, as demonstrated for tissue inhibitor of metalloproteinases (Kaczorek *et al.*, 1987) but this requires firstly the use of denaturants to solubilize the protein followed by a controlled refolding step. Not all heterologous proteins expressed intracellularly in yeast form insoluble material. Superoxide dismutase appears to be found as a completely soluble active protein after expression in yeast (Hallewell *et al.*, 1987).

One of the first recombinant DNA products to reach the market is a hepatitis B vaccine based on hepatitis B surface antigen expressed intracellularly in yeast. When expressed in yeast hepatitis B surface antigen forms 22-nm viral-like particles which are similar to those found in human serum. (Valenzuela *et al.*, 1982; Hitzemann *et al.*, 1983; Miyanohara *et al.*, 1983). When expressed the yeast-derived material is produced as a high-molecular-mass aggregate which is held together by non-covalent forces and can be dissociated by 2% sodium dodecyl sulphate without reducing agents (Wampler *et al.*, 1985). During *in vitro* purification of the antigen, disulphide bonds form spontaneously resulting in a particle of monomers and disulphide-bonded dimers. Treatment with thiocyanate further converts this material into a fully disulphide-bonded form which has been used to prepare a vaccine (Wampler *et al.*, 1985). Protection from infection with this vaccine has been seen (McAleer *et al.*, 1984). Work with hepatitis B surface antigen has led to the development of this system as a basis for the presentation of novel antigens.

The particles produced from hepatitis B surface antigen are highly immuno-genic and co-expression of another antigen may result in the production of hybrid antigen particles. This has been achieved with herpes simplex glycoprotein D (Valenzuela *et al.*, 1985). Recently yeast has been used to produce human immunodeficiency virus proteins which were antigenic *in vivo* (Barr *et al.*, 1987).

Expression in yeast can also result in post-translational modification. Glycosylation only takes place when proteins enter the secretory pathway and is well documented in yeast (see section 4.4). Other post-translational modifications can also take place. For example human superoxide dismutase produced in yeast is efficiently acetylated at the amino terminus to produce a protein identical to material derived from human tissue (Hallewell *et al.*, 1987). Natural yeast superoxide dismutase is not N-acetylated, yet the yeast cell appears able to recognize the signal for acetylation of the human enzyme. Phosphorylation can also take place. Miyamoto *et al.* (1985) demonstrated phosphorylation of the human c-myc protein produced in yeast which was localized to the nuclear fraction. Also attachment of lipid (palmitylation) may also take place for human Ha-ras DNA encoded p21 (Clark *et al.*, 1985). It is also interesting that heterologous proteins produced in yeast can take up cofactors to produce active molecules. Rat liver cytochrome P-450 has been expressed in yeast (Oeda *et al.*, 1985). This protein is normally an integral endoplasmic reticulum membrane protein and contains a haem group. Oeda *et al.* (1985) demonstrated that approximately 50% of the material produced in yeast took up haem and was assembled as a functional membrane-bound enzyme. The same group have also expressed rat liver NADPH: cytochrome P-450 reductase in yeast (Murakami *et al.*, 1986). This enzyme requires two flavin groups for activity and when produced in yeast alongside cytochrome P-450, the two enzymes formed a functional electron transport chain which demonstrated the same specificity towards drug monooxygenation as the natural rat enzyme. This work is being extended to produce chimaeric cytochrome P-450 enzymes from different cytochrome P-450 genes expressed in yeast (Sakaki *et al.*, 1987).

4.4 Secretion

The presence of a well-defined secretory system in yeast is one of the major reasons that yeast has found popularity as a host for recombinant protein production. Secretion into the growth medium allows the production of soluble active protein away from the majority of yeast proteins which are retained within the cell. This can mean that the recovery and purification of proteins produced in this manner is relatively straightforward and therefore inexpensive.

Another advantage is that it is possible to produce glycosylated proteins. Many mammalian proteins are glycosylated and may require glycosylation for activity or stability. Glycosylation occurs within the secretory pathway in the endoplasmic reticulum of the cell and in the Golgi apparatus. Inner-core N-linked glycosylation occurs in the endoplasmic reticulum where a dolichol-linked carbohydrate is attached to an asparagine residue at an appropriate glycosylation site. After transfer to the Golgi apparatus outer core glycosylation takes place. The nature of outer-core glycosylation is different in yeast and mammalian cells. In yeast the outer core contains mostly mannose residues added stepwise to the inner core (Schekman and Novick, 1982) whereas in mammalian cells outer glycosylation is often complex and branched, with several different sugars added to an inner-core which may be trimmed prior to their addition (Dumphy and Rothman, 1985). It is worth noting that the sites for N-linked glycosylation recognized in yeast and mammalian cells are the same (see below) such that sugar is attached to the same sites on the protein; however, the nature of the carbohydrate attached to the protein produced in yeast will be different from that of the natural mammalian protein or the protein produced in mammalian cell culture.

The secretory pathway in *S. cerevisiae* is very similar to that in higher eukaryotes and is shown diagrammatically in Figure 4.1. Proteins and peptides to be secreted usually have an N-terminal signal sequence which is responsible for the translocation of the protein or peptide into the lumen of the endoplasmic reticulum (Novick *et al.*, 1981). The signal sequence is usually composed of about twenty amino acid residues containing a large number of hydrophobic residues. The signal sequences of yeast and mammalian proteins are very similar and indeed many mammalian signal sequences function in

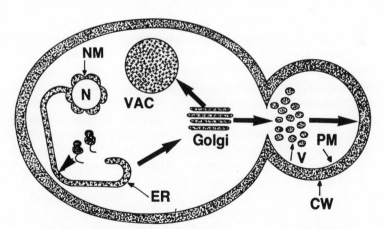

Figure 4.1 Cartoon of the secretory pathway in *S. cerevisiae*. Proteins are directed into the endoplasmic reticulum, transported from the endoplasmic reticulum to the Golgi apparatus, and from the Golgi to secretory vesicles and thus to the cell surface. An alternate pathway from the Golgi to the vacuole also exists. Abbreviations: *N*, nucleus; *NM*, nuclear membrane; *ER*, endoplasmic reticulum; *VAC*, vacuole; *V*, secretory vesicles; *PM*, plasma membrane; *CW*, cell wall.

Table 4.1 Signal sequences used to secrete heterologous proteins from yeast.

Derivation of signal sequence	Examples of recombinant proteins secreted from yeast	Reference
α-Factor	Epidermal growth factor	Brake *et al.* (1984)
	Interleukin-2, GM-CSF	Price *et al.* (1987)
Killer toxin	Cellulose	Skipper *et al.* (1985)
Invertase	α-1-Antitrypsin	Moir-Dumais (1987)
Acid phosphatase	α-Interferon	Hinnen *et al.* (1983)
Amyloglucosidase	Gastric lipase	Yarranton *et al.* (unpublished)
Natural (heterologous to yeast)	α-Interferon	Hitzemann *et al.* (1983)
	Wheat-α-amylase	Rothstein *et al.* (1984)

yeast (see below). Translocation of proteins into the endoplasmic reticulum is usually, but not always, co-translational (Rothblatt *et al.*, 1987), and the signal sequence is removed by a luminal signal peptidase (Waters *et al.*, 1988). The order of events in the yeast secretory pathway has been worked out largely by the use of mutants by Schekman and colleagues (Novick *et al.*, 1981; Schekman and Novick, 1982). After inner-core glycosylation in the endoplasmic reticulum the protein is transferred to the Golgi apparatus. Outer-core glycosylation takes place in the Golgi apparatus where proteins are also sorted into secretory proteins and those destined for the yeast vacuole. The mechanism behind this process is not understood, but for carboxypeptidase Y the information may reside in the N-terminal propeptide (Valls *et al.*, 1987; Johnson *et al*; 1987). Proteins and peptides to be secreted are then packaged into secretory vesicles which are then transported to the cell surface where fusion with the cell membrane takes place releasing the contents from the cell. This process requires energy and many different gene products which have been defined by mutations (Novick *et al.*, 1981).

    S. cerevisiae naturally secretes very few proteins. Enzymes such as invertase and acid phosphatase are secreted into the periplasmic space and do not pass into the culture medium, whereas the sex pheromone α-factor and yeast killer toxin are secreted into the culture medium. The signal sequences from all four of these secreted proteins have been used to direct secretion of heterologous proteins with varying degrees of success. Table 4.1 shows some of the systems used to achieve secretion from yeast.

4.4.1 *Prepro-α-factor directed secretion*
By far the most extensively used system for the secretion of foreign proteins from yeast is that using the prepro region of the *MFα1* gene (Kurjan and Herskowitz, 1982). Alpha factor is a 13-amino-acid sex pheromone which is secreted into the growth medium by haploid mating type alpha cells of *S. cerevisiae*. It is produced as a 165-amino-acid precursor containing four copies

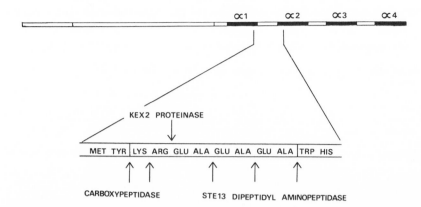

Figure 4.2 Processing of prepro-α-factor in *S. cerevisiae*. The prepro-α-factor precursor containing four copies of the alpha-factor peptide is cleaved by the *KEX2* encoded endoproteinase after lys-arg sequence. The C-terminus of α-factor is generated by the action of a carboxypeptidase encoded by the *KEX1* gene and the N-terminus by the action of the dipeptidyl aminopeptidase encoded by the *STE13* gene.

of the alpha factor peptide separated by spacer sequences (Julius *et al.*, 1983). The pre region is a signal sequence which is sufficient to allow translocation into the endoplasmic reticulum but is not sufficient for secretion of α-factor peptides. Proteolytic processing takes place in the Golgi apparatus to release the 13-amino-acid peptides from the precursor (Julius *et al.*, 1983; 1984). This process has been described in detail elsewhere in this volume (Wolf, 1988), and is summarized in Figure 4.2. Three yeast proteinases are required for the release of mature α-factor. The initial cleavage of the pro-α-factor precursor is achieved through an endoproteinase encoded by the *KEX2* gene (proteinase yscF) with specificity after a pair of basic residues, lys-arg. The amino terminus of mature α-factor is then generated by sequential removal of pairs of residues by a dipeptidyl aminopeptidase encoded by the *STE13* gene (dipeptidyl aminopeptidase ysc IV). The carboxyl terminus of mature α-factor is generated through the action of a carboxypeptidase (carboxypeptidase ysc α) which removes the lys and arg residues left by the action of the *KEX2* proteinase. This carboxypeptidase is encoded by the *KEX1* gene (Dmochowska *et al.*, 1987).

Correct processing of a heterologous polypeptide is obviously essential to obtain an authentic product. During expression at high levels processing of the prepro-α-factor-polypeptide precursor may become limiting to efficient secretion. To investigate processing at high expression levels the entire prepro-α-factor gene has been overexpressed using a multicopy plasmid (Smith *et al.*, 1986). The resulting α-factor peptides in the medium were then subjected to analysis by FAB-mass spectrometry and N-terminal sequence analysis. Alpha factor peptides accumulated to very high levels in the medium during a fed-

Table 4.2 Polypeptides secreted from yeast using the MFα prepro sequence to direct secretion.

Protein	Reference
Epidermal growth factor	Brake et al. (1984)
β-Endorphin	Bitter et al. (1984)
Interferon	Singh et al. (1984)
	Zsebo et al. (1986)
Interleukin-2	Miyajima et al. (1985)
	Price et al. (1987)
GM-CSF	Miyajima et al. (1986)
	Shaw et al. (1988)
Calcitonin	King, Walton and Yarranton, unpublished
	Bitter et al. (1984)
Insulin	Thin et al. (1986)
Phospholipase A$_2$	Van den Bergh et al. (1987)
Hirudin	Loison et al. (1988)
Somatomedin C	Ernst et al. (1986)
Somatostatin	Green et al. (1986)
Prochymosin	Moir et al. (1984)
Atrial natriuretic factor	Vlasuk et al. (1986)
α-Amylase	Astolfi-Filho et al. (1986)
β-Lactamase	Roggenkamp et al. (1986)
Invertase	Emr et al. (1983)

batch fermentation yet only approximately 10% of the peptides secreted were authentic α-factor. The majority of the material seen contained extra glu-ala or asp-ala dipeptides at the amino terminus indicating incomplete processing by the dipeptidyl aminopeptidase encoded by *STE13*. Thus during overexpression of α-factor, processing by this enzyme was found to be limiting to the secretion of authentic α-factor peptide.

A large number of heterologous proteins and peptides have now been secreted from yeast using the alpha-factor system. Some of these are listed in Table 4.2. One of the first heterologous polypeptides to be studied was epidermal growth factor (Brake *et al.*, 1984). Epidermal growth factor (EGF) was expressed and secreted to high levels, with over 95% of the material produced, 5–10-μg/ml or 7% of total cell protein, being secreted into the growth medium. The sequence preceding the amino terminus of EGF was varied in attempts to achieve correct processing. When a sequence similar to the α-factor spacer peptide was used, lys-arg-glu-ala-glu-ala-glu-ala-EGF, cleavage after the lys-arg was complete. However, the glu-ala dipeptides were not efficiently removed. A second lys-arg dipeptide was inserted immediately N-terminal to EGF. Again cleavage at the first lys-arg was complete, and 80% of the material was cleaved at the second lys-arg dipeptide, resulting in only 20% of incorrectly processed material. When most of the spacer peptide was deleted leaving only one lys-arg immediately before the EGF sequence the material secreted was 100% mature EGF indicating good activity by the *KEX2* encoded proteinase even in the absence of the remainder of the spacer sequence. The levels of EGF secreted in this case were about half of those found when the complete spacer sequence was used. With all of these different plasmids, the EGF secreted contained a proportion of dimer and trimer, probably formed by incorrect disulphide pairing.

In some cases use of prepro-α-factor gene fusions has resulted in not only cleavage at the N-terminus but also internal cleavages in the protein of interest. When β-endorphin, a 31-amino-acid peptide, was secreted using an α-factor gene fusion two cleavages inside the β-endorphin peptide took place at the carboxyl side of two lysines (Bitter et al., 1984). The 20-kDa protein α-interferon was also efficiently secreted by the α-factor system. In this case some internal cleavage was also observed though to a much lower extent (Bitter et al., 1984). Low internal cleavage of α-interferon might be due to protection by the folding of the molecule. In this work also it was noted that the STE13 encoded dipeptidyl aminopeptidase was rate-limiting to efficient processing.

We have used the prepro leader sequence of α-factor to direct the secretion of several heterologous polypeptides including calcitonin-glycine. Calcitonin is a 32-amino-acid peptide hormone containing a 1, 7-disulphide bridge and a C-terminal proline amide residue. Calcitonin lowers blood calcium levels and is used in the treatment of Paget's Disease and osteoporosis. Because of the natural C-terminal amide group we have made a calcitonin-glycine precursor

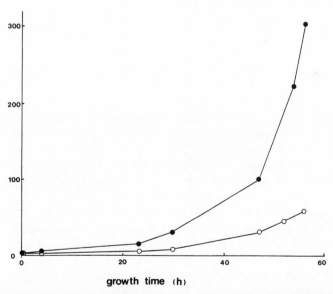

growth time (h)

Figure 4.3 The secretion of human calcitonin from S. cerevisiae directed by the prepro-α-factor leader sequence. A fed-batch fermentation of yeast secreting human calcitonin to high levels. ○, calcitonin secreted (mg per litre culture); ●, growth of yeast (g dry weight per litre culture).

which can be converted to calcitonin *in vitro* using a pituitary amidating enzyme.

Calcitonin-glycine was expressed and secreted by fusing the peptide coding region to the prepro-α-factor leader and expressing using either the α-factor promoter or the PGK promoter. In both cases high levels of calcitonin were found in the growth medium with little or no difference in expression or secretion levels. Greater than 97% of the total calcitonin-glycine produced was recovered in the growth medium. The production of calcitonin-glycine increased in proportion to yeast biomass (Figure 4.3).

Two expression plasmids were made to direct secretion of calcitonin-glycine. The plasmid pYC 207 contained two lys-arg dipeptides for cleavage in the sequence:

lys-arg-glu-ala-glu-ala-glu-ala-
ser-lys-arg-calcitonin-glycine.

Material from yeast containing this plasmid was subjected to purification by reverse-phase HPLC and the purified peptides were analysed by sequence analysis and FAB-mass spectrometry. All of the material recovered was cleaved at the first *KEX2* cleavage site; however, we found no cleavage at all at the second lys-arg immediately preceding the calcitonin sequence. In a proportion of the material processing by the *STE13* dipeptidyl aminopeptidase had taken place, again suggesting that the action of this enzyme might be limiting during overexpression. This result was surprising given the results of Brake *et al.* (1984) described above. With calcitonin-glycine several effects might prevent cleavage of the second *KEX2* site, including (i) the conformation of the protein precursor might be such as to prevent access of the *KEX2* proteinase to the second site, and (ii) the amino acid sequence surrounding the second *KEX2* site might result in a poor substrate for the proteinase.

A second expression plasmid, pYC51 was made in which most of the spacer sequence above was deleted. This now contained one lys-arg dipeptide immediately before the calcitonin-glycine sequence. This material was again purified and analyzed. In this case, correctly processed calcitonin-glycine was recovered.

The generation of an authentic C-terminus is also important for the production of heterologous products. In some cases heterologous C-terminii have been reported with peptides secreted by the α-factor system. Vlasuk *et al.* (1986) found that human atrial natriureic peptide (ANP) was efficiently secreted but only a small proportion of the peptide recovered had an intact C-terminus. The majority of the ANP lacked two residues from the C-terminus. This may be a result of the action of the carboxypeptidase normally involved in α-factor processing, particularly as the carboxypeptidase cleavage site of the natural α-factor peptide (tyr-lys) is similar to the bond cleaved to shorten ANP (aromatic-basic, phe-arg). In this work the disulphide bond of ANP was formed correctly, with no evidence of mispairing. Correct disulphide bond formation has also been found in many other cases, for example human interferon (Zsebo et al. 1986) and porcine phospholipase A_2 where all seven disulphide bridges were correctly formed (Van den Bergh et al., 1987).

Both N-linked and O-linked glycosylation can take place during secretion and glycosylation sites used are often identical to those used on the natural protein (Miyajima et al., 1986; Ernst et al., 1987).

The secretion of larger proteins using the α-factor leader sequence has proved difficult. A consensus interferon was secreted but remained trapped in the periplasmic space/cell wall (Zsebo et al., 1986). Also the prepro-α-factor sequence was inefficiently removed. With human gastric lipase, a 50-kDa glycoprotein, we have also found poor secretion and inefficient removal of the leader sequence (Yarranton, King and Walton, unpublished). However, a very large Epstein-Barr virus envelope glycoprotein (gp 350) has been secreted as a 400-kDa glycoprotein (Schultz et al., 1987). The expression of this protein was toxic to the cell, so that regulated expression was required to achieve good levels of the protein. About 50% of the material produced was secreted into the culture medium. On analysis only the most highly glycosylated form was secreted, and these authors suggested a role for N-linked carbohydrate in directing secretion. Tunicamycin, an inhibitor of N-linked glycosylation, prevented section of interleukin-2 (Miyajima et al., 1985). A role for carbohydrate in secretion of proteins from animal cells has also been proposed (Guan et al., 1985; Kelly, 1985). However, carbohydrate is not involved in the direction of yeast proteins to the vacuole (Valls et al., 1987; Johnson et al., 1987) and many non-glycosylated polypeptides have now been secreted from yeast as described in this chapter.

4.4.2 Other yeast signal sequences used for secretion

The leader sequence from yeast killer toxin has also been used for heterologous protein secretion. A bacterial cellulose (endo-1,4-β glucanase) has been secreted using a gene fusion of the S. cerevisiae killer toxin leader sequence to the cellulose gene (Skipper et al., 1985). Also the leader sequence of Kluy-veromyces lactis killer toxin, which is structurally and functionally distinct from that of S. cerevisiae (Sugisaki et al., 1983) has been used to direct secretion of an active fragment of interleukin 1β (Baldari et al., 1987). This 17-kDa protein was secreted into the growth medium as a 22-kDa glycosylated form which was correctly processed at the N-terminus and showed biological activity.

The signal sequence of yeast acid phosphatase has also been used to secrete proteins from yeast, for example, α-interferon (Hinnen et al., 1983). However, it was not able to direct the secretion of tissue plasminogen activator (Lemontt et al., 1985).

The signal sequence of invertase has been used to secrete human interferon α-2 with efficient removal of the leader sequence during secretion to gene-rate a mature N-terminus (Chang et al., 1986). In this case production of interferon α-2 was increased when the invertase signal sequence was used compared to the natural human signal sequence. Approximately 50% of the material produced was secreted into the growth medium. Human α1-antitrypsin has also been secreted using the invertase leader sequence (Moir and Dumais, 1987). Yeast glycosylated the three asparagines which are glycosylated in the natural protein. Twenty per cent of the α1-antitrypsin produced was secreted and this was all fully glycosylated. The material within the cell was partially glycosylated and the rate-limiting step to secretion appeared to be transport from the endoplasmic reticulum to the Golgi apparatus as defined by the yeast SEC 18 mutation.

Smith *et al.* (1985) reported the secretion of calf prochymosin using the invertase leader sequence; however, less than 10% of the material produced was secreted. Most of the intracellular invertase-prochymosin was located in the vacuole of the cell. A rate-limiting step in secretion might direct protein to the vacuole or the heterologous protein might be misdirected. Immunoglobulin chains expressed in yeast were also directed to the vacuole with only low-level secretion (Wood, 1985). The expression level of prochymosin could be increased by substituting the *SUC2* promoter with the stronger *TPI* (triose phosphate isomerase) promoter but no increase in secretion efficiency was seen (Smith *et al.*, 1985). Integration of the gene fusion into the chromosome surprisingly increased the efficiency of secretion from 1–2% to 8–10%. Integration may result in a more stable expression unit and abolish the need to grow cells under selection to maintain the plasmid, but the number of gene copies will be much lower than those achieved with a multicopy plasmid (see section 4.2.4). Smith *et al.* found that strains containing several integrated copies produced about the same amount of prochymosin as strains bearing a multicopy plasmid, but secreted about four times as much.

In attempts to increase secretion efficiency further Smith *et al.* (1985) screened for and isolated yeast mutants with improved secretion efficiency. These supersecreting mutants which arose at very low frequency produced about the same amount of prochymosin as the wild-type strain, but secreted up to ten times the amount. Mutations in at least four genes were suggested to lead to supersecretion, and two genes, designated SSC1 and SSC2, were particularly strong. It was hoped that these mutants would be of general utility for increasing the secretion of a number of heterologous proteins, and improved secretion was noted for bovine growth hormone (Smith *et al.*, 1985). However, the general utility of these mutants remains to be proved, as no further reports have appeared in the literature.

Human gastric lipase has been used to assess secretion using the yeast DEX leader sequence. The *DEX1* gene encodes a dextrinase, which is a large protein of approximately 150 000 Mr and is heavily glycosylated (Meaden *et al.*, 1985). This enzyme is secreted into the growth medium and allows the use of starch as sole carbon source by *S. diastaticus*. Using the DEX leader sequence we have obtained the production of active lipase to 0.5–1% of total cell protein with no

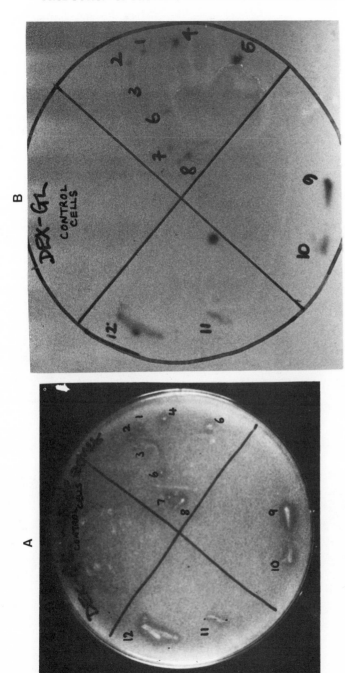

Figure 4.4 The secretion of human gastric lipase from *S. cerevisiae* directed by the DEX signal sequence. *A*: An agar plate assay for detection of yeast colonies secreting hGL. An agar plate containing 0.75% tributyrin shows zones of clearing around colonies of yeast secreting lipase. Tributyrin is cleaved by human gastric lipase. *B*: An autoradiogram of a nitrocellulose filter overlaid over the same plate, probed with anti-human gastric lipase antiserum and iodinated protein A.

evidence for any inactive enzyme being produced. Of this 30–40% was secreted, most of which appeared to be present in the periplasm/cell wall fraction of the cell released by zymolyase treatment. Some lipase was secreted into the growth medium, as measured by a plate assay and by assaying concentrated growth medium for enzymic activity (Figure 4.4). The lipase produced was glycosylated, as shown by a molecular mass of approx. 50 000 (three major glycosylation variants were found in cell lysates), and by tunicamycin experiments in which formation of the high-molecular mass bands was inhibited and accumulation of a 40 000 band was observed. The secreted material corresponded to the most highly glycosylated variant.

4.4.3 *Secretion of heterologous proteins directed by their own signal sequence*
In many cases the natural signal sequence of the heterologous protein has been used in attempts to achieve secretion. The results of this approach have been variable with some proteins secreted at high levels and others not secreted at all. Some examples of this approach are listed in Table 4.3.

Calf prochymosin has been studied by two groups and in neither case has secretion from a preprochymosin gene been observed (Mellor *et al.*, 1983; Goff *et al.*, 1984). Attempts to secrete human tissue plasminogen activator

Table 4.3 Polypeptides expressed in yeast with their natural signal sequence.

Protein	Source	Secretion	Reference
α-Interferon	Human	+	Hitzemann *et al.* (1983)
Prochymosin	Bovine	−	Mellor *et al.* (1983)
			Goff *et al.* (1984)
α1-Antitrypsin	Human	−	Cabezon *et al.* (1984)
α-Amylase	Wheat	+	Rothstein *et al.* (1984)
α-Amylase	Murine	+	Thomsen (1983)
α-Amylase	*Bacillus*	+	Ruohonen *et al.* (1987)
β-Lactamase	*E. coli*	−	Roggenkamp *et al.* (1985)
Phospholipase A₂	Bovine	+	Tanaka *et al.* (1988)
Pancreatic Secretory Trypsin inhibitor	Human	+	Izumoto *et al.* (1987)
Lysozyme	Chicken	+	Oberto and Davison (1985)
Lysozyme	Human	+	Jigami *et al.* (1986)
Tissue inhibitor of metalloproteases	Human	−	Kaczorek *et al.* (1987)
Immunoglobulin chains λ and μ	Murine	+	Wood *et al.* (1985)
Tissue plasminogen activator	Human	−	Lemontt *et al.* (1985)
Haptoglobin	Human	−	Van der Straken *et al.* (1986)
Gastric lipase	Human	+	Yarranton *et al.*, unpublished
Thaumatin	Plant	+	Edens *et al.* (1984)
Glucoamylase	*Aspergillus*	+	Innis *et al.* (1985)
Cellobiohydrolase	*Trichoderma*	+	Penttila *et al.* (1988)

(Lemontt et al., 1985), human tissue inhibitor of metalloproteinases (Kaczorek et al., 1987), human α1-antitrypsin (Cabezon et al., 1984) and bacterial β-lactamase (Roggenkamp et al., 1985) have also resulted in no secretion into the medium being observed. In the case of tPA, the acid phosphatase signal sequence was also unable to direct secretion (Lemontt et al., 1985). With both signal sequences, glycosylated tPA was produced with some fibrinolytic activity and this appeared to be associated with large cellular debris. These results indicate entry into the secretory pathway, but further analysis was not carried out to determine whether localization to the cell wall was taking place.

With human α1-antitrypsin expression levels were higher when no secretion signal was present, 1% versus 0.3% of total cell protein (Cabezon et al., 1984), as has also been observed for prochymosin (Mellor et al., 1983). No α1-antitrypsin activity could be detected in cells expressing the preprotein although high activity was seen when α1-antitrypsin was expressed intra-cellularly without a secretory signal, particularly in strains carrying the pep 4-3 mutation (Cabezon et al., 1984). The α1-antitrypsin pre-sequence was not removed although some glycosylation appeared to take place.

Hitzemann et al. (1983) were the first to report that a heterologous signal sequence could be functional for heterologous protein secretion in yeast. Interferons α1 and α2 were expressed and secreted using their natural human leader sequences. A relatively small proportion (10–20%) of the material synthesized was secreted, and the amino terminus of the material produced was heterogeneous with some of the pre-sequence still remaining on a proportion of the secreted protein. Human salivary α-amylase has been secreted to higher levels, with about half of the enzyme produced being secreted into the culture medium (Nakamura et al., 1986; Sato et al., 1986). Three different glycosylated forms of α-amylase were produced and the amino-terminal sequence appeared to be non-heterogeneous and authentic. The N-terminal residue is converted to pyroglutamate as is the case with the natural human material. The C-terminal residue was also as expected from the DNA sequence, with no apparent C-terminal degradation taking place. Experiments with tunicamycin suggested that glycosylation was not necessary for efficient secretion. Also, the α-amylase gene expressed without the signal sequence was neither secreted nor glycosylated, again suggesting glycosylation is taking place during the secretory process (Sato et al., 1986).

The human signal sequence of pancreatic secretory trypsin inhibitor has been used to secrete the 56-amino-acid trypsin inhibitor (Izumoto et al., 1987). For this polypeptide the efficiency of secretion was very high (95%) with 1.6 $\mu g\,mL^{-1}$ recovered in the culture medium. The signal sequence was efficiently cleaved, the recovered material having an identical specific activity to natural material. Human haptoglobulin, however, is expressed very poorly, and when made as the prepro form the haptoglobin produced is partially glycosylated, yet none is secreted into the growth medium (Van der Straten et al., 1986). Other mammalian signal sequences have also been shown to function in yeast. The mouse immunoglobulin pre-sequence for both λ and μ chains directs the secretion of antibody chains, although at low levels (Wood et al., 1985). In this work the assembly of a small amount of functional antibody was seen. The chicken lysozyme signal sequence has been used to secrete both chicken lysozyme (Oberto and Davison, 1985) and human lysozyme (Jigami et al., 1986). Chicken lysozyme was expressed at 1.5% of yeast protein, and two-thirds of this was secreted into the growth medium as correctly processed material. A similar amount of human lysozyme was also secreted, 55–65%, when the cells were allowed to grow to stationary phase. However, during exponential growth the activity was higher in the periplasmic space (Jigami et al., 1986). This may suggest that lysozyme accumulates in the periplasm during rapid synthesis in exponential growth and is then slowly released into the medium. Mouse α-amylase is also readily secreted from yeast when expressed as the preprotein (Thomsen, 1983). In this case α-amylase was expressed at approximately 0.1% of total yeast protein, with 90% of the material found in the culture medium.

Bovine pancreatic phospholipase A_2 has been expressed in yeast as a proenzyme, with the dog signal sequence from the same enzymes being used for secretion (Tanaka et al., 1988). The dog pre-sequence resulted in the secretion of almost all of the material produced and was efficiently removed. In about 80% of the phospholipase A_2 produced the pro-sequence was not removed such that about 20% of the material was secreted as active enzyme. The proenzyme could be activated to phospholipase A_2 by trypsin treatment.

Secretory signals for plant proteins can also work in yeast. Wheat α-amylase has been secreted using its wheat signal peptide (Rothstein et al., 1984; Rothstein et al., 1986). The plant protein thaumatin has also been secreted when expressed as preprothaumatin (Edens et al., 1984). Fungal sequences have also been used, for example, in the secretion of an Aspergillus glucoamylase (Innis et al., 1985) and for Trichoderma reesei cellobiohydrolases (Penttila et al., 1988). In both cases, high levels of secretion were observed, with the active enzymes being recovered from the growth medium.

A prokaryotic signal sequence has also been shown to be functional in S. cerevisiae. The signal peptide of Bacillus amyloliquefaciens α-amylase led to efficient secretion of this enzyme (75% of total) and the complete 31-amino-acid signal sequence was required to achieve this (Ruohonen et al., 1987). When a partial signal sequence of 16 amino acids was used, which was predicted to resemble a yeast signal sequence, no secretion was observed, suggesting a stringent signal sequence requirement.

The function of the signal sequence may be only to insert the protein into the endoplasmic reticulum of the cell and thus into the secretory pathway. After this event, the nature of the protein itself may be more important for secretion. Widely differing results have been obtained for protein secretion from S. cerevisiae, and this may depend on the nature of the signal sequence, the nature of the protein, the type and level of expression or a complex interplay of factors. The requirements for signal peptide function in yeast may not be as stringent as has previously been thought. Kaiser *et al.* (1987) have shown that many random sequences can replace the signal peptide of yeast invertase. Results using different signal peptides in attempts to achieve secretion of prochymosin have shown that the natural calf sequence is non-functional, whereas secretion can be achieved using either the invertase or α-factor signal sequence (Mellor *et al.*, 1983; Goff *et al.*, 1984; Smith *et al.*, 1985). Work on human gastric lipase has revealed that the natural human pre-sequence can direct secretion as well as the yeast DEX signal sequence. The use of the α-factor signal sequence can also lead to glycosylated lipase, indicating entry into the secretory pathway, but in this case the α-factor pro-sequence is inefficiently removed. With the current state of knowledge, an empirical approach needs to be taken with each protein to determine the best system to use for efficient secretion. In general, the α-factor system is very efficient for peptides and small proteins, but for larger proteins the situation is more complex.

4.5 Conclusions

S. cerevisiae is an attractive host for the production of recombinant polypeptides. It can be easily manipulated genetically, and as a primitive eukaryote it contains a secretory system with many features in common with that in higher eukaryotes. This system can be used to secrete heterologous polypeptides to high levels and thus produce active soluble products away from contaminating cellular proteins, greatly facilitating their subsequent recovery and purification. The factors affecting secretion efficiency are poorly understood; the nature of the signal sequence, the protein itself, and many other factors may all be important. Also, the design of recovery processes to take advantage of secretion requires accumulation of the product to a high level in the culture broth. Secretion also introduces the need for a substantial concentration step to reduce processing volumes, usually by ultrafiltration or a specific adsorption technique. Also, the protein of interest must be stable to the harsh extracellular environment of the fermenter, and cell lysis should be kept to a minimum to reduce contamination with other proteins and to prevent proteinase release which may degrade the secreted polypeptide.

Intracellular expression is also useful, allowing high levels of product to be produced. It is worth noting that one of the first recombinant DNA products on the market is a hepatitis B vaccine produced intracellularly in yeast.

References

Ammerer, G., Hunter, C.P., Rothman, J.H., Saari, G.C., Valls, L.A. and Stevens, T.H. (1986) *PEP4* gene of *Saccharomyces cerevisiae* encodes proteinase A, a vacuolar enzyme required for processing of vacuolar precursors. *Mol. Cell Biol.* **6**: 2490.

Astolfi-Filho, S., Galembeck, E.V., Faria, J.B. and Scheberg Franscino, A.C. (1986) Stable yeast transformants that secrete functional α-amylase encoded by cloned mouse pancreatic cDNA. Bio/technology 4: 311.

Baldari, C., Murray, J.A.H., Ghiara, P., Cesarani, G. and Galeotti, C.L. (1987) A novel leader peptide which allows efficient secretion of a fragment of human interleukin 1β in Saccharomyces cerevisiae. EMBO J. 6: 229.

Barr, P.J., Steimer, K.S., Sabin, E.A., Parkes, D., George-Nascimento, C., Stephens, J.C., Powers, M.A., Gyenes, A., Van Nest, G.A., Miller, E.T., Higgins, K.W. and Luciw, P.A. (1987) Antigenicity and immunogenicity of domains of the human immunodeficiency virus (HIV) envelope polypeptide expressed in the yeast Saccharomyces cerevisiae. Vaccine 5: 90.

Bebbington, C. and Hentschel, C. (1985) The expression of recombinant DNA products in mammalian cells. Trends in Biotechnol. 3: 314.

Beggs. J.D. (1978) Transformation of yeast by a replicating hybrid plasmid. Nature 275: 104.

Beggs, J.D. (1981) Multiple-copy yeast plasmid vectors. In Molecular Genetics in Yeast, eds. Von Wettstein, D., Friis, J., Kielland-Bradt, M. and Stenderup, A., Alfred Benzon Symp. 16: 383.

Bitter, G.A., Chen, K.A., Banks, A.R. and Lai, P.H. (1984) Secretion of foreign proteins from Saccharomyces cerevisiae directed by α-factor gene fusions. Proc. Natl. Acad. Sci. USA 81: 5330.

Bodmer, M., Angal, S., Yarranton, G.T., Harris, T.J.R., Lyons, A., King, D.J., Pieroni, G., Riviere, C., Verger, R. and Lowe, P.A. (1987) Molecular cloning of a human gastric lipase and expression of the enzyme in yeast. Biochim. Biophys. Acta 909: 237.

Boeke, J.D., Xu, H. and Fink, G.R. (1988) A general method for the chromosomal amplification of genes in yeast. Science 239: 280.

Botstein, D., Falco, S.C., Stewart, S.E., Brennan, M., Scherer, S., Stinchcomb, D.T., Struhl, K. and Davis, R.W. (1979) Sterile host yeasts (SHY) a eukaryotic system of biological containment for recombinant DNA experiments. Gene 8: 17.

Brake, A.J., Merryweather, J.P., Coit, D.G., Heberlein, V.A., Masiarz, F.R., Mullenbach, G.T., Urdea, P., Valenzuela, P. and Barr, P.J. (1984) α-factor directed synthesis and secretion of mature foreign proteins in Saccharomyces cerevisiae. Proc. Natl. Acad. Sci. USA 81: 4642.

Broach, J.R. and Hicks, J.B. (1980) Replication and recombination functions associated with the yeast plasmid, 2μ circle. Cell 21: 501.

Broach, J.R., Strathern, J.N. and Hicks, J.B. (1979) Transformation in yeast: development of a hybrid cloning vector and the isolation of the CAN1 gene. Gene 8: 121.

Broach. J.R., Guarascio, V.R. and Jayaram, M. (1982) Recombination within the 2μ circle is site-specific. Cell 29: 227.

Broach, J.R., Li, Y.-Y, Wu, L.C.-C. and Jayaram, M. (1983) Vectors for high-level inducible expression of cloned genes in yeast. In Experimental Manipulation of Gene expression, ed. Inouye, M., Academic Press, New York.

Cabezon, T., de Wilde, M., Herion, P., Loriau, R. and Bollen, A. (1984) Expression of human α-1 antitrypsin cDNA in the yeast Saccharomyces cerevisiae. Proc. Natl. Acad. Sci. USA 81: 6594.

Campbell, J.L. (1983) Yeast DNA replication. In Genetic Engineering Principles and Methods, eds. Setlow, J.K. and Hollaender, A., vol. 5, Plenum, New York, 109.

Chang. C.N., Matteucci, M., Perry, L.J., Wulf, J.J., Chen, C.Y. and Hitzemann, R.A. (1986) Saccharomyces cerevisiae secretes and correctly processes human interferon hybrid proteins containing yeast invertase signal peptides. Mol. Cell. Biol. 6: 1812.

Clark, S.G., McGrath, J.P. and Levinson, A.D. (1985) Expression of normal and activated human Ha-ras cDNAs in Saccharomyces cerevisiae. Mol. Cell. Biol. 5: 2746.

Clarke, L. and Carbon, J. (1980) Isolation of yeast centromere and construction of functional small chromosomes. Nature 257: 504.

Clarke-Walker, G.D. and Miklos, G.L.D. (1984) Localisation and quantification of circular DNA in yeast. Eur. J. Biochem. 41: 359.

Cousens, L.S., Shuster, J.R., Gallegos, C., Ku, L., Stempien, M.M., Urdea, M.S., Sanchez-Pescador, R., Taylor, A. and Tekamp-Olson, P. (1987) High level expression proinsulin in the yeast Saccharomyces cerevisiae. Gene 61: 265.

Dmochowska, A., Dignard, D., Henning, D., Thomas, D.Y. and Bussey, H. (1987) Yeast KEX1 gene encodes a putative protease with a carboxypeptidase B-like function involved in killer toxin and α-factor precursor processing. Cell 50: 573.

Dunphy, W.G. and Rothman, J.E. (1985) Compartmental organization of the Golgi stack, Cell 42: 13.

Edens, L., Bom, I., Ledeboer, A.M., Maat, J., Toonan, M.Y., Visser, C. and Verrips, C.T. (1984) Synthesis and processing of the plant protein thaumatin in yeast. *Cell* **37**: 629.

Emr, S.D., Schekman, R., Flessel, M.C. and Thorner, J. (1983) An MF alpha 1-SUC2 (alpha factor-invertase) gene fusion for study of protein localisation and gene expression in yeast. *Proc. Natl. Acad. Sci. USA* **80**: 7080.

Ernst, J.F. (1986) Improved secretion of heterologous proteins by *Saccharomyces cerevisiae*: effects of promoter substitution in alpha-factor fusions. *DNA* **5**: 483.

Etcheverry, T., Forrester, W. and Hitzemann, R. (1986) Regulation of the chelatin promoter during the expression of human serum albumin or yeast phosphoglycerate kinase in yeast. *Bio/technology* **4**: 726.

Forman, H.J. and Fridovich, I. (1973) On the stability of bovine superoxide dismutase. *J. Biol. Chem.* **248**: 2645.

Futcher, A.B. (1986) Copy number amplification of the 2-μm circle plasmid of *Saccharomyces cerevisiae*. *J. Theoret. Biol.* **119**: 197.

Gerbaud, C. and Guerineau, M. (1980) 2-μm plasmid copy number in different yeast strains and repartition of endogenous and 2-μm chimeric plasmids in transformed strains. *Curr. Genet.* **1**: 219.

Goff, C.G., Moir, D.T., Kohno, T., Gravius, T.C., Smith, R.A., Yamasaki, E. and Taunton-Rigby, A. (1984) Expression of calf prochymosin in *Saccharomyces cerevisiae*. *Gene* **27**: 35.

Goodey, A.R., Doel, S.M., Piggott, J.R., Watson, M.E.E., Zealey, G.R., Cafferkey, R. and Carter, B.L.A. (1986) The selection of promoters for the expression of heterologous genes in the yeast *Saccharomyces cerevisiae*. *Mol. Gen. Genet.* **204**: 505.

Green, R., Schaber, M.D., Shields, D. and Kramer, R. (1986) Secretion of somatostatin by *Saccharomyces cerevisiae*. *J. Biol. Chem.* **261**: 7558.

Gritz, L. and Davies, J. (1983) Plasmid encoded hygromycin B resistance: the sequence of hygromycin B phosphotransferase and its expression in *Escherichia coli* and *Saccharomyces cerevisiae*. *Gene* **25**: 179.

Guan, J.L., Machames, L.E. and Rox, J.K. (1985) Glycosylation allows cell surface transport of an anchored secretory protein. *Cell* **42**: 489.

Gunge, N. (1983) Yeast DNA plasmids. *Ann. Rev. Microbiol.* **37**: 253.

Hadfield, C., Cashmore, A.M. and Meacock, P.A. (1986) An efficient chloramphenicol-resistance marker for *Saccharomyces cerevisiae* and *Escherichia coli*. *Gene* **45**: 149.

Hallewell, R.A., Mills, R., Tekamp-Olson, P., Blacher, R., Rosenberg, S., Otting, F., Maziarz, F.R. and Scandella, C.J. (1987) Amino terminal acetylation of authentic human Cu, Zn superoxide dismutase produced in yeast. *Bio/technology* **5**: 363.

Harris, T.J.R. (1983) Expression of eukaryotic genes in *E. coli*. In *Genetic Engineering*, ed. Williamson, R. Academic Press, London, 127.

Hartley, J.L. and Donelson, J.E. (1980) Nucleotide sequence of the yeast plasmid. *Nature* **286**: 860.

Henderson, R.C.A., Cox, B.S. and Tubb, R.S. (1985) The transformation of brewing yeast with a plasmid containing the gene for copper resistance. *Curr. Genet.* **9**: 133.

Hinnen, A., Hicks, J.B. and Fink, G.R. (1978) Transformation of yeast. *Proc. Natl. Acad. Sci. USA* **75**: 1929.

Hinnen, A. Meyhack, B. and Tsapis, R. (1983) High expression and secretion of foreign proteins in yeast. In *Gene Expression in Yeast*, vol. 1, eds. Korhola, M. and Vaisanen, E., Foundation for Biotechnical and Industrial Fermentation Research, Kauppakirjapaino, Helsinki, 157.

Hitzemann, R.A., Hagie, F.E., Levine, H.L., Goeddel, D.V., Ammerer, G. and Hall, B.D. (1981) Expression of a human gene for interferon in yeast. *Nature* **293**: 717.

Hitzemann, R.A., Leung, D.W., Perry, L.J., Kohr, W.J., Levine, H.L. and Goeddel, D.V. (1983*a*) Secretion of human interferons by yeast. *Science* **219**: 620.

Hitzemann, R.A., Chen, C.Y., Hagie, F.E., Patzer, E.J., Liu, C.C., Estell, D.A., Miller, J.V., Yaffe, A., Kleid, D.G., Levinson, A.D. and Opperman, H. (1983*b*) Expression of hepatitis B virus surface antigen in yeast. *Nucleic Acids Res.* **11**: 2745.

Innis, M.A., Holland, M.J., McCabe, P.C., Cole, G.E., Wittman, V.P., Tal, R., Watt, K.W.K., Gelfand, D.H., Holland, J.P. and Meade, J.H. (1985) Expression, glycosylation and secretion of an Aspergillus glucoamylase by *Saccharomyces cerevisiae*. *Science* **228**: 21.

Izumoto, Y., Sato, T., Yamamoto, T., Yoshida, N., Kikuchi, N., Ogawa, M. and Matsubara, K. (1987) Expression of human pancreatic secretory trypsin inhibitor in *Saccharomyces cerevisiae*. *Gene.* **59**: 151.

Jigami, Y., Muraki, M., Harada, N. and Tanaka, H. (1986) Expression of synthetic human-lysozyme gene in *Saccharomyces cerevisiae*: use of a synthetic chicken-lysozyme signal sequence for secretion and processing. *Gene* **43**: 273.

Jimenez, A. and Davies, J. (1980) Expression of a transposable antibiotic resistance element in *Saccharomyces*. *Nature* **287**: 869.

Johnson, L.M., Bankaitis, V.A. and Emr, S.D. (1987) Distinct sequence determinants direct intracellular sorting and modification of a yeast vacuolar protease. *Cell* **48**: 875.

Johnston, M. and Davis, R.W. (1984) Sequences that regulate the divergent *GAL1-GAL10* promoter in *Saccharomyces cerevisiae*. *Mol. Cell. Biol.* **4**: 1440.

Julius, D.J., Blair, L.C., Brake, A.J., Sprague, G.F. and Thorner, J. (1983) Yeast alpha-factor is processed from a larger precursor polypeptide: the essential role of a membrane bound dipeptidyl aminopeptidase. *Cell* **32**: 839.

Julius, D.J., Schekman, R. and Thorner, J. (1984) Glycosylation and processing of prepro-α-factor through the yeast secretory pathway. *Cell* **36**: 309.

Kaczorek, M., Honore, N., Ribes, V., Dehoux, P., Cornet, P., Cartwright, T. and Streeck, R.E. (1987) Molecular cloning and synthesis of biologically active human tissue inhibitor of metalloproteinases in yeast. *Bio/technology* **5**: 595.

Kaiser, C.A., Preuss, D., Grisafi, P. and Botstein, D. (1987) Many random sequences functionally replace the secretion signal sequence of yeast invertase. *Science* **235**: 312.

Kelly, R.B. (1985) Pathways of protein secretion in eukaryotes. *Science* **230**: 25.

Kitano, K., Nakao, M., Hoh, Y. and Fujisawa, Y. (1987) Recombinant hepatitis B virus surface antigen P31 accumulates as particles in *Saccharomyces cerevisiae*. *Bio/technology* **5**: 281.

Kniskern, P.J., Hagiopian, A., Montgomery, D.L., Burke, P., Dunn, N.R., Hofmann, K.J., Miller, W.J. and Ellis, R.W. (1986) Unusually high-level expression of a foreign gene (hepatitis B core antigen) in *Saccharomyces cerevisiae*. *Gene* **46**: 135.

Kurjan, J. and Herskowitz, I. (1982) Structure of a yeast pheromone (MFα). A putative α-factor precursor contains four tandem copies of mature alpha-factor. *Cell* **30**: 933.

Lemontt, J.F., Wei, C. and Dackowski, W.R. (1985) Expression of active human uterine tissue plasminogen activator in yeast. *DNA* **4**: 419.

Loison, G., Findeli, A., Bernard, S., Nguyen-Juilleret, M., Marquet, M., Riehl-Bellon, N., Carvallo, D., Guerra-Santos, L., Brown, S.W., Courtney, M., Roitsch, C. and Lemoine, Y. (1988) Expression and secretion in *S. cerevisiae* of biologically active leech hirudin. *Bio/technology* **6**: 72.

Marston, F.A.O. (1986) The purification of eukaryotic polypeptides expressed in *E coli*. *Biochem. J.* **240**: 1.

McAlleer, W.J., Buynak, E.B., Maigetter, R.Z., Wampler, D.E., Miller, W.J. and Hilleman, M.R. (1984) Human hepatitis B vaccine from recombinant yeast. *Nature* **307**: 178.

McNiel, J.B. and Friesen, J. (1981) Expression of *Herpes simplex* virus thymidine kinase gene in *Saccharomyces cerevisiae*. *Mol. Gen. Genet.* **184**: 386.

Meaden, P., Ogden, K., Bussey, H. and Tubb, R.S. (1985) A DEX gene conferring production of extracellular amyloglucosidase on yeast. *Gene* **34**: 325.

Mellor, J., Dobson, M.J., Roberts, N.A., Tuite, M.F., Emtage, J.S., White, S., Lowe, P.A., Patel, T., Kingsman, A.J. and Kingsman, S.M. (1983) Efficient synthesis of enzymatically active calf chymosin in *Saccharomyces cerevisiae*. *Gene* **24**: 1.

Mellor, J., Dobson, M.J., Roberts, N.A., Kingsman, A.J. and Kingsman, S.M. (1985). Factors affecting heterologous gene expression in *Saccharomyces cerevisiae*. *Gene* **33**: 215.

Miyajima, A., Miyajima, I., Arai, K. and Arai, N. (1984) Expression of plasmid R388-encoded type II dihydrofolate reductase as a dominant selective marker in *S. cerevisiae*. *Mol. Cell. Biol.* **4**: 407.

Miyajima, A., Bond, M.W., Otsu, K., Arai, K. and Arai, N. (1985) Secretion of mature mouse interleukin-2 by *Saccharomyces cerevisiae*: use of a general secretion vector containing promoter and leader sequences of the mating pheomone α-factor. *Gene* **37**: 155.

Miyajima, A., Otsu, K., Schreurs, J., Bond, M.W., Abrams, J.S. and Arai, K. (1986) Expression of murine and human granulocyte macrophage colony stimulating factors in *S. cerevisiae*: mutagenesis of the potential glycosylation sites. *EMBO J.* **5**: 1193.

Mujamoto, C., Chizzonite, R., Crowl, R., Rupprecht, K., Kramer, R., Schaber, M., Kumar, G., Poonian, M. and Ju, G. (1985) Molecular cloning and regulated expression of the human c-myc gene in *Escherichia coli* and *Saccharomyces cerevisiae*: comparison of the protein products. *Proc. Natl. Acad. Sci. USA* **82**: 7232.

Mujanohara, A., Toh-e, A., Nozaki, C., Hamada, F., Ohtomo, N. and Matsubara, K. (1983) Expression of hepatitis B surface antigen gene in yeast. *Proc. Natl. Acad. Sci. USA* **80**: 1.

Moir, D., Duncan, M., Kohno, T., Mao, J. and Smith, R. (1984) Production of calf chymosin by the yeast *S. cerevisiae*. In *The World Biotech Report* 1984 Online, Pinner, 189.

Moir, D.T. and Dumais, D.R. (1987) Glycosylation and secretion of human alpha-1-antitrypsin by yeast. *Gene* **56**: 209.

Murakami, H., Yabusaki, Y. and Ohkawa, H. (1986) Expression of rat NADPH: cytochrome P-450 reductase cDNA in *Saccharomyces cerevisiae. DNA* **5**: 1.

Murray, A.W. and Szostak, J.W. (1983) Pedigree analysis of plasmid segregation in yeast. *Cell* **34**: 911.

Nagai, K., Perutz, M.F. and Poyast, C. (1985) Oxygen binding properties of human mutant haemoglobins synthesised in *Escherichia coli. Proc. Natl. Acad. Sci. USA* **82**: 7252.

Nakamura, Y., Sato, T., Emi, M., Miyanohara, A., Nishide, T. and Matsubara, K. (1986) Expression of human salivary α-amylase gene in *Saccharomyces cerevisiae* and its secretion using the mammalian signal sequence. *Gene* **50**: 239.

Novick, P., Ferro, S. and Schekman, R. (1981) Order of events in the yeast secretory pathway. *Cell* **25**: 461.

Oberto, J. and Davison, J. (1985) Expression of chicken egg white lysozyme by *Saccharomyces cerevisiae. Gene* **40**: 57.

Oeda, K., Sakaki, T. and Ohkawa, H. (1985) Expression of rat liver cytochrome P-450MC cDNA in *Saccharomyces cerevisiae. DNA* **4**: 203.

Parent, S.A., Fenimore, C.M. and Bostian, K.A. (1985) Vector systems for the expression, analysis and cloning of DNA sequences in *S. cerevisiae. Yeast* **1**: 83.

Penttila, M.E., Andre, L., Lehtovaara, P., Bailey, M., Teeri, T.T. and Knowles, J.K.C. (1988) Efficient secretion of two fungal cellobiohydrolases by *Saccharomyces cerevisiae. Gene* **63**: 103.

Price, V., Mochizuki, D., March, C.J., Cosman, D., Deeley, M.C., Klinke, R., Clevenger, W., Gillis, S., Baker, P. and Urdal, D. (1987) Expression, purification and characterization of recombinant murine granulocyte-macrophage colony-stimulating factor and bovine interleukin-2 from yeast. *Gene* **55**: 287.

Roggenkamp, R., Dargatz, H. and Hollenberg, C.P. (1985) Precursor of β-lactamase is enzymatically inactive. Accumulation of the preprotein in *Saccharomyces cerevisiae. J. Biol. Chem.* **260**: 1508.

Rosenberg, S., Barr, P.J., Majoriam, R.C. and Hallewell, R.A. (1984) Synthesis in the yeast *Saccharomyces cerevisiae* of a functional oxidation resistant mutant of human α-1-antitrypsin. *Nature* **312**: 77.

Rothblatt, J.A., Webb, J.R., Ammerer, G. and Meyer, D.I. (1987) Secretion in yeast: structural features influencing the post-translational translocation of prepro-α-factor *in vitro. EMBO J.* **6**: 3455.

Rothstein, R.J. (1983) One-step gene disruption in yeast. *Meth. Enzymol.* **101**: 202.

Rothstein, S.J., Lazarus, C.M., Smith, W.E., Baulcombe, D.C. and Gatenby, A.A. (1984) Secretion of a wheat α-amylase expressed in yeast. *Nature* **308**: 662.

Rothstein, S.J., Lahner, K.N., Lazarus, C.M., Baulcombe, D.C. and Gatenby, A.A. (1987) Synthesis and secretion of wheat α-amylase in *Saccharomyces cerevisiae. Gene* **55**: 353.

Ruohonen, L., Hackman, P., Lehtovaara, P., Knowles, J.K.C. and Keranen, S. (1987) Efficient secretion of *Bacillus amyloliquefaciens* α-amylase by its own signal peptide from *Saccharomyces cerevisiae* host cells. *Gene* **59**: 161.

Sakaki, T., Shibata, M., Yabusaki, Y. and Ohkawa, H. (1987) Expression in *Saccharomyces cerevisiae* of chimaeric cytochrome P-450 cDNAs for rat cytochrome P-450c and P-450d. *DNA* **6**: 31.

Sato, T., Tsunasawa, S., Nakamura, Y., Emi, M., Sakiyama, F. and Matsubara, K. (1986) Expression of the human salivary α-amylase gene in yeast and characterization of the secreted protein. *Gene* **50**: 247.

Schekman, R. and Novick, P. (1982) The secretory process and yeast cell surface assembly. In *The Molecular Biology of the Yeast Saccharomyces. Metabolism and Gene Expression*, eds. Strathern, J.H., Jones, E.W. and Broach, J.R., Cold Spring Harbor, New York, 361.

Schultz, L.D., Tanner, J., Hofmann, K.J., Emini, E.A., Condra, J.H., Jones, R.E., Kieft, E. and Ellis, R.W. (1987) Expression and secretion in yeast of a 400-kDa envelope glycoprotein derived from Epstein-Barr virus. *Gene* **54**: 113.

Shaw, K.J., Frommer, B.R., Anagnost, J.A., Narsula, S. and Liebowitz, P.J. (1988) Regulated secretion of MuGM-CSF in *Saccharomyces cervisiae* via GALI: MFα1 prepro sequences. *DNA* **7**: 117.

Singh, A., Lugovoy, J.M., Kohr, W.J. and Perry, L.J. (1984). Synthesis, secretion and processing of α-factor-interferon fusion proteins in yeast. *Nucleic Acids Res.* **12**: 8927.

Skipper, N., Sutherland, M., Davies, R.W., Kilburn, D., Miller, R.C., Warren, A. and Wong, R. (1985) Secretion of a bacterial cellulase by yeast. *Science* **230**: 958.

Smith, R.A., Duncan, M.J. and Moire, D.T. (1985) Heterologous protein secretion from yeast. *Science* **230**: 1219.

Smith, B.W., Walton, E.F., Caulcott, C.A., King, D.J. and Yarranton, G.T. (1986) The secretion of homologous proteins from yeast. *Proc. 14th Int. Congr. of Microbiology*, Manchester.

Som, T., Armstrong, K.A., Volkert, F.C. and Broach, J.R. (1988) Autoregulation of 2-µm circle gene expression provides a model for maintenance of stable plasmid copy levels. *Cell* **52**: 27.

Struhl, K., Stinchcomb, D.T., Scherer, S. and Davis, R.W. (1979) High-frequencing transformation in yeast: Autonomous replication of hybrid DNA molecules. *Proc. Natl. Acad. Sci. USA.* **76**: 1035.

Struhl, K. (1983) Direct selection for gene replacement events in yeast. *Gene* **26**: 231.

Sugisaki, Y., Gumage, N., Sakaguchi, K., Yamasaki, M. and Tamwa, G. (1983) *Kluyveromyces lactis* killer toxin inhibitors adenylate cyclase of sensitive yeast cells. *Nature* **304**: 464.

Tanaka, T., Kimura, S. and Ota, Y. (1988) Secretion of the proenzyme and active bovine pancreatic phospholipase A₂ enzyme by *Saccharomyces cerevisiae*: design and use of a synthetic gene. *Gene* **64**: 257.

Thim, L., Hansen, M.T., Norris, K., Hoegh, I., Boel, E., Forstrom, J., Ammerer, G. and Fiil, N.P. (1986) Secretion and processing of insulin precursors in yeast. *Proc. Natl. Acad. Sci. USA* **83**: 6766.

Thomsen, K.K. (1983) Mouse α-amylase synthesised by *Saccharomyces cerevisiae* is released into the culture medium. *Carlsberg Res. Commun.* **48**: 545.

Tuite, M.F., Dobson, M.J., Roberts, N.A., King, R.M., Burke, D.C., Kingsman, S.M. and Kingsman, A.J. (1982) Regulated high efficiency expression of human interferon-alpha in *Saccharomyces cerevisiae*. *EMBO J.* **1**: 603.

Valenzuela, P., Medina, A., Rutter, W.J., Ammerer, G. and Hall, B.D. (1982) Synthesis and assembly of hepatitis B virus surface antigen particles in yeast. *Nature* **298**: 347.

Valenzuela, P., Coit, D., Medina-Selby, A., Kuo, C.H., Van Nash, G., Burke, L., Bull, P., Urdea, M.S. and Graves, P.V. (1985) Antigen engineering in yeast: Synthesis and assembly of hybrid hepatitis B surface antigen—Herpes simplex 1-gD particles. *Bio/technology* **3**: 323.

Valls, L.A., Hunter, C.P., Rothman, J.H. and Stevens, T.H. (1987) Protein sorting in yeast: The localisation determinant of yeast vacuolar carboxypeptidase γ resides in the propeptide. *Cell* **48**: 887.

Van den Bergh, C.J., Bakkers, A.C.A.P.A., De Geus, P., Verheij, H.M. and de Haas, G.H. (1987) Secretion of biologically active porcine prophospholipase A₂ by *Saccharomyces cerevisiae*. *Eur. J. Biochem.* **170**: 241.

Van der Straken, A., Falque, J.C., Loriau, R., Bollen, A. and Cabezon, T. (1986) Expression of cloned haptoglobulin and α₁-antitrypsin complementary DNAs in *Saccharomyces cerevisiae*. *DNA* **5**: 129.

Vlasuk, G.P., Bencen, G.H., Scarborough, R.M., Tsai, P.K., Whang, J.L., Maack, T., Camargo, M.J.F., Kirsher, S.W. and Abraham, J.A. (1986) Expression and secretion of biologically active human atrial natriuretic peptide in *Saccharomyces cerevisiae*. *J. Biol. Chem.* **261**: 4789.

Wampler, D.E., Lehman, E.D., Boger, J., McAleer, W.J. and Scolnick, E.M. (1985) Multiple chemical forms of hepatitis B surface antigen produced in yeast. *Proc. Natl. Acad. Sci. USA* **82**: 6830.

Waters, M.G., Evans, E.A. and Blobel, G. (1988) Prepro-α-factor has a cleavable signal sequence. *J. Biol. Chem.* **263**: 6209.

Webster, T.D. and Dickson, R.C. (1983) Direct selection of *Saccharomyces cerevisiae* resistant to the antibiotic G418 following transformation with a DNA vector carrying the kanamycin resistance gene of Tn903. *Gene* **26**: 243.

Wood, C.R., Boss, M.A., Kenten, J.H., Calvert, J.E., Roberts, N.A. and Emtage, J.S. (1985) The synthesis and *in vivo* assembly of functional antibodies in yeast. *Nature* **314**: 446.

Woolford, C.A., Daniels, L.B., Park, F.J., Jones, E.W., Van Arsdell, J.N., Innis, M.A. (1986) The *PEP4* gene encodes an aspartyl protease implicated in the post-translational regulation of *Saccharomyces cerevisiae* vacuolar hydrolases. *Mol. Cell. Biol.* **6**: 2500.

Zhu, J., Contreras, R., Gheysen, D., Ernst, J. and Fiers, W. (1985) A system for dominant transformation and plasmid amplification in *Saccharomyces cerevisiae*. *Bio/technology* **3**: 451.

Zsebo, K.M., Lu, H.S., Fieschko, J.C., Goldstein, L., Davis, J., Duker, K., Suggs, S.V., Lai, P.H. and Bitter, G.A. (1986) Protein secretion from *Saccaromyces cerevisiae* directed by the prepro-α-factor leader region. *J. Biol. Chem.* **261**: 5858.

Zubenko, G.S., Park, F.J. and Jones, E.W. (1982) Genetic properties of mutations at the *PEP4* locus in *Saccharomyces cerevisiae*. *Genetics* **102**: 679.

5 Yeast (*Saccharomyces cerevisiae*) proteinases: structure, characteristics and function

HANS H. HIRSCH, PAZ SUAREZ RENDUELES
and DIETER H. WOLF*

5.1 Introduction

Proteolysis as a fundamental mechanism in the biology of organisms is currently receiving increasing attention. A multitude of biological functions have been shown to involve catalytic action of proteinases (Pine, *et al.*, 1972; Goldberg and Dice, 1974; Miller, 1975; Holzer *et al.*, 1975; Goldberg and St John, 1976; Wolf, 1980; Holzer and Heinrich, 1980; North, 1982; Docherty and Steiner, 1982; Achstetter and Wolf, 1985; Wolf, 1986; Pontremoli and Melloni, 1986; Suarez Rendueles and Wolf, 1988). The basic process of hydrolytic cleavage of peptide bonds in proteins catalysed by proteinases appears costly and potentially detrimental to an organism if not properly controlled. Corresponding to the respective functions, various ways of controlling proteolytic events have evolved. The desired limits to proteolytic action are achieved through the primary and secondary specificity of proteinases, by compartmentalization of proteinases and their substrates within the cell, through modification of the substrates allowing recognition by the respective proteinases, by regulation of proteinase activity via zymogen activation, and the presence or absence of specific inhibitors, as well as through the regulation of proteinase gene expression. The combination of these factors creates the reliability and accuracy of proteolytic processes, hence making proteolysis a general principle exerting biological control.

In the advent of modern life sciences in the last century, yeast was chosen to serve as model organism of cellular biology, establishing the 'enzyme principle' by reconstitution of 'biochemical' reactions in the test tube (Berthelot, 1860; Buchner, 1897). It was as early as 1898 that proteolytic activities were detected in yeast (Hahn, 1898). The amenability of the yeast *Saccharomyces cerevisiae* to modern biochemical and molecular biology techniques, in conjunction with its intrinsic biological and genetic properties, made it a truly biotechnological tool from today's point of view. For these very reasons *Saccharomyces cerevisiae* became a pace-maker in the study of eukaryotic cell functions, and in particular, in the study of the role of proteinases in the biology of the cell.

In the past, intracellular proteolytic events were thought to be exclusively associated with the lysosomal compartment of the cell. Today, however, we

*To whom correspondence should be addressed.

have come to realize that proteolytic events most probably take place in every cellular compartment, including nucleus, mitochondria, cytoplasm, secretory pathway and periplasm.

The proteolytic enzymes described to date in *Saccharomyces cerevisiae* are summarized in Table 5.1. These enzymes are listed according to their endo- or exopeptidolytic specificity. Their structural genes are indicated, as well as their cellular localization and the cellular roles that have been found or ascribed to them. The nomenclature used throughout this review was introduced previously (Achstetter *et al.*, 1984*a*).

5.2 Vacuolar proteinases

The vauole of the yeast *Saccharomyces cerevisiae* contains major hydrolytic enzymes such as RNase, phosphatase and proteinase activities. It therefore represents a functional analogue to the mammalian lysosome. The vacuole also serves as a storage compartment for polyphosphates, amino acids, carbohydrates and Ca^{2+}-ions (Matile and Wiemken, 1967; Wiemken *et al.*, 1979; Wiemken and Dürr, 1974; Messenguy *et al.*, 1980).

Storage contents as well as hydrolytic activities appear to be regulated in a growth phase-and-nutrient-dependent fashion (Saheki *et al.*, 1974; Hansen *et al.*, 1977; Lille and Pringle, 1980). The role of the vacuole as a digestive organelle involved in the turnover of cellular proteins has been well documented (Wolf and Ehmann, 1979, 1981; Mechler and Wolf, 1981; Zubenko and Jones, 1981; Teichert *et al.*, 1987). This function is accomplished largely by a set of highly active proteinases, characterized initially using unspecific substrates like collagen derivatives, casein or hemoglobin (Hata *et al.*, 1967*a*; Lenney and Dalbec, 1967; Lenney *et al.*, 1974).

5.2.1 *Proteinase yscA*

Proteinase yscA is a soluble aspartic acid endopeptidase (Hata *et al.*, 1967; Lenney and Dalbec, 1967; Hata *et al.*, 1967*b*; Meussdoerffer *et al.*, 1980; Magni *et al.*, 1982). The enzyme is synthesized as a 52-kDa glycosylated precursor, and the mature, active form of the protein resides in the vacuole as a 42-kDa glycoprotein (Meussdoerffer *et al.*, 1980; Mechler *et al.*, 1982*a*). The primary structure of the mature protein has been determined by amino acid sequencing. Proteinase yscA consists of 329 amino acid residues and contains 8.5% neutral sugar and 1% glucosamine. The carbohydrate is attached to asparagine residues in positions 67 and 267 (Dreyer *et al.*, 1986). This finding confirms previous results which showed that two glycosylation reactions were prevented when tunicamycin, an inhibitor of N-glycosylation, was added to cells (Mechler *et al.*, 1982*a*; Elbein, 1981).

The enzyme is active against acid denatured hemoglobin or casein, with pH optima of 3 and 6, respectively. Proteinase yscA can also be measured using the fluorogenic peptide succinyl-Arg-Pro-Phe-His-Leu-Leu-Val-Tyr-7-

amino-4-methyl-coumarin as substrate at about pH 5. Hydrolysis occurs at the Leu-Val bond (Yokosawa *et al.*, 1983). The enzyme is inhibited by pepstatin, diazoacetyl-D, L-norleucine methylester and by 1, 2-epoxy-3-(4-nitrophenoxy) propane. Proteinase yscA activity is partly cryptic in whole cell extracts. This is due to a proteinase yscA-specific cytoplasmic inhibitor protein (I_3^A) of M_r of about 7.7 kDa (Saheki *et al.*, 1974; Lenney *et al.*, 1974; Matern *et al.*, 1974a; Lenney, 1975; Nunez de Castro and Holzer, 1976). Incubation of cell extracts at low pH releases proteinase yscA activity from inhibition most likely due to digestion of I_3^A by proteinase yscB (Matern *et al.*, 1974a; Beck *et al.*, 1980). Activity levels of proteinase yscA increase several-fold *in vivo* when cells enter diauxic growth and stationary phase, grow on acetate or are transferred on to nitrogen-free medium (Hansen *et al.*, 1977; Klar and Halvorson, 1975; Betz and Weiser, 1976; Holzer *et al.*, 1977).

Two classes of mutations that abolish proteinase yscA activity have been described. Mutations in the structural gene (*PRA1*) specifically affect proteinase yscA; *pra1* mutants are defective in proteinase yscA activity but show normal levels of other vacular enzyme activities (Mechler and Wolf, 1981; Mechler *et al.*, 1982b). In contrast, a number of *pep4* mutations have been reported (of which *pep4-3* is the best characterized allele) resulting in a pleiotropic phenotype affecting not only proteinase yscA activity but also other vacuolar hydrolases. Cloning of the *PEP4* gene revealed that it codes for proteinase yscA. Complementation studies showed that the *PRA1* and *PEP4* gene are identical. The structural gene of proteinase yscA is located on chromosome XIV (Jones, 1977; Jones *et al.*, 1982; Rothman *et al.*, 1986; Ammerer *et al.*, 1986; Woolford *et al.*, 1986; Mechler *et al.*, 1987; Strathern *et al.*, 1983).

The gene codes for a polypeptide of 405 amino acids (Figure 5.1). Proteolytic processing of the proproteinase yscA zymogen occurs between amino acid residues Glu_{76}-Gly_{77}, resulting in the mature 329-amino-acid residues containing vacuolar form. Proteinase yscA exhibits extensive amino acid homology with mammalian aspartic proteases, such as human lysosomal cathepsin D (46%) and human pepsin (40%). Interestingly, pro-cathepsin D and pro-proteinase yscA share the identical zymogen processing site (Thr_{75}-Glu_{76}-Gly_{77}), suggesting a similar activation mechanism.

5.2.2 *Proteinase yscB*

Proteinase yscB is a soluble serine-, sulphydryl endopeptidase exhibiting an activity optimum in the neutral pH range (Fujishiro *et al.*, 1980; Kominami *et al.*, 1981a, b). Studies on the biosynthesis of proteinase yscB identified a 73-kDa precursor form of the enzyme which is transferred into a 42-kDa precursor form (Mechler *et al.*, 1988). This precursor is subsequently converted into a protein of 33 kDa (Mechler *et al.*, 1982a) identical in molecular mass to the purified, active proteinase yscB. The mature 33-kDa proteinase yscB is a glycoprotein containing about 10% carbohydrate (Kominami *et al.*, 1981a).

```
                                                                                              -728 GAATTCAT
-720  CTATTCATAACTTTTTGGGTTTTTATTACATATAGAGTGACGTTTTGATACGATTTACATTTTGAAATATATAGATTTTTGTATATTAATAAGTTTTGATTTTCAGTTTTAGTTATGCTT
-600  AATGAATTTTTGAATCAGAAAAAGAAGTTCTGAAAAAAACATAATCCTGCTTGATGTGGTACAACAAGCTCATAAATAATTTTAAAAAGTTTGTTAATCCGTTTTCAATATCTTGAGCTC
-480  CTCAATTGTATTTGCTGAGGTCTGATTATTTCTATAACCAAAAGCGGTTATTGAATCTATGGAGAGGCTGTAACCCGTCTTATGCCTTCCGGGTACTATATTTCATTTGCGGGTGTCGAT
-360  GGATTAAGGGGCGGAGGCGGCCCTTTTTAGGATTTATATAAAAAGCCATACTTCCGTACTTCGTAACCTCTTATCAACTGGTTAAGGGAACAGAGTAAAGAAGTTTGGGTAATTCGCTGC
-240  TATTTATTCATTCCACCTTCTTCTTTTTTGAGCGAAGCCTTTATAATCAAATTTTAGTGGTCTTTTCTATTTTTATTTGAAGCCTACCACGTAAGGGAAGAATAACAAAAAGTATATC
-120  TCACCCTACTGTATTCATAAAAAGTTTTTTCTATTAGAATTCTATAAGAAAAGAAAAAAAAAAGCCTAGTGACCTAGTATTTAATCCAAATAAAATTCTAAACAAAAACCAAAACTAAC

  1   ATG TTC AGC TTG AAA GCA TTA TTG CCA TTG GCC TTG TTG TTG GTC AGC GCC AAC CAA GTT GCT GCA AAA GTC CAC AAG GCT AAA ATT TAT
  1   Met Phe Ser Leu Lys Ala Leu Leu Pro Leu Ala Leu Leu Leu Val Ser Ala Asn Gln Val Ala Ala Lys Val His Lys Ala Lys Ile Tyr

 91   AAA CAC GAG TTG TCC GAT GAG ATG AAA GAA GTC ACT TTC GAG CAA CAT TTA GCT CAT TTA GGC CAA AAG TAC TTG ACT CAA TTT GAG AAA
 31   Lys His Glu Leu Ser Asp Glu Met Lys Glu Val Thr Phe Glu Gln His Leu Ala His Leu Gly Gln Lys Tyr Leu Thr Gln Phe Glu Lys

181   GCT AAC CCC GAA GTT GTT TTT TCT AGG GAG CAT CCT TTC TTC ACT GAA GGT GGT CAC GAT GTT CCA TTG ACA AAT TAC TTG AAC GCA CAA
 61   Ala Asn Pro Glu Val Val Phe Ser Arg Glu His Pro Phe Phe Thr Glu Gly Gly His Asp Val Pro Leu Thr Asn Tyr Leu Asn Ala Gln

271   TAT TAC ACT GAC ATT ACT TTG GGT ACT CCA CCT CAA AAC TTC AAG GTT ATT TTG GAT ACT GGT TCT TCA AAC CTT TGG GTT CCA AGT AAC
 91   Tyr Tyr Thr Asp Ile Thr Leu Gly Thr Pro Pro Gln Asn Phe Lys Val Ile Leu Asp Thr Gly Ser Ser Asn Leu Trp Val Pro Ser Asn

361   GAA TGT GGT TCC TTG GCT TGT TTC CTA CAT TCT AAA TAC GAT CAT GAA GCT TCA TCA AGC TAC AAA GCT AAT GGT ACT GAA TTT GCC ATT
121   Glu Cys Gly Ser Leu Ala Cys Phe Leu His Ser Lys Tyr Asp His Glu Ala Ser Ser Ser Tyr Lys Ala Asn Gly Thr Glu Phe Ala Ile

451   CAA TAT GGT ACT GGT TCT TTG GAA GGT TAC ATT TCT CAA GAC ACT TTG TCC ATC GGG GAT TTG ACC ATT CCA AAA CAA GAC TTC GCT GAG
151   Gln Tyr Gly Thr Gly Ser Leu Glu Gly Tyr Ile Ser Gln Asp Thr Leu Ser Ile Gly Asp Leu Thr Ile Pro Lys Gln Asp Phe Ala Glu

541   GCT ACC AGC GAG CCG GGC TTA ACA TTT GCA TTT GGC AAG TTC GAT GGT ATT TTG GGT TTG GGT TAC GAT ACC ATT TCT GTT GAT AAG GTG
181   Ala Thr Ser Glu Pro Gly Leu Thr Phe Ala Phe Gly Lys Phe Asp Gly Ile Leu Gly Leu Gly Tyr Asp Thr Ile Ser Val Asp Lys Val

631   GTC CCT CCA TTT TAC AAC GCC ATT CAA CAA GAT TTG TTG GAC GAA AAG AGA TTT GCC TTT TAT TTG GGA GAC ACT TCA AAG GAT ACT GAA
211   Val Pro Pro Phe Tyr Asn Ala Ile Gln Gln Asp Leu Leu Asp Glu Lys Arg Phe Ala Phe Tyr Leu Gly Asp Thr Ser Lys Asp Thr Glu

721   AAT GGC GGT GAA GCC ACC TTT GGT GGT ATT GAC GAG TCT AAG TTC AAG GGC GAT ATC ACT TGG TTA CCT GTT CGT CGT AAG GCT TAC TGG
241   Asn Gly Gly Glu Ala Thr Phe Gly Gly Ile Asp Glu Ser Lys Phe Lys Gly Asp Ile Thr Trp Leu Pro Val Arg Arg Lys Ala Tyr Trp

811   GAA GTC AAG TTT GAA GGT ATC GGT TTA GGC GAC GAG TAC GCC GAA TTG GAC AGC CAT GGT GCC GCC ATC GAT ACT GGT ACT TCT TTG ATT
271   Glu Val Lys Phe Glu Gly Ile Gly Leu Gly Asp Glu Tyr Ala Glu Leu Glu Ser His Gly Ala Ala Ile Asp Thr Gly Thr Ser Leu Ile

901   ACC TTG CCA TCA GGA TTA GCT GAA ATG ATT AAT GCT GAA ATT GGG GCC AAG AAG GGT TGG ACC GGT CAA TAT ACT CTA GAC TGT AAC ACC
301   Thr Leu Pro Ser Gly Leu Ala Glu Met Ile Asn Ala Glu Ile Gly Ala Lys Lys Gly Trp Thr Gly Gln Tyr Thr Leu Asp Cys Asn Thr

991   AGA GAC AAT CTA CCT GAT CTA ATT TTC AAC TTC AAT GGC TAC AAC TTC ACT ATT GGG CCA TAC GAT TAC ACG CTT GAA GTT TCA GGC TCC
331   Arg Asp Asn Leu Pro Asp Leu Ile Phe Asn Phe Asn Gly Tyr Asn Phe Thr Ile Gly Pro Tyr Asp Tyr Thr Leu Glu Val Ser Gly Ser

1081  TGT ATC TCT GCA ATT ACA CCA ATG GAT TTC CCA GAA CCT GTT GGC CCA CTG GCC ATC GTT GGT GAT GCC TTC TTG CGT AAA TAC TAT TCT
361   Cys Ile Ser Ala Ile Thr Pro Met Asp Phe Pro Glu Pro Val Gly Pro Leu Ala Ile Val Gly Asp Ala Phe Leu Arg Lys Tyr Tyr Ser

1171  ATT TAC GAT TTG GGC AAC AAT GCG GTT GGT TTG GCC AAA GCA ATT TGA GCTAAACTTTTCTTACTTCTCCGCCCTATCCTTTTCTGCCATCTAGAGAGCTTTTA
391   Ile Tyr Asp Leu Gly Asn Asn Ala Val Gly Leu Ala Lys Ala Ile End

1275  TAAGTAGATAACAATAAAAAAAACTATAGTATATTTAAAAAAAAAAAACAAGACAAACCATCTTGTCCTCAGTTTTAGAATCCATTGTTCTATGCTGCTGCCCATAATGTCATTATATGC

1395  GGGTAGCCCGATGATGCGGCTCGAG
```

Figure 5.1 The structural gene of proteinase ysCA (*PEP4/PRA1*) (Ammerer *et al.*, 1986).

Tunicamycin interferes with the glycosylation of the 42-kDa precursor, whereas, however, the authentic proteinase yscB molecule remains unchanged. These data support the idea that an asparagine-linked carbohydrate is part of the pro-sequence which is cleaved off during maturation, and a tunicamycin-insensitive linkage of carbohydrate to the mature protein. Also, endoglycosidase H treatment of mature proteinase yscB *in vitro* did not result in deglycosylation of the protein (Mechler *et al.*, 1982a).

The enzyme can be assayed with the chromophoric collagen derivatives Azocoll (Ulane and Cabib, 1976; Cabib and Ulane, 1973; Saheki and Holzer, 1974) and Hide Powder Azure (Ulane and Cabib, 1976). In accordance with its low specificity, proteinase yscB splits urea-denatured proteins, gelatin hemoglobin and casein. The chromogenic peptide succinyl-Ala-Ala-Pro-Phe-4-nitroanilide (Achstetter *et al.*, 1981) as well as ester substrates of trypsin and chymotrypsin (Kominami *et al.*, 1981b) are also cleaved. The enzyme is strongly inhibited by the serine modifying agents diisopropylfluorophosphate and phenylmethylsulphonyl fluoride (Hata *et al.*, 1967*a*; Kominami *et al.*, 1981*a*). Mercury compounds are also strong inhibitors of proteinase yscB as is the actinomycete inhibitor chymostatin (Hata *et al.*, 1967*a*; Ulane and Cabib, 1976; Kominami *et al.*, 1981*a*). Proteinase yscB activity is cryptic in whole cell extracts. This phenomenon is brought about by a proteinase yscB-specific cytoplasmic inhibitor protein (I_2^B) of M_r of about 8.5-kDa (Lenney, 1975; Lenney and Dalbec, 1969; Bünning and Holzer, 1977; Maier *et al.*, 1979). Incubation of cell extracts at low pH (Saheki *et al.*, 1974) or with 0.05 to 0.65% sodium dodecyl sulphate at neutral pH or with pepsin (Jones, 1977; Wolf and Ehmann, 1978*b*; Schwencke, 1981) releases proteinase yscB activity from inhibition. Activity levels of proteinase yscB increase several-fold *in vivo* when cells enter diauxic growth and stationary phase, grow on acetate or are transferred on to nitrogen-free medium (Hansen *et al.*, 1977; Klar and Halvorson, 1975; Betz and Weiser, 1976).

The structural gene of proteinase yscB (*PRB1*) is located on chromosome V (H.H. Hirsch and D.H. Wolf, unpublished results). *prb1* mutants are defective in proteinase yscB activity, but show normal levels of proteinase yscA and carboxypeptidase yscY activities (Wolf and Ehmann, 1978*b*, 1979; Mechler *et al.*, 1982*b*; Zubenko *et al.*, 1979, 1980). The proteinase yscB structural gene has been cloned by complementation of the *prb1* mutation (Mechler *et al.*, 1988; Moehle *et al.*, 1987*b*). The open reading frame of *PRB1* extends over 635 amino acids, predicting the high-molecular-mass precursor found recently (Figure 5.2). The *in vivo* pulse-chase data predict two processing events to yield mature proteinase yscB. N-terminal amino acid sequencing of purified proteinase yscB, together with the proteinase yscB sequence, suggested as the first event removal of a N-terminal extension of 31.5 kDa by a yet unknown protease yielding the 42-kDa precursor. As a second event, a carboxyterminal 'pro'-peptide is cleaved in a proteinase yscA-dependent fashion, resulting in the mature 33-kDa proteinase yscB. The C-terminal peptide would have to contain the single site which is N-glycosylated. Proteinase yscB shares significant homology with subtilisins. The homology to the fungal member of this family, proteinase K, is 43%. The proteinase yscB sequence shows no homology to those of the trypsinchymotrypsin family of serine proteinases (Mechler *et al.*, 1982*a*; Moehle *et al.*, 1987*a,b*; Mechler *et al.*, 1988).

5.2.3 *Carboxypeptidase yscY*

Carboxypeptidase yscY is the most extensively studied peptidase in *Saccharomyces cerevisiae*. It is a soluble serine exopeptidase, active in the slightly acidic to neutral pH range with a M_r of 61-kDa. The enzyme has been sequenced (Martin *et al.*, 1982; Svendsen, 1982) and found to consist of one polypeptide chain of 430 amino acids. The enzyme carries four asparagine-linked oligosaccharide chains, three of which have an average composition of $GlcNAc_2Man_{16}$ and carry phosphate groups (Hasilik and Tanner, 1976,

AACACACCCG CGATAAAGAG CGCGATGAAT ATAAAAGGGG GCCAATGTTA CGTCCCGTTA TATTGGAGTT CTTCCCATAC AAACTTAAGA GTCCAATTAG CTTCATCGCC
AATAAAAAAA CAAACTAAAC CTAATTCTAA CAAGCAAAG

ATG AAG TTA GAA AAT ACT CTA TTT ACA CTC GGT GCC CTA GGG AGC ATC TCT GCT GCT TTG GTC ATC CCA AAT CTT GAA AAT GCC GCC GAC
MET Lys Leu Glu Asn Thr Leu Phe Thr Gly Ala Leu Gly Ser Ile Ser Ala Ala Leu Val Ile Pro Asn Leu Glu Asn Ala Ala Asp -251

CAC CAC GAA CTG ATT AAC AAG GAA GAT CAC CAC GAG AGA CCC AGA AAA GTG GAA TTC ACT AAG GAC GAT GAT GAG GAG CCA TCT GAC TCT
His His Glu Leu Ile Asn Lys Glu Asp His His Glu Arg Pro Arg Lys Val Glu Phe Thr Lys Asp Asp Asp Glu Glu Pro Ser Asp Ser -221

GAA GAT AAA GAA CAT GGA AAG TTC CAT AAG AAG GGC CGC AAG GGC CAA GAC AAG GAG TCT CCG GAA TTC AAC GGT AAA CGT GCA AGT GGC
Glu Asp Lys Glu His Gly Lys Phe His Lys Lys Gly Arg Lys Gly Gln Asp Lys Glu Ser Pro Glu Phe Asn Gly Lys Arg Ala Ser Gly -191

TCT CAT GGG AGC GCC CAC GAG GGA GGA AAG GGC ATG AAG CCT AAG CAT GAA AGT TCC AAT GAT GAT GAT AAT GAT GAT AAG AAG AAG AAG
Ser His Gly Ser Ala His Glu Gly Gly Lys Gly Met Lys Pro Lys His Glu Ser Ser Asn Asp Asp Asp Asn Asp Asp Lys Lys Lys Lys -161

CCT CAC CAT AAG GGT GGC TGC CAC GAA AAT AAG GTG GAG GAG AAG AAG ATG AAA GGT AAG GGC AAG GTG AAG GTC AAG GGC AAG CAC GAA AAG
Pro His His Lys Gly Gly Cys His Glu Asn Lys Val Glu Glu Lys Lys Met Lys Gly Lys Gly Lys Val Lys Gly Lys Lys His His Glu Lys -131

⌢⌢
ACG TTG GAG AAA GGG AGG CAC CAC AAC AGG CTG GCT CCT CTC GTG TCC ACT GCA CAA TTC AAC CCA GAC GCG ATC TCC AAG ATC ATC CCC
Thr Leu Glu Lys Gly Arg His His Asn Arg Leu Ala Pro Leu Val Ser Thr Ala Gln Phe Asn Pro Asp Ala Ile Ser Lys Ile Ile Pro -101

AAC CGC TAC ATT ATA GTC TTC AAG AGA GGT GCC CCT CAA GAA GAG ATC GAT TTC CAC AAG GAA AAC GTC CAG CAG GCA CAA TTA CAA CTT
Asn Arg Tyr Ile Ile Val Phe Lys Arg Gly Ala Pro Gln Glu Glu Ile Asp Phe His Lys Glu Asn Val Gln Gln Ala Gln Leu Gln Ser -71

◆
GTA GAG AAC TTA TCT GCC GAA GAC GCT TTC TTC ATT TCT ACT AAA GAC ACC TCC TTG TCC ACC TCT GAG GCT GGC GGT ATC CAG GAC TCA
Val Glu Asn Leu Ser Ala Asp Ala Phe Phe Ile Ser Thr Lys Asp Thr Ser Leu Ser Thr Ser Glu Ala Gly Gly Ile Gln Asp Ser -41

TTC AAC ATC GAT AAT CTT TTC TCC GGT TAC ATC GGT TAC TTC ACC TAC GTA GAC TTG ATA CGT CAA AAC CCA TTG AAC GAC TTT
Phe Asn Ile Asp Asn Leu Phe Ser Gly Tyr Ile Gly Tyr Phe Thr Tyr Val Asp Leu Ile Arg Gln Asn Pro Leu Val Asp Phe -11

GTT GAG AGA GAC TCT ATT GTC GAG GCT ACA GAA TTT GAC ACT CAA AAT AGC GCC CCA TGG GGG TTG GCC CGT ATT TCC CAC AGA GAG CGC
Val Glu Arg Asp Ser Ile Val Glu Ala Thr⌐Glu Phe Asp Thr Gln Asn Ser Ala Pro Trp Gly Leu Ala Arg Ile Ser His Arg Glu Arg 20

                                                                                        ★
CTC AAC CTG GGG TCC TTC AAC AAG TAT CTC TAC GAT GAT GAT GCC GGT CGC GGT GTC ACG TCC TAT GTT ATT GAC ACG GGT GTC AAC ATC
Leu Asn Leu Gly Ser Phe Asn Lys Tyr Leu Tyr Asp Asp Asp Ala Gly Arg Gly Val Thr Ser Tyr Val Ile Asp Thr Gly Val Asn Ile 50

                                                                                        ★
AAC CAC AAG GAC TTC GAA AAG AGA GCC ATT TGG GGG AAA ACC ATC CCA CTT AAC GAC GAA GAT CTC GAC GGT AAC GGC CAC GGT ACC CAC
Asn His Lys Asp Phe Glu Lys Arg Ala Ile Trp Gly Lys Thr Ile Pro Leu Asn Asp Glu Asp Leu Asp Gly Asn Gly His Gly Thr His 80

                                                                                        ◆
TGT GCC GGT ACT ATC GCT TCC AAA CAC TAC GGT GTC GCT AAA AAT GCC AAC GTT GTT GCG GTG AAA GTC TTG GGA TCA AAG GGC TCT GGT
Cys Ala Gly Thr Ile Ala Ser Lys His Tyr Gly Val Ala Lys Asn Ala Asn Val Val Ala Val Lys Val Leu Arg Ser Asn Gly Ser Gly 110

⌢⌢
ACC ATG TCT GAT GTC GTC AAA GGT GTC GAA TAT GCC GCA AAG GCG CAC CAA AAA GAA GCC CAA GAA AAG AAA AAG GGG TTC AAA GGT TCC
Thr Met Ser Asp Val Val Lys Gly Val Glu Tyr Ala Ala Lys Ala His Gln Lys Glu Ala Gln Glu Lys Lys Lys Gly Phe Lys Gly Ser 140

ACA GCC AAT ATG TCG CTT GGT GGT GGC AAG TCC CCA GCT TTG GAC TTG GCT GTT AAT GCA GCC GTT GAA GTC GGT ATT CAC TTT GCC GTG
Thr Ala Asn Met Ser Leu Gly Gly Gly Lys Ser Pro Ala Leu Asp Leu Ala Val Asn Ala Ala Val Glu Val Gly Ile His Phe Ala Val 170

                                                                                        ◆
GCT GCT GGT AAC GAA AAC CAA GAC GCT TGT AAT ACC TCC CCA GCT TCT GCT GAC AAG GCC ATC ACC GTC GGA GGT TCC ACG TTG AGC GAT
Ala Ala Gly Asn Glu Asn Gln Asp Ala Cys Asn Thr Ser Pro Ala Ser Ala Asp Lys Ala Ile Thr Val Gly Ala Ser Thr Leu Ser Asp 200

                                                                    ⌢⌢
GAC AGA GCC TAC TTT TCC AAC TGG GGT AAG TGT GTC GAC GTT TTC GCC CCA GGT TTA AAT ATT TCC ACC TAC ATT GGA AGT GAT GAT GAC
Asp Arg Ala Tyr Phe Ser Asn Trp Gly Lys Cys Val Asp Val Phe Ala Pro Gly Leu Asn Ile Leu Ser Thr Tyr Ile Gly Ser Asp Asp 230

                        ★   ⌢⌢
GCC ACC GCC ACT TTA TCG GGT ACC TCA ATG GCT TCC CCT CAC GTT GCT GGT TTG TTG ACC TAC TTT TTG TCT TTG CAA CCA GGT TCT GAT
Ala Thr Ala Thr Leu Ser Gly Thr Ser Met Ala Ser Pro His Val Ala Gly Leu Leu Thr Tyr Phe Leu Ser Leu Gln Pro Gly Ser Asp 260

AGT GAA TTT TTC GAA TTG GGC CAG GAT TCT TTG ACT CCT CAG CAA TTG AAA AAG AAG CTA ATT CAT TAC AGT ACA AAG GAC ATT TTG TTC
Ser Glu Phe Phe Glu Leu Gly Gln Asp Ser Leu Thr Pro Gln Gln Leu Lys Lys Lys Leu Ile His Tyr Ser Thr Lys Asp Ile Leu Phe 290

                                                                                ◆
GAT ATC CCT GAA GAC ACT CCA AAT GTT TTA ATC TAC AAC GGT GGT GGT CAA GAT TTG TCC GCT TTC TGG AAT GAT ACC AAA AAG TCC CAT
Asp Ile Pro Glu Asp Thr Pro Asn Val Leu Ile Tyr Asn Gly Gly Gly Gln Asp Leu Ser Ala Phe Trp Asn Asp Thr Lys Lys Ser His 320

TCA TCT GGT TTC AAG CAA GAA TTG AAC ATG GAT GAA TTC ATT GGA TCA AAG ACC GAT TTA ATC TTT GAT CAA GTG AGA GAT ATT CTT GAT
Ser Ser Gly Phe Lys Gln Glu Leu Asn Met Asp Glu Phe Ile Gly Ser Lys Thr Asp Leu Ile Phe Asp Gln Val Arg Asp Ile Leu Asp 350

AAA TTG AAT ATT ATT TAA
Lys Leu Asn Ile Ile

TTCTTCATTT AGAAAAATTT CAGCTGCTTT TTTTTTTCTT TTTCTTTCCT TAGGCGTCTC GAGGTTACAA GTCGGAGTCC CTCTTCACTA TCGTTTGTCC ACTTTTTTTA
TATCCCCATT ATTTTCAATC TGAATTTCAT TTTTTTTTTT TAATTCATGA AATTTATATG TCCCACGTAT TACTACATAT TTGCGTTTTT AATTAAATAA ATAACTGTTA
CTTTTATTAT ATCTTATTTG CAGATCACTT ATCTGATCAA ATGTTTTCGT TTTCGTGTGT GGTGACGATG TATTAGGTAC GCGAAATAAA CAAAACAAAC AAACAAGGCC

Figure 5.2 The structural gene of proteinase yscB (*PRB1*) (Moehle *et al.*, 1987*b*).

1978*a*, *b*; Trimble and Maley, 1977; Hashimoto *et al.*, 1981; Trimble *et al.*, 1983). The fourth chain is phosphate-free and has an average composition of GlcNAc$_2$Man$_{10}$. *In vivo* studies revealed that carboxypeptidase yscY is synthesized as a precursor of two different species varying only in their carbohydrate content. The species P$_1$ of M_r 67-kDa appears in the endoplasmic reticulum, whereas the 69-kDa species P$_2$ is found in the Golgi apparatus. The final maturation step consists of scission of an N-terminal peptide yielding the 61-kDa protein (Hasilik and Tanner, 1978*a*, *b*; Trimble *et al.*, 1983; Hemmings *et al.*, 1981; Stevens *et al.*, 1982; Müller and Müller, 1981).

The enzyme exhibits a broad specificity against peptides, aminoacyl esters, amides and the chromogenic substrate benzoyl-Tyr-4-nitroanilide. It also attacks proteins (Doi *et al.*, 1967; Hayashi, 1976). The enzyme is strongly inhibited by the serine residue modifying agents, diisopropyl-fluorophosphate

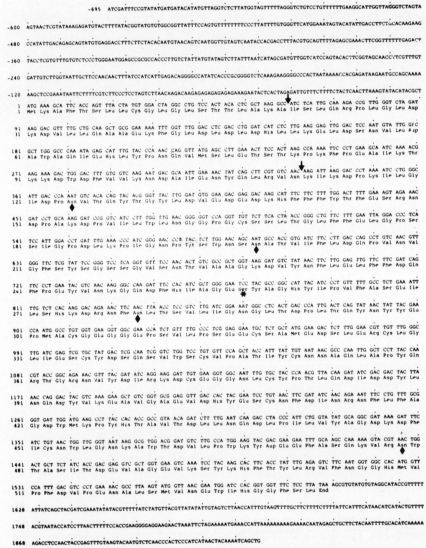

Figure 5.3 The structural gene of carboxypeptidase yscY (*PRC1*) (Rothman *et al.*, 1986).

and phenylmethylsulphonyl fluoride. Benzyloxy-carbonyl-L-phenylalanine chloromethyl ketone inactivates the enzyme by alkylating a single histidine residue necessary for activity (Hayashi *et al.*, 1975). Mercurials inhibit carboxypeptidase yscY activity by binding to a single SH-group, which is, however, not part of the active centre. A specific polypeptide inhibitor (I^C) of carboxypeptidase yscY, of M_r about 25-kDa, is located in the cytoplasm of the yeast cell (Lenney, 1975; Matern *et al.*, 1974*b*). Activity levels of carboxypeptidase yscY increase several-fold *in vivo*, when cells enter diauxic growth and stationary phase, grow on acetate, grow on a peptide, which is a substrate for the enzyme, or are transferred on to nitrogen-free medium (Hansen *et al.*, 1977; Klar and Halvorson, 1975; Wolf and Ehmann, 1978*a*).

The structural gene for carboxypeptidase yscY is *PRC1* (Wolf and Fink, 1975; Wolf and Weiser, 1977); *prc1* mutants are devoid of carboxypeptidase yscY activity but have normal levels of other vacuolar hydrolases. The structural gene for carboxypeptidase yscY has been cloned by complementation of the *prc1* mutation, and was found to encode a precursor of 532 amino acids, 111 amino acids larger than the mature vacuolar protein (Figure 5.3). Zymogen cleavage occurs at the $Asn_{111}-Lys_{112}$ bond (Stevens *et al.*, 1986; Valls *et al.*, 1987).

5.2.4 *Carboxypeptidase yscS*

In a mutant lacking carboxypeptidase yscY a second vacuolar carboxypeptidase, called carboxypeptidase yscS, was detected. The enzyme cleaves benzyloxy-carbonyl-glycyl-leucine with high efficiency. The metal ion chelating agent EDTA is a potent inhibitor of carboxypeptidase yscS, indicating that the enzyme is a metal ion-dependent peptidase. Zn^{2+} ions completely restore activity. Activity levels of carboxypeptidase yscS are regulated by the nitrogen source. Highest activity can be achieved when cells are grown on benzyloxy-carbonyl-glycyl-leucine as sole nitrogen source (Wolf and Ehmann, 1978*a*; Wolf and Weiser, 1977; Emter and Wolf, 1984).

The putative structural gene for carboxypeptidase yscS is *CPS1; cps1* mutants are deficient in carboxypeptidase yscS activity and they show normal levels of other vacuolar peptidases (Wolf and Ehmann, 1981).

5.2.5 *Aminopeptidase yscI*

The vacuolar aminopeptidase yscI is a soluble metallo-exopeptidase of M_r of 640-kDa (Frey and Röhm, 1978; Matile *et al.*, 1971). Aminopeptidase yscI activity can be measured in the slightly basic pH range using Leucine-4-nitroanilide or Alanine-4-nitroanilide. The enzyme is also active on a variety of peptides such as Leu-Gly, Leu-Leu, Leu-Val, Leu-Gly-Gly, Ala-Thr-Ala and leucine-amide. It is strongly inhibited by metal chelating agents (Frey and Röhm, 1978).

Chloride ions are allosteric activators of the enzyme. The structural gene for aminopeptidase yscI is probably *LAP4; lap4* mutants specifically lack this vacuolar aminopeptidase. The *LAP4* gene was located on chromosome XI. (Hetz and Röhm, 1987; Trumbly and Bradley, 1983).

5.2.6 *Aminopeptidase yscCo*

Aminopeptidase yscCo is a soluble metallo-exopeptidase of M_r of about 100 kDa. The enzyme is active only in the presence of small amounts of Co^{2+} ions. The enzyme cleaves Arginine-4-nitroanilide and Lysine-4-nitroanilide with high efficiency. pH optimum is in the moderately alkaline range. The enzyme is strongly inhibited by EDTA and Zn^{2+} ions. The actinomycete inhibitor bestatin, which inhibits most yeast aminopeptidases tested so far, is not effective on aminopeptidase yscCo. Aminopeptidase yscCo activity dramatically increases when cells reach stationary growth phase (Emter and Wolf, 1984; Achstetter *et al.*, 1982, 1983, 1984).

5.2.7 *Dipeptidyl aminopeptidase yscV*

Dipeptidyl aminopeptidase yscV is an expopeptidase associated with the vacuolar membrane. The enzyme cleaves Alanyl-prolyl-4-nitroanilide at the 4-nitroanilide bond in the neutral pH range. In general, the enzyme specifically hydrolyses peptide bonds involving the carboxyl group of prolyl residues penultimate to unprotected amino termini, unless arginine is the N-terminal amino acid. The enzyme is strongly inhibited by phenylmethylsulphonyl fluoride, indicating a serine residue in the active centre. Heavy metal ions were also found to inhibit the enzyme (Bordallo *et al.*, 1984; Garcia Alvarez *et al.*, 1985).

The structural gene for dipeptidyl aminopeptidase yscV is *DAP2*. A gene called *DPP2*, potentially encoding this vacuolar membrane protein, was isolated on the basis of its ability to suppress the mating defect of a *MATα ste13* mutant at a high copy number. Allelism tests performed with a strain harbouring a deletion in the genomic copy of *DPP2* and *dap2* mutant showed that *DPP2* indeed encodes dipeptidyl aminopeptidase yscV (Suarez Rendueles and Wolf, 1987; Pohlig *et al.*, 1986; Julius *et al.*, 1983a).

5.2.8 *Other vacuolar proteinases*

Cell fractionation studies using a mutant devoid of four vacuolar enzymes, the two endoproteinases yscA and yscB and the two carboxypeptidases yscY and yscS, uncovered additional proteolytic activity in the vacuole. Most pronounced activity was found against the two chromogenic peptide substrates benzoyl-Ile-Glu-Gly-Arg-4-nitroanilide and tosyl-Gly-Pro-Arg-4-nitroanilide splitting at a site other than the 4-nitroanilide bond (Emter and Wolf, 1984). The number of proteinases represented by this activity is unknown.

5.3 Mitochondrial proteinases

5.3.1 Matrix-located proteinase–proteinase yscMpI

The matrix-located proteinase, for reasons of a unifying nomenclature here called proteinase yscMpI (Mp, mitochondria (precursor) processing), is a soluble metallo-endopeptidase of M_r about 115-kDa. The enzyme is active only against precursor forms of nuclear coded proteins which are imported into the mitochondrion. Precursors of proteins destined for the mitochondrial matrix, inner membrane and intermembrane space are cleaved. The proteinase activity has a pH optimum in the neutral range and is inhibited by the metal chelators 1, 10-phenanthroline and EDTA. Zn^{2+}, Co^{2+}, and Mn^{2+} ions restore the enzyme activity (McAda and Douglas, 1982; Böhni et al., 1983; Cerletti et al., 1983).

Using a variety of fluorogenic synthetic substrates, endopeptidase and aminopeptidase activities were detected to be associated with the mitochondrial matrix fraction (Yasuhara and Ohashi, 1987). The activities were assayed at pH 7.5 and shown to be sensitive to o-phenanthroline but insensitive to iodoacetate and phenylmethylsulphonyl fluoride. One of the endopeptidase activities preferentially cleaves at the amino side of paired basic amino acid residues in the peptides benzyloxy-carbonyl-Ala-Arg-Arg-4-methylcoumaryl-7-amide and butoxy-carbonyl-Gln-Arg-Arg-4-methylcoumaryl-7-amide. Another endopeptidase could be assayed directly with succinyl -Ala-Pro-Ala-4-methylcoumaryl-7-amide. The aminopeptidases found were active against Arg-(4-methylcoumaryl-7-amide) and Leu-(4-methylcoumaryl-7-amide), and were inhibited by bestatin (Yasuhara and Ohashi, 1987).

5.3.2 Other mitochondrial proteinases

Three proteolytic activities of M_r of about 17-kDa located in the inner mitochondrial membrane and active against cytochrome c as substrate have been described (Zubatov et al., 1984). Hydrolytic activity against cytochrome c is sensitive to inhibition by phenylmethylsulphonyl fluoride, p-chloromercuriphenyl sulphonate and leupeptin. In addition, the presence of three proteins active against benzoyl-DL-Arginine-β-naphthylamide and benzoyl-DL-Arginine-4-nitroanilide was reported. Whether one or more activities can be indentified with the above described matrix-associated peptidases remains to be determined.

5.4 Periplasmic space proteinases

5.4.1 Aminopeptidase yscII

Aminopeptidase yscII is a soluble metallo-exopeptidase most active in the neutral pH range (Frey and Röhm, 1978, 1979b). Forty percent of aminopeptidase yscII activity was detected on intact cells, suggesting an external location,

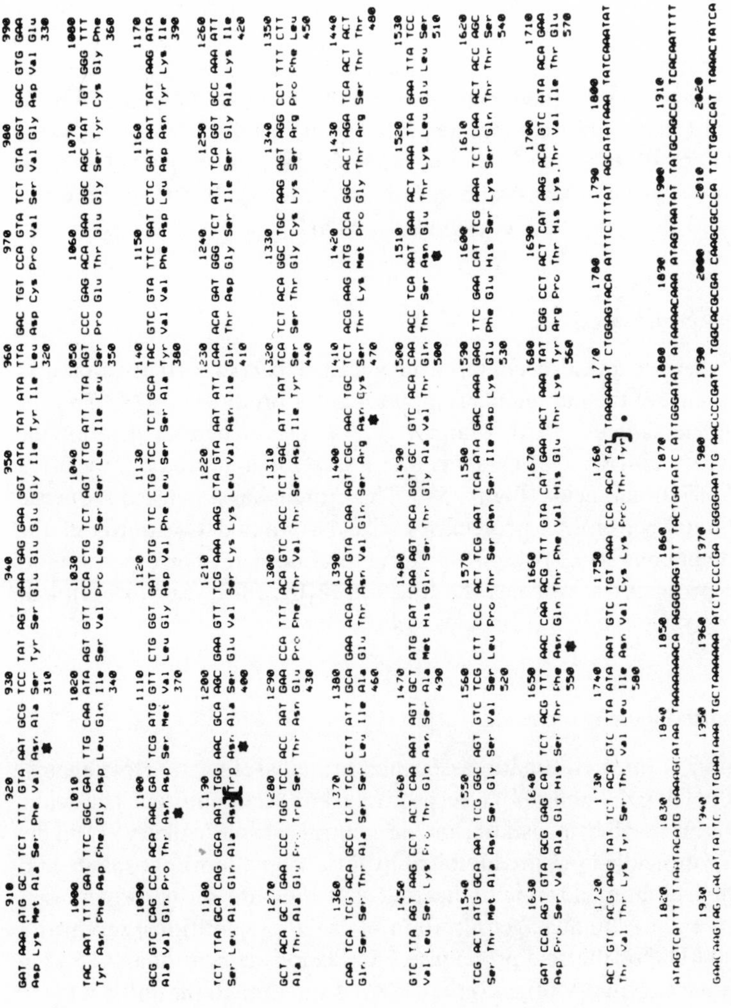

Figure 5.4 The structural gene of 'barrier activity' (BAR1) (MacKay et al., 1988).

F

the residual amount being located internally. The M_r of the enzyme protein was reported to range from 85 to 140-kDa, the differences being attributed to varying degrees of glycosylation. The enzyme activity can be assayed using a variety of different aminoacyl-4-nitroanilides, aminoacyl-β-naphthylamides, dipeptides, tripeptides and tetrapeptides. The enzyme is strongly inhibited by EDTA, bestatin and Zn^{2+} and Hg^{2+} ions at high concentrations (Frey and Röhm, 1978b, 1979, Knüver, 1979; Achstetter et al., 1983; Hirsch et al., 1988).

The structural gene for aminopeptidase yscII is APE2, and ape2 mutants selected on the basis of their inability to hydrolyse Lys-β-naphthylamide externally also lack internal aminopeptidase yscII activity. The APE2 gene was located on chromosome XI at a position similar to the reported locus of LAPI, indicating that APE2 and LAPI are most likely the same genes (Trumbly and Bradley, 1983; Hirsch et al., 1988).

5.5 Secreted proteinases

MATa cells secrete an extracellular protein, called 'barrier' activity, that acts as an antagonist of the mating pheromone α-factor produced by MATα cells (Hicks and Herskowitz, 1976; Manney, 1983). The structural gene of the barrier activity (BARI, chromosome IX) codes for a primary translation product of 587 amino acids (Figure 5.4). The protein has a marked sequence homology with pepsin-like proteinases. The N-terminal two-thirds of the barrier protein consists of two distinct structural domains corresponding to pepsin-like proteinases, whereas the function of the third, carboxyterminal domain is as yet unknown (MacKay et al., 1988).

5.6 Chromatin-bound proteinases

The detection of three chromatin-associated proteinases most active against histones at pH 4, pH 7 and pH 11, respectively, has been reported. The acidic proteinase is inhibited by pepstatin, and the neutral enzyme (called yB) and the alkaline enzyme (called yC) are inhibited by diisopropylfluorophosphate and phenylmethylsulphonylfluoride, indicating that they are serine proteinases. Molecular weight estimates (gel filtration of crude preparations) revealed a value of 100-kDa for the acid proteinase, 35-kDa for the neutral and 25-kDa for the alkaline enzyme (Motizuki et al., 1986). Even though the authors try to rule out the possibility that the neutral and the alkaline enzyme are vacuolar contaminations (no inhibition of yB by chymostatin, M_r of yC very different from carboxypeptidase yscY), their statement that the acidic proteinase might be identical with proteinase yscA leaves doubts. This matter needs to be investigated more thoroughly.

5.7 Proteinases of unknown localization—soluble proteinases

5.7.1 Proteinase yscD

Proteinase yscD is a metallo-sulphydryl endopeptidase. It consists of a single polypeptide chain of M_r of about 83 kDa. The enzyme is not localized in the vacuole of the yeast cell. Proteinase yscD cleaves the Pro-Phe bond of the synthetic peptide substrate benzoyl-Pro-Phe-Arg-4-nitroanilide and the Ala-Ala-bond of acetyl-Ala-Ala-Pro-Ala-4-nitroanilide, acetyl-Ala-Ala-Pro-Phe-4-nitroanilide and methoxysuccinyl-Ala-Ala-Pro-Met-4-nitroanilide with high efficiency. pH optima for cleavage range from slightly acidic to neutral. Mercurials and EDTA are potent inhibitors of proteinase yscD (Emter and Wolf, 1984; Achstetter *et al.*, 1985).

The structural gene for proteinase yscD is *PRD1*. Strains carrying the *prd1* mutation have lost the ability to split the Pro-Phe as well as the Ala-Ala bonds of the above mentioned substrates (Garcia Alvarez *et al.*, 1987).

5.7.2 Proteinase yscE

Proteinase yscE is a sulphydryl endopeptidase of M_r about 600-kDa. The enzyme seems to be composed of one type of subunit of molecular mass 70-kDa, and a series of 10–12 subunits of between 38 and 22 kDa. The enzyme is not located in the vacuole of the yeast cell. Activity can be assayed with the synthetic peptide substrates benzyloxycarbonyl-Gly-Gly-Leu-4-nitroanilide or succinyl-Phe-Leu-Phe-4-nitroanilide in the moderate alkaline pH range of 8–8.5. Cleavage occurs at the 4-nitroanilide bond. The enzyme exhibits a small activity against ^{3}H-methylcasein. Mercurials were found to be strong inhibitors of the enzyme activity (Achstetter *et al.*, 1984a; Emter and Wolf, 1984).

5.7.3 Other soluble proteinases

In addition to the well-characterized proteinases yscD and yscE described above, a variety of additional soluble endopeptidases has been found, but these enzymes have not yet been extensively characterized.

As with proteinases yscD and yscE, detection of a variety of proteinases was only possible in strains lacking the vacuolar endoproteinases and carboxypeptidases of broad specificity. In addition, these strains enabled the detection of three distinct ^{3}H-methylcasein-splitting activities which together comprise less than 1% of the total ^{3}H-methylcasein-splitting activity detectable in wild-type cells. One of these activities, proteinase yscK, can be activated by ATP as well as polyvalent-ions (Achstetter *et al.*, 1981; 1984b).

An endopeptidase of M_r 43-kDa, able to cleave the Leu-enkephaline derivative Tyr-Gly-Gly-Phe-Leu-Arg-Arg-Phe-Ala-NH$_2$ between the Arg-Arg pair at an optimum pH in the neutral range, has been reported. Due to its specificity cleaving at pairs of basic amino acid residues, the signals of pro-hormone processing, the enzyme was called pro-pheromone convertase Y.

Activity is strongly inhibited by the serine residue modifying agents diiso-propylfluorophosphate and phenylmethylsulphonyl fluoride (Mizuno and Matsuo, 1984).

In addition to the vacuolar aminopeptidases yscI and yscCo and the periplasmic aminopeptidase yscII a variety of soluble aminopeptidase activities have been found (Masuda et al., 1975; Frey and Röhm, 1978; Achstetter et al., 1983; Trumbly and Bradley 1983). An enzyme activity named aminopeptidase II of either 30-kDa (Frey and Röhm, 1978) or 34-kDa (Achstetter et al., 1983) molecular mass has been described. The different substrate specificity reported, however, may point to two different enzymes. Four leucine aminopeptidases (LAPI to LAPIV) have been described, previously. LAPIV can be identified with certainty as one of the known aminopeptidases, the vacuolar aminopeptidase yscI. LAPI is most likely identical with aminopeptidase yscII (Hirsch et al., 1988). LAPI and LAPII are metallo-exopeptidases of M_r 85-kDa, splitting leucine- and lysine-bound organic chromophores and are subject to EDTA inhibition; LAPIII, a protein of 95-kDa molecular weight is not inhibited by EDTA. LAP3 was located on chromosome XIV (Trumbly and Bradley, 1983).

LAPII and LAPIII are most probably among the additional eleven aminopeptidase activities (aminopeptidase yscIV to aminopeptidase yscXIV) that have been detected by Achstetter et al. (1983). Some of these activities have been characterized to some extent. Aminopeptidase yscV is a sulphydryl exopeptidase of about 170-kDa molecular mass, is sensitive to Hg^{2+} ions and insensitive to nitrilotriacetic acid. Aminopeptidase yscVIII is a metallo-exopeptidase of about 165-kDa molecular mass, insensitive to Hg^{2+} ions, but strongly inhibited by nitrilotriacetic acid. Aminopeptidase yscXI is a sulphydryl metallo-exopeptidase of about 110-kDa molecular mass, sensitive to both Hg^{2+} ions and nitrilotriacetic acid. Aminopeptidase yscXII is a sulphydryl metallo-exopeptidase of about 97-kDa molecular mass, more sensitive to Hg^{2+} ions than to nitrilotriacetic acid. Aminopeptidase yscXII is present preferentially in exponentially growing cells, whereas aminopeptidase yscXI is present in both growing and stationary-phase cells. Evidence from the characterization of remaining aminopeptidase activities in ape2 mutants suggests that LAPII and aminopeptidase yscXII are most likely identical activities. All these characterized aminopeptidases are sensitive to the actinomycete inhibitor bestatin (Achstetter et al., 1983; Hirsch et al., 1988).

Three additional dipeptidyl aminopeptidases have also been described: dipeptidyl aminopeptidase yscI, dipeptidyl aminopeptidase yscII and dipeptidyl aminopeptidase yscIII (Achstetter et al., 1983). Dipeptidyl aminopeptidase yscII has been characterized in some detail. It is a metallo-exopeptidase of about 78-kDa molecular mass, active at a slightly acidic pH. It cleaves the chromogenic peptide Val-Ala-4-nitroanilide. The enzyme is sensitive to EDTA and Zn^{2+} ions.

A dipeptidase of M_r 120-kDa, dipeptidase yscI, has been identified. Its activity is dependent on metal ions. The enzyme cleaves only dipeptides composed of L-amino acids, Gly-Leu, Gly-Gly and Ala-Ile being among the most efficient substrates (Röhm, 1974; Frey and Röhm, 1978).

Three additional carboxypeptidases have been detected in a mutant strain lacking the activities of carboxypeptidase yscY and carboxypeptidase yscS. Carboxypeptidase yscγ is a metallo-exopeptidase splitting benzyloxycarbonyl-Ser-Phe; carboxypeptidase yscδ is a high-molecular-mass serine exopeptidase hydrolysing benzyloxycarbonyl-Ala-Phe; and carboxypeptidase yscε is a low-molecular-mass serine exopeptidase splitting benzyloxycarbonyl-Ala-Phe as well (Wolf and Ehmann, 1981).

5.8 Membrane-bound proteinases

The enzymes described in this section have been identified in the whole membrane fraction of yeast cells (for references see below).

5.8.1 Proteinase yscF
Proteinase yscF, the product of the *KEX2* gene (chromosome XIV), is a membrane-bound endopeptidase. Low concentrations of Triton X-100 (1%) activate the enzyme. Proteinase yscF can be solubilized by Triton X-100 (1%) or Zwittergent 3–14 (0.5%). Dithiothreitol leads to enzyme inactivation. The enzyme is active in the neutral pH range and completely dependent on Ca^{2+} ions. No other ion can substitute for Ca^{2+}. The metal chelating agents EGTA and EDTA, but not 1, 10-phenanthroline, inhibit the enzyme strongly. Activity of proteinase yscF is also dependent on intact sulphydryls. The activity is abolished by heavy metals such as Zn^{2+} and Hg^{2+} ions as well as by iodoacetate, iodoacetamide and N-ethyl-male imide. The enzyme is not inhibited by tosyl-lysyl-chloromethylketone but is sensitive to Ala-Lys-Arg-chloromethylketone. High concentrations of phenylmethylsulphonyl fluoride do not affect enzyme activity but high concentrations of diisopropylfluorophosphate are inhibitory. The enzyme contains a domain with significant homology to a class of bacterial serine proteinases, the subtilisins, indicating that proteinase yscF might be a novel type of serine proteinase. The enzyme has a molecular weight of about 135–150-kDa. Proteinase yscF has been identified by its ability to cleave synthetic chromogenic peptide substrates at the C-terminal side of a pair of basic amino acid residues: benzyloxycarbonyl-Tyr-Lys-Arg-4-nitroanilide and butoxycarbonyl-Gln-Arg-Arg-7-amido-4-methylcoumarin. Chromogenic peptides containing only a pair of basic amino acid residues as total amino acid sequence or typical trypsin substrates containing only one basic amino acid at the cleavage site are not recognized by proteinase yscF, pointing to very specific requirements for substrate recognition (Julius *et al.*, 1984a; Achstetter and Wolf, 1985; Fuller *et al.*, 1985, 1988; Wagner *et al.*, 1987).

The *KEX2* gene has been cloned by complementation of the *kex2* mutation. The nucleotide sequence of the *KEX2* gene predicts a primary sequence of 814 amino acids with M_r for the peptide chain of 91 kDa. The nucleotide sequence predicts a hydrophobic aminoterminal signal sequence and a second hydrophobic sequence nearer to the carboxy terminus, which most likely represents the membrane spanning domain. The enzyme has several sites for N- and O-glycosylation. Localization of proteinase yscF has to be assumed in the late Golgi- and (or) in secretory vesicle membranes (Fuller *et al.*, 1985; 1987).

5.8.2 *Carboxypeptidase yscα*

Carboxypeptidase yscα has been found in a mutant strain lacking the two vacuolar carboxypeptidases yscY and yscS. It is a membrane-associated exopeptidase most active between pH 6 and 7.5, with high specificity towards C-terminal basic amino acid residues of synthetic peptides such as benzyloxycarbonyl-Tyr-Lys-Arg and benzyloxycarbonyl-Tyr-Lys (Achstetter and Wolf, 1985; Wagner *et al.*, 1987).

The enzyme is strongly inhibited by phenylmethylsulphonyl fluoride and micromolar amounts of mercury-containing compounds. The inhibition pattern of carboxypeptidase yscα strongly resembles that of the vacuolar carboxypeptidase yscY. A membrane-associated peptidase with high homology with carboxypeptidase yscY has been inferred from the sequence of the *KEX1* gene (Figure 5.5). Carboxypeptidase yscα activity is missing in *kex1* mutant cells. These facts together indicate that *KEX1* is the structural gene of carboxypeptidase yscα (Dmochowska *et al.*, 1987; Wagner and Wolf, 1987; Bussey, 1988).

5.8.3 *Dipeptidyl aminopeptidase yscIV*

Membrane-bound dipeptidyl aminopeptidase activity had been first described by Suárez Rendueles *et al.* (1981). It was later found that this activity was due to two enzymes differing in their sensitivity to heat, the heat-resistant enzyme called dipeptidyl aminopeptidase A and the heat-labile enzyme called dipeptidyl aminopeptidase B. The heat labile enzyme was identified as the vacuolar dipeptidyl aminopeptidase yscV (see section on dipeptidyl aminopeptidase yscV). Dipeptidyl aminopeptidase A (for reasons of consistency with the yeast proteinase nomenclature this is called dipeptidyl aminopeptidase yscIV) is a serine exopeptidase active in the neutral pH range. The activity can be assayed using the chromogenic peptide substrate Ala-Pro-4-nitroanilide. The enzyme is inhibited by phenylmethylsulphonyl fluoride. It is remarkably heat stable at 60°C. (Achstetter *et al.*, 1983; Julius *et al.*, 1983a; Bordallo *et al.*, 1984; Garcia Alvarez *et al.*, 1985; Suarez Rendueles and Wolf, 1987).

Strains carrying mutations in the *STE13* gene have low levels of dipeptidyl aminopeptidase yscIV. The structural gene is *STE13*, for it is a

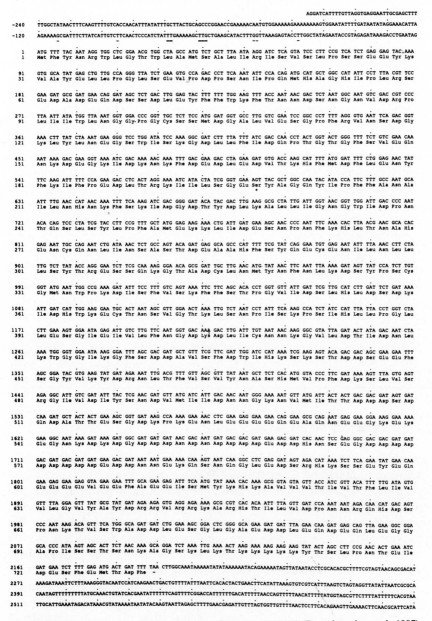

Figure 5.5 The structural gene of carboxypeptidase yscα (*KEX1*) (Dmochowska *et al.*, 1987).

dosage-sensitive locus. The *STE13* gene was cloned. The nucleotide sequence predicts a polypeptide of 931 amino acid residues (Fuller *et al.*, 1988).

5.8.4 *Other membrane-bound proteinases*

In addition to proteinase yscF, two membrane-bound metallo-endopeptidases have been found – proteinase yscG and proteinase yscH. Proteinase yscG has been detected by its ability to split the synthetic peptide substrate benzyloxycarbonyl-Ala-Ala-Leu-4-nitroanilide at a bond different from the 4-nitroanilide bond. The enzyme is inhibited by EDTA and is sensitive to treatment with the reducing agent dithiothreitol (T. Achstetter and D.H. Wolf, unpublished observations). In contrast to proteinase yscF, Hg^{2+} ions do not inhibit the enzyme. Proteinase yscH can be assayed with the synthetic peptide substrate acetyl-Ala-Ala-Pro-Ala-4-nitroanilide which is cleaved at a bond different from the 4-nitroanilide bond. The enzyme is inhibited by EDTA, but is insensitive to Hg^{2+} ions as well as to treatment with dithiothreitol (Achstetter and Wolf, 1985; and unpublished observations).

Aminopeptidase yscP is a metallo-exopeptidase active in the neutral pH range. The enzyme can be assayed using the chromogenic peptide substrate Leu-4-nitroanilide. Chelating agents and the actinomycete inhibitor bestatin are potent inhibitors of aminopeptidase yscP (Achstetter *et al.*, 1983; 1984*b*).

Ion exchange chromatography of the salt wash of a membrane fraction containing mostly peripheral and low concentrations of soluble proteins revealed the existence of a unique aminopeptidase activity named aminopeptidase yscXV. The activity is able to cleave only Leu-4-nitroanilide from all aminoacyl 4-nitroanilides tested including Lys-4-nitroanilide, Phe-4-nitroanilide and Ala-4-nitroanilide (T. Achstetter and D.H. Wolf, unpublished results). This enzyme is detectable only in exponentially growing cells, and is active in the neutral pH range.

A benzyloxycarbonyl-Phe-Leu splitting membrane associated carboxy-peptidase (preliminarily called carboxypeptidase yscλ) was found which is sensitive to inhibition by mercurials (Wagner *et al.*, 1987).

Earlier observations ascribed two endoproteolytic activities named proteinase M and proteinase P to a membrane fraction in yeast. Later results, obtained with rigorously salt-washed membranes, could not confirm this location. Thus, these activities represent soluble proteins which are contained in the membrane fraction previously obtained and which, under conditions of higher ionic strength, were completely removed (Achstetter *et al.*, 1981; 1984*b*).

Recently, a Leu-Lys cleaving endopeptidase was detected in the medium after cross-linking *MAT*α yeast cells. The enzyme cleaves α-factor pheromone between Leu^6-Lys^7 at a pH optimum of 4.0, yielding two distinct peptides. A similar enzyme in *MAT*a cells has been reported by other groups. As suggested by the substrate specificity, the activity shares features of the above-mentioned proteinase yscP (Maness and Edelmann, 1978; Ciejek and Thorner, 1979; Finkelstein and Strausberg, 1979; Okada *et al.*, 1987).

5.9 Proteolysis and Function

The specificity of a proteinase is assumed to be closely related to its physiological function, as is its cellular location or its mode of expression. Generally, proteinases of rather broad specificity should be involved in degradation of substrate protein down to the amino acid level. Therefore, non-specific proteinases have to be extremely well-controlled in order to protect the cell from random degradation of other than correct target proteins. This may be achieved by restriction of non-specific proteinases into secluded cellular compartments such as the lysosome or the vacuole. It appears necessary as well as conceivable that control elements exist which select dispensable proteins in accordance with cellular needs and initiate their degradation in the lysosomal compartment.

Alternatively, non-lysosomal proteinases of broad cleavage specificity should be controlled by the requirement for an additional and selective signal to indicate a correct substrate protein. In this case too, a machinery detecting and marking the respective substrate proteins should exist. Taken together, proteolysis carried out by non-specific proteinases leading to degradation of proteins in the cell is by no means non-selective. Here the function of recognition and selection, however, resides in a separate entity rather than the one arising from the immediate proteinase–substrate interaction.

In contrast, the selectivity of limited proteolysis appears to reside more directly in the proteinase–substrate interaction. Limited proteolysis as an activation, inactivation or modification event may derive its selectivity from the proteolytic enzyme which recognizes only specific amino acid target sequences. On the other hand, the selectivity of limited proteolysis may come about by the selective exposure of a 'processing site' under certain conditions, such as those at pH, ionic strength, or secondary modifications, thus allowing an otherwise non-specific proteinase to catalyze a highly specific event. The maturation of vacuolar hydrolases in *Saccharomyces cerevisiae* gives an example of the latter kind.

The elucidation of cellular functions involving a biochemically character-ized proteinase is almost impossible unless at least a hint is given by the location, the regulation or the substrate specificity of the enzyme under investigation. On the other hand, certain cell functions may be known to involve proteolytic events for which, however, the respective catalysts remain to be found. An unequivocal answer to the involvement of a certain proteinase in a certain cellular event is, however, possible only when the catalyst of the reaction is switched off *in vivo* and the resulting alterations are analysed. Thus, the biochemical and molecular analysis is the ultimate tool to provide the answers about proteinase function and cellular control.

Protein degradation is a process known to occur in growing and stationary-phase yeast cells. It is well known that metabolic conditions greatly influence the degradation rate. Degradation rates range from 0.5–1% per hour when cells grow on glucose, 2% per hour when cells grow on ethanol, 2–3% per hour when cells starve for nitrogen, to 3% per hour when cells starve for carbon at low phosphate concentrations. Concerning protein degradation in growing cells, two classes of proteins can be distinguished: the bulk of the yeast protein being slowly degraded ($t_{1/2}$ = 160 h) and a small portion of the protein being degraded rather quickly ($t_{1/2}$ = 0.8–2.4 h) (Halvorson, 1958a, b; Bakalkin et al., 1976; Betz, 1976; López and Gancedo, 1979; Gancedo et al., 1982). Most of the evidence available indicates that in higher eukaryotes the degradation of the long-lived proteins which occurs under a variety of metabolic conditions takes place in the lysosome (Hershko and Ciechanover, 1982). This seems to be the situation also in yeast under sporulation conditions where vacuolar proteinases are responsible for the bulk of protein degradation (Wolf and Ehmann, 1979; Zubenko et al., 1979; Mechler and Wolf, 1981; Zubenko and Jones, 1981; Teichert et al., 1987) (see section on function of vacuolar proteinases). In contrast, selective turnover of intracellular proteins in the cytoplasm under normal metabolic conditions is ATP-dependent and nonlysosomal in higher eukaryotes (Hershko and Ciechanover, 1982; Hershko, 1983; Finley and Varshavsky, 1985; Pontremoli and Melloni, 1986).

Recent biochemical and genetic evidence indicates that covalent conjugation of ubiquitin to short-lived intracellular proteins is essential for their selective degradation. Thus, ubiquitination might be one signal triggering degradation by proteinases which recognize their protein substrates once conjugated to ubiquitin. On the other hand, it cannot be excluded that recognition for degradation of a protein occurs upon exposure of a signal different from ubiquitin, and that ubiquitin has its function in perturbing the protein structure, by destabilizing it and thereby facilitating degradation (Goldstein et al., 1975; Ciechanover et al., 1984; Finley et al., 1984).

By analogy to the signal sequences that confer on proteins the ability to enter distinct cellular compartments, it has been suggested that proteins might contain sites of specific amino acid sequences that alone or in combination would act to determine the half-life of each protein in vivo. In fact, by means of site-directed mutagenesis, it has recently been possible to expose different amino acid residues at the amino-termini of an otherwise identical protein (β-galactosidase) produced in vivo in Saccharomyces cerevisiae. Strikingly, depending on the nature of the N-terminal residue, the resulting β-galactosidase proteins show a range of half-lives from more than 20 hours to less than three minutes. From these results it has been concluded that the set of individual amino acids can be ordered with respect to the half-lives that they confer on a non-compartmentalized intracellular protein when exposed at its amino-terminus. Protein-stabilizing residues are Met, Ser, Ala, Thr, Val and Gly, whereas proteins with either Arg, Lys, Phe, Leu or Asp at the amino-terminus have very short half-lives. Interestingly enough, the unblocked amino-terminal residues found in metabolically stable, non-compartmentalized proteins from both prokaryotes and eukaryotes are exclusively of the stabilizing class. In addition, it has been hypothesized that this mechanism could also account for the degradation of long-lived proteins. In this case, the rate-limiting step might be a slow stepwise aminopeptidase cleavage that eventually exposes a destabilizing residue at the amino-terminus of the protein, followed by rapid degradation (Blobel, 1980; Schatz and Butow, 1983; Rogers et al., 1986; Bachmair et al., 1986).

Inactivation of selected enzymes upon glucose addition to yeast cells growing on a non-fermentable carbon source, a process called catabolite inactivation (Holzer, 1976), might represent another example of unspecific proteolysis affecting proteins that are modified following a metabolic signal. The enzymes investigated most thoroughly are fructose-1,6-bisphosphatase (Gancedo, 1971), cytoplasmic malate dehydrogenase (Witt et al., 1966) and phosphoenolpyruvate carboxykinase (Haarasilta and Oura, 1975). Another process, carbon-starvation-induced inactivation of NADP-dependent glutamate dehydrogenase, was studied (Mazon, 1978). The inactivation of these enzymes was thought to be due to proteolysis as the inactivation processes were found to be irreversible and the enzyme proteins cross-reacting with specific antibodies disappeared at the same time as enzyme activities (Holzer 1976; Neeff et al., 1978; Mazon and Hemmings, 1979; Funayama et al., 1980; Müller et al., 1981).

It has been shown for fructose-1,6-bisphosphatase that catabolite inactivation is a two-step process: phosphorylation of the enzyme, which inactivates it partly (Müller and Holzer, 1981; Tortora et al., 1981; Purwin et al., 1982; Mazon et al., 1982a), is followed by proteolysis (Funayama et al., 1980). There is some indirect evidence implying that phosphoenolpyruvate carboxykinase and cytoplasmic malate dehydrogenase might also be marked by phosphorylation prior to proteolysis (Tortora et al., 1983).

In contrast to the above described proteolytic processes, the proteolytic events described below in the section on protein maturation and secretion of proteins are good examples of highly specific proteolytic events.

5.10 Cellular functions of known proteinases

The isolation of mutants lacking one or several proteolytic activities, as well as cloning some of their genes, has enabled the elucidation of the intracellular functions of a variety of the known proteinases.

5.10.1 Function of vacuolar proteinases: their role in protein degradation and protein maturation

Based on studies using mutant strains lacking proteinase yscA (pra1), proteinase yscB (prb1), carboxypeptidase yscY (prc1) and carboxypeptidase yscS (cps1), the major function of these enzymes was found to reside in the differentiation process of sporulation which is triggered by nitrogen starvation. Diploid MATa/MATα strains, homozygous for the absence of proteinase yscA or yscB, showed a reduced protein degradation rate of 30% and 40–50%, respectively, as compared to diploid $MATa/MAT\alpha$ wild-type cells. Sporulation rate in diploid strains lacking one or the other endoproteinase was considerably reduced (Wolf and Fink, 1975; Wolf and Ehmann, 1979; 1981; Mechler and Wolf, 1981; Zubenko and Jones, 1981; Wolf, 1981; Teichert et al., 1987).

Diploid $MATa/MAT\alpha$ double mutant strains homozygous for the absence of both endoproteinases yscA and yscB have lost 85% of the protein degradation capacity under sporulation conditions as compared to wild-type $MATa/MAT\alpha$ cells, and completely cease to sporulate (Teichert et al., 1987).

Diploid $MATa/MAT\alpha$ cells devoid of the activities of carboxypeptidase yscY and/or yscS do not show any considerably impaired production of spores (Wolf and Fink et al., 1975; Wolf and Ehmann, 1981; Zubenko and Jones, 1981). However, diploid $MATa/MAT\alpha$ double mutant strains lacking one of the endoproteinases (proteinase yscB) in addition to carboxypeptidase yscY show only a rate of 10% of sporulation visible under the microscope as compared to wild-type diploids. Diploid $MATa/MAT\alpha$ triple mutant strains lacking the two carboxypeptidases yscY and yscS in addition to proteinase yscB nearly cease to sporulate (sporulation rate 3% as compared to wild type) (Wolf, 1981; Wolf and Ehmann, 1981). Thus, the vacuolar endoproteinases yscA and yscB and the two vacuolar carboxypeptidases yscY and yscS are enzymes needed for the differentiation process of sporulation. Here, they most likely provide amino acids at the expense of unneeded (vegetative) cell protein for new protein synthesis for the spores to be formed. Earlier biochemical data showed that proteinase yscA and yscB (Klar and Halvorson, 1975; Betz and Weiser, 1976), as well as carboxypeptidase yscY increase considerably in activity upon shift of diploid cells to sporulation medium. This fits quite well the results concerning the function of the above four peptidases in sporulation obtained by means of genetic analysis. However, as defects in proteinase yscA or combined defects in proteinase yscA and proteinase yscB abolish processing and activation of other vacuolar enzymes (Ammerer et al., 1986; Woolford et al., 1986; Mechler et al., 1987) (see below), it cannot be excluded that absence of other vacuolar activity which also occurs in the mutants, is in addition involved in the sporulation event.

Proteinase yscA activity is necessary for cell survival under nitrogen starvation conditions. In the absence of nitrogen (but in the presence of glucose) a defect in proteinase yscA (allele *pep4-3*) leads to nearly 100% cell death within about three to five days of cultivation. The absence of proteinase yscB does not affect cellular life under these conditions (U. Teichert and D.H. Wolf, unpublished). It may not necessarily be the missing intracellular nitrogen supply alone which leads to cell death. As proteinase yscA is the trigger in the activation and maturation event of a variety of vacuolar enzymes, it might be the missing activity of one or more of the enzymes concerned which is not compatible with cell survival.

Vacuolar proteinases are involved in highly non-specific protein and peptide degradation. In support of this idea, it has been found that the capacity of cell extracts of a quadruple mutant strain lacking the two endoproteinases yscA and yscB, as well as the two carboxypeptidases yscY and yscS, to degrade an 'unspecific' substrate such as ^{3}H-methylcasein is reduced by 95–99% as compared to a wild-type strain harbouring these enzymes. Thus, one can assume that these enzymes represent the major unspecific peptidases of the yeast cell. However, recent experiments showed clearly that the vacuolar endoproteinases yscA and yscB are not the only enzymes involved in protein degradation in yeast: while their absence in starving and sporulating cells reduces protein degradation by 85%, reduction in growing cells and cells synthesizing false proteins amounts to only 30–40% (Achstetter et al., 1984b; Teichert et al., 1987). Thus under these conditions, more than 60% of the protein degradative capacity must be due to some proteolytic system(s) different from the vacuolar proteinases yscA and yscB. The recently detected ubiquitin system in yeast and the degradation pathway of different species of β-galactosidase in yeast suggest that a ubiquitin-dependent proteolytic system might be operating under those conditions (Özkaynak et al., 1984; 1987; Bachmair et al., 1986; Wilkinson et al., 1986; Jentsch et al., 1987; Finley et al., 1987).

Recent findings uncovered that the two vacuolar endoproteinases yscA and yscB involved in such an unspecific process as is protein degradation have in addition a highly specific function. They are the catalysts triggering the maturation of vacuolar enzymes by limited proteolysis on the route from the site of synthesis (the endoplasmic reticulum) to their site of action (the vacuole). This will be described in the section on protein maturation and secretion (Ammerer et al., 1986; Woolford et al., 1986; Mechler et al., 1987).

5.10.2 *Function of mitochondrial proteinases: protein import, protein degradation*

The mitochondrial matrix proteinase (proteinase yscMpI) was purified and characterized on the basis of its high specificity for the catalysis of mitochondrial precursor protein processing. The enzyme cleaves precursors in vitro to proteins imported into the mitrochondrial matrix and the mitochondrial inner membrane (McAda et al., 1982; Böhni et al., 1983; Cerletti et al., 1983). The enzyme also appears to be involved in the two-step processing of precursors to proteins destined for the outer face of the inner mitochondrial membrane and for the intermembrane space. It cleaves these precursors to intermediate forms. The second cleavage, converting the intermediate forms to the mature forms, is mediated by a second proteinase which has not been identified as yet (Daum et al., 1982; Reid et al., 1982). The mitochondrial matrix proteinase (proteinase yscMpI) shows a unique degree of specificity, it cleaves mitochondrial precursors only. Cytochrome c oxidase subunit V is shown to be correctly processed by the enzyme. Thus, the biochemical data leave little doubt that this enzyme is really involved in these processes *in vivo*, even though no mutants of the enzyme have been studied as yet (McAda *et al.*, 1982; Böhni *et al.*, 1983; Cerletti *et al.*, 1983).

The situation is completely different with the mitochondrial proteinases of low molecular mass acting on benzoyl-D, L-Arginine-β-naphthylamide and benzoyl-D, L-Arginine-4-nitroanilide: these enzymes are thought to be involved in the breakdown of mitochondrial translation products. Recent fractionation of mitochondrial proteinase activities suggests that a set of endo- and aminopeptidases exists, allowing degradation of proteins to the amino acid level within mitochondria. However, until mutants are available to test this hypothesis, this proposal must remain tentative (Zubatov *et al.*, 1984; Yasuhara and Ohashi, 1987).

5.10.3 *Function of periplasmic proteinases*

Mutants devoid of aminopeptidase yscII (*ape2*) were isolated. The enzyme is not involved in any process vital for vegetative growth on mineral or complete medium. Neither mating nor the differentiation process of sporulation is disturbed. The enzyme might be involved in amino acid assimilation from exogenously supplied peptides (Hirsch *et al.*, 1988).

5.10.4 *Function of membrane-bound proteinases: their role in hormone precursor processing and secretion*

Three membrane-bound enzymes—proteinase yscF, dipeptidyl aminopeptidase yscIV and carboxypeptidase yscα—were shown to be involved in the biosynthesis of the mating pheromone α-factor (Julius *et al.*, 1983a; 1984a; Achstetter and Wolf, 1985; Wagner *et al.*, 1987; Fuller *et al.*, 1988; Bussey, 1988). In the yeast *Saccharomyces cerevisiae* α-factor, one of the two oligopeptide pheromones triggering sexual conjugation of the haploid cell types *MATa* and *MATα* (Betz *et al.*, 1981; Thorner, 1981) is synthesized as a high-molecular-mass precursor protein which must undergo proteolytic maturation to yield the active pheromone (Emter *et al.*, 1983; Julius *et al.*, 1984b). The cloned α-factor precursor gene (*MFα1*) revealed that α-factor is present in four repeats in the precursor molecule. The repeats are flanked by peptide spacers, each starting with the amino acid sequence Lys-Arg and followed by -(Glu-Ala)$_2$-, -(Glu-Ala)$_3$- or -Glu-Ala-Asp-Ala-Glu-Ala-sequences (Kurjan and Herskowitz, 1982) (Figure 5.6). How does processing of this molecule proceed? Sequences of two basic amino acids have been found in many mammalian hormone precursor molecules and are thought to represent the initial processing sites. Based on the finding that chromogenic peptide substrates represent an extremely powerful tool in uncovering proteinases in yeast (Achstetter *et al.*, 1981; 1984a; 1984b; 1985), the chromogenic peptide substrate benzyloxycarbonyl-Tyr-Lys-Arg-4-nitroanilide harbouring the carboxy-terminal tyrosine residue of the α-factor and the consecutive Lys-Arg-sequence of the spacer peptide was introduced into a search for the processing

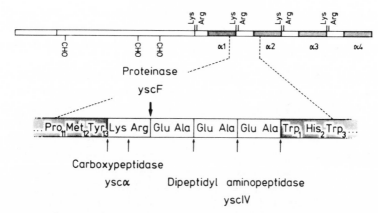

Figure 5.6 Maturation of the sex hormone α-factor and the enzymes involved. For details see text. (Achstetter and Wolf, 1985).

endopeptidase (Achstetter and Wolf, 1985). In another approach the chromogenic peptide butoxy-carbonyl-Gln-Arg-Arg-7-amido-4-methylcoumarin was used (Julius *et al.*, 1984*a*). The search resulted in the detection of the Ca^{2+} ion dependent proteinase yscF (Achstetter and Wolf, 1985) or 'lysine-arginine-cleaving endopeptidase' (Julius *et al.*, 1984*a*) which is strictly specific for splitting these peptides after the Lys-Arg or Arg-Arg bond. Strong evidence that this enzyme is the maturase responsible for the initial cut of the α-factor precursor molecule comes from a mutation (*kex2*) which causes a defect in killer toxin secretion (the phenotype found after mutant selection), and which leads to α-sterility (Leibowitz and Wickner, 1976; Bussey, 1981), and to accumulation of an overglycosylated α-factor precursor (Julius *et al.*, 1983*b*). Proteinase yscF activity is absent in this mutant. Cloning of the *KEX2* gene restores mating ability of *MATα kex2* mutants and identifies the structural gene of the enzyme (Julius *et al.*, 1984*a*; Achstetter and Wolf, 1985).

Recognition and cleavage of pairs of basic amino acid residues in the maturation of peptide hormones, and of some plasma proteins in mammals, may involve closely related enzymes. This was demonstrated *in vitro* through the correct processing of human pro-albumin to albumin at the carboxyl side of Arg-Arg-residues by the yeast *KEX2* encoded proteinase yscF. Proteinase yscF was selectively inhibited by the mutant variant 'Pittsburgh' of the plasma serine protease inhibitor $α_1$-antitrypsin, whereas normal $α_1$-antitrypsin was ineffective. Since pro-albumin and $α_1$-antitrypsin are both known to proceed through the secretory pathway of liver cells, unprocessed pro-albumin found in the circulation of carriers of the mutant $α_1$-antitrypsin ('Pittsburgh') is thought to be due to inhibition of the relevant enzyme which according to its specificity is closely related to proteinase yscF (Bathurst *et al.*, 1987).

Another maturase, called pro-pheromone convertase Y, has been proposed to be the catalyst of this initial processing step (Mizuno and Matsuo, 1984). However, no data relevant to the processing *in vivo* were presented, leaving severe doubts as to the correctness of this hypothesis.

Splitting of the α-factor precursor molecule after the Lys-Arg amino acid pair leaves this amino acid sequence carboxy-terminally connected to the α-factor molecules and -(Glu-Ala)$_2$-, -(Glu-Ala)$_3$- or Glu-Ala-Asp-Ala-Glu-Ala sequences connected to the amino terminus of the pheromone (Figure 5.6). Carboxypeptidase yscα is responsible for further maturation at the carboxyterminal cleavage site of α-factor (Figure 5.6). The enzyme is highly specific for the carboxyterminal cleavage of basic amino acid residues from synthetic sequences which mimic the cleavage site of the α-factor precursor molecule (Achstetter and Wolf, 1985; Wagner *et al.*, 1987). A *kex1* mutant strain defective in killer toxin production (as are *kex2* mutants) (Leibowitz and Wickner, 1976; Wickner and Leibowitz, 1976), secretes α-factor activity which is greatly reduced compared to wild type. This activity can be markedly activated by incubation with carboxypeptidase B, indicating that the *kex1* mutant has lost the capacity to remove the basic amino acids at the carboxyterminus of α-factor. The facts that *KEX1* encodes a serine exopeptidase with high homology to the vacuolar carboxypeptidase yscY (Dmochowska *et al.*, 1987; Bussey 1988), that carboxypeptidase yscα exhibits an identical inhibition spectrum to carboxypeptidase yscY (Wagner *et al.*, 1987), and that the activity is missing in *kex1* mutant cells strongly suggest that carboxypeptidase yscα is encoded by the *KEX1* gene. The aminoterminal extensions remaining on α-factor after the initial cut by proteinase yscF are removed by dipeptidyl aminopeptidase yscIV (also called dipeptidyl aminopeptidase A) (Julius *et al.*, 1983*a*) (Figure 5.6). According to the specificity of the enzyme, removal of amino acids occurs pairwise, as has been found in the maturation process of the bee venom melittin (Kreil *et al.*, 1980). Strong evidence that dipeptidyl aminopeptidase yscIV (dipeptidyl aminopeptidase A) is responsible for this maturation step comes from a mutant lacking the *STE13* gene product activity. The *STE13* gene is required for fertility of *MATα* cells but not of *MATa* cells (Sprague *et al.*, 1981). *MATα* mutant cells defective in this gene secrete biologically inactive α-factor molecules because the aminoterminal extension remains after the initial cut of the precursor by proteinase yscF, and they lack dipeptidyl aminopeptidase yscIV activity. Cloning of the *STE13* gene in *MATα ste13* mutants restores dipeptidyl aminopeptidase yscIV activity and reverses sterility (Julius *et al.*, 1983*a*). The scheme for α-factor processing is depicted in Figure 5.6.

5.10.5 More on protein maturation and secretion
The secretory pathway in yeast has been well defined by mutants temperature sensitive for growth and the export of secretory proteins (*sec*) (Novick *et al.*, 1980; 1981; Schekman, 1985*a,b*). Based on studies with these mutants, the

secretory pathway has been defined as either:

(i) Endoplasmic reticulum → Golgi → vesicles → secretion
(ii) Endoplasmic reticulum → Golgi → vacuoles.

The mutants allow a temperature-sensitive blockage of the delivery of secretory proteins, either into the endoplasmic reticulum, from the endoplasmic reticulum to the Golgi complex, from the Golgi complex to secretory vesicles (or, for vacuolar enzymes, from the Golgi complex to the vacuole), and from the secretory vesicles to the external space. By these means, the mutants cause the accumulation of secretory proteins in the respective cell compartment from which delivery is blocked and allow analysis of the molecular forms of the secretory proteins retained in this compartment. Studies on these mutants demonstrated the existence of α-factor as a high-molecular-mass precursor and allowed tracing of its proteolytic processing along the secretory pathway (Emter *et al.*, 1983; Julius *et al.*, 1984*b*).

Invertase, the enzyme encoded by the *SUC2* gene and required for growth of yeast cells on sucrose as sole carbon source, is secreted via the yeast secretory pathway (Schekman, 1985*a*). The enzyme bears a signal sequence characteristic for secretory proteins (Perlman *et al.*, 1982; Carlson *et al.*, 1983). Comparison of the molecular mass of the enzyme protein, as deduced from the DNA sequence with the protein moiety accumulating in *sec18* cells blocked in the delivery step from the endoplasmic reticulum to the Golgi as well as with the protein moiety of external invertase, shows that the signal sequence is removed upon transfer of the protein into the endoplasmic reticulum. Also, acid phosphatase, which is secreted to the cell surface, contains a signal peptide which is cleaved off during the secretion process (Emr *et al.*, 1983; Haguenauer-Tsapsis and Hinnen, 1984).

The α-factor precursor and the killer toxin precursor also encode what appear to be typical signal sequences and both are secreted via the secretory pathway. Evidence has been accumulated, which indicates that the signal sequences are not removed from these proteins as a separate entity and that no cleavages occur prior to the Golgi step (Bussey, 1981; Kurjan and Herskowitz, 1982; Bussey *et al.*, 1983; Julius *et al.*, 1984*b*; Hanes *et al.*, 1986).

However, *in vitro* experiments have shown that the signal sequence of pre-pro-α-factor may be removed (Tillman *et al.*, 1987).

Interestingly, substitution of the authentic leader peptide of pre-pro-killer toxin with the signal sequence of repressible acid phosphatase (*PHO5*) leads to signal sequence cleavage still conferring the expression of immunity phenotype, but resulting in lower levels of secreted toxin than expected from the strength of the *PHO5* promoter (Hanes *et al.*, 1986). The role of the α-factor precursor peptide and of the retained, non-cleaved signal sequence in secretion and processing of the pheromone remain to be clarified.

The enzyme involved in signal peptide processing, signal peptidase, should be located in the endoplasmic reticulum. On statistical grounds, the specificity and the processing site of signal peptidase were predicted (Von Heinje, 1983) to contain small neutral residues in positions -1 and -3 of a potential recognition site extending from position -5 to -1 of the N-terminus of the secretory protein to be processed. The specificity of signal peptidase was tested systematically *in vivo*, introducing point mutations of all 19 amino acids at the -1 position of the processing site of the *PHO5* acid phosphatase signal sequence. In agreement with von Heinje's prediction, signal sequence cleavage was observed only when Ala at position -1 was changed to either Gly, Cys, Ser or Thr, but no processing was found when Ala at position -1 was substituted with large aliphatic, aromatic or charged residues. The unprocessed *PHO5* species of the latter class appeared to be largely retained in the endoplasmic reticulum. Only reduced amounts of unprocessed enzyme appeared to be secreted (Monod *et al.*, 1986; Monod *et al.*, 1987).

The vacuole of *Saccharomyces cerevisiae* contains numerous hydrolytic enzymes (Wiemken *et al.*, 1979). The biosynthesis and transport of vacuolar enzymes also follows part of the secretory pathway, as has been shown with carboxypeptidase yscY (Stevens *et al.*, 1982). Like externally secreted proteins, the vacuolar located carboxypeptidase yscY also contains a signal sequence, which is cleaved off during synthesis and transport into the endoplasmic reticulum (Valls *et al.*, 1987). Sorting from secreted proteins occurs in the Golgi body. All the soluble vacuolar proteinases characterized so far are glycoproteins that are synthesized as higher-molecular-mass zymogens. As previously summarized (see section on vacuolar proteinases), proteinase yscA is synthesized as a precursor of M_r 52 kDa, proteinase yscB as precursors of M_r 73 kDa ('super-pro-proteinase yscB') and M_r 42 kDa (pro-proteinase yscB) (Mechler *et al.*, 1982*a*; 1988) and carboxypeptidase yscY as a precursor of M_r 67–69 kDa (Hasilik and Tanner, 1978*b*; Müller and Müller 1981). The precursor forms are cleaved to yield the active, mature enzymes of M_r 42 kDa (proteinase yscA), M_r 33 kDa (proteinase yscB) (Mechler *et al.*, 1982*a*; 1988), and M_r 61 kDa (carboxypeptidase yscY) (Hasilik and Tanner 1978*b*; Müller and Müller 1981; Stevens *et al.*, 1982; Mechler *et al.*, 1982*a*). Maturation of proteinase yscA and carboxypeptidase yscY is due to removal of an N-terminal peptide (Hemmings *et al.*, 1981; Ammerer *et al.*, 1986; Woolford *et al.*, 1986; Valls *et al.*, 1987). Using conditional secretion deficient mutants, it was shown that maturation of carboxypeptidase yscY occurs either in transit from the Golgi body to the vacuole or after the pro-enzyme arrives in the vacuole (Stevens *et al.*, 1982). Also, maturation of the precursor form of proteinase yscA occurs after passage through the endoplasmic reticulum, most likely in the same compartment where maturation of carboxypeptidase yscY takes place (Mechler *et al.*, 1988).

Proteinase yscB is synthesized as a high-molecular-mass precursor of M_r 73-kDa (super-pro-proteinase yscB) which is cleaved into the low-molecular-mass precursor of M_r 42-kDa (pro-proteinase yscB). Cleavage of the 73-kDa precursor occurs before or during transit into the endoplasmic reticulum. The processing catalyst is unknown. The 42-kDa precursor is enzymatically active against the collagen derivative Azocoll. As with the precursor forms of proteinase yscA and carboxypeptidase yscY, maturation of the 42-kDa precursor of proteinase yscB occurs after passage through the endoplasmic reticulum, most likely in transit from the Golgi body to the vacuole or in the vacuole. As one may predict from the sequence of the proteinase yscB gene, transfer of the 73-kDa precursor to the 42-kDa precursor occurs most probably via scission of an N-terminal peptide; processing of the 42-kDa precursor to the mature 33-kDa enzyme is most probably due to cleavage of a C-terminal peptide (Moehle *et al.*, 1987*b*; Mechler *et al.*, 1988).

Considerable progress has recently been made in the identification of the proteolytic enzymes responsible for maturation of the precursor forms of the vacuolar proteinases. Mutants with defects in the *PEP4* gene have been isolated and shown to be defective for a variety of vacuolar hydrolase activities (Jones, 1977; Hemmings *et al.*, 1981; Jones *et al.*, 1982; Zubenko *et al.*, 1983). Using immunochemical techniques, it was shown that *pep4–3* cells accumulate the precursor form of carboxypeptidase yscY and the 42-kDa precursor of proteinase yscB. No precursor protein of proteinase yscA could be found (Hemmings *et al.*, 1981; Mechler *et al.*, 1982*b*). The pleiotropic nature of the *pep4* mutation led to the hypothesis that the *PEP4* gene encodes a proteinase required for the processing of the vacuolar protein precursors (Jones, 1984). Cloning of the *PEP4* and the proteinase yscA gene as well as complementation studies between *pep4-3* cells and proteinase yscA mutant cells defective in the structural gene of proteinase yscA (*pra1*) (Ammerer *et al.*, 1986; Rothman *et al.*, 1986; Woolford *et al.*, 1986; Mechler *et al.*, 1987) uncovered the identity of the two genes and thus the central role of the vacuolar proteinase yscA in processing of vacuolar proteinase precursors. The extensive sequence homology between pro-proteinase yscA and pepsinogen led to the suggestion of an activation mechanism for pro-proteinase yscA similar to the one for pepsinogen (Bustin and Conway-Jacobs, 1971; James and Sielecki 1986). The model predicts that pro-proteinase yscA undergoes autoactivation upon transport to the low-pH environment of the vacuole (Ammerer *et al.*, 1986; Woolford *et al.*, 1986). However, activation and processing of proteinase yscA by an enzyme as yet unknown cannot be completely excluded. Proteinase yscA is a central protein in processing and activation of other vacuolar enzymes. *In vitro* studies show that pro-carboxypeptidase yscY cannot be activated by purified proteinase yscA (Mechler *et al.*, 1987). However, vacuole lysates containing proteinase yscA (but deficient in active proteinase yscB) or vacuole lysates devoid of intrinsic proteinase yscA and proteinase yscB, but with added purified proteinase yscA, lead to processing and activation of pro-carboxypeptidase yscY. The optimum pH for the processing event is 5. The additional vacuolar principle needed for activation can be mimicked by polyphosphate. Processing of pro-carboxypeptidase yscY by proteinase yscA leads to a carboxypeptidase yscY molecule of somewhat higher molecular

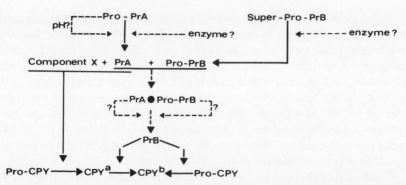

Figure 5.7 Model of maturation reactions yielding active carboxypeptidase ycsY. (Pro-) PrA = (pro-) proteinase yscA, super-pro-PrB = 72-kD 'precursor of proteinase yscB, (Pro-) PrB = (42-kD precursor of) proteinase yscB, (Pro-) CPY = (pro-) carboxypeptidase yscY. Dotted arrows: hypothetical reactions; Solid arrows: reactions based on experimental results.

mass (CPY[a]) than the authentic, mature enzyme, which is, however, fully active (Figure 5.7). This intermediate-molecular-mass form of carboxypeptidase yscY can be transferred by proteinase yscB into its apparently authentic, mature form (CPY[b]). Proteinase yscB is also able to directly mature pro-carboxypeptidase yscY to the apparently mature, authentic form *in vitro* (Figure 5.7). The actual additional participation of proteinase yscB in processing and maturation of vacuolar proteinase precursors *in vivo* is strongly indicated by the fact that pro-carboxypeptidase yscY is processed and activated in a strain carrying a *pra1* mutant allele that prevents activation of carboxypeptidase yscY by proteinase yscA but allows activation of proteinase yscB. In addition, the molecular mass differences in matured carboxypeptidase yscY brought about by proteinase yscA action and alternatively by proteinase yscB action *in vitro* can also be found *in vivo* (Hasilik *et al.*, 1978b; Mechler *et al.*, 1987).

The 42-kDa low-molecular-mass precursor of proteinase yscB accumulates in mutant cells deficient in proteinase yscA (allele *pep4-3*). This precursor is active against the proteinase yscB substrate Azocoll. One may speculate that a function of the peptide cleaved off the 73-kDa high-molecular-mass precursor of proteinase yscB is that of an inhibitor of the 42-kDa low-molecular-mass precursor activity to prevent damage during traffic of this molecule through the secretory pathway. Accumulation of the 42-kDa low-molecular-mass precursor of proteinase yscB in proteinase yscA (allele *pep4-3*)-deficient mutant cells points to the involvement of the proteinase yscA protein in the maturation process to yield the mature 33-kDa enzyme. *In vitro* processing of the lower-molecular-mass precursor to mature proteinase yscB is initiated by purified proteinase yscA. The maturation process, however, is not only inhibited by the addition of the proteinase yscA-specific inhibitor pepstatin but also by the addition of chymostatin or I_2^B, two specific inhibitors of proteinase yscB and its 42-kDa precursor. As the activity of proteinase yscB is, in addition, dependent on the gene dosage of proteinase yscA, the currently most likely explanation for maturation of the 42-kDa proteinase yscB precursor is complex formation of this precursor with proteinase yscA, followed by proteolytic steps involving proteinase yscA activity and the activity of the 42-kDa proteinase yscB precursor itself (Jones *et al.*, 1982; Woolford *et al.*, 1986; Mechler *et al.*, 1988).

Thus, the maturation of vacuolar hydrolases appears to come about by proteolytic processes, involving both proteinase yscA and proteinase yscB or pro-proteinase yscB (42-kDa precursor) (Figure 5.7). The activation of the vacuolar alkaline phosphatase (*PHO8*) also follows this pattern. Whereas *pra1* mutants containing some active proteinase yscB, or *prb1* mutants containing active proteinase yscA, show wild-type activity of alkaline phosphatase against α-naphthylphosphate, *pep4-3* mutants (containing active 42-kDa proteinase yscB precursor) possess reduced levels of alkaline phosphatase (Jones *et al.*, 1982; H.H. Hirsch and D.H. Wolf, unpublished). In contrast, no alkaline phosphatase activity was detected in double mutants *pra1 prb1* or *pep4-3 prb1*, completely deficient in any proteinase yscB activity (H.H. Hirsch and D.H. Wolf, unpublished).

As stated above, vacuolar and secreted proteins together traverse a portion of the same pathway in yeast. Sorting and direction to their respective target location occurs in the Golgi body (Stevens *et al.*, 1982; Schekman 1985*a*, *b*). It has been shown that glycosylation does not play a role in vacuolar targeting (Schwaiger *et al.*, 1982) and that yeast cells do not utilize the mannose-6-phosphate recognition marker as a mechanism of vacuolar protein sorting (Stevens *et al.*, 1982). Recent studies on the sorting mechanism of carboxypeptidase yscY showed that the localization determinant for the vacuole resides in the propeptide (Valls *et al.*, 1987). Thus, the aminoterminal extension of carboxypeptidase yscY serves three different functions: it mediates translocation across the endoplasmic reticulum (followed by proteolytic cleavage of the 20-amino-acid signal peptide by signal peptidase), and the remaining propeptide targets the carboxypeptidase yscY to the vacuole and renders the enzyme inactive during transit until it is cleaved off in a proteinase yscA and proteinase yscB triggered event.

Similar to the mating pheromone of *MATα* cells, the pair of mating pheromones secreted by *MATa* cells, *a*-factor, consists of small peptides. They contain sequences of 12 amino acids which differ only in the amino acid at position 6 (Val, Leu). They contain a S-alkylated carboxyterminal residue (Betz *et al.*, 1987). Two DNA stretches have been found in *MATa* cells, which potentially encode *a*-factor precursor polypeptides of 36 and 38 amino acids, respectively, containing a single copy of *a*-factor. The *a*-factor sequences are preceded by N-terminal peptide extensions containing a pair of basic amino acid residues (Lys–Lys) which may provide the signal for the initial proteolytic processing step as is the case for α-factor (Brake *et al.*, 1985; Fuller *et al.*, 1988).

Table 5.1 Proteinases of *Saccharomyces cerevisiae*.

Enzymes	Structural gene(s)	Characteristics	Cellular localization	Cellular role(s)
Proteinase yscA	PRA1/PEP4	Aspartic acid endopeptidase	Vacuole (soluble)	Protein degradation, nitrogen metabolism, vacuolar hydrolase precursor processing.
Proteinase yscB	PRB1	Serine sulphydryl endopeptidase	Vacuole (soluble)	Protein degradation, nitrogen metabolism, vacuolar hydrolase precursor processing.
Proteinase yscD	PRD1	Metallo (sulphydryl) endopeptidase	Non-vacuolar (soluble)	Unknown
Proteinase yscE	Unknown	Sulphydryl endopeptidase	Non-vacuolar (soluble)	Unknown
Proteinase yscF	KEX2	Serine sulphydryl endopeptidase (Ca^{2+}-dependent)	Membrane fraction (Golgi?)	Precursor processing, α-factor, killer factor
Proteinase yscG	Unknown	Metallo endoproteinase	Membrane fraction	Unknown
Proteinase yscH	Unknown	Metallo endoproteinase	Membrane fraction	Unknown
Proteinase yscK	Unknown	Unknown	Unknown (soluble)	Unknown
Proteinase yscMpI	Unknown	Metallo endoproteinase (Zn^{2+})	Mitochondrial matrix (soluble)	Precursor processing of nuclear coded mitochondrial proteins
3 Mitochondrial proteinases	Unknown	Serine sulphydryl endopeptidases	Inner mitochondrial membrane	Unknown
Propherome convertase Y	Unknown	Serine endopeptidase	Unknown (soluble)	Unknown
Aminopeptidase yscI	LAP4	Metallo sulphydryl exopeptidase (Zn^{2+})	Vacuole (soluble)	Unknown
Aminopeptidase yscII	APE2/LAP1	Metallo sulphydryl exopeptidase (Zn^{2+})	Periplasm	Nitrogen metabolism
Aminopeptidase yscIII	Unknown	Unknown	Unknown (soluble)	Unknown
13 Aminopeptidases (including probably LAPII and LAPIII): amino-peptidases yscIV-XV, P	Unknown	Exopeptidases	Unknown	Unknown

Enzyme	Gene	Type	Localization	Function
Aminopeptidase yscCo	Unknown	Metallo sulphydryl exopeptidase (Co^{2+})	Vacuole (soluble)	Unknown
Dipeptidyl aminopeptidase yscI	Unknown	Unknown	Unknown (soluble)	Unknown
Dipeptidyl aminopeptidase yscII	Unknown	Metallo exopeptidase	Unknown (soluble)	Unknown
Dipeptidyl aminopeptidase yscIII	Unknown	Unknown	Unknown (soluble)	Unknown
Dipeptidyl aminopeptidase yscIV	STE13	Serine exopeptidase	Membrane fraction (Golgi?)	Precursor processing, α-factor
Dipeptidyl aminopeptidase yscV	DAP2	Serine exopeptidase	Vacuole (membrane)	Unknown
Carboxypeptidase yscY	PRC1	Serine sulphydryl exopeptidase	Vacuole (soluble)	Protein degradation, nitrogen metabolism.
Carboxypeptidase yscS	CPS1	Metallo exopeptidase (Zn^{2+})	Vacuole (soluble)	Protein degradation, nitrogen metabolism.
Carboxypeptidase yscα	KEX1	Serine exopeptidase	Membrane fraction	Precursor processing α-factor, killer factor
Carboxypeptidase yscγ	Unknown	Metallo exopeptidase	Unknown (soluble)	Unknown
Carboxypeptidase yscδ	Unknown	Serine exopeptidase	Unknown (soluble)	Unknown
Carboxypeptidase yscε	Unknown	Serine exopeptidase	Unknown (soluble)	Unknown
Carboxypeptidase yscλ	Unknown	Sulphydryl exopeptidase	Membrane fraction	Unknown
Dipeptidase yscI	Unknown	Metallo exopeptidase (Mg^{2+})	Unknown (soluble)	Unknown

For detailed characteristics of the proteinases and for references see text.

Proteinase yscF, the α-factor maturase, does not seem to be responsible for this initial processing step of the a-factor precursor, as $MATa$ cells of mutants defective in the structural gene for proteinase yscF ($KEX2$) are not sterile (Leibowitz and Wickner, 1976). The catalysts for liberation of mature a-factor from the precursor have yet to be found.

Killer yeast cells secrete a heterodimeric α, β-subunit protein toxin, which is lethal to sensitive yeasts, yet cells secreting the toxin are immune to its action. Secretion of the toxin occurs via the normal secretory pathway (Bussey et al., 1983). The killer determinant is a cytoplasmically inherited double-stranded RNA (the M_1 double-stranded RNA) that is encapsidated in a virus-like particle (Tipper et al., 1984). A cDNA clone of the killer precursor gene has been sequenced and expressed in yeast (Bostian et al., 1984; Lolle et al., 1984; Skipper et al., 1984). The precursor has a prepro-toxin structure: it consists of an N-terminal leader, followed by the two toxin subunits α and β, which are separated by a central glycosylated γ-peptide (γ-domain). The pro-toxin is a glycosylated protein of M_r 42–44-kDa, consisting of a protein portion of M_r 34–35-kDa (Bostian et al., 1983). A minimum of three cleavage events must occur to yield the mature toxin (Bostian et al., 1984). The proposed cleavage sites of the toxin precursors are the Arg_{44}-Glu_{45} bond between the leader and the α-subunit of the toxin, the Trp_{130}-Gly_{131} bond between the α-subunit and the γ-domain, and the Arg_{233}-Tyr_{234} bond (the Arg is part of a basic amino acid pair of the sequence Lys-Arg-Tyr) between the γ-domain and the β-subunit of the toxin (Bostian et al., 1984; Tipper et al., 1984).

Processing of the precursor to killer toxin is abnormal in $kex2$ mutants (Bussey, 1981; Bussey et al., 1983; Tipper et al., 1984). $kex2$ mutants lack proteinase yscF, an enzyme specific for cleavage after pairs of basic amino acid residues and responsible for maturation of the mating pheromone α-factor (Julius et al., 1984a; Achstetter and Wolf 1985). From its high specificity for cleaving after pairs of basic amino acid residues, but its broad specificity for the amino acid following the basic amino acid pair, one might predict that proteinase yscF is responsible for cleavage of the Arg-Tyr bond in the precursor between the γ-domain and the β-subunit of the toxin. By this split the β-subunit of the toxin becomes liberated.

$kex1$ mutants are defective in production of active killer toxin (Leibowitz and Wickner, 1976; Wickner and Liebowitz 1976). A gene which complements $kex1$ mutants and maps at the $KEX1$ locus has been isolated and sequenced. Analysis of the nucleotide sequence of the cloned DNA indicates an open reading frame coding for a protein of 80-kDa that has a considerable homology with the vacuolar serine peptidase carboxypeptidase yscY (Dmochowska et al., 1987). Carboxypeptidase yscα is absent in $kex1$ mutant cells (Wagner and Wolf, 1987) indicating that, besides its action in α-factor precursor processing, this enzyme is also involved in killer factor processing. Recent re-evaluation of the α-cleavage site of the protoxin indicates processing at a pair of basic residues (Arg_{148}-Arg_{149}), a likely proteinase yscF cleavage site, instead of the Trp_{130}-Gly_{131} bond (reviewed in Bussey, 1988; Fuller et al.,

1988). Carboxypeptidase yscα may be responsible for removal of these basic residues after proteinase yscF cleavage.

Even though many proteolytic enzymes have now been detected in yeast, for a variety of processes of proteolytic nature the catalysts have yet to be uncovered.

5.11 Proteolytic processes requiring proteinases

5.11.1 Protein synthesis

Eukaroytic protein synthesis is initiated by methionine. This N-terminal methionine is removed from cytoplasmic proteins when alanine, serine, glycine, proline, threonine or valine is the penultimate amino acid. Excision does not occur when methionine precedes residues of arginine, asparagine, aspartic acid, glutamine, glutamic acid, isoleucine, leucine, lysine or methionine (Tsunasawa et al., 1985; Huang et al., 1987) (see also section on proteolysis and heterologous gene expression). The aminopeptidase responsible has yet to be identified. Studies on mutants devoid of aminopeptidase yscII exclude involvement of this enzyme in the N-terminal methionine removal (Hirsch et al., 1988). It is interesting to note that the methionine-removing aminopeptidase cleaves methionine off only those aminotermini where methionine is followed by an amino acid, which, according to the hypothesis of Bachmair et al. (1986), does not render the protein accessible to rapid proteolysis.

5.11.2 Catabolite inactivation

Addition of glucose to cells growing on a non-fermentable carbon source leads to specific inactivation of a very limited set of enzymes, a process called catabolite inactivation (for a review see Holzer, 1976). The enzymes analysed best, and for which disappearance of the protein has clearly been shown, are the gluconeogenic enzymes fructose-1, 6-bisphosphatase (Gancedo, 1971; Funayama et al., 1980), cytoplasmic malate dehydrogenase and phosphoenoylpyruvate carboxykinase. Also, for the maltose and galactose uptake systems, uridine nucleosidase and aminopeptidase yscI, glucose-triggered inactivation has been reported (Witt et al., 1966; Haarasilta and Oura, 1975; Neeff et al., 1978; Müller et al., 1981; Ferguson et al., 1967; Gancedo and Schwerzmann, 1976; Müller and Holzer, 1981; Robertson and Halvorson, 1957; Görts, 1969; Matern and Holzer, 1977; Magni et al., 1977; Frey and Röhm, 1979a).

Studies with mutants indicated that the vacuolar proteinases yscA and yscB are not required for inactivation, ruling out their involvement in this process which had been previously proposed on the basis of biochemical data.

In the case of fructose-1, 6-bisphosphatase it was later shown that cAMP-dependent phosphorylation leads to partial inactivation of the enzyme and precedes disappearance of its protein moiety concomitant with further disappearance of the activity (Tortora et al., 1981; Müller and Holzer, 1981;

Purwin *et al.*, 1982; Mazon *et al.*, 1982*a, b*). Due to this finding the inactivation process of fructose-1, 6-bisphosphatase could then be divided into two distinct events, a non-proteolytic primary inactivation step and a secondary proteolytic step. Thus, the statement that the vacuolar endoproteinases yscA and yscB are not involved in inactivation of fructose-1, 6-bisphosphatase was clearly valid for the primary inactivation step (Wolf and Ehmann, 1979; Zubenko and Jones, 1979; Mechler and Wolf, 1981).

The data had to be re-evaluated for the second inactivation step, which however, is of proteolytic nature. Analysis of the data found with strains carrying the proteinase yscA mutation (allele *pra1–1*) or the proteinase yscB mutation (allele *prb1–1* and *prb1–9*) shows that the extent of inactivation of fructose-1, 6-bisphosphatase ranges from 90 to 95% in these mutants, a value also found for wild-type cells (Wolf and Ehmann, 1979; Zubenko and Jones, 1979; Mechler and Wolf, 1981). This indicates that neither of the two endopeptidases is involved in the second, proteolytic inactivation step of fructose-1, 6-bisphosphatase. Recently, it has been reported that a *pep4–3* mutant strain defective in the proteinase yscA gene and also lacking the activities of proteinase yscB, carboxypeptidase yscY and other hydrolases (Jones *et al.*, 1982; Ammerer *et al.*, 1986; Woolford *et al.*, 1986), was defective in the proteolytic step of fructose-1, 6-bisphosphatase inactivation. If this were so, the phenotype observed could be explained by the proposal that the combined action of the two endopeptidases yscA and yscB or some other hydrolase not activated in this strain was needed for fructose-1, 6-bisphosphatase inactivation. However, the results obtained by Funagama *et al.* (1985) are not in agreement with the findings of Teichert *et al.*, (U. Teichert, B. Mechler and D.H. Wolf, unpublished). The *pep4–3* mutant was found to inactivate and (by using specific antibodies against fructose-1, 6-bisphosphatase) to degrade the enzyme in a similar fashion to wild type. The reason for the discrepancy in the results is unknown.

5.11.3 *Starvation-initiated inactivation*
NADP-dependent glutamate dehydrogenase is inactivated during carbon starvation of cells paralleled by disappearance of its immunologically detectable protein (Mázon, 1978; Mázon and Hemmings, 1979). The vacuolar endopeptidases, proteinases yscA and yscB, are not involved in this process. The catalyst has yet to be found. The catalyst for the degradation of the phosphorylated NAD-dependent glutamate dehydrogenase generated under carbon starvation of cells is also yet to be discovered (Wolf and Ehmann, 1979; Hemmings *et al.*, 1980; Hemmings, 1980; Mechler and Wolf, 1981).

5.11.4 *Cell growth*
Budding and cell division of yeast cells requires the synthesis of chitin which is deposited in the primary septum between the mother cell and the emerging daughter cell (Bacon *et al.*, 1969; Cabib, 1976).

Until recently, synthesis of chitin was thought to be accomplished by a single enzyme, chitin synthetase, present as an inactive zymogen in the plasma membrane (Duran et al., 1975). In vitro experiments suggested that proteinase yscB may be responsible for the activation of this enzyme in vivo (Cabib and Ulane, 1973; Cabib, 1976; Ulane and Cabib, 1976). Genetic evidence seemed to have ruled out this proposal: strains deficient in proteinase yscB synthesize chitin and grow normally (Wolf and Ehmann, 1979; Zubenko et al., 1979). Recent isolation and disruption of the structural gene (CHS1) which codes for chitin synthetase (now called chitin synthetase I) led to the unexpected finding that cells carrying the disrupted gene are not only viable, but in addition exhibit normal amounts of chitin, despite the absence of chitin synthetase activity I in the biochemical assay (Bulawa et al., 1986).

Subsequent studies on the chitin synthetase I disrupted strain led to the discovery of a second enzyme activity, called chitin synthetase II, of sufficiently different characteristics to allow detection in wild-type cells also. Chitin synthetase II catalyses a reaction identical to that catalysed by chitin synthetase I, and also appears to be located on the cell surface. Controversial reports exist, however, concerning the effect on chitin synthetase II of limited proteolysis in general and of trypsin in particular. Initial investigation suggested that chitin synthetase II is also present as a zymogen. In contrast to chitin synthetase I, proteinase yscB was not capable of activating chitin synthetase II in vitro, whereas various other proteases including trypsin were effective. A second report indicates, however, that proteolysis is not only unnecessary for chitin synthetase II but in fact decreases its activity. The differences in activation requirements of both enzymes may be related to their respective roles in the physiological state of the yeast cell. At present, no physiological condition is known in which chitin synthetase I is essential and is activated by an as yet unknown proteolytic process. As only growth and chitin synthesis, events that take place also in the absence of chitin synthetase I, have been tested in proteinase yscB mutants in order to elucidate proteinase yscB participation in chitin synthetase I activation, the question as to the involvement of this proteinase in chitin synthetase I activation is again open (Wolf and Ehmann, 1979; 1981; Sburlati and Cabib, 1986; Orlean, 1987).

5.11.5 Pheromone response
Hormones (pheromones) are signals triggering the induction of cellular metabolic changes. The two mating pheromones, α-factor and a-factor, secreted by the two different haploid mating cell types of Saccharomyces cerevisiae, have different amino acid sequences, but the effects they induce on their specific target cells are very similar. The responses to the pheromone include an accumulation of specific mRNAs, arrest of the cell cycle in the G_1 phase, the appearance of cell surface agglutinins, and a morphological change resulting in 'shmoo' formation (Betz et al., 1981; Thorner, 1981; Sprague et al., 1983).

α-factor has been shown to bind to a specific receptor on $MATa$ cells (see Chapter Two). Once the α-factor is bound to its receptor, the complexes enter the cell probably by receptor-mediated endocytosis (Riezman *et al.*, 1986). The internalized α-factor is then degraded rapidly (Chvatchko *et al.*, 1986). Degradation most likely occurs in the yeast vacuole, and a vacuolar proteinase(s) should be responsible for degradation. Whether one of the well-characterized vacuolar proteinases or an as yet unidentified enzyme is responsible has yet to be elucidated.

Non-mating haploids can recover from the effect of the pheromones and resume vegetative growth. Recovery from division arrest has been shown for $MATα$ cells to be due to inactivation of α-factor by a cell-associated proteinase. *sst1* (*bar1*) mutants are supersentive to α-factor and are deficient in an activity which inactivates α-factor. The $BAR1$ gene has recently been cloned and sequenced. It encodes a pepsin-like proteinase. The protein is only expressed in $MATa$ cells. Its exact mode of action on α-factor is still unknown (Sprague and Herskowitz, 1981; Chan and Otte, 1982*a*; 1982*b*; MacKay *et al.*, 1988).

5.11.6 *Proteolysis in mutagenesis and repair*

In *Escherichia coli* DNA damage leads to a pleiotropic SOS response which is due to DNA repair, prophage induction, filamentation and to the induction of a proteinase, the *recA* protein, which is believed to be essential for the regulation of SOS functions. It has been speculated that proteolytic activities may act upon different repressors giving rise to the pleiotropic SOS response. Also, in eukaryotes the involvement of proteolytic processes in epigenetic and genetic events triggered by radiation or chemical carcinogenesis has been indicated. In *Saccharomyces cerevisiae*, of the five proteinases measured (proteinase yscA, proteinase yscB, carboxypeptidase yscY, aminopeptidase yscI and aminopeptidase yscII), the vacuolar proteinase yscB showed a significant increase in UV-irradiated cells. It was proposed that this increase in the proteinase yscB level might be related to the triggering of the mutagenic DNA repair. It is completely unknown, however, how the enzyme might participate in this repair process. Nothing is known about a probable involvement of more specific proteinases in the induction of mutagenic DNA repair (Witkin, 1976; Troll *et al.*, 1978; Hanawalt *et al.*, 1979; Radman, 1980; Gottesman, 1981; Schwencke and Moustacchi, 1982*a*, *b*).

5.11.7 *Degradation of aberrant proteins*

The cell must ensure that aberrant and non-fuctional protein is quickly degraded to avoid accumulation of protein waste and disturbance of cell metabolism. In contrast to starvation conditions, in which cells carrying mutations in the vacuolar proteinase yscA and proteinase yscB genes have lost 85% of their degradative capacity, growing mutant cells and mutant cells synthesizing false protein due to application of canavanine retain 50–70% of their degradative potential (Teichert *et al.*, 1987). Aberrant proteins and protein fragments due to mutation have been found to be greatly reduced in cells. A defect in processing of the precursor to the 25 S RNA of the 60 S subunit of the ribosome due to mutation leads to rapid degradation of the unassembled proteins of the 60 S ribosomal subunit (Gorenstein and Worner, 1977; Bigelis and Burridge, 1978; Bigelis *et al.*, 1981; Donahue and Henry, 1981). Mutations in either one of the two *FAS* genes which code for the two subunits of the fatty acid synthetase complex lead to absence of immunologically detectable fatty acid synthetase complex (Dietlein and Schweizer 1975; Schweizer *et al.*, 1978). One possible explanation for this phenotype is the rapid degradation of aberrant and unassembled subunits. Unassembled mitochondrial proteins are also thought to be degraded (Galkin *et al.*, 1980). A mutationally altered carboxypeptidase yscY protein was shown to be rapidly degraded (Mechler *et al.*, 1982*b*). Mutants defective in the vacuolar endoproteinases yscA and yscB as well as in carboxypeptidase yscS revealed that these enzymes are not involved in the degradation process of the carboxypeptidase yscY mutant protein (Teichert *et al.*, 1987). This clearly indicates that yeast must contain additional degradative potential besides the known vacuolar proteinases.

Ubiquitin, a polypeptide of 8.5-kDa, has been found in yeast (Goldstein *et al.*, 1975), and, among other proteolytic systems, yeast may contain the ATP- and ubiquitin-dependent degradative system found in several eukaryotic cell types. An energy-requiring process leads to conjugation of ubiquitin to proteins which are subsequently degraded. (Hershko and Ciechanover, 1982; Hershko, 1983; Ciechanover *et al.*, 1984; Finley *et al.*, 1984; Jentsch *et al.*, 1987).

It has recently been found that *Saccharomyces cerevisiae* contains four distinct loci encoding ubiquitin (Özkaynak *et al.*, 1984; 1987; Finley 1987). *UBI1*, *UBI2* and *UBI3* each contain a single ubiquitin coding sequence, followed by open reading frames encoding 50–76 amino acid residues ('tail sequences'). Thus, *UBI1*, *UBI2* and *UBI3* encode hybrid proteins, the aminoterminal 76-amino-acid residues of which are identical to mature yeast ubiquitin. The fourth ubiquitin gene, *UBI4*, contains five ubiquitin encoding repeats in a head-to-tail arrangement, and thus encodes a polyubiquitin precursor protein. The ubiquitin molecules are joined directly via Gly-Met peptide bonds between the last and the first residue of mature ubiquitin, respectively (Özkaynak *et al.*, 1984; 1987). A processing proteinase splitting this Gly-Met bond and thus liberating mature ubiquitin from the precursor had to be postulated, and evidence for its existence has been found. Expression of a chimeric gene encoding a ubiquitin-β-galactosidase fusion protein leads to β-galactosidase molecules from which ubiquitin is cleaved off, yielding a deubiquibinated β-galactosidase. The ubiquitin-β-galactosidase junction encoded by the chimeric gene is identical to the junctions between adjacent repeats in the polyubiquitin precursor protein (Bachmair *et al.*, 1986). It is highly probable that the same proteinase is responsible for both processes, conversion of polyubiquitin into mature ubiquitin and cleavage of the fusion protein.

There is strong evidence for involvement of ubiquitin in the degradation of abnormal proteins. Mutants in the *UBI4* locus are highly sensitive to amino acid analogues which lead to the accumulation of abnormal proteins (Finley *et al.*, 1987).

5.11.8 *Proteolysis and heterologous gene expression*

Amenability to recombinant DNA technology and longstanding experience in industrial handling triggered a widespread biotechnological interest in *Saccharomyces cerevisiae* as expression system for the production of foreign proteins (Kingsman *et al.*, 1987; Sturley and Young, 1987). Also, given the biochemical and genetic potential, *S. cerevisiae* may also prove an useful alternative as an *in vivo* and *in vitro* test system for certain heterologous proteins of which certain molecular modes of action may be difficult to investigate in a conventional host. The attractiveness of yeast for heterologous gene expression resides in the eukaryotic modus of co- and post-translational modification and the eukaryotic secretory apparatus, providing advantageous features not found in prokaryotic expression systems but sometimes essential to the biological activity of the heterologous gene product. Modifications of foreign proteins observed in yeast involve N-terminal acetylation, fatty acid acylation, phosphorylation, disulphide bond formation, N- and O-glycosylation and proteolytic processing (Kramer *et al.*, 1985; Miyamoto *et al.*, 1985; Takatsuji *et al.*, 1986; Wen and Schlesinger, 1986; Barr *et al.*, 1987; Jabbar and Nayak, 1987; Kornbluth *et al.*, 1987).

The amount of expressed foreign protein may not only depend on the copy number of the expression vector, the quality of the promoter and terminator, but will also be affected by intrinsic properties of the protein product, by the subcellular location, by the growth phase, and the respective properties of the host strain. It is conceivable that the 'defence strategies' of the host strain in order to respond to the foreign 'load' do not only involve plasmid loss, down regulation of transcription or translation and gene rearrangement, but also include enhanced proteolytic degradation of the foreign protein.

In Table 5.2, an arbitrary compilation of heterologous proteins expressed in *S. cerevisiae* is given under the aspect of proteolytic processing and degradation. Not unexpectedly, specific processing events mainly associated with secretion of the foreign protein processing, e.g. signal sequence cleavage or proteinase yscF (*KEX2* gene product), are often characterized in more detail. In contrast, the stability or degradation kinetics of foreign proteins and the likely proteolytic system involved are rarely documented (Jabbar and Nayak, 1987). Some heterologous gene expression is carried out in proteinase-deficient strains, but it remains unclear whether this simply represents a prophylactic measure. Is degradation taking place already *in vivo*, or instead during extraction procedures after disintegration of cells *in vitro*? Targeting of foreign proteins into the secretory pathway and secretion into the culture medium was expected to facilitate the isolation of the product, circumventing unspecific degradation upon cell disruption.

Even though the secretion efficiency of foreign proteins could be improved in some cases by using yeast signal sequences instead of authentic heterologous 'pre'-sequences, the yields of the secreted products are often still up to 10-fold lower than those obtained from expression without the signal sequence, retaining the proteins in the cytoplasm (Hitzemann *et al.*, 1983; Mellor *et al.*, 1983; Chabezon *et al.*, 1984; Smith *et al.*, 1985; Cramer *et al.*, 1987). Whether this difference is due to cytoplasmic precipitation of the highly expressed protein (Wood *et al.*, 1985; Huang *et al.*, 1987), to degrading proteinases during or after secretion, or to a rate-limiting step along the secretory pathway, remains to be determined (Bitter *et al.*, 1984; Smith *et al.*, 1985; Ernst, 1986; Wen and Schlesinger, 1986).

Secretion of heterologous proteins was investigated using either 'pre'-sequences, yeast signal sequences or fusions to the α-factor pheromone precursor. As indicated in Table 5.2, no prediction for efficiency can be made. Either the 'pre'- or the yeast signal sequence functioned in some construction, but did not in others. Generally, however, the yeast signal sequences appeared to be somewhat more efficient than the authentic foreign 'pre'-sequences. Correct signal sequence cleavage was observed in most cases at the predicted site yielding a mature N-terminus ($+1$). Heterogeneous N-termini, however, were obtained from *preD* or *preD/A* interferon α1 processing (Hitzemann *et al.*, 1983; Chang *et al.*, 1986): more than half of the interferon species were processed at sites (-14), (-8) or (-3), different from that of the expected residue ($+1$). The reasons for the changed proteolytic specificity are unknown, but could involve the high number of 23-amino-acid residues of the 'pre'-sequence, the occurrence of a charged residue (lysine) at position (-7) after the sufficiently long stretch of 16 uncharged residues in the signal sequence, or inhibition by incorrect disulphide bonding.

Two defined consecutive proteolytic processing events were observed in pre-glucoamylase expressed in yeast (Innis *et al.*, 1985). There, the initial signal sequence cleavage occurred at the correct site leaving Asn at position $+1$, but was followed by a proteinase yscF (*KEX2* gene product) dependent processing step at the carboxy side of Lys($+5$)-Arg($+6$), yielding Ala($+7$) as N-terminus. No significant impairment of the activity of glucoamylase by this additional processing step was observed.

Possibly in a similar manner, the glycosylated degradation products of the yeast acid phosphatase PHO5 (1–256)-phaseolin fusion protein (ungly-cosylated M_r 69 000) of about M_r 54–58 000 (Cramer *et al.*, 1987) could be explained. Speculatively, these products result from cleavage at a Lys(240)-Arg(241) site of PHO5, now accessible in the hybrid protein.

In the secretory pathway, the accessibility to Lys-Arg sites is critical to processing, as documented by two examples. Fusing the α-factor leader

Table 5.2 Proteolysis and heterologous gene expression in *Saccharomyces cerevisiae*

Host strain	Expression vector	Promoter	Protein expressed — Signal sequence	Proteolytic events and product characterization — Processing event	Identification by means of	Yield	Degradation	Localization of product	Ref.
AB103 (α leu2 ura3 his4 pep4-3)	pC11 (LEU2 2μ)	MFα1	αF(1-89)-epidermal growth factor (hEGF)	yscF-cleavage of extracellular material, N-terminus (Glu-Ala)$_3$ extension	1. Antibody 2. Sequencing 3. Bioassay	7%	n.d.	s.p., p, m	1
			αF(1-83)-Lys-Arg-hEGF	yscF-cleavage of extracellular material, intracellular material αF-hEGF fusion or hEGF mono-, di-, trimer?					
GM3C-2 (α leu2 trp1 his4 cyc1 cyp3)	pGT41 (LEU2 2μ)	MFα1	αF(1-89) β-endorphin (human)	yscF-cleavage, extracellular material with 1 or 2 N-terminal (Glu-Ala) extensions; internal trypsin-like degradation products	1. Antibody 2. HPLC 3. Sequencing	0.5 mg/l	equal in pep4-3	m	2
20B-12 (α trp1 pep4-3)	pYE (TRP1 2μ)	MFα1	αF(1-89) interferon α1	yscF-cleavage, extracellular material with 1 or 2 N-terminal (Glu-Ala) extensions; 10% as αF-interferon fusion	1. Antibody 2. Activity	2 × 10^8 U/l	n.d.	s.p., p, m	3
			αF-(1-83)-Lys-Arg-interferon α1	yscF-cleavage of 50% of extracellular material; 50% αF-interferon fusion					
			αF(1-83)-Lys-Arg-β endorphin	n.d.	Antibody	1 mg/l	n.d.	m	
			αF(1-83)-Lys-Arg-calcitonin	n.d.	Antibody	1 mg/l	n.d.	m	
20B-12 (α trp1 pep4-3)	pYE (TRP1 2μ)	MFα1	αF(1-89)-somatostatin	yscF-cleavage, extracellular material contains N-terminal (Glu-Ala) extensions;	1. Antibody 2. HPLC 3. Sequencing	0.2 mg/l	n.d.	(c?, s.p., p) m	4
		PHO5	αF(1-89)-somatostatin						
		PHO5	αF(1-83)-Lys-Arg-somatostatin	intracellular material as αF-somatostatin fusion; yscF-cleavage of extracellular material, N-terminus (+1)					

Strain	Plasmid	Promoter	Product	Result	Analysis	Yield			Ref
BJ1991 (α ura3 leu2 trp1 prb1 pep4-3)	pJDB207 (LEU2 2μ) pJD-CEN6 (TRP1 ARS1 CEN6 p446/1 (URA3 2μ)	MFα1 ACT CYC1	αF(1-83)-Lys-Arg-somatomedin	yscF-cleavage of extracellular material; intracellular material n.d.	Antibody	30 mg/l	n.d.	(s.p., p)m	5
MT556 (α leu2 pep4-3)	pMT544 (LEU2 2μ)	TPI	αF(1-89)-glucagon (human)	yscF-cleavage of extracellular material, 1/3 N-terminus (+1), 2/3 N-terminal (Glu-Ala) extensions	1. Antibody 2. HPLC 3. Sequencing	0.9 μM	n.d.	m	6
SEY2101 (a ura3 leu2 ade2)	pSEY210 (URA3 2μ)	MFα1	αF(1-89)/pre invertase (SUC2)	yscF-cleavage of extracellular product; intracellular material present	1. Activity 2. Antibody	n.d.	n.d.	(s.p.)p	7
			αF(1-83)-Lys-Arg-glucagon	yscF-cleavage of extracellular material, N-terminus (+1); intracellular material n.d.		0.2 μM		m	
MT615 tpi/tpi pep4/pep4	pMT612 (LEU2 2μ POT)		αF(1-83)-pro-insulin (human)	yscF-cleavage of extracellular material, 90% N-terminal (Glu-Ala) extension, 10% N-terminus (+1)	1. Antibody 2. HPLC 3. Sequencing 4. Bioassay	2 μM	n.d.	m	8
			αF(1-83)-Lys-Arg-proX-insulin	90% N-terminus (+1), also yscF-cleavage at internal site(s); mostly incorrect disulphide bridging (X = variation of C-peptide spacer)					
			αF(1-83)-Lys-Arg-(B-chain)-Arg-Arg-Leu-Gln-Lys-Arg-(A-chain)-insulin	Extracellular material contains equal amounts of correctly yscF-cleaved and disulphide bridged insulin and C-terminally processed (B-chain).					9
W301-18Aμ (α leu2 trp1 ura3 ade2 his3 can1)	pCV7 (LEU2 2μ)	MFα1	αF(1-83)-Lys-Arg-hANP (atrial natriuretic peptide)	yscF-cleavage of extracellular material, N-terminus (+1); 2/3 show C-terminal removal of 2 amino acids.	1. Antibody 2. HPLC 3. Sequencing	0.1 mg/l	n.d.	m	10

Table 5.2 (Contd.)

Host strain	Expression vector	Promoter	Protein expressed / Signal sequence	Processing event	Identification by means of	Yield	Degradation	Localization of product	Ref.
				Proteolytic events and product characterization					
TGY1sp4 (α ura3 his3)	pTG881 (dLEU2 2µ URA3)	PGK MFα1	αF(1-83)-Lys-Arg-hirudin V2	yscF-cleavage of extracellular material, N-terminus (+1)	1. Activity 2. HPLC 3. Sequencing	1%	n.d.	m	11
			αF(1-89)-hirudin V2	yscF-cleavage of extracellular material, 50% contains N-terminal (Glu-Ala)$_2$ extension					
BJ1991 (α leu2 trp1 ura3 pep4-3 prb1)	pUR (LEU2 2µ)	GAPDH	αF(1-87)-prophospholipase A2 (human)	yscF-cleavage of extracellular material, retained N-terminal extension; trypsin-activatable	1. Antibody 2. Activity 3. Bioassay 4. Sequencing	0.5 mg/l	equal in pep4-3 or prb1	m	12
GRF 18 (α leu2 his3)			Pre-prophospholipase A2	No activity or antigen detectable			unstable?		
AB103.1 (α leu2 ura3 pep4-3 cir⁰)	pAB18 (URA3 2µ)	MFα1	αF(1-83)-Lys-Arg-CTAP III (human connective tissue activating peptide)	yscF-cleavage of extracellular material, N-terminus (+1); inactive degradation products?	1. Bioassay 2. HPLC 3. Sequencing	3×10^4 U/l	n.d.	m	13
20B-12 (α trp1 pep4-3)	pTRP584 (TRP1 ARS1 2µ)	MFα1	αF(1-83)-Lys-Arg-interleukin2 (mouse)	yscF-cleavage of extracellular material, N-terminus (+1); intracellular material as glycosylated αF-interleukin fusion	1. Antibody 2. Bioassay 3. HPLC 4. Sequencing	10 µg/l	n.d.	s.p, m	14
			αF(1-83)-Lys-Arg-mGM/CSF (murine granulocyte macrophage colony stimulating factor)	yscF-cleavage of extracellular material, N-terminus (+5), removal of Ala-Pro-Thr-Arg; heterogeneously glycosylated	(As above)				15
XV2181 (trp1/trp1)	pαADH2	ADH2	αF(1-83)-Lys-Arg-mGM/CSF	yscF-cleavage of extracellular material, 50% N-terminus (+1), 50% N-terminus (+3), removal of Ala-Pro; glycosylated	(As above)	50 mg/l	n.d.	m	16

Strain	Plasmid	Promoter	Signal	Gene product	Processing	Analysis	Yield	pep4 effect	Location	Ref
RH218 (trp1)	pFRS (TRP1 ARS1)	ADH1		αF(1-83)-Lys-Arg-interleukin 2 (bovine)	yscF-cleavage of extracellular material, N-terminus (+1) i.e. retained Ala-Pro-Thr-Gln; unglycosylated	(As above)	60 mg/l	n.d.	m	17
20B-12 (α trp1 pep4-3)	YEp1PT (TRP1 2μ)	PGK	pre D	interferon αD (human)	n.d.	1. Antibody	0.4%	n.d.	c	18
				interferon α1	sp-cleavage: heterogeneous N-terminus	2. Bioassay	0.3%	equal in pep4-3	s.p., p>m	
				interferon α1	n.d.	3. Sequencing	1.0%		c	
GMC-2 (a trp1 leu2 his4 cys1 cps1) or	pYE (TRP1 2μ)	PH05	pre D/A	interferon α2	sp-cleavage: heterogeneous N-terminus	1. Antibody	0.2%	n.d.	s.p., p>m	19
			pre D	interferon α2	n.d.	2. Bioassay	1.0%		c	
			pre	interferon α2	n.d.	3. Sequencing	n.d.		s.p., p>m	
				interferon γ	n.d.		n.d.		s.p., p>m	
				interferon γ	n.d.		n.d.		c	
W301-18A (α trp1 leu2 ade2 ura3 his3)				interferon αD	n.d.	Bioassay	0.2%		c	
n.d.	pJDP207 (LEU2 2μ)	PH05	PH05/pre	interferon αD	n.d.	Bioassay	0.5%*		n.d.	20
				interferon αD	n.d.		5.0%*		c	
20B-12 (α trp1 pep4-3)	YEpIPT (TRP1 2μ)	PGK	SUC2	interferon α2	sp-cleavage: N-terminus (+1)	1. Antibody	29×10^6 U/l	n.d.	p, m	21
			SUC2	Met-interferon α2	sp-cleavage: N-terminus (+1)	2. Bioassay	22×10^6 U/l	n.d.	m	
			pre	interferon α2	sp-cleavage: heterogeneous N-terminus	3. Sequencing	12×10^6 U/l	n.d.	(s.p.)	
S150-2B (leu2 ura3 trp1 his3 cir+)	YEpsec1 (dLEU2 2μ URA3)	GAL/CYC1	pre	(K. lactis killer toxin)-interleukin1β (human)	sp-cleavage: N-terminus (+1)	1. Antibody 2. Bioassay	2 mg/l	n.d.	m	22
(pep4)	pADH040-Z (LEU2 2μ)	ADH1	pre	β-lactamase (E. coli)	sp-cleavage: product active unprocessed: product inactive	1. Antibody 2. Activity	2%	reduced in pep4-3	c < s.p.	23
KK4 (α leu2 ura3 his trp1 gal180)	YEp51 (LEU2 2μ)	GAL10	'pre'	lysozyme (human)	sp-cleavage: N-terminus (+1)	Activity	1.5 mg/l	n.d.	s.p. < p, m	24
				lysozyme	n.d.			n.d.	c	

Table 5.2 (Contd.)

Host strain	Expression vector	Promoter	Signal sequence	Protein expressed	Proteolytic events and product characterization					Ref.
					Processing event	Identification by means of	Yield	Degradation	Localization of product	
Mc16 (α leu2 his4 ade2 lys2 SUF2)	pMA230 (LEU2 2μ)	PGK	pre	α-amylase (wheat) α-amylase	~ sp-cleavage	1. Antibody 2. Activity	n.d.	n.d.	m (c)	25
DB746 (α trp1 leu2 ura3 his3)	pMS (TRP1 2μ)	ADC1	pre	α-amylase (mouse)	~ sp-cleavage	1. Activity 2. Affinity	0.1%	n.d.	m	26
AH22 (α leu2 his4 can1 cir⁺)	pAM82 (LEU2 ARS1 2μ)	PHO5	pre	α-amylase (human) α-amylase	~ sp-cleavage and yscF-cleavage. C-terminus correct, glycosylated; n.d.; unglycosylated	1. Antibody 2. Activity 3. Sequencing	900 × 1 ×	n.d.	m c	16
C468 (α leu2 his3 mal)	pAC1 (LEU2 2μ)	ENO1	pre	glucoamylase (Aspergillus)	~ sp-cleavage and yscF-cleavage, N-terminus (+7);	1. Activity 2. Growth on starch 3. Antibody 4. Sequencing	25 U/l	n.d.	m	27
(leu2 kex2)					sp-cleavage: N-terminus (+1)					
MT302/1C (α leu2 his3 ade1 arg5 pep4-3)	pMA27 (LEU2 2μ)	PGK	pre	pro-chymosin (calf) pro-chymosin Met-chymosin	~ sp-cleavage: activity = 0.05%* n.d., activity = 0.1%* n.d., activity not detectable	1. Antibody 2. Activity	1.0%* 5.0%* 0.1%	n.d. n.d. n.d.	(s.p.) c c	28
CGY339 (α his4 ura3 pep4-3) or super secretion derivatives	pCGS40 (URA3 2μ) or YIp5	GAL1 URA3 GAL1 TPI SUC2 GAL1 GAL1 MFα1 GAL1	pre SUC2 SUC2 SUC2 PHO5 α(1-89) pre	pro-chymosin (calf) URA3-pro-chymosin pro-chymosin growth harmone (bovine)	n.d. n.d. ~ sp-cleavage n.d.	1. Antibody 2. Activity Antibody	0.25%* 0.07%* 0.25%* 0.25%* 0.05%* 0.3%* 0.3%* 0.2%* 0.3% 0.3%	n.d. n.d.	c c s.p., v?, p.m. s.p., p, m	29

Host	Plasmid	Promoter	Signal	Product	Processing	Assay	Yield		Localization	Ref
n.d.	pYEGF-2	GAPDH		epidermal growth factor (human)	n.d.	Bioassay	15 µg/l	n.d.	c	30
AH22 (a leu 2 his4 can1 cir+)	pUR (LEU2 2µ)	GAPDH	pre	pro-thaumatin	sp-cleavage: N-terminus (+1)	1. Antibody 2. Sequencing	100 ×	n.d.	(s.p.)	31
			pre	pro-thaumatin thaumatin	n.d. ~ sp-cleavage		0.1 × 10 ×		(c) (s.p.)	32
AB103.1 (a leu2 ura3 pep4-3 his4 cir0)	pC1/1 (LEU2 2µ)	PHO5		Met-α_1-antitrypsin (human) Met-α_1-antitrypsin Val-358	n.d.	1. Activity 2. Antibody	2%	reduced in pep4-3	c	33
Ic1697d (arg J+ leu2)	pRIT (LEU2 2µ)	ARG3	pre	α_1-antitrypsin Met-α_1-antitrypsin	~ sp-cleavage; glycosylated n.d.; unglycosylated	1. Activity 2. Antibody	0.3%* 1.0%*	reduced in pep4-3	s.p. c	34
AB110 (leu2 ura3 his4 pep4-3 cir0)	pC1/1 (LEU2 2µ)	GAPDH		Met-α_1-antitrypsin (human) Met-α_1-antitrypsin Val-385	N-terminus (+1)	1. Activity 2. Protease binding 3. Antibody 4. Sequencing	10%* 10%*	s.above	c c	35
CGY339 (α ura3 leu2 his4 pep4-3 or mnn9; sec18; alg1 strains)	pCGS (URA3 2µ)	TPI	SUC2	α-antitrypsin Glu-α_1-antitrypsin	~ sp-cleavage, glycosylated n.d.; unglycosylated	Antibody	0.5%* n.d.	n.d. n.d.	s.p., p, m c	36
AH22 (a leu2 his4 can1 cir+)	pAM82 (LEU2 ARS1 2µ)	PHO5	pre	pancreatic secretory trypsin inhibitor (human)	sp-cleavage: N-terminus (+1)	1. Activity 2. Antibody 3. Sequencing	1.7 mg/l	n.d.	p, m	37
OL1 (α ura3 leu2 his3)	pPV2 (URA3 2µ)	PHO5	pre	tissue inhibitor of metalloproteases (human)	no sp-cleavage, insoluble	1. Antibody 2. Activity	0.4%	n.d.	c	38
HT246 (a leu2 prb1)	pJDB207 (LEU2 2µ)	PHO5 PHO5		tissue plasminogen activator (human TPA) PHO5-Lys-Arg-hTPA	sp-cleavage: N-terminus (+1) sp-cleavage, no yscF processing	1. Antibody 2. Activity 3. Sequencing	0.01%	reduced in prb1 extracts	(s.p.)	39

Table 5.2 (Contd.)

Host strain	Expression vector	Promoter	Signal sequence	Protein expressed	Proteolytic events and product characterization					Ref.
					Processing event	Identification by means of	Yield	Degradation	Localization of product	
ABYS (leu2 pra1 prb1 prc1 cps1)	pJDB207 (LEU2 2μ)	PHO5 GAPDH	PHO5	hirudin-V1	sp-cleavage: N-terminus (+1) C-terminus 10% heterogeneous	1. Activity 2. HPLC 3. Sequencing	60 mg/l	reduced in pra1 prb1 prc1 cps1 strains	m	40, 41
X4003-5B (a leu2 ura3 his4 met2 trp5 gal1)	pMA91 (LEU2 2μ) or pLG89 (URA3 2μ)	PGK	pre pre	immunoglobulin λ (mouse) immunoglobulin λ immunoglobulin μ immunoglobulin μ	~ sp-cleavage; (75% insoluble product) n.d. ~ sp-cleavage n.d.	1. Antibody 2. Hapten binding	0.5%* n.d. 0.5%* n.d.	n.d.	s.p., v?, p, m c, v? s.p., v?, p, m c, v?	42
W301-18A (α trp1 leu2) or 20B-12 (α trp1 pep4-3)	pYE7 (TRP1 ARS1 2μ)	PHO5	pre PHO5 PHO5/pre PHO5	phaseolin phaseolin phaseolin phaseolin PHO5(1-256)phaseolin	n.d. ~ sp-cleavage partial sp-cleavage no cleavage ~ sp-cleavage, degradation products	Antibody	3.0%* 2.0%* 2.6%* 2.0%* 1.7%*	equal in both strains	(c) (s.p.) (c, s, p.) (c) (c, s.p)	43, 44
N1B (α cyc1 leu2 his4 ura3)	YEp (LEU2 2μ) YCp (URA3 ARS1 CEN4)	ADC1	(mito pre)	cytochrome c (rat)	n.d. cyc1 complementation no complementation	1. Antibody 2. Growth on lactate	0.6% 0.1%	n.d.	mito	45
JRY438 (α ura3 leu2 his4)	pCGS109 (URA3 2μ)	GAL1	(mito pre)	OTCase (human) (ornithine transcarbamoylase)	~ mito-pre cleavage	1. Antibody 2. Activity 3. arg3 complementation	25 U/mg	n.d.	mito (matrix)	46
RP11 (a arg3 ura3 ade2)			(mutant) Arg-Gly23 pre OTCase		no mito-pre cleavage				mito (mem)	

182

Host strain	Vector	Promoter	Protein expressed	pre	Processing / growth	Assay	Amount	Stability	Localization	Ref
20B-12 (α trp1 pep4-3)	pWY4 (TRP1 2µ)	ADH1	hemagglutinin of human influenza virus (HA 565)		n.d., normal growth	1. Antibody 2. Sequencing	n.d.	n.d.	c	47
			HA 500	pre	~ sp-cleavage; slow growth; sp-cleavage: N-terminus (+1)		0.3%	$t_{1/2} = 2.5$ h	s.p., mem	48
			HA 484	pre	n.d.			$t_{1/2} = 0.5$ h	(c?)	
			HA 325	pre	sp-cleavage: N-terminus (+1) hyperglycosylated			$t_{1/2}$ 2.5 h	s.p., mem, m	
			HA 308		n.d.			$t_{1/2}$ 0.5 h	(c?)	
YM722 (a leu2 ura3 his3 lys2 ade2) or sec1; sec7; sec18; sec53 strains	pYMS-2 (LEU2 2µ)	GAL1	sindbis virus 265 protein precursor		autoproteolytic? capsid protein processing (pE2pE1 secreted and processed) E1 glycosylated, pE2p unglycosylated, pE2 glycosylated, E2 glycosylated	Antibodies	2.5%	$t_{1/2}$ 0.5 h in extracts	c (s.p.) (s.p.) (s.p.)	49
			vesicular stomatitis virus G protein		n.d., reduced growth	Antibody	3.0%	n.d.	(s.p., mem)	
TD4 (a trp1 leu2 his4 ura3)	pYT (TRP1 2µ)	ADC1	acetylcholine receptor, α subunit (T. calif)		n.d.	1. Antibody 2. Ligand binding	1.0% +	n.d.	mem	50
W303-1A (leu2 ura trp his ade)	YEp51 (LEU2 2µ)	GAL10	viral p60-src cellular p60-src		N-terminal Met removal and myristoylation	1. Antibody 2. Kinase activity	n.d.	n.d.	n.d.	51
XV610-8C (α trp1 leu2 ade lys1 can1)	pMA56 (TRP1 2µ)	ADH1	HBsAg (S gene) (hepatitis B virus)		n.d., monomer or particle non-glycosylated	1. Antibody 2. Sedimentation 3. Electron microscopy 4. Antigenicity	25 µg/l	n.d.	(c?, mem?)	52
20B-12 (α trp1 pep4-3)	pYe (TRP1 ARS1 2µ)	PGK					50 µg/l			18, 53
AH22 (a leu2 his4 can1 cir⁺)	pAM82 (LEU2 ARS1 2µ)	PHO5	HBsAg (S gene)		n.d.		2.5 mg/l			54

Table 5.2 (Contd.)

Host strain	Expression vector	Promoter	Signal sequence	Protein expressed	Processing event	Proteolytic events and product characterization				Ref.
						Identification by means of	Yield	Degradation	Localization of product	
AH22 R⁻ (a leu2 his3 can1 cir⁺ pho80)	pSH19 (LEU2 ARS1 2μ)	GAPDH PHO5	pre	HBsAg (Pre-S2-S gene)	n.d. glycosylated	5. Polymerized albumin receptor	n.d.	increased in extracts	(s.p., mem?)	55
OL1 (α ura3 leu2 his3)	pPV2 (URA3 2μ)	PHO5		HBsAg (PreS1-PreS2-S gene) HBsAg (S gene)	n.d. unglycosylated unglycosylated		0.1%	n.d.	c (mem) c	56
ATCC (α ilv2 tmp1 typ1 5'dTMP⁻)	YEp3 (LEU2 2μ)	GAPDH		thymidine kinase of herpes simplex virus	n.d. as αF-invertase fusion	1. Activity 2. Growth on thymidine	n.d.	n.d.	(c)	57
20B-12 (α trp1 pep4-3)	pYE7 (TRP1 ARS1 2μ)	PHO5		human c-myc oncogene	n.d. phosphorylation	Antibody	n.d.	n.d.	n	58
AH22 (a leu his4 can1 cir⁺)	pAM82 (LEU2 ARS1 2μ)	PHO5		reverse transcriptase (cauliflower mosaic virus)	n.d.	Activity	1.2 U/mg	n.d.	c?	59
20B-12 (α trp1 pep4-3)	pYE72 (TRP1 2μ)	PHO5		reverse transcriptase (gag pol of HTLVIII)	p56 processing by retroviral protease into p24, p14, p16	Antibody	n.d.	n.d.	c?	60
AB110 (a leu2 ura3 his4 pep4-3)	pAB24 (LEU2 URA3 2μ)	GAPDH		reverse transcriptase (HIV-SF2)	p66 processing C-terminally by yeast? protease into p51	1. Antibody 2. Activity	18. U/mg	n.d.	c?	61
AB110 (a leu2 ura3 his4 pep4-3)	pDP34 (URA3 dLEU2)	PHO5		protease inhibitor eglin c (H. medic.)	N-terminal Met-removal, partial acetylation	1. Activity 2. HPLC 3. Sequencing	n.d.	n.d.	c	40, 62

184

HH103 (α leu2 ura3 ape2)			(as above)			n.d.	n.d.	c	63
BB25-1d (α leu2 gal2)	pJDP207 (dLEU2 2μ)	PGK	Met1-X(+2)-thaumatin (X(+2) = permutated amino acid residue)	N-terminal Met-removal, acetylation	Sequencing 5-10%	n.d.	c	64

pre = authentic signal sequence of heterlogous protein; *PH05* = yeast promoter or part of open reader frame, e.g. signal sequence of acid phosphatase; αF(1-89) = αfactor pheromone leader, amino acid residues e.g. 1 to 89; sp = signal peptidase; yscF = proteinase yscF (*KEX2* gene product); % = percent of total protein; %* = percent of total soluble protein; | | = other quantification; |10 ×| = relative quantification, e.g. 10-fold; c = cytoplasm; s.p. = secretory pathway; n = nucleus; v = vacuole; p = periplasm, m = medium; mem = membrane; mito = mitochondria; ~ = most likely; n.d. = not documented; % t = percentage of total membrane protein.

(1) Brake et al. (1984); (2) Bitter et al. (1984); (3) Zsebo et al. (1986); (4) Green et al. (1986); (5) Ernst (1986); (6) Moody et al. (1987); (7) Emr et al. (1983); (8) Thim et al. (1986); (9) Thim et al. (1987); (10) Vlasuk et al. (1986); (11) Loison et al. (1986); (12) Van den Bergh et al. (1987); (13) Mullenbach et al. (1986); (14) Miyajima et al. (1985); (15) Miyajima et al. (1986); (16) Price et al. (1987); (17) Hitzemann et al. (1981); (18) Hitzemann et al. (1983); (19) Kramer et al. (1984); (20) Hinnen et al. (1983); (21) Chang et al. (1986); (22) Baldari et al. (1987); (23) Roggenkamp et al. (1985); (24) Jigami et al. (1986); (25) Rothstein et al. (1984); (26) Thomsen (1983); (27) Innis et al. (1985); (28) Mellor et al. (1983); (29) Smith et al. (1985); (30) Urdea et al. (1983); (31) Edens et al. (1984); (32) Edens et al. (1985); (33) Rosenberg et al. (1984); (34) Chabezon et al. (1984); (35) Travis et al. (1985); (36) Moir and Dunais (1987); (37) Izumoto et al. (1987); (38) Kaczorek et al. (1987); (39) Heim et al. (1985); (40) Meyhack et al. (1987); (41) Grossenbacher et al. (1987); (42) Wood et al. (1985); (43) Cramer et al. (1985); (44) Cramer et al. (1987); (45) Scarpulla and Nye (1986); (46) Chang et al. (1987); (47) Jabbar et al. (1985); (48) Jabbar et al. (1987); (49) Wen et al. (1986); (50) Fujita et al. (1986); (51) Kornbluth et al. (1987); (52) Valenzuela et al. (1982); (53) Wampler et al. (1985); (54) Miyanohara et al. (1983); (55) Itoh et al. (1986); (56) Dehoux et al. (1986); (57) Zhu et al. (1984); (58) Miyamoto et al. (1985); (59) Takatsuji et al. (1986); (60) Kramer et al. (1985); (61) Barr et al. (1987); (62) Meyhack et al. (1986); (63) Hirsch et al. (1988); (64); Huang et al. (1987).

αF(1–89) which includes a (Glu-Ala)$_2$ spacer sequence at the C-terminus to interferon α con1, more than 90% of the extracellular interferon was processed at the Lys($+$84)-Arg($+$85) site. The N-terminus of interferon, however, contained two, one or no Glu-Ala extensions. When the proteinase yscF (*KEX2* gene product) processing site preceded mature interferon α con1 without any spacer, only 50% mature interferon α con1 was obtained, and 50% was detected as α-factor-interferon α con1 fusion protein (Bitter *et al.*, 1984; Zsebo *et al.*, 1986).

The influence of neighbouring sequences on accessibility to the pairs of basic amino acid residues flanking the 'C-peptide spacer' of human insulin to the proteinase yscF (*KEX2* gene product) was examined in α-factor-(1–85)-insulin fusions (Thim *et al.*, 1986; 1987). Size reduction of the C-peptide spacer to a single pair of basic residues drastically reduced the proteinase yscF (*KEX2* gene product) processing, but, interestingly, increased the amount of correctly disulphide-bonded insulin.

As described above, the (Glu-Ala)$_2$ spacer sequence was deleted in order to obtain homogenous N-termini, since the removal of the (Glu-Ala)$_2$ extension by dipeptidyl aminopeptidase yscIV (A) (*STE13* gene product) proved to be inefficient. However, deletion of the (Glu-Ala)$_2$ spacer sequence may not necessarily always result in the projected N-terminus. In the case of murine granulocyte/macrophage colony stimulating factor, two and four N-terminal amino acid residues of the mature glycosylated protein were found to be removed (Bitter *et al.*, 1984; Brake *et al.*, 1984; Green *et al.*, 1986; Loison *et al.*, 1986; Miyajima *et al.*, 1986; Moody *et al.*, 1987; Price *et al.*, 1987; Thim *et al.*, 1986; Van den Bergh *et al.*, 1987).

C-Terminal processing of heterologous proteins is rarely documented. In some reports, C-terminal removal of two or three amino acid residues from secreted proteins was detected. Whether the carboxypeptidase yscα (*KEX1* gene product) is involved in this process remains to be shown (Sato *et al.*, 1986; Vlasuk *et al.*, 1986; Grossenbacher *et al.*, 1987; Meyhack *et al.*, 1987; Thim *et al.*, 1987).

The removal of the C-terminus of the p66 reverse transcriptase from human immunodeficiency virus (HIV-SF2) expressed in the yeast cytoplasm could be mimicked by an as yet undefined yeast proteinase (Barr *et al.*, 1987), since the coding region of the retroviral proteinase was apparently not included in the cloned DNA fragment. Speculatively, the acid proteinase presumably encoded by the similarly organized endogenous yeast Ty element in the *TYB* gene (Mellor *et al.*, 1986) could be an interesting candidate for the processing of the C-terminal peptide. In contrast, N-terminal processing appeared to be strictly dependent on the presence of the viral proteinase (Kramer *et al.*, 1985).

Finally, N-terminal methionine removal by a specific aminopeptidase was observed on several cytoplasmically located foreign proteins (Meyhack *et al.*, 1986; Barr *et al.*, 1987; Huang *et al.*, 1987; Hirsch *et al.*, 1988). Using mature Met-X-thaumatin, the predicted influence of the penultimate amino acid residue X on methionine removal and acetylation was investigated (Huang *et al.*, 1987). The thaumatin mutant proteins fell into four categories:

(i)   Methionine removal, no acetylation when X = Pro, Val, Cys
(ii) Methionine removal, acetylation when X = Gly, Ala, Ser, Thr
(iii) No methionine removal, no acetylation when X = Ile, Leu, Met, Phe, Tyr, Trp, Lys, His, Arg, Gln
(iv) No methionine removal, acetylation when X = Asp, Asn, Glu.

Exceptions to these general classifications, including partial acetylation, are observed (Travis *et al.*, 1985; Tsunasawa *et al.*, 1985; Meyhack *et al.*, 1986; Kornbluth *et al.*, 1987; Hirsch *et al.*, 1988).

Acknowledgements

The authors thank H. Gottschalk and W. Fritz for excellent help during the preparation of the manuscript. This work was supported by grants from the Deutsche Forschungsgemeinschaft, Bonn, and the Fonds der Chemischen Industrie, Frankfurt.

References

Achstetter, T., Ehmann, C. and Wolf, D.H. (1981) New proteolytic enzymes in yeast. *Arch. Biochem. Biophys.* **207**: 445.

Achstetter, T., Ehman, C. and Wolf, D.H. (1982) Aminopeptidase Co, a new yeast peptidase. *Biochem. Biophys. Res. Commun.* **109**: 341.

Achstetter, T., Ehmann, C. and Wolf, D.H. (1983) Proteolysis in eukaryotic cells: Aminopeptidases and dipeptidyl aminopeptidases of yeast revisited. *Arch. Biochem. Biophys.* **226**: 292.

Achstetter, T., Ehmann, C., Osaki, A. and Wolf, D.H. (1984*a*) Proteolysis in eukaryotic cells. Proteinase yscE, a new yeast peptidase. *J. Biol. Chem.* **259**: 13344.

Achstetter, T., Emter, O., Ehmann, C. and Wolf, D.H. (1984*b*) Proteolysis in eukaryotic cells. Identification of multiple proteolytic enzymes in yeast. *J. Biol. Chem.* **259**: 13344.

Achstetter, T. and Wolf, D.H. (1985*a*) Proteinases, proteolysis and biological control in the yeast *Saccharomyces cerevisiae. Yeast* **1**: 139.

Achstetter, T. and Wolf, D.H. (1985) Hormone processing and membrane-bound proteinases in yeast. *EMBO J.* **4**: 173.

Achstetter, T., Ehmann, C. and Wolf, D.H. (1985) Proteinase yscD. Purification and characterization of a new yeast peptidase. *J. Biol. Chem.* **260**: 4585.

Aibara, S., Hayashi, R. and Hata, T. (1971) Physical and chemical properties of yeast proteinase C. *Agr. Biol. Chem.* **35**: 658.

Ammerer, G., Hunter, C.P. Rothman, J.H., Saari, G.C., Valls, L.A. and Stevens, T.H. (1986) *PEP4* gene of *Saccharomyces cerevisiae* encodes proteinase A, a vacuolar enzyme required for processing of vacuolar precursors. *Mol. Cell. Biol.* **6**: 2490.

Bachmair, A., Finley, D. and Varshavsky, A. (1986) *In vivo* half-life of a protein is a function of its amino-terminal residue. *Science* **234**: 179.

Bacon, J.S.D., Farmer, V.C., Jones, D. and Taylor, I.F. (1969) The glucan components of the cell wall of baker's yeast (*Saccharomyces cerevisiae*) considered in relation to its ultrastructure. *Biochem. J.* **114**: 557.

Bai, Y. and Hayashi, R. (1975) A possible role for a single cysteine residue in carboxypeptidase Y. *FEBS Lett.* **56**: 43.

Bai, Y. and Hayashi, R. (1979) Properties of the single sulfhydryl group of carboxypeptidase Y. Effects of alkyl and aromatic mercurials on activities towards various synthetic substrates. *J. Biol. Chem.* **254**: 8473.

Bakalkin, G.Y., Kalnov, S.L., Zubatov, A.S. and Luzikov, V.N. (1976) Degradation of total cell protein at different stages of *Saccharomyces cerevisiae* yeast growth. *FEBS Lett.* **63**: 218.

Baldari, C., Murray, J.A.H., Ghiara, P., Cesareni, G. and Galeotti, C.L. (1987) A novel leader peptide which allows secretion of a fragment of human interleukin 1 in *Saccharomyces cerevisiae*. *EMBO J.* **6**: 229.

Barr, P.J., Power, M.D., Lee-Ng, C.T., Gibson, H.L. and Luciw, P.A. (1987) Expression of active human immunodeficiency virus reverse transcriptase in *Saccharomyces cerevisiae*. *Biotechnology* **5**: 486.

Bathurst, I.C., Brennan, S.O., Carell, R.W., Cousens, L.S., Brake, A. and Barr, P.J. (1987) Yeast KEX2 protease has the properties of a human proalbumin converting enzyme. *Science* **235**: 348.

Beck, I., Fink, G.R. and Wolf, D.H. (1980) The intracellular proteinases and their inhibitors in yeast. A mutant with altered regulation of proteinase A inhibitor activity. *J. Biol. Chem.* **255**: 4821.

Berthelot, M. (1860) Sur la fermentation glycosique du sucre de canne. *C. R. Acad. Sci. Paris.* **50**: 980.

Betz, H. and Weiser, U. (1976) Protein degradation and proteinases during yeast sporulation. *Eur. J. Biochem.* **62**: 65.

Betz, H. (1976) Inhibition of protein synthesis stimulates intracellular protein degradation in growing yeast cells. *Biochem. Biophys. Res. Commun.* **72**: 121.

Betz, R., Manney, T.R. and Duntze, W. (1981) Hormonal control of gametogenesis in the yeast *Saccharomyces cerevisiae*. *Gamete Res.* **4**: 571.

Betz, R., Crabb, J.W., Meyer, H.E., Wittig, R. and Duntze, W. (1987) Amino acid sequences of a-factor mating peptides from *Saccharomyces cerevisiae*. *J. Biol. Chem.* **262**: 546.

Bigelis, R. and Burridge, K. (1978) The immunological detection of yeast nonsense termination fragments on sodium dodecylsulfate-polyacrylamide gels. *Biochem. Biophys. Res. Commun.* **82**: 322.

Bigelis, R., Keesey, J.K. and Fink, G.R. (1981) The yeast his4 multifunctional protein. Immunochemistry of the wild-type protein and altered forms. *J. Biol. Chem.* **256**: 5144.

Bitter, G.A., Chen, K.K., Banks, A.R. and Lai, P.H. (1984) Secretion of foreign proteins from *Saccharomyces cerevisiae* directed by α-factor gene fusions. *Proc. Natl. Acad. Sci. USA* **81**: 5330.

Blobel, G. (1980) Intracellular protein topogenesis. *Proc. Natl. Acad. Sci. USA* **77**: 1496.

Böhni, P.C., Daum, G. and Schatz, G. (1983) Import of proteins into mitochondria. Partial purification of a matrix-located protease involved in cleavage of mitochondrial precursor polypeptides. *J. Biol. Chem.* **258**: 4937.

Bordallo, C., Schwencke, J. and Suarez Rendueles, M.P. (1984) Localization of the thermosensitive X-prolyl dipeptidyl aminopeptidase in the vacuolar membrane of *Saccharomyces cerevisiae*. *FEBS Lett.* **173**: 199.

Bostian, K.A., Jayachandran, S. and Tipper, D.J. (1983) A glycosylated protoxin in killer yeast: models for its structure and maturation. *Cell* **32**: 169.

Bostian, K.A., Elliott, Q., Bussey, H., Burn, V., Smith, A. and Tipper, D.J. (1984) Sequence of the preprotoxin dsRNA gene of type I killer yeast: multiple processing events produce a two-component toxin. *Cell* **36**: 741.

Brake, A.J., Merryweather, J.P., Coit, D.G., Heberlein, U.A., Masiarz, F.R., Mullenbach, G.T., Urdea, M.S., Valenzuela, P. and Barr, P.J. (1984) α-factor-directed synthesis and secretion of mature foreign proteins in *Saccharomyces cerevisiae*. *Proc. Natl. Acad. Sci. USA* **81**: 4642.

Brake, A.J., Brenner, C., Najarian, R., Laybourn, P. and Merryweather, J. (1985) Structure of genes encoding precursors of the yeast peptide mating pheromone a-factor. In *Current Communications in Molecular Biology. Protein Transport and Secretion*, ed. Gething, M.J., Cold Spring Harbor Laboratory, New York, 103.

Buchner, E. (1897) Alkoholische Garung ohne Hefezellen. *Ber. Dt. Chem. Ges.* **30**: 117.

Bulawa, C., Slater, M., Cabib, E., Au-Young, J., Sburlati, A., Adair, L.W. Jr., and Robbins, P.W. (1986) The *S. cerevisiae* structural gene for chitin synthetase is not required for chitin synthesis *in vivo*. *Cell* **46**: 213.

Bunning, P. and Holzer, H. (1977) Natural occurrence and chemical modification of proteinase B inhibitors from yeast. *J. Biol. Chem.* **252**: 5316.

Bussey, H. (1981) Physiology of killer factor in yeast. *Adv. Microbial Physiol.* **22**: 93.

Bussey, H., Saville, D., Greene, D., Tipper, D.J. and Bostian, K.A. (1983) Secretion of *Saccharomyces cerevisiae* killer toxin: processing of the glycosylated precursor. *Mol. Cell. Biol.* **3**: 1362.

Bussey, H. (1988) Proteases and the processing of precursors to secreted proteins in yeast. *Yeast* **4**: 17.

Bustin, M. and Conway-Jacobs, A. (1971) Intramolecular activation of procine pepsinogen. *J. Biol. Chem.* **246**: 615.

Cabib, E. and Ulane, R. (1973) Chitin synthetase activating factor from yeast, a protease. *Biochem. Biophys. Res. Commun.* **50**: 186.

Cabib, E. (1976) The yeast primary septum: a journey into three-dimensional biochemistry. *Trends Biochem. Sci.* **1**: 275.

Carlson, M., Taussig, R., Kustu, S. and Botstein, D. (1983) The secreted form of invertase in *Saccharomyces cerevisiae* is synthesized from mRNA encoding a signal sequence. *Mol. Cell. Biol.* **3**: 439.

Cerletti, N., Böhni, P.C. and Suda, K. (1983) Import of proteins into mitochondria. Isolated yeast mitochondria and a solubilized matrix protease correctly process cytochrome c oxidase subunit V precursor at the NH_2 terminus. *J. Biol. Chem.* **258**: 4944.

Chabezon, T., De Wilde, M., Herion, P., Loriau, R. and Bollen,A. (1984) Expression of human α_1-antitrypsin cDNA in the yeast *Saccharomyces cerevisiae*. *Proc. Natl. Acad. Sci. USA* **81**: 6594.

Chan, R.K. and Otte, C.A. (1982a) Isolation and genetic analysis of *Saccharomyces cerevisiae* mutants supersensitive to G1 arrest by *a*-factor and α-factor pheromones. *Mol. Cell. Biol.* **2**: 11.

Chan, R.K. and Otte, C.A. (1982b) Physiological characterization of *Saccharomyces cerevisiae* mutants supersensitive to G1 arrest by *a*-factor and α-factor pheromones. *Mol. Cell. Biol.* **2**: 21.

Chang, C.N., Matteucci, M., Perry, J., Wulf, J.J., Chen, C.Y. and Hitzemann, R.A. (1986) *Saccaromyces cerevisiae* secretes and correctly processes human interferon hybrid proteins containing yeast invertase signal peptides. *Mol. Cell. Biol.* **6**: 1812.

Cheng, M.Y., Pollock, R.A., Hendrick, J.P. and Horwich, A.L. (1987) Import and processing of human ornithine transcarbamoylase precursor by mitochondria from *Saccharomyces cerevisiae*. *Proc. Natl. Acad. Sci. USA* **84**: 4063.

Chvatchko, Y., Howald, I. and Riezman, H. (1986) Two yeast mutants defective in endocytosis are defective in pheromone response. *Cell* **46**: 355.

Ciechanover, A., Finley, D. and Varshavsky, A. (1984) Ubiquitin dependence of selective protein degradation demonstrated in the mammalian cell cycle mutant ts85. *Cell* **37**: 57.

Ciejek, E. and Thorner, J. (1979) Recovery of *S. cerevisiae* a cells from G1 arrest by α-factor pheromone requires endopeptidase action. *Cell* **18**: 623.

Cramer, J.H., Lea, K. and Slightom, J.L. (1985) Expression of phaseolin cDNA genes in yeast under control of natural plant DNA sequences. *Proc. Natl. Acad. Sci. USA* **82**: 334.

Cramer, J.H., Lea, K., Schaber, M.D. and Kramer, R.A. (1987) Signal peptide specificity in posttranslational processing of the plant protein phaseolin in *Saccharomyces cerevisiae*. *Mol. Cell. Biol.* **7**: 121.

Daum, G., Gasser, S.M. and Schatz, G. (1982) Import of proteins into mitochondria. Energy-dependent, two-step processing of the intermembrane space enzyme cytochrome by isolated yeast mitochondria. *J. Biol. Chem.* **257**: 13075.

Dehoux, P., Ribes, V., Sobczak, E. and Streeck, R.E. (1986) Expression of the hepatitis b_2 virus large envelope protein in *Saccharomyces cerevisiae*. *Gene* **48**: 155.

Dietlein, G. and Schweizer, E. (1975) Control of fatty-acid-synthetase biosynthesis in *Saccharomyces cerevisiae*. *Eur. J. Biochem.* **58**: 177.

Dmochowska, A., Dignard, D., Henning, D., Thomas, D.Y. and Bussey, H. (1987) Yeast *KEX1* gene encodes a putative protease with a carboxypeptidase B-like function involved in killer toxin and α-factor precursor processing. *Cell* **50**: 573.

Docherty, K. and Steiner, D.F. (1982) Post-translational proteolysis in polypeptide hormone biosynthesis. *Ann. Rev. Physiol.* **44**: 625.

Doi, E., Hayashi, R. and Hata, T. (1967) Purification of yeast proteinases. Part III. Purification and some properties of yeast proteinase C. *Agr. Biol. Chem.* **31**: 160.

Donahue, T.F. and Henry, S.A. (1981) Myo-Inositol-1-phosphate synthase. Characteristics of the enzyme and identification of its structural gene in yeast. *J. Biol. Chem.* **256**: 7077.

Dreyer, T., Halkier, B., Svendsen, I. and Ottesen, M. (1986) Primary structure of the aspartic proteinase A from *Saccharomyces cerevisiae*. *Carlsberg Res. Commun.* **51**: 27.

Duran, A., Bowers, B. and Cabib, E. (1975) Chitin synthetase zymogen is attached to the yeast plasma membrane. *Proc. Natl. Acad. Sci. USA* **72**: 3952.

Edens, L., Bom, I., Ledeboer, A.M., Maat, J., Toonen, M. Y., Visser, C. and Verrips. C.T. (1984) synthesis and processing of the plant protein thaumatin in yeast. *Cell* **37**: 629.

Edens, L. and Van der Wel, H. (1985) Microbial synthesis of the sweet-tasting plant protein thaumatin. *Trends Biotechnol.* **3**: 61.

Elbein, A.D. (1981) The tunicamycins–useful tools for studies on glycoproteins. *Trends Biochem. Sci.* **6**: 219.

Emr, S.D., Schekman, R., Flessel, M.C. and Thorner, J. (1983) An MFα 1-SUC2 (α-factor-invertase) gene fusion for study of protein localization and gene expression in yeast. *Proc. Natl. Acad. Sci. USA* **80**: 7080.

Emter, O., Mechler, B., Achstetter, T., Müller, H. and Wolf, D.H. (1983) Yeast pheromone α-factor is synthesized as a high molecular weight precursor. *Biochem. Biophys. Res. Commun.* **116**: 822.

Emter, O. and Wolf, D.H. (1984) Vacuoles are not the sole compartment of proteolytic enzymes in yeast. *FEBS Lett.* **166**: 321.

Ernst, J.F. (1986) Improved secretion of heterologous proteins by *Saccharomyces cerevisiae*: effects of promotor substitution in alpha-factor fusions. *DNA* **5**: 483.

Ferguson, J.J. Jr., Boll, M. and Holzer, H. (1967) Yeast malate dehydrogenase: enzyme inactivation in catabolite repression. *Eur. J. Biochem.* **1**: 21.

Finkelstein, D. and Strausberg, S. (1979) Metabolism of α-factor by a-mating type cells of *Saccharomyces cerevisiae*. *J. Biol. Chem.* **254**: 796.

Finley, D., Ciechanover, A. and Varshavsky, A. (1984) Thermolability of ubiquitin-activating enzyme from the mammalian cell cycle mutant ts85. *Cell* **37**: 43.

Finley, D. and Varshavsky, A. (1985) The ubiquitin system: functions and mechanisms. *Trends Biochem. Sci.* **10**: 343.

Finley, D., Ozkaynak, E. and Varshavsky, A. (1987) The yeast polyubiquitin gene is essential for resistance to high temperature, starvation and other stresses. *Cell* **48**: 1035.

Frey, J. and Röhm, K.H. (1978) Subcellular localization and levels of aminopeptidases and dipeptidases in *Saccharomyces cerevisiae*. *Biochim. Biophys. Acta* **527**: 31.

Frey, J. and Röhm, K.H. (1979*a*) The glucose-induced inactivation of aminopeptidase I in *Saccharomyces cerevisiae*. *FEBS Lett.* **100**: 261.

Frey, J. and Röhm, K.H. (1979*b*) External and internal forms of yeast aminopeptidase II. *Eur. J. Biochem.* **97**: 169.

Fujishiro, K., Sanada, I.Y., Tanaka, H. and Katunuma, N. (1980) Purification and characterization of yeast protease B. *J. Biochem.* **87**: 1321.

Fujita, N., Nelson, N., Fox, T.D., Claudio, T., Lindstrom, J., Riezman, H. and Hess, G.P. (1986) Biosynthesis of the *Torpedo californica* acetylcholine receptor α-subunit in yeast. *Science* **231**: 1284.

Fuller, R.S., Brake, A.J., Julius, D.J. and Thorner, J. (1985) The KEX2 gene product required for processing of yeast prepro-α-factor is a calpain-like endopeptidase specific for cleaving at pairs of basic residues. In *Current Communications in Molecular Biology. Protein Transport and Secretion*, ed. Gething, M.J., Cold Spring Harbor Laboratory, New York, 97.

Fuller, R.S., Sterne, R.E. and Thorner, J. (1988) Enzymes required for yeast prohormone processing. In *Recombinant DNA to Study Neuropeptide Processing*, ed. Herbert, E., *Ann. Rev. Physiol.* **50** (in press).

Funayama, S., Gancedo, J.M. and Gancedo, C. (1980) Turnover of yeast fructose-bisphosphatase in different metabolic conditions. *Eur. J. Biochem.* **109**: 61.

Funayama, T., Toyoda, Y. and Sy, J. (1985) Catabolite inactivation of fructose-1,6-bisphosphatase and cytoplasmic malate dehydrogenase in yeast. *Biochem. Biophys. Res. Commun.* **130**: 467.

Galkin, A.V., Tsoi, T.V. and Luzikov, V.N. (1980) Regulation of mitochondrial biogenesis. Occurrence of nonfunctioning components of the mitochondrial respiratory chain in *Saccharomyces cerevisiae* grown in the presence of proteinase inhibitors: Evidence for proteolytic control over assembly of the respiratory chain. *Biochem. J.* **190**: 145.

Gancedo, C. (1971) Inactivation of fructose-1, 6-bisphosphatase by glucose in yeast. *J. Bacteriol.* **107**: 401.

Gancedo, C. and Schwerzmann, K. (1976) Inactivation by glucose of phosphoenolpyruvate carboxykinase from *Saccharomyces cerevisiae*. *Arch. Microbiol.* **109**: 221.

Gancedo, J.M., Lopez, S. and Ballesteros, F. (1982) Calculation of half-lives of proteins *in vivo*. Heterogeneity in the rate of degradation of yeast proteins. *Mol. Cell. Biochem.* **43**: 89.

Garcia Alvarez, N., Bordallo, C., Gascon, S. and Suarez Rendueles, P. (1985) Purification and characterization of a thermosensitive X-prolyl dipeptidyl amino peptidase (dipeptidyl aminopeptidase yscV) from *Saccharomyces cerevisiae*. *Biochim. Biophys. Acta* **832**: 119.

Garcia Alvarez, N., Teichert, U. and Wolf, D.H. (1987) Proteinase yscD mutants of yeast. Isolation and characterization. *Eur. J. Biochem.* **163**: 339.

Goldberg, A.L. and Dice, J.F. (1974) Intracellular protein degradation in mammalian and bacterial cells. *Ann. Rev. Biochem.* **43**: 835.

Goldberg, A.L. and St. John, A.C. (1976) Intracellular protein degradation in mammalian and bacterial cells: Part 2. *Ann. Rev. Biochem.* **45**: 747.

Goldstein, G., Scheid, M., Hammerling, E., Boyse, E.A., Schlesinger, D.H. and Naills, H.D. (1975) Isolation of a polypeptide that has lymphocyte-differentiating properties and is probably represented universally in living cells. *Proc. Natl. Acad. Sci. USA* **72**: 11.

Gorenstein, C. and Warner, J.R. (1977) Synthesis and turnover of ribosomal proteins in the absence of 60 S subunit assembly in *Saccharomyces cerevisiae*. *Mol. Gen. Genet.* **157**: 327.

Görts, C.P.M. (1969) Effect of glucose on the activity and the kinetics of the maltose-uptake system and of α-glucosidase *Saccharomyces cerevisiae*. *Biochim. Biophys. Acta* **184**: 299.

Gottesman, S. (1981) Genetic control of the SOS system in *E. coli*. *Cell* **23**: 1.

Green, R., Schaber, M.D., Shields, D. and Kramer, R. (1986) Secretion of somatostatin by *Saccharomyces cerevisiae*. *J. Biol. Chem.* **261**: 7558.

Grossenbacher, H., Auden, J.A.L., Bill, K., Liersch, M. and Maerki, W. (1987) Isolation and characterization of recombinant desulfatohirudin from yeast. A highly selective thrombin inhibitor. *Thromb. Res. Suppl.* **VII**: 34.

Haarasilta, S. and Oura, E. (1975) On the activity and regulation of anaplerotic and gluconeogenetic enzymes during the growth process of baker's yeast. *Eur. J. Biochem.* **52**: 1.

Haguenauer-Tsapis, R. and Hinnen, A. (1984) A deletion that includes the signal peptidase cleavage site impairs processing, glycosylation, and secretion of cell surface yeast acid phosphatase. *Mol. Cell. Biol.* **4**: 2668.

Hahn, M. (1898) Das proteolytische Enzym des Hefepressaftes. *Ber. Dt. Chem. Ges.* **31**: 200.

Halvorson, H. (1958a) Studies on protein and nucleic acid turnover in growing cultures of yeast. *Biochim. Biophys. Acta* **27**: 227.

Halvorson, H. (1958b) Intracellular protein and nucleic acid turnover in resting yeast cells. *Biochim. Biophys. Acta* **27**: 255.

Hanawalt, P.C., Cooper, P.K., Ganesan, A.K. and Smith, C.A. (1979) DNA repair in bacteria and mammalian cells. *Ann. Rev. Biochem.* **48**: 783.

Hanes, S.D., Burn, V.E., Sturley, S.L., Tipper, D.J., Bostian, K.A. (1986) Expression of a cDNA derived from the yeast killer preprotoxin gene: implications for processing and immunity. *Proc. Natl. Acad. Sci. USA* **83**: 1675.

Hansen, R. J., Switzer, R.L., Hinze, H. and Holzer, H. (1977) Effects of glucose and nitrogen source on the levels of proteinases, peptidases, and proteinase inhibitors in yeast. *Biochim. Biophys. Acta* **496**: 103.

Hashimoto, C., Cohen, R.E., Zhang, W.J. and Ballou, C.E. (1981) Carbohydrate chains on yeast carboxypeptidase Y are phosphorylated. *Proc. Natl. Acad. Sci. USA* **78**: 2244.

Hasilik, A. and Tanner, W. (1976) Inhibition of the apparent rate of synthesis of the vacuolar glycoprotein carboxypeptidase Y and its protein antigen by tunicamycin in *Saccharomyces cerevisiae*. *Antimicrob. Agents Chemother.* **10**: 402.

Hasilik, A. and Tanner, W. (1978a) Carbohydrate moiety of carboxypeptidase Y and perturbation of its biosynthesis. *Eur. J. Biochem.* **91**: 567.

Hasilik, A. and Tanner, W. (1978a) Biosynthesis of the vacuolar yeast glycoprotein carboxypeptidase Y. Conversion of precursor into the enzyme. *Eur. J. Biochem.* **85**: 599.

Hata, T., Hayashi, R. and Doi, E. (1967) Purification of yeast proteinases. Part III. Isolation and physicochemical properties of yeast proteinase A and C. *Agr. Biol. Chem.* **31**: 357.

Hayashi, R., Aibara, S. and Hata, T. (1970) A unique carboxypeptidase activity of yeast proteinase C. *Biochim. Biophys. Acta* **212**: 359.

Hayashi, R., Moore, S. and Stein, W.H. (1973) Carboxypeptidase from yeast. Large scale preparation and the application to COOH-terminal analysis of peptides and proteins. *J. Biol. Chem.* **248**: 2296.

Hayashi, R., Bai, Y. and Hata, T. (1975) Evidence for an essential histidine in carboxypeptidase Y. Reaction with the chloromethyl ketone derivative of benzyloxycarbonyl-L-phenylalanine. *J. Biol. Chem.* **250**: 5221.

Hayashi, R. (1976) Carboxypeptidase Y. In: Proteolytic Enzymes. *Meth. Enzymol.* **45B**: 568.

Heim, J., Treichler, H.J., Hanni, C., Meyhack, B. and Hinnen, A. (1985) Biochemical characterization of recombinant mature t-PA and acid phosphatase t-PA hybrids in yeast. *EMBO Workshop. Amalfi.*

Hemmings, B.A. (1980) Phosphorylation and proteolysis regulate the NAD-dependent glutamate dehydrogenase from *Saccharomyces cerevisiae*. *FEBS Lett.* **122**: 297.

Hemmings, B.A., Zubenko, G.S. and Jones, E.W. (1980) Proteolytic inactivation of the NADP-dependent glutamate dehydrogenase in proteinase-deficient mutants of *Saccharomyces cerevisiae*. *Arch. Biochem. Biophys.* **202**: 657.

Hemmings, B.A., Zubenko, G.S., Hasilik, A. and Jones, E.W. (1981) Mutant defective in processing of an enzyme located in the lysosome-like vacuole of *Saccharomyces cerevisiae*. *Proc. Natl. Acad. Sci. USA* **78**: 435.

Hershko, A. and Ciechanover, A. (1982) Mechanisms of intracellular protein breakdown. *Ann. Rev. Biochem.* **51**: 335.

Hershko, A. (1983) Ubiquitin: Roles in protein modification and breakdown. *Cell* **34**: 11.

Hicks, J.B. and Herskowitz, I. (1976) Evidence for a new diffusible element of mating pheromones in yeast. *Nature* **260**: 246.

Hinnen, A., Meyhack, B. and Tsapis, R. (1983) High expression and secretion of foreign proteins in yeast. In *Gene Expression in Yeast, Proc. Alko Yeast Symp. Helsinki 1983*, Vol. 1, eds. Korhola, M. and Vaisanen, E., Foundation for Biotechnical and Industrial Fermentation Research, 157.

Hirsch, H., Suarez Rendueles, M.P., Aschstetter, T. and Wolf, D.H. (1988) Aminopeptidase yscII of yeast: isolation of mutants and their biochemical and genetic analysis. *Eur. J. Biochem.* (1988) **173**: 589.

Hitzemann, R.A., Hagie, F.E., Levine, H.L., Goeddel, D.V., Ammerer, G. and Hall, B.D. (1981) Expression of a human gene for interferon in yeast. *Nature* **293**: 717.

Hitzemann, R.A., Chen, C.Y., Hagie, F.E., Patzer, E.J., Lui, C.C., Estell, D.A., Miller, J.V., Yaffe, A., Kleid, D.G., Levison, A.D. and Oppermann, H. (1983b) Expression of hepatitis B virus surface antigen in yeast. *Nucleic. Acids Res.* **11**: 2745.

Hitzemann, R.A., Leung, D.W., Perry, L.J., Kohr, W.J., Levine, H.L. and Goeddel, D.V. (1983a) Secretion of human interferons by yeast. *Science* **219**: 620.

Holzer, H., Betz, H. and Ebner, E. (1975) Intracellular proteinases in microorganisms. *Curr. Top. Cell. Reg.* **9**: 103.

Holzer, H. (1976) Catabolite inactivation in yeast. *Trends Biochem. Sci.* **1**: 178.

Holzer, H., Bünning, P. and Meussdoerffer, F. (1977) Characteristics and functions of proteinase A and its inhibitors in yeast. In *Acid Proteases*, ed. Tang, J., Plenum, New York, 271.

Holzer, H. and Heinrich, P.C. (1980) Control of proteolysis *Ann. Rev. Biochem.* **49**: 63.

Huang, S., Elliott, R.C., Liu, P.S., Koduri, R.K., Weickmann, J.L., Lee, J.H., Blair, L.C., Ghosh-Dastidar, P., Bradshaw, R.A., Bryan, K.M., Einarson, B., Kendall, R.L., Kolacz, K.H. and Saito, K. (1987) Specificity of Cotranslational Amino-Terminal Processing of Proteins in Yeast. *Biochemistry* **26**: 8242.

Innis, M.A., Holland, M.J., McCabe, P.C., Cole, G.E., Wittman, V.P., Tal, R., Watt, K.W.K., Gelfand, D.H., Holland, J.P. and Meade, J.H. (1985) Expression, glycosylation, and secretion of an *Aspergillus* glucoamylase by *Saccharomyces cerevisiae*. *Science* **228**: 21.

Itoh, Y., Hayakawa, T. and Fujisawa, Y. (1986) Expression of hepatitis B virus surface antigen P31 gene in yeast. *Biochem. Biophys. Res. Commun.* **138**: 268.

Jabbar, M.A., Sivasubramanian, N. and Nayak, D.P. (1985) Influenza viral (A/WSN/33) hemagglutinin is expressed and glycosylated in the yeast *Saccharomyces cerevisiae*. *Proc. Natl. Acad. Sci. USA* **82**: 2019.

Jabbar, M.A. and Nayak, D.P. (1987) Signal processing glycosylation, and secretion of mutant hemagglutinins of a human influenza virus by *Saccharomyces cerevisiae*. *Mol. Cell. Biol.* **7**: 1476.

James, M.N.G. and Sielecki, A.R. (1986) Molecular structure of an aspartic proteinase zymogen, porcine pepsinogen, at 1.8 A resolution. *Nature* **319**: 33.

Jenness, D.D., Burkholder, A.C. and Hartwell, L.H. (1983) Binding of α-factor pheromone to yeast a cells. Chemical and genetic evidence for an α-factor receptor. *Cell* **35**: 521.

Jentsch, S., McGrath, J.P. and Varshavsky, A. (1987) The yeast DNA repair gene RAD6 encodes a ubiquitin-conjugating enzyme. *Nature* **329**: 131.

Jigami, Y., Muraki, M., Harada, N. and Tanaka, H. (1986) Expression of synthetic human-lysozyme gene in *Saccharomyces cerevisiae*: use of a synthetic chicken-lysozyme signal for secretion and processing. *Gene* **43**: 273.

Johansen, J.T., Breddam, K. and Ottesen, M. (1976) Isolation of carboxypeptidase Y by affinity chromatography. *Carlsberg Res. Commun.* **41**: 1.

Jones, E.W. (1977) Proteinase mutants of *Saccharomyces cerevisiae*. *Genetics* **85**: 23.

Jones, E.W., Zubenko, G.S. and Parker, R.R. (1982) *PEP4* gene function is required for expression of several vacuolar hydrolases in *Saccharomyces cerevisiae*. *Genetics* **102**: 665.

Jones, E.W. (1984) The synthesis and function of proteases in *Saccharomyces cerevisiae*: genetic approaches. *Ann. Rev. Genet.* **18**: 233.

Julius, D., Blair, L.C., Brake, A.J., and Thorner, J. (1983*b*) Yeast pheromone biosynthesis: over-glycosylation of prepro-α-factor in kex2 mutants. In *The Molecular Biology of Yeast*, eds. Hicks, J.B., Klar, A. and Strathern, J.N., Cold Spring Harbor Laboratory, New York, 351.

Julius, D., Blair, L., Brake, A., Sprague, G. and Thorner, J. (1983*a*) Yeast α-factor is processed from a larger precursor polypeptide: the essential role of a membrane-bound dipeptidyl aminopeptidase. *Cell* **32**: 839.

Julius, D., Brake, A., Blair, L., Kunisawa, R. and Thorner, J. (1984*a*) Isolation of the putative structural gene for the lysine-arginine-cleaving endopeptidase required for processing of yeast prepro-α-factor. *Cell* **37**: 1075.

Julius, D., Schekman, R. and Thorner, J. (1984*b*) Glycosylation and processing of prepro-α-factor through the yeast secretory pathway. *Cell* **36**: 309.

Kaczorek, M., Honore, N., Ribes, V., Dehoux, P., Cornet, P., Cartwright, T. and Streeck, R.E. (1987) Molecular cloning and synthesis of biologically active human tissue inhibitor of metalloproteinases in yeast. *Biotechnology* **5**: 595.

Kingsman, S.M., Kingsman, A.J. and Mellor, J. (1987) The production of mammalian proteins in *Saccharomyces cerevisiae*. *Trends in Biotechnol.* **5**: 53.

Klar, A.J.S. and Halvorson, H.O. (1975) Proteinase activities of *Saccharomyces cerevisiae* during sporulation. *J. Bacteriol.* **124**: 863.

Knuver, J. (1979) Isolierung und Charakterisierung der Aminopeptidase II aus Hefe (*Saccharomyces cerevisiae*). Diplomarbeit, Fachbereich Chemie der Philipps-Universität Marburg/Lahn.

Kominami, E., Hoffschulte, H., Leuschel, L., Maier, K. and Holzer, H. (1981*b*) The substrate specificity of proteinase B from baker's yeast. *Biochim. Biophys. Acta.* **661**: 136.

Kominami, E., Hoffschulte, H. and Holzer, H. (1981) Purification and properties of proteinase B from yeast. *Biochim. Biophys. Acta* **661**: 124.

Kornbluth, S., Jove, R. and Hanafusa, H. (1987) Characterization of avian and viral p60 src proteins expressed in yeast. *Proc. Natl. Acad. Sci. USA* **84**: 4455.

Kramer, R.A., DeChiara, T.M., Schaber, M.D. and Hilliker, S. (1984) Regulated expression of a human interferon gene in yeast: control by phosphate concentration or temperature. *Proc. Natl. Acad. Sci. USA* **81**: 367.

Kramer, R.A., Schaber, M.D., Skalka, A.M., Ganguly, K., Wong-Staal, F. and Reddy, E.P. (1985) HTLV-III gag protein is processed in yeast cells by the virus pol-protease. *Science* **231**: 1580.

Kreil, G., Haiml, L. and Suchanek, G. (1980) Stepwise cleavage of the pro part of promelittin by dipeptidyl-peptidase IV. Evidence of a new type of precursor—product conversion. *Eur. J. Biochem.* **111**: 49.

⫣ Kuhn, R.W., Walsh, K.A. and Neurath, H. (1974) Isolation and partial characterization of an acid carboxypeptidase from yeast. *Biochemistry* **13**: 3871.

Kuhn, R.W., Walsh, K.A. and Neurath, H. (1976) Reaction of yeast carboxypeptidase C with group-specific reagents. *Biochemistry* **22**: 4481.

Kurjan, J. and Herskowitz, I. (1982) Structure of a yeast pheromone gene (MFα): a putative α-factor precursor contains four tandem copies of mature α-factor. *Cell* **30**: 933.

Leibowitz, M.J. and Wickner, R.B. (1976) A chromosomal gene required for killer plasmid expression, mating, and spore maturation in *Saccharomyces cerevisiae*. *Proc. Natl. Acad. Sci. USA* **73**: 2061.

Lenney, J.F. and Dalbec, J.M. (1967) Purification and properties of two proteinases from *Saccharomyces cerevisiae*. *Arch. Biochem. Biophys.* **120**: 42.

Lenney, J.F. and Dalbec, J.M. (1969) Yeast proteinase B: Identification of the inactive forms as an enzyme-inhibitor complex. *Arch. Biochem. Biophys.* **129**: 407.

⫣ Lenney, J.F., Matile, P., Wiemken, A., Schellenberg, M. and Meyer, J. (1974) Activities and cellular localization of yeast proteases and their inhibitors. *Biochem. Biophys. Res. Commun.* **60**: 1378.

Lenney, J.F. (1975) Three yeast proteins that specifically inhibit yeast proteases A, B and C. *J. Bacteriol.* **122**: 1265.

Lille, S.H. and Pringle, J.L. (1980) Reserve carbohydrate metabolism in *Saccharomyces cerevisiae*. Responses to nutrient limitation. *J. Bacteriol.* **143**: 1384.

Loison, G., Findeli, A., Bernard, A., Nguyen-Juilleret, M., Marquet, M., Riehl-Bellon, N., Carvallo, D., Guerra-Santos, L., Brown, S.W., Courtney, M., Roitsch, C. and Lemoine, Y. (1986) Expression and secretion in *Saccharomyces cerevisiae* of biologically active leech hirudin. *Biotechnology* **6**: 72.

Lolle, S., Skipper, N., Bussey, H. and Thomas, D.Y. (1984) The expression of cDNA clones of yeast M1 double-stranded RNA in yeast confers both killer and immunity phenotypes. *EMBO J.* **3**: 1383.

Lopez, S. and Gancedo, J.M. (1979) Effect of metabolic conditions on protein turnover in yeast. *Biochem. J.* **178**: 769.

MacKay, V.L., Welch, S.K., Insley, M.Y., Manney, T.R., Holly, J., Saari, G.C. and Parker, M.L. (1988) The *Saccharomyces cerevisiae* BAR1 gene encodes an exported protein with homology to pepsin. *Proc. Natl. Acad. Sci. USA* **85**: 55.

Magni, G., Santarelli, I., Natalini, P., Ruggieri, S. and Vita, A. (1977) Catabolite inactivation of baker's-yeast uridine nucleosidase. Isolation and partial purification of a specific proteolytic inactivase. *Eur. J. Biochem.* **75**: 77.

Magni, G., Natalini, P. Santarelli, I. and Vita, A. (1982) Baker's yeast protease A. Purification and enzymatic and molecular properties. *Arch. Biochem. Biophys.* **213**: 426.

Maier, K., Müller, H. and Holzer, H. (1979) Purification and molecular characterization of two inhibitors of yeast proteinase B. *J. Biol. Chem.* **254**: 8491.

Maness, P. and Edelmann, G. (1978) Inactivation and chemical alteration of mating factor by cells and spheroblasts of yeast. *Proc. Natl. Acad. Sci. USA.* **75**: 1304.

Manney, T.R. (1983) Expression of the BAR1 gene in *Saccharomyces cerevisiae*: induction by the α-mating pheromone of an activity associated with a secreted protein. *J. Bacteriol.* **155**: 291.

Martin, B.M., Svendsen, I., Viswanatha, T. and Johansen, J.T. (1982) Amino acid sequence of carboxypeptidase Y.I. Peptides from cleavage with cyanogen bromide. *Carlsberg Res. Commun.* **47**: 1.

Masuda, T. Hayashi, R. and Hata, T. (1975). Aminopeptidase in the acidic fraction of the yeast autolysate. *Agr. Biol. Chem.* **39**: 499.

Matern, H., Hoffmann, M. and Holzer, H. (1974) Isolation and characterization of the carboxypeptidase Y inhibitor from yeast. *Proc. Natl. Acad. Sci. USA* **71**: 4874.

Matern, H., Betz, H. and Holzer, H. (1974a) Compartmentation of inhibitors of proteinase A and B and carboxypeptidase Y in yeast. *Biochem. Biophys. Res. Commun.* **60**: 1051.

Matern, H. and Holzer, H. (1977) Catabolite inactivation of the galactose uptake system in yeast. *J. Biol Chem.* **252**: 6399.

Matile, P. and Wiemken, A. (1967) The vacuole as the lysosome of the yeast cell. *Arch. Microbiol.* **56**: 148.

Matile, P., Wiemken, A. and Guyer, W. (1971) Alysosomal aminopeptidase isoenzyme in differentiating yeast cells and protoplasts. *Planta* **96**: 43.

Mazon, M.J. (1978) Effect of glucose starvation on the nicotinamide adenine dinucleotide phosphate-dependent glutamate dehydrogenase of yeast. *J. Bacteriol.* **133**: 780.

Mazon, M.J. and Hemmings, B.A. (1979) Regulation of *Saccharomyces cerevisiae* nicotinamide adenine dinucleotide phosphate-dependent glutamate dehydrogenase by proteolysis during carbon starvation. *J. Bacteriol.* **139**: 686.

Mazon, M.J., Gancedo, J.M. and Gancedo, C. (1982a) Inactivation of yeast fructose-1, 6-bisphosphatase. *In vivo* phosphorylation of the enzyme. *J. Biol. Chem.* **257**: 1128.

Mazon, M.J., Gancedo, J.M. and Gancedo, C. (1982b) Phosphorylation and inactivation of yeast fructose-bisphosphatase *in vivo* by glucose and by proton ionophores. *Eur. J. Biochem.* **127**: 605.

McAda, P.C. and Douglas, M.G. (1982) A neutral metallo endoprotease involved in the processing of an F-ATPase subunit precursor in mitochondria. *J. Biol. Chem.* **257**: 3177.

Mechler, B. and Wolf, D.H. (1981) Analysis of proteinase A function in yeast. *Eur. J. Biochem.* **121**: 47.

Mechler, B., Müller, M., Müller, H., Meussdoerffer, F. and Wolf, D.H. (1982a) *In vivo* biosynthesis of the vacuolar proteinases A and B in the yeast *Saccharomyces cerevisiae*. *J. Biol. Chem.* **257**: 11203.

Mechler, B., Müller, M., Müller, H. and Wolf, D.H. (1982b) *In vivo* biosynthesis of vacuolar proteinases in proteinase mutants of *Saccharomyces cerevisiae*. *Biochem. Biophys. Res. Commun.* **107**: 770.

Mechler, B., Müller, H. and Wolf, D.H. (1987) Maturation of vacuolar (lysosomal) enzymes in yeast: proteinase yscA and proteinase yscB are catalysts of the processing and activation event of carboxypeptidase yscY. *EMBO J.* **6**: 2157.

Mechler, B., Hirsch, H.H., Müller, H. and Wolf, D.H. (1988) Biogenesis of the yeast lysosome (vacuole): biosynthesis and maturation of proteinase yscB. *EMBO J.* **7**: 1705.

Mellor, J., Dobson, M.J., Roberts, N.A., Tuite, M.F., Emtage, J.S., White, S., Lowe, P.A., Patel, T., Kingsman, A.J. and Kingsman, S.M. (1983) Efficient synthesis of enzymatically active calf chymosin in *Saccharomyces cerevisiae. Gene* **24**: 1.

Mellor, J., Kingsman, A.J. and Kingsman, S.M. (1986) Ty, an endogenous retrovirus of yeast? *Yeast.* **2**: 145.

Messenguy, F., Lolin, D. and ten Have, J.P. (1980) Regulation of compartmentation of amino acid pools in *Saccharomyces cerevisiae* and its effects on metabolic control. *Eur. J. Biochem.* **108**: 439.

Meussdoerffer, F., Tortora, P. and Holzer, H. (1980) Purification and properties of proteinase A from yeast. *J. Biol. Chem.* **255**: 12087.

Meyhack, B., Marki, W., Heim, J. and Hinnen, A. (1986) High level expression of a proteinase inhibitor in yeast. *Experientia* **42**: 644.

Meyhack, B., Heim, J., Rink, H., Zimmerman, W. and Marki, W. (1987) Desulfatohirudin, a specific thrombin inhibitor: expression and secretion in yeast. *Thromb. Res. Suppl.* **VII**: 33.

Miller, G.C. (1975) Peptidases and proteases of *Escherichia coli* and *Salmonella typhimurium. Ann. Rev. Microbiol.* **29**: 485.

Miyajima, A., Bond, M.W., Otsu, K., Arai, K. and Arai, N. (1985) Secretion of mature mouse interleukin-2 by *Saccharomyces cerevisiae*: use of a general secretion vector containing promoter and leader sequences of the mating pheromone α-factor. *Gene* **37**: 155.

Miyajima, A., Otsu, K., Schreurs, J., Bond, M.W., Abrams, J.S. and Arai, K. (1986) Expression of murine and human granulocyte-macrophage colony-stimulating factors in *Saccharomyces cerevisiae*: mutagenesis of the potential glycosylation sites. *EMBO J.* **6**: 1193.

Miyamoto, C., Chizzonite, R., Crowl, R., Rupprecht, K., Kramer, R., Schaber, M., Kumar, G., Poonian, M. and Ju, G. (1985) Molecular cloning and regulated expression of the human c-myc gene in *Escherichia coli* and *Saccharomyces cerevisiae*: comparison of the protein products. *Proc. Natl. Acad. Sci. USA* **82**: 7232.

Miyanohara, A., Toh-E, A., Nozaki, C., Hamada, F., Ohtomo, N. and Matsubara, K. (1983) Expression of hepatitis B surface antigen in yeast. *Proc. Natl. Acad. Sci. USA* **80**: 1.

Mizuno, K. and Matsuo, H. (1984) A novel protease from yeast with specificity towards paired basic residues. *Nature* **309**: 558.

Moehle, C.M., Aynardi, M.W., Kolodny, M.R., Park, F.J. and Jones, E.W. (1987*a*) Protease B of *Saccharomyces cerevisiae*. Isolation and regulation of the *PRB1* structural gene. *Genetics* **115**: 255.

Moehle, C.M., Tizard, R., Lemmon, S.K., Smart, J. and Jones, E.W. (1987*b*) Protease B of the lysosome-like vacuole of the yeast *Saccharomyces cerevisiae* is homologous to the subtilisin family of serine proteases. *Mol. Cell. Biol.* **7**: 4390.

Moir, D.T. and Dumais, D.R. (1987) Glycosylation and secretion of human alpha-1-antitrypsin by yeast. *Gene* **56**: 209.

Monod, M., Rauseo-Koenig, I., Silve, S., Haguenauer-Tsapis, R. and Hinnen, A. (1986) Phenotype associated with *in vitro* constructed signal sequence mutations in the yeast *PHO5* gene. In *Genetics of Industrial Microorganisms. Proc. Int. Symp.*, eds. Alacevic, M., Hranueli, D. and Toman, A., 193.

Monod, M., Silve, S., Haguenauer-Tsapis, R. and Hinnen, A. (1987) *In vivo* and *in vitro* studies of *PHO5* mutants at the signal peptidase cleavage site. *Found. Biotechnol. Indust. Ferment. Res.* **5**: 149.

Moody, A.J., Norris, F., Norris, K., Hansen, M.T. and Thim, L. (1987) The secretion of glucagon by transformed yeast strains. *FEBS Lett.* **212**: 302.

Motizuki, M., Mitsui, K., Endo, Y. and Tsurugi, K. (1986) Detection and partial characterization of the chromatin-associated proteases of yeast *Saccharomyces cerevisiae. Eur. J. Biochem.* **158**: 345.

Mullenbach, G.T., Tabrizi, A., Blacher, R.W. and Steimer, K.S. (1986) Chemical synthesis and expression in yeast of a gene encoding connective tissue activating peptide-III. *J. Biol. Chem.* **261**: 719.

Müller, M. and Holzer, H. (1981) Proteolysis and catabolite inactivation in yeast. In *Proteinases*

196   MOLECULAR AND CELL BIOLOGY OF YEASTS

and Their Inhibitors. Structure, Function and Applied Aspects, eds. Turk, V. and Vitale, L., Pergamon, Oxford, 57.

Müller, M., Müller, H. and Holzer, H. (1981) Immunochemical studies on catabolite inactivation of phosphoenoylpyruvate carboxykinase in *Saccharomyces cerevisiae. J. Biol. Chem.* **256**: 723.

Müller, D. and Holzer, H. (1981) Regulation of fructose-1, 6-bisphosphatase in yeast by phosphorylation/dephosphorylation. *Biochem. Biophys. Res. Commun.* **103**: 926.

Müller, M. and Müller, H. (1981) Synthesis and processing of the *in vitro* and *in vivo* precursors of the vacuolar yeast enzyme carboxypeptidase Y. *J. Biol. Chem.* **256**: 11962.

Neeff, J., Hagele, E., Nauhaus, J., Heer, U. and Mecke, D. (1978) Evidence of catabolite degradation in the glucose-dependent inactivation of yeast cytoplasmic malate dehydrogenase. *Eur. J. Biochem.* **86**: 489.

North, M.J. (1982) Comparative biochemistry of the proteinases of eukaryotic microorganisms. *Microbiol. Rev.* **46**: 308.

Norvick, P., Field, C. and Schekman, R. (1980) Identification of 23 complementation groups required for post-translational events in the yeast secretory pathway. *Cell* **21**: 205.

Novick, P., Ferro, S. and Schekman, R. (1981) Order of events in the yeast secretory pathway. *Cell* **25**: 461.

Nunez de Castro, I. and Holzer, H. (1976) Studies on proteinase-A inhibitor I from yeast. *Hoppe-Seyler's Z. Physiol. Chem.* **357**: 727.

Okada, T., Sonomoto, K., Tanaka, A. (1987) Novel Leu-lys-specific peptidase (LEULYSIN) produced by gel entrapped yeast cells. *Biochem. Biophys. Res. Commun.* **145**: 316.

Orlean, P. (1987) Two chitin synthetases in *Saccharomyces cerevisiae. J. Biol. Chem.* **262**: 5732.

Özkaynak, E., Finley, D. and Varshavsky, A. (1984) The yeast ubiquitin gene: head-to-tail repeats encoding a polyubiquitin precursor protein. *Nature* **312**: 663.

Özkaynak, E., Finley, D., Solomon, M.J. and Varshavsky, A. (1987) The yeast ubiquitin genes: a family of natural gene fusions. *EMBO J.* **6**: 1429.

Perlman, D., Halvorson, H.O. and Cannon, E.L. (1982) Presecretory and cytoplasmic invertase polypeptides encoded by distinct mRNAs derived from the same structural gene differ by a signal sequence. *Proc. Natl. Acad. Sci. USA* **79**: 781.

Pine, M.J. (1972) Turnover of intracellular proteins. *Ann. Rev. Microbiol.* **26**: 103.

Pohlig, G., Roberts, C.J., Rothman, J.H. and Stevens, T.H. (1986) Biochemical and genetic analysis of vacuolar membrane protein localization. In *13th Int. Conf. on Yeast Genetics and Molecular Biology, Yeast* **2**: S306.

Pontremoli, S. and Melloni, E. (1986) Extralysosomal protein degradation. *Ann. Rev. Biochem.* **55**: 455.

Price, V., Mochizuki, D., March, C.J., Cosman, D., Deeley, M.C., Klinke, R., Clevenger, W., Gillis, S., Baker, P. and Urdal, D. (1987) Expression, purification and characterization of recombinant murine granulocyte-macrophage colony-stimulating factor and bovine interleukin-2 from yeast. *Gene* **55**: 287.

Purwin, C., Leidig, F. and Holzer, H. (1982) Cyclic AMP-dependent phosphorylation of fructose-1, 6-bisphosphatase in yeast. *Biochem. Biophys. Res. Commun.* **107**: 1482.

Radman, M. (1980) Is there SOS induction in mammalian cells? *Photochem. Photobiol.* **32**: 823.

Reid, G.A., Yonetani, T. and Schatz, G. (1982) Import of proteins into mitochondria. Import and maturation of the mitochondrial intermembrane space enzymes cytochrome b_2 and cytochrome c peroxidase in intact yeast cells. *J. Biol. Chem.* **257**: 13068.

Riezman, H., Chvatchko, Y. and Dulic, V. (1986) Endocytosis in yeast. *Trends Biochem. Sci.* **11**: 325.

Robertson, J.J. and Halvorson, H.O. (1957) The components of maltozymase in yeast and their behavior during deadaptation. *J. Bacteriol.* **73**: 186.

Rogers, S., Wells, R. and Rechsteiner, M. (1986) Amino acid sequences common to rapidly degraded proteins: the PEST hypothesis. *Science* **234**: 364.

Roggenkamp, R., Dargatz, H. and Hollenberg, C.P. (1985) Precursor of β-lactamase is enzymatically inactive. *J. Biol. Chem.* **260**: 1508.

Röhm, K.H. (1974) Properties of a highly purified dipeptidase from brewer's yeast. *Hoppe-Seyler's Z. Physiol. Chem.* **355**: 675.

Rosenberg, S., Barr, P.J., Najarian, R.C. and Hallewell, R.A. (1984) Synthesis in yeast of a functional oxidation-resistant mutant of human α-antitrypsin. *Nature* **312**: 77.

Rothman, J.H., Hunter, C.P., Valls, L.A. and Stevens, T.H. (1986) Overproduction-induced mislocalization of a yeast vacuolar protein allows isolation of its structural gene. *Proc. Natl. Acad. Sci.* **83**: 3248.

Rothstein, S.J., Lazarus, C.M., Smith, W.E., Baulcombe, D.C. and Gatenby, A.A. (1984) Secretion of a wheat α-amylase expressed in yeast. *Nature* **308**: 662.

Saheki, T., Matsuda, Y. and Holzer, H. (1974) Purification and characterization of macromolecular inhibitors of proteinase A from yeast. *Eur. J. Biochem.* **47**: 325.

Saheki, T. and Holzer, H. (1974) Comparison of the tryptophan synthase inactivation enzymes with proteinases from yeast. *Eur. J. Biochem.* **42**: 621.

Sato, T., Tsunasawa, S., Nakamura, Y., Emi, M., Sakiyama, F. and Matsubara, K. (1986) Expression of the human salivary α-amylase gene in yeast and characterization of the secreted protein. *Gene* **50**: 247.

Sburlati, A. and Cabib, E. (1986) Chitin synthetase 2, a presumptive participant in septum formation in *Saccharomyces cerevisiae. J. Biol. Chem.* **261**: 15147.

Scarpulla, R.C. and Nye, S.H. (1986) Functional expression of rat cytochrome c in *Saccharomyces cerevisiae. Proc. Natl. Acad. Sci. USA* **83**: 6352.

Schatz, G. and Butow, R.A. (1983) How are proteins imported into mitochondria? *Cell* **32**: 316.

Schekman, R. (1985a) The secretory pathway in yeast. *Trends Biochem. Sci.* **7**: 243.

Schekman, R. (1985b) Protein localization and membrane traffic in yeast. *Ann. Rev. Cell Biol.* **1**: 115.

Schwaiger, H., Hasilik, A., Von Figura, K., Wiemken, A. and Tanner, W. (1982) Carbohydrate-free carboxypeptidase Y is transferred into the lysosome-like yeast vacuole. *Biochem. Biophys. Res. Commun.* **104**: 950.

Schweizer, E., Werkmeister, K. and Jain, M. (1978) Fatty acid biosynthesis in yeast. *Mol. Cell. Biochem.* **21**: 95.

Schwencke, J. (1981) Measurement of proteinase B activity in crude yeast extracts: A novel procedure of activation using pepsin. *Anal. Biochem.* **118**: 315.

Schwencke, J. and Moustacchi, E. (1982a) Proteolytic activities in yeast after UV irradiation. I. Variation in proteinase levels in repair proficient RAD strains. *Mol. Gen. Genet.* **185**: 290.

Schwencke, J. and Moustacchi, E. (1982b) Proteolytic activities in yeast after UV irradiation. II. Variation in proteinase levels in mutants blocked in DNA-repair pathways. *Mol. Gen. Genet.* **185**: 296.

Skipper, N., Thomas, D.Y. and Lau, P.C.K. (1984) Cloning and sequencing of the preprotoxin-coding region of the yeast M1 double-stranded RNA. *EMBO J.* **3**: 107.

Smith, R.A., Duncan, M.J. and Moir, D.T. (1985) Heterologous protein secretion from yeast. *Science* **229**: 1219.

Sprague, G.F. Jr., Rine, J. and Herskowitz, I. (1981) Control of yeast cell type by the mating type locus II. Genetic interactions between MAT α and unlinked α-specific STE genes. *J. Mol. Biol.* **153**: 323.

Sprague, G.F. Jr. and Herskowitz, I. (1981) Control of yeast cell type by the mating type locus. I. Identification and control of expression of the a-specific gene BAR1. *J. Mol. Biol.* **153**: 305.

Sprague, G.F. Jr., Blair, L.C. and Thorner, J. (1983) Cell interactions and regulation of cell type in the yeast *Saccharomyces cerevisiae. Ann. Rev. Microbiol.* **37**: 623.

Stevens, T., Esmon, B. and Schekman, R. (1982) Early stages in the yeast secretory pathway are required for transport of carboxypeptidase Y to the vacuole. *Cell* **30**: 439.

Stevens, T.H., Rothman, J.H. Payne, G.S. and Schekman, R. (1986) Gene dosage-dependent secretion of yeast vacuolar carboxypeptidase yscY. *J. Cell. Biol.* **102**: 1551.

Strathern, J.N., Jones, E.W. and Broach, J.R. (eds.) (1983) Metabolism and gene expression. In Appendix 1: *The Molecular Biology of the Yeast Saccharomyces cerevisiae*, Cold Spring Harbor Laboratory, New York, 639.

Sturley, S.L. and Young, T.W. (1987) Production of foreign proteins by yeast. (ed. Russel, G.E.) *Biotech. Genet. Eng. Rev.* **4**: 1.

Suarez Rendueles, M.P., Scwencke, J., Garcia Alvarez, N. and Gascon, S. (1981) A new X-prolyl-dipeptidyl aminopeptidase from yeast associated with a particulate fraction. *FEBS Lett.* **131**: 296.

Suarez Rendueles, P. and Wolf, D.H. (1987) Identification of the structural gene for dipeptidyl aminopeptidase yscV (DAP2) in *Saccharomyces cerevisiae. J. Bacteriol.* **169**: 4041.

Suarez Rendueles, P. and Wolf, D.H. (1988) Proteinase function in yeast: biochemical and genetic approaches to a central mechanism of post-translational control in the eukaryote cell. *FEMS Microbiol. Rev.* **54**: 17.

Svandsen, I., Martin, B.M., Vishwanatha, T. and Johansen, J.T. (1982) Amino acid sequence of carboxypeptidase Y. II. Peptides from enzymatic cleavages. *Carlsberg Res Commun.* **47**: 15.

Takatsuji, H., Hirochika, H., Fukushi, T. and Ikeda, J.E. (1986) Expression of cauliflower mosaic virus reverse transcriptase in yeast. *Nature* **319**: 240.

Teichert, U., Mechler, B., Müller, H. and Wolf, D.H. (1987) Protein degradation in yeast. *Biochem. Soc. Trans.* **15**: 811.

Thim, L., Hansen, M.T., Norris, K., Hoegh, I., Boel, E., Forstron, J., Ammerer, G. and Fiil, N.P. (1986) Secretion and processing of insulin precursors in yeast. *Proc. Natl. Acad. Sci. USA* **83**: 6766.

Thim, L., Hansen, M.T., Ørensen, A.R. (1987) Secretion of human insulin by a transformed yeast cell. *FEBS Lett.* **212**: 307.

Thomsen, K.K. (1983) Mouse α-amylase synthesized by *Saccharomyces cerevisiae* is released into the culture medium. *Carlsberg Res. Commun.* **48**: 545.

Thorner, J. (1981) Pheromonal regulation of development in *Saccharomyces cerevisiae*. In *The Molecular Biology of the Yeast Saccharomyces. Life Cycle and Inheritance*, eds. Strathern, J.N., Jones, E.W. and Broach, J.R., Cold Spring Harbor Laboratory, New York, 143.

Tillmann, U., Günther, R., Schweden, J. and Bause, E. (1987) Subcellular location of enzymes involved in the N-glycosylation and processing of asparagine-linked oligosaccharides in *Saccharomyces cerevisiae*. *Eur. J. Biochem.* **162**: 635.

Tipper, D.J. and Bostian, K.A. (1984) Double-stranded ribonucleic acid killer system in yeasts. *Microbiol. Rev.* **48**: 125.

Tortora, P., Birtel, M., Lenz, A.G. and Holzer, H. (1981) Glucose-dependent metabolic interconversion of fructose-1, 6-bisphosphatase in yeast. *Biochem. Biophys. Res. Commun.* **100**: 688.

Tortora, P., Burlini, N., Leoni, F. and Guerritore, A. (1983) Dependence on cyclic AMP of glucose-induced inactivation of yeast gluconeogenetic enzymes. *FEBS Lett.* **155**: 39.

Travis, J., Owen, M., George, P., Carrell, R., Rosenberg, S., Hallewell, R.A. and Barr, P.J. (1985) Isolation and properties of recombinant DNA produced variants of human α_1-proteinase inhibitor. *J. Biol. Chem.* **260**: 4384.

Trimble, R.B. and Maley, F. (1977) The use of endo-β-N-acetylglucosaminidase H in characterizing the structure and function of glycoproteins. *Biochem. Biophys. Res. Commun.* **78**: 935.

Trimble, R.B., Maley, F. and Chu, F.K. (1983) Glycoprotein biosynthesis in yeast. Protein conformation affects processing of high mannose oligosaccharides on carboxypeptidase Y and invertase. *J. Biol. Chem.* **258**: 2562.

Troll, W., Meyn, M.S. and Rossman, T.G. (1978) Mechanism of protease action in carcinogenesis. In *Carcinogenesis*, Vol. 2, eds. Slaga, T.J., Sivak, A. and Boutwell, R.K., Raven Press, New York, 301.

Trumbly, R.J. and Bradley, G. (1983) Isolation and characterization of aminopeptidase mutants of *Saccharomyces cerevisiae*. *J. Bacteriol.* **156**: 36.

Tsunasawa, S., Stewart, J.W. and Sherman, F. (1985) Aminoterminal processing of mutant forms of yeast iso-1-cytochrome c. The specificities of methionine aminopeptidase and acyltransferase. *J. Biol. Chem.* **260**: 5382.

Tzumoto, Y., Sato, T., Yamamoto, T., Yoshida, N., Kikuchi, N., Ogawa, M. and Matsubara, K. (1987) Expression of human pancreatic secretory trypsin inhibitor in *Saccharomyces cerevisiae*. *Gene* **59**: 151.

Ulane, R.E. and Cabib, E. (1976) The activating system of chitin synthetase from *Saccharomyces cerevisiae*. *J. Biol. Chem.* **251**: 3367.

Urdea, M.S., Merryweather, J.P., Mullenbach, B.T., Coit, D., Heberlein, U., Valenzuela, P. and Barr, P.J. (1983) Chemical synthesis of a gene for human epidermal growth urogastrone and its expression in yeast. *Proc. Natl. Acad. Sci. USA* **80**: 7461.

Valenzuela, P., Medina, A., Rutter, W.J., Ammerer, G. and Hall, B.D. (1982) Synthesis and assembly of hepatitis B virus surface antigen particles in yeast. *Nature* **298**: 347.

Valls, L.A., Hunter, C.P., Rothman, J.H. and Stevens, T.H. (1987) Protein sorting in yeast: The localization determinant of yeast vacuolar carboxypeptidase Y resides in the propeptide. *Cell* **48**: 887.

Van den Bergh, C.J., Bekkers, C.A.P.A., De Geus, P., Verheij, H.M. and De Haas, G.H. (1987)

Secretion of biologically active porcine prophospholipase A_2 by *Saccharomyces cerevisiae*. Use of the prepro sequence of the α-mating factor. *Eur. J. Biochem.* **170**: 241.

Vlasuk, G.P., Bencen, G.H., Scarborough, R.M., Tsai, P.K., Whang, J.L., Maack, T., Camargo, M.J.F., Krisher, S.W. and Abraham, J.A. (1986) Expression and secretion of biologically active human atrial natriuretic peptide in *Saccharomyces cerevisiae*. *J. Biol. Chem.* **261**: 4789.

Von Heinje, G. (1983) Patterns of amino acids near signal-sequence cleavage sites. *Eur. J. Biochem.* **133**: 17.

Wagner, J.C. and Wolf, D.H. (1987) Hormone (Pheromone) processing enzyme in yeast. *FEBS Lett.* **221**: 423.

Wagner, J.C., Escher, C. and Wolf, D.H. (1987) Some characteristics of hormone (pheromone) processing enzymes in yeast. *FEBS Lett.* **216**: 31.

Wampler, D.E., Lehman, E.D., Boger, J., McAleer, W.J. and Scolnik, E. (1985) Multiple chemical forms of hepatitis B surface antigen produced in yeast. *Proc. Natl. Acad. Sci. USA* **82**: 6830.

Wen, D. and Schlesinger, M.J. (1986) Regulated expression of Sindbis and vesicular stomatitis virus glycoproteins in *Saccharomyces cerevisiae*. *Proc. Natl. Acad. Sci. USA* **83**: 3639.

Wickner, R.B. and Leibowitz, M.J. (1976) Two chromosomal genes required for killing expression in killer strains of *Saccharomyces cerevisiae*. *Genetics* **82**: 429.

Wiemken, A. and Durr, M. (1974) Characterization of amino acid pools in the vacuolar compartment of *S. cerevisiae*. *Arch. Microbiol.* **101**: 45.

Wiemken, A., Schellenberg, M. and Urech, K. (1979) Vacuoles: the sole compartments of digestive enzymes in yeast (*Saccharomyces cerevisiae*)? *Arch. Microbiol.* **123**: 23.

Wilkinson, K.D., Cox, M.J., O'Connor, L.B. and Shapira, R. (1986) Structure and activities of a variant ubiquitin sequence from bakers yeast. *Biochemistry* **25**: 4999.

Witkin, E.M. (1976) Ultraviolet mutagenesis and inducible DNA repair in *Escherichia coli*. *Bacteriol. Rev.* **40**: 869.

Witt, I., Kronau, R. and Holzer, H. (1966) Repression von Alkoholdehydrogenase, Malatdehydrogenase, Isocitratlyase und Malatsynthase in Hefe durch Glucose. *Biochim. Biophys. Acta* **118**: 522.

Wolf, D.H. and Fink, G.R. (1975) Proteinase C (carboxypeptidase Y) mutant of yeast. *J. Bacteriol.* **123**: 1150.

Wolf, D.H. and Weiser, U. (1977) Studies on a carboxypeptidase Y mutant of yeast and evidence for a second carboxypeptidase activity. *Eur. J. Biochem.* **73**: 553.

Wolf, D.H. and Ehmann, C. (1978a) Carboxypeptidase S from yeast: Regulation of its activity during vegetative growth and differentiation. *FEBS Lett.* **91**: 59.

Wolf, D.H. and Ehmann, C. (1978b) Isolation of yeast mutants lacking proteinase B activity. *FEBS Lett.* **92**: 121.

Wolf, D.H. and Ehmann, C. (1979) Studies on a proteinase B mutant of yeast. *Eur. J. Biochem.* **98**: 375.

Wolf, D.H. (1980) Control of metabolism in yeast and other lower eukaryotes through action of proteinases. *Adv. Microbial Physiol.* **21**: 267.

Wolf, D.H. (1981) Proteinases and sporulation in yeast. In *The Fungal Spore: Morphogenetic Controls*, eds. Turian, G. and Hohl, H.R., Academic Press, London, New York, 355.

Wolf, D.H. and Ehmann, C. (1981) Carboxypeptidase Y- and carboxypeptidase S-deficient mutants of *Saccharomyces cerevisiae*. *J. Bacteriol.* **147**: 418.

Wolf, D.H. (1982) Proteinase action *in vitro* versus proteinase function *in vivo*: mutants shed light on intracellular proteolysis in yeast. *Trends Biochem. Sci.* **7**: 35.

Wolf, D.H. (1986) Cellular control in the eukaryotic cell through action of proteinases: The yeast *Saccharomyces cerevisiae* as a model organism. *Microbiol. Sci.* **3**: 107.

Wood, C.R., Boss, M.A., Kenten, J.H., Calvert, J.E., Roberts, N.A. and Emtage, J.S. (1985) The synthesis and *in vivo* assembly of functional antibodies in yeast. *Nature* **314**: 446.

Woolford, C.A., Daniels, L.B., Park, F.J., Jones, E.W., Van Arsdell, J.N. and Innis, M.A. (1986) The *PEP4* gene encodes an aspartyl protease implicated in the posttranslational regulation of *Saccharomyces cerevisiae* vacuolar hydrolases. *Mol. Cell. Biol.* **6**: 2500.

Yasuhara, T. and Ohashi, A. (1987) New chelator-sensitive proteases in matrix of yeast mitochondria. *Biochem. Biophys. Res. Commun.* **144**: 277.

Yokosawa, H., Ito, H., Murata, S. and Ishii, S.I. (1983) Purification and fluorometric assay of proteinase A from yeast. *Anal. Biochem.* **134**: 210.

Zhu, X.L., Ward, C. and Weissbach, A. (1984) Control of *Herpes simplex* virus thymidine kinase gene expression in *Saccharomyces cerevisiae* by a yeast promoter sequence. *Mol. Gen. Genet.* **194**: 31.

Zsebo, K.M., Lu, H.S., Fieschko, J.C., Goldstein, L., Davis, J., Duker, K., Suggs, S.V., Lai, P.H. and Bitter, G.A. (1986) Protein secretion from *Saccharomyces cerevisiae* directed by the prepro-α-factor leader region. *J. Biol. Chem.* **261**: 5858.

Zubatov, A.S., Mikhailova, A.E. and Luzikova, V.N. (1984) Detection, isolation and some properties of membrane proteinases from yeast mitochondria. *Biochim. Biophys. Acta* **787**: 188.

Zubenko, G.A. and Jones, E.W. (1979) Catabolite inactivation of gluconeogenic enzymes in mutants of yeast deficient in proteinase B. *Proc. Natl. Acad. Sci. USA* **76**: 4581.

Zubenko, G.S., Mitchell, A.P. and Jones, E.W. (1979) Septum formation, cell division, and sporulation in mutants of yeast deficient in proteinase B. *Proc. Natl. Acad. Sci. USA* **76**: 2395.

Zubenko, G.S., Mitchell, A.P. and Jones, E.W. (1980) Mapping of the proteinase B structural gene PRB1, in *Saccharomyces cerevisiae* and identification of nonsense alleles within the locus. *Genetics* **96**: 137.

Zubenko, G.S. and Jones, E.W. (1981) Protein degradation, meiosis and sporulation in proteinase-deficient mutants of *Saccharomyces cerevisiae*. *Genetics* **97**: 45.

Zubenko, G.S., Park, F.J. and Jones, E.W. (1983) Mutations in PEP4 locus of *Saccharomyces cerevisiae* block final step in maturation of two vacuolar hydrolases. *Proc. Natl. Acad. Sci. USA* **80**: 510.

6 Signal transduction by GTP-binding proteins in *Saccharomyces cerevisiae*

KUNIHIRO MATSUMOTO*, KOZO KAIBUCHI, KEN-ICHI ARAI,
MASATO NAKAFUKU and YOSHITO KAZIRO

6.1 Introduction

One family of guanine nucleotide-binding proteins, G proteins, serves as modulators or transducers in various transmembrane signalling systems that consist of three distinct components; receptors, G proteins, and amplifiers. The receptors detect extracellular signals while G proteins functionally couple receptor stimulation to the regulation of amplifiers (Gilman, 1987). The entire chain of events comprises signal transduction. Figure 6.1 depicts the mechanism by which G proteins transduce signals between receptor and amplifier

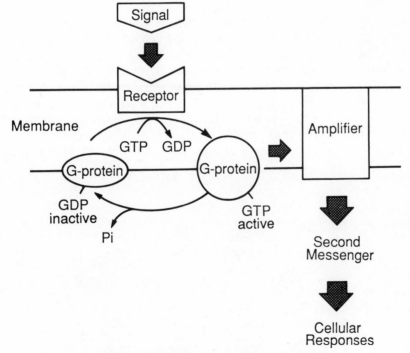

Figure 6.1 Role of G proteins in signal transduction.

*To whom correspondence should be addressed.

proteins. The binding of extracellular signals to the receptor protein alters the conformation of the receptor, enabling it to activate the G protein. The G protein binds GTP. Binding of GTP changes the conformation of the G protein so that the GTP-bound form activates the amplifier protein to produce a second messenger molecule. Hydrolysis of bound GTP terminates the activation of the amplifier.

There are several different types of G proteins (Figure 6.2). Two G proteins, Gs and Gi, are involved in hormonal regulation of the adenylate cyclase activity; the former activates the cyclase in response to the β-adrenergic stimuli while the latter mediates inhibition of the enzyme (Gilman, 1987). Transducin, Gt1, which is present in retinal rods, regulates the activity of cyclic GMP-phosphodiesterase and mediates visual signal transduction (Stryler and Bourne, 1986). Gt2 is thought to play the analogous role in cones (Stryler and Bourne, 1986). Another G protein, Go, which is abundunt in brain tissues (Sternweis and Robishaw, 1984), may be involved in neuronal responses, but its precise function has not yet been clarified. Furthermore, recent evidence suggests the presence of other molecular species of G proteins which may be involved in the activation of phospholipase C (Ohta et al., 1985) and phospholipase A_2 (Nakamura and Ui, 1985), and the gating of K^+ (Pfaffinger et al., 1985) and Ca^{2+} channels (Holz et al., 1986).

Each G protein consists of three different subunits, α ($M_r = 39\,000$–$45\,000$), β ($M_r = 35\,000$–$36\,000$) and γ ($M_r = 8000$–$10\,000$) (Figure 6.2). The α-subunits contain a guanine nucleotide binding site and determine the specificity of the proteins for its receptor and amplifier. Recently cDNA sequences for α-subunits of several G proteins have been determined; for example, Gs from bovine (Nukada et al., 1986a; Robishaw et al., 1986), rat (Itoh et al., 1986), and human tissues (Bray et al., 1986; Mattera et al., 1986); Gi from bovine (Nukada et al., 1986b), rat (Itoh et al., 1986), mouse (Sullivan et al., 1986), and human tissues (Bray et al., 1987; Suki et al., 1987); Go from rat

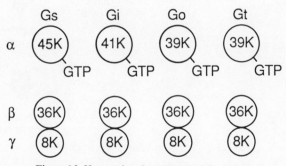

Figure 6.2 Heterotrimeric structure of G proteins.

(Itoh *et al.*, 1986) and bovine brain (Meurs *et al.*, 1987); and Gt1 and Gt2 from bovine retina (Lochrie *et al.*, 1985; Medynski *et al.*, 1985; Tanabe *et al.*, 1985; Yatsunami and Khorana, 1985). The amino acid sequence of Gsα from rat brain (Itoh *et al.*, 1986) is almost identical (about 99% homologous) to that of bovine adrenal Gsα (Robishaw *et al.*, 1986) and to that of bovine brain Gsα (Nukada *et al.*, 1986*a*).

There are three distinct forms of Giα, Gi1α, Gi2α and Gi3α (Jones and Reed, 1987). The sequence of rat brain Gi2α (Itoh *et al.*, 1986) is about 89% homologous with the bovine brain Gi1α sequence (Nukada *et al.*, 1986*b*). These studies have revealed that the nucleotide and the deduced amino acid sequences are highly homologous among different mammalian species.

In view of the strong conservation of the amino acid sequences of each G protein species among different organisms, Nakafuku and colleagues searched for G protein homologous genes in yeast and isolated two different genes, *GPA1* and *GPA2*, from *Saccharomyces cerevisiae* (Nakafuku *et al.*, 1987, 1988). Yeast cells offer several advantages over mammalian cells because of the ease of applying genetic approaches. Studies of the function of G proteins in yeast are expected to shed more light on their role in signal transduction in mammalian cells.

Another family of GTP-binding proteins, the *ras* family, is also widely distributed and is highly conserved among eukaryotes. The *ras* gene products, like α-subunits of G proteins, bind guanine nucleotides, show GTPase activity, and are associated with the cytoplasmic surface of the plasma membrane. The similarities between the *ras* family and G proteins suggest that the *ras* proteins would be components of signal transduction through membrane signalling systems (Gilman, 1987). In *S. cerevisiae* there are two *ras* genes, *RAS1* and *RAS2* (Defeo-Jones *et al.*, 1983; Powers *et al.*, 1984), and four *ras*-related genes, *RHO1* and *RHO2* (Madaule *et al.*, 1987), *SEC4* (Salminen and Novick, 1987), and *YPT1* (Gallwitz *et al.*, 1983). In this chapter, we describe the roles of yeast GTP-binding proteins in the yeast signal transduction systems.

6.2 GTP-binding proteins in yeast

6.2.1 *The G protein family*

GPA1 and GPA2. Two genes, *GPA1* and *GPA2*, have been isolated from yeast *Saccharomyces cerevisiae* by cross-hybridization with cDNAs that code for the α-subunit of rat brain Giα and Goα (Nakafuku *et al.*, 1987, 1988). *GPA1* encodes a protein (GP1α) of 472 amino acids with a calculated molecular weight of 54 075. *GPA2* encodes a protein (GP2α) of 449 amino acids with a molecular weight of 50 516. The molecular sizes of the predicted yeast GP1α and GP2α are considerably larger than the mammalian G protein α subunits. GP1α is involved in mating factor-mediated signal transduction in yeast haploid cells (Dietzel and Kurjan, 1987; Miyajima *et al.*, 1987). GP2α may be involved in the regulation of the cyclic AMP level (Nakafuku *et al.*, 1988). The functions of GP1α and GP2α in the yeast signal transduction systems are described below.

Comparison of amino acid sequences of yeast GP1α and GP2α with those of mammalian G protein α subunits. The deduced amino acid sequences of yeast

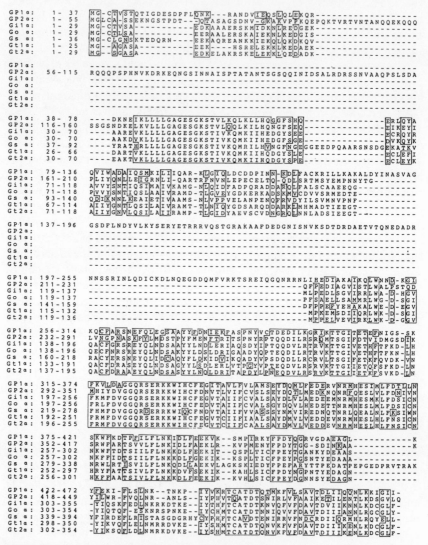

Figure 6.3 Comparison of amino acid sequences of yeast GP1α and GP2α, rat Gsα, Giα and Goα, and bovine Gt1α and Gt2α. Gaps, indicated by hyphens, were introduced to obtain maximum homology. Sets of identical or conservative residues are enclosed within solid lines.

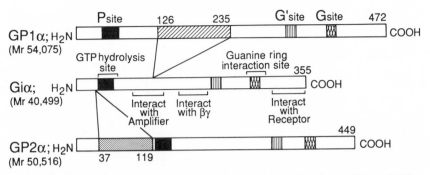

Figure 6.4 Schematic diagram of the structural and functional domains of yeast GP1α and GP2α, and rat Giα polypeptides. See text and Figure 6.6 for discription of P, G, and G' site.

GP1α and GP2α revealed remarkable homology with those of mammalian G protein α-subunits (Figure 6.3). When GP1α and GP2α are aligned with rat brain Giα to obtain maximal homology, the extra sequences of 110 and 83 amino acids which are unique to GP1α and GP2α, respectively, are found in the N-terminal portion (Figures 6.3, 6.4). The extra sequence of GP1α is inserted between Gly-118 and Met-119 of Giα, while that of GP2α is found closer to the N-terminus, between Lys-29 and Ala-30 of Giα. The NH$_2$-terminal region of mammalian G proteins is assumed to be the site interacting with an amplifier protein (Bourne, 1984). The sequences unique to GP1α and GP2α may play an important role in the interaction with their putative amplifiers. The overall homology in nucleotide and amino acid sequences of yeast GP1α, yeast GP2α, rat Giα, and rat Gsα is summarized in Figure 6.5. Disregarding the unique sequences present in GP1α (residues 126–235) and GP2α (residues 37–119), they are 60% homologous if the conservative amino acid substitutions are considered to be homologous. The homology is smaller than that between rat Giα and Goα (85%) but is comparable to that between rat Giα and Gsα (60%).

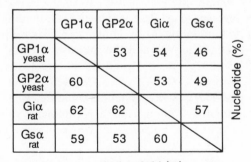

	GP1α	GP2α	Giα	Gsα	
GP1α yeast		53	54	46	
GP2α yeast	60		53	49	
Giα rat	62	62		57	Nucleotide (%)
Gsα rat	59	53	60		

Amino Acid (%)

Figure 6.5 Summary of comparison between the nucleotide and amino acid sequences of yeast GP1α and GP2α, and rat Gsα and Giα.

P site

```
GP1α  :  43-  58:  K L L L L G A G E S G K S T V L
GP2α  :125-140:  K V L L L G A G E S G K S T V L
Giα   :  35-  50:  K L L L L G A G E S G K S T I V
Goα   :  35-  50:  K L L L L G A G E S G K S T I V
Gsα   :  42-  57:  R L L L L G A G E S G K S T I V
Gt1α  :  31-  46:  K L L L L G A G E S G K S T I V
Gt2α  :  35-  50:  K L L L L G A G E S G K S T I V
H-ras :   5-  20:  K L V V V G A G G V G K S A L T
EF-Tu :  13-  28:  N V G T I G H V D H G K T T L T
```

G site

```
GP1α  :381-397:  T P - F I L F L N K I D L F E E K V
GP2α  :358-374:  T S - V V L F L N K I D L F A E K L
Giα   :263-279:  T S - I I L F L N K K D L F E E K I
Goα   :263-279:  T S - I I L F L N K K D L F G E K I
Gsα   :284-290:  I S - V I L F L N K Q D L L A E K V
Gt1α  :258-274:  T S - I V L F L N K K D V F S E K I
Gt2α  :262-278:  T S - I V L F L N K K D L F E E K I
H-ras :109-125:  V P - M V L V G N K C D L A A R T V
EF-Tu :127-144:  V P Y I I V F L N K C D M V D D E E
```

G'site

```
GP1α  :319-335:  D A G G Q R S E R K K W I H C F E
GP2α  :296-312:  D V G G Q R S E R K K W I H C F D
Giα   :201-217:  D V G G Q R S E R K K W I H C F E
Goα   :201-217:  D V G G Q R S E R K K W I H C F E
Gsα   :224-240:  D V G G Q R D E R R K W I Q C F N
Gt1α  :196-212:  D V G G Q R S E R K K W I H C F E
Gt2α  :200-216:  D V G G Q R S E R K K W I H C F E
```

Figure 6.6 Amino acid sequence homology in three regions (P, G and G′ sites) between yeast GP1α and GP2α, rat Gsα, Giα and Goα, and bovine Gt1α and Gt2α, *ras* protein, and EF-Tu. Regions of exact homology or with conservative substitutions are boxed.

As illustrated in Figure 6.6, there are three regions in which the homologies are especially pronounced.

(i) GTP hydrolysis site (P site). The P site (amino acid residues 43–58 of GP1α, 125–140 of GP2α, and 35–50 of Giα) is conserved in all G proteins and also in other GTP-binding proteins such as EF-Tu and the *ras* family. This site is suggested to be involved in GTPase activity of GTP-binding proteins, based on the following facts: mutation of Gly-12 of mammalian *ras* protein results in the decrease in GTPase activity and the increase in transforming activity (Gibbs *et al.*, 1984; Seeburg *et al.*, 1984) and the amino acid residue Val-20 of *E. coli* EF-Tu, corresponding to Gly-12 of *ras*, is located in close proximity to the phosphoryl group of GTP in X-ray crystallographic analysis (Jurnak, 1985). In the P site, 12 contiguous amino acids are completely identical, and 16 contiguous amino acids are homologous between yeast G proteins and mammalian G proteins.

(ii) Guanine ring interacting site (G site). The G site (amino acid residues 382–397 of GP1α, 359–374 of GP2α, and 264–279 of Giα) is also conserved in all G proteins and other GTP-binding proteins. This site is thought to be responsible for the guanine ring interaction since Cys-137 of *E. coli* EF-Tu was protected by GDP or GTP against modification with N-ethylmaleimide (Arai *et al.*, 1980). The recent results of X-ray crystallographic analysis of EF-Tu by Jurnak (1985) indicate clearly that the sequence Asn-135 to Asp-138 is situated close to the guanine ring of the bound GTP/GDP. In the G site, 12 out of 16 amino acids are homologous between yeast G proteins and mammalian G proteins. The consensus sequence Asn-Lys-Xaa-Asp is found in all families of GTP-binding proteins.

(iii) A region unique to G proteins (G' site): The G' site is strongly conserved in all G proteins (amino acid residues 319–335 of GP1α, 296–312 of GP2α, and 201–217 of Giα) and is unique to the G protein family. In the G' site, identical amino acid sequences containing 12 contiguous amino acids are shared by yeast and mammalian G proteins.

Other regions of homology among mammalian G proteins are the sites for ADP-ribosylation of Gsα and Gtα by cholera toxin and of Giα, Goα, and Gtα by pertussis toxin. The cystein residue which is ADP-ribosylated by pertussis toxin (a cystein residue located in the fourth position from the COOH-terminus) is replaced by isoleucine and serine in yeast GP1α and GP2α, respectively. This indicates that yeast G proteins are probably refractory to the modification by islet-activating protein. On the other hand, the sequences around Arg-297 of yeast GP1α and Arg-273 of yeast GP2α are highly homologous with the sequence around Arg-201 of Gsα, which is thought to be the site of ADP-ribosylation by cholera toxin (Figure 6.3). The sequence around Arg-201 of Gsα is very homologous not only to transducin, which is ADP-ribosylated by the cholera-toxin, but also to Gi1α and Goα which are not (Figure 6.3).

In mammalian cells, G proteins are comprised of three subunits, α, β and γ. No β- or γ-subunits have been identified in yeast at this time, though the *CDC4* gene product has weak homology with a β-subunit of transducin (Fong *et al.*, 1986). However, a remarkable conservation between the structures of mammalian and yeast Gα suggests that yeast may posses β- and γ-subunits of G proteins.

6.2.2 The ras family

RAS1 and *RAS2*. The *ras* family also consists of several closely related proteins. Members of this family were first identified as the transforming genes of sarcoma viruses (Ha-, Ki, and N-*ras*) and subsequently were found in the genomes of various eukaryotes (Shilo and Weinberg 1981; Defeo-Jones *et al.*, 1983; Powers *et al.*, 1984; Reymond *et al.*, 1984; Fukui and Kaziro, 1985).

Two genes, *RAS1* and *RAS2*, have been isolated from *S. cerevisiae* by using the viral *ras* gene as a hybridization probe (Defeo-Jones *et al.*, 1983; Powers *et al.*, 1984). They code for proteins of 309 and 322 amino acids respectively. The yeast *RAS* genes are about twice the size of the mammalian *ras* protein of

188 or 189 amino acids. Yeast *RAS* proteins are 90% homologous to the mammalian *ras* protein in their first 90 (amino-terminal) amino acids, and about 50% homologous in their next 80 amino acids. All the proteins then have a variable region and finally terminate in a short conserved sequence (Cys-A-A-X; where A is an aliphatic amino acid). Yeast strains containing a disruption in either *RAS1* or *RAS2* are viable, but disruptions in both are lethal (Kataoka *et al.*, 1984; Tatchell *et al.*, 1984). Thus, *RAS* functions are essential for yeast cell viability and proliferation. Expression of human *ras* protein can complement this lethality (Defeo-Jones *et al.*, 1985; Kataoka *et al.*, 1985). Moreover, chimeric yeast-mammalian *ras* genes and a modified yeast *RAS1* gene that contains a deletion of the nonhomologous region near the carboxy terminus can transform mouse NIH 3T3 cells in gene transfer assays (Defeo-Jones *et al.*, 1985). These results imply that the amino acid homology extends to the functional level. It has been shown that RAS2 protein stimulates adenylate cyclase (Broek *et al.*, 1985; Toda *et al.*, 1985) while RAS1 protein seems to negatively affect inositol phospholipid turnover (Kaibuchi *et al.*, 1986b). The effects of yeast RAS proteins on adenylate cyclase and inositol phospholipid turnover are described below.

6.2.3 The ras-related family

RHO1 and RHO2. The *rho* genes comprise an evolutionarily conserved family with significant homology to the *ras* gene family (Madaule and Axel, 1985). Two genes, *RHO1* and *RHO2*, were isolated from *S. cerevisiae* by using the *rho* gene of the marine snail *Aplysia* as a hybridization probe (Madaule *et al.*, 1987). The predicted sizes for the products of *RHO1* and *RHO2* are 23 000 and 21 000 dalton respectively. The yeast genes *RHO1* and *RHO2* are 70% and 57% homologous, respectively, to the *rho* gene of *Aplysia*, and they are 53% homologous to each other. The products of *RHO1* and *RHO2* share only 25–32% homology with Ha-*ras* proteins; however, the identities are clustered into several discrete highly conserved domains implicated in GTP binding and hydrolysis. Genetic experiments demonstrate that *RHO1* is an essential gene in yeast, while *RHO2* is not required for cell viability. The lethality associated with inactivation of *RHO1* was not suppressed by overproduction of cyclic AMP-dependent protein kinase, suggesting that *RHO1* is not required to regulate adenylate cyclase activity. The function of the *RHO1* and *RHO2* proteins may be to link a GTP-dependent amplifier pathway to the cell surface.

SEC4. The *SEC4* gene was originally isolated by its ability to partially complement the *sec15* growth defect (Salminen and Novick, 1987). *SEC4* protein plays an essential role in the transport of secretory protein from the Golgi to the cell surface. *SEC4* encodes a protein of 23 500 dalton and shares 32% homology with the Ha-*ras* gene. The regions of best homology are those involved in the binding and hydrolysis of GTP. More extensive homology, 48%, is seen with another yeast *ras*-related gene, *YPT1* (described below). *SEC4* is essential for growth. Salminen and Novick (1987) proposed that the *SEC4* product is a GTP-binding protein that plays an essential role in controlling a late stage of the secretory pathway.

YPT1. *YPT1*, located between the actin and the β-tublin structural genes and first identified as an open reading frame (Gallwitz *et al.*, 1983), encodes a 206-amino-acid protein (23 213 dalton) and is essential for cell growth (Schmitt *et al.*, 1986; Segev and Botstein, 1987). *YPT1* protein was shown to specifically bind and hydrolyse GTP. It seems to be involved in microtubule organization and function (Schmitt *et al.*, 1986; Wagner *et al.*, 1987).

6.3 The role of yeast GTP-binding proteins in yeast signal transduction

S. cerevisiae has two different signal transduction systems at the G1 phase of the cell cycle (Figure 6.7). One is mediated by nutrients, such as glucose, which positively regulate the early G1 phase; the other is mediated by mating

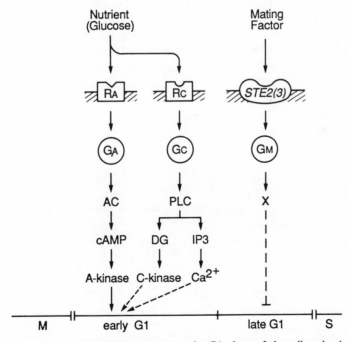

Figure 6.7 Yeast signal transduction systems at the G1 phase of the cell cycle. A terminal arrowhead indicates stimulation; a terminal bar indicates inhibition. Symbols: R, receptor; G, G protein; AC, adenylate cyclase; PLC, phospholipase C; DG, diacylglycerol; IP$_3$, inositol triphosphate; A-kinase, cyclic AMP-dependent protein kinase; C-kinase, protein kinase C.

H

pheromones, which negatively regulate the late G1 phase (Reed, 1980). In the first case, glucose serves as an extracellular signal for the regulation of adenylate cyclase and inositol phospholipid turnover (Kaibuchi *et al.*, 1986*b*).

6.3.1 *Mating pheromone-mediated signal transduction at late G1 phase*

S. cerevisiae has three distinct cell types: two haploid cell types, *a* and α, and an *a*/α diploid cell type. The *a* and α cell types can mate to yield *a*/α diploids through a multistep mating process. This mating process is initiated by peptide pheromones secreted by each haploid cell type. α cells secrete α-factor which acts on *a* cells, whereas *a* cells produce *a*-factor that acts on α cells. These responses ultimately cause cell cycle arrest of the target cells at late G1 phase (Sprague *et al.*, 1983). It is believed that both mating factors generate an intracellular signal by interacting with their specific membrane receptors on the target cells. Genetic and biochemical studies have suggested that *STE2* and *STE3* encode the receptors for α-factor and *a*-factor, respectively (Jenness *et al.*, 1983; Hagen *et al.*, 1986). *STE2* and *STE3* proteins have been shown to share a common signal transduction pathway by interacting with a common or interchangeable protein within the target cells (Bender and Sprague, 1986; Nakayama *et al.*, 1987). The *STE2* and *STE3* gene products are predicted to have seven potential transmembrane domains (Burkholder and Hartwell, 1985; Nakayama *et al.*, 1985) as in the case of rhodopsin (Nathans and Hogness, 1984) and β-adrenergic receptor (Dixon *et al.*, 1986) which interact with mammalian G proteins in the signal transduction process (Stryer and Bourne, 1986; Gilman, 1987). Therefore, G protein(s) may interact with *STE2* and *STE3* proteins to transmit mating factor signals in an analogous manner to mammalian G proteins (Figure 6.8).

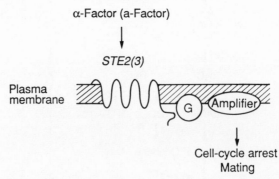

α-Factor (a-Factor)

↓

STE2(3)

Plasma
membrane

Cell-cycle arrest
Mating

Figure 6.8 Schematic representation of interaction that leads to mating of yeast cells.

The following five key observations suggest that GP1α is involved in mating factor-mediated signal transduction in yeast haploid cells (Miyajima *et al.*, 1987):

(i) The *GPA1* transcript is found in haploid cells but is not detectable in a/α diploids.

(ii) Disruption of *GPA1* is lethal in haploid cells but not in a/α diploid cells indicating that *GPA1* is a haploid-specific gene essential for cell growth.

(iii) Upon regulation of expression of *GPA1* by the galactose-inducible *GAL1* promoter, the loss of the GP1α function leads to cell cycle arrest at the same stage in the cell cycle as do the mating factors..

(iv) Mutants (*sgp1* and *sgp2*) that suppress the lethality of the *gpa1* disruption mutation show a sterile phenotype that is not cell-type-specific. These mutants produce but do not respond to mating factors.

(v) Loss of GP1α function can relieve the sterility caused by a *ste2* mutation.

Dietzel and Kurjan (1987) have isolated a gene (*SCG1*) which, when present on a multicopy plasmid, suppresses the supersensitivity of *sst2* mutant cells to mating factors. Comparison of the sequence of *SCG1* with *GPA1* revealed only five differences, in both the nucleotide and amino acid sequences, suggesting that they are the same gene from different strains of yeast. This result supports the view that GP1α is involved in the mating factor-mediated signal transducing pathway; a model of which is shown in Figure 6.9. This model is based on the reaction mechanism of GTP-binding protein originally proposed for translational elongation factors and extrapolated to other systems (Kaziro, 1978). The reaction mechanism consists of three basic concepts:

(i) The protein has two qualitatively different conformations defined by the bound ligand, GDP or GTP.

(ii) The GTP-bound form is an active conformation that is able to activate a biological amplifier, while the GDP-bound form is inactive.

(iii) Hydrolysis of the bound GTP is, therefore, required to shut off the signal.

According to this model (Figure 6.9), haploid cell arrest in the presence of mating factors is mediated by the activity of GTP-bound GP1α in a manner similar to mammalian G protein in hormonal signal transduction. In the absence of mating factor, GP1α binds GDP and restrains a putative amplifier molecule that is responsible for generation of a cell-cycle arrest signal. Binding of mating factor to its receptor promotes the exchange of the bound GDP with GTP. The GTP-bound form of GP1α activates the amplifier which is then able to transmit a signal to arrest the cell cycle at late G1 phase. Based on this model, the lethality of the cell with disrupted *GPA1* can be explained by the uncoupling of the amplifier from GP1α (see *gpa1::HIS3* in Figure 6.9). The amplifier, which may be unlocked from the mating factor receptor complex, may elicit a constitutive signal for cell-cycle arrest regardless of the presence of mating factors. Loss of the GP1α function can suppress the sterile phenotype of *STE2* disruption. This supports the idea that *GPA1* disruption results in continuous production of a cell-cycle arrest signal and in promotion of conjugation in the absence of a mating factor signal. Suppressor mutations

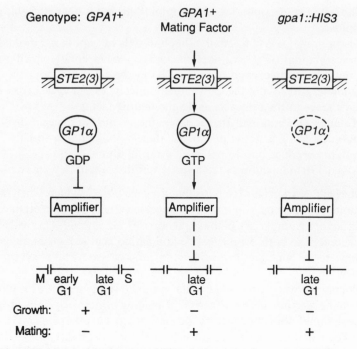

Figure 6.9 Model of the function of GP1α in the mating factor signaling of yeast.

(*sgp1* and *sgp2*) of the *gpa1* disruption mutation can be used to define further components involved in this system. One possibility is that the suppressor mutation may occur in the amplifier. In this case, the mutant amplifier may be altered so as not to generate the arrest signal. Therefore, this type of mutation may be able to suppress the lethality of the *gpa1* disruption mutation and may show a cell-type non-specific sterile phenotype. Both biochemical and genetic studies are necessary to clarify the real reaction mechanisms of GP1α during signal tranduction. For example, based on existing biochemical observations on the mechanism of signal transduction for mammalian hormones, it will be interesting to test for an *in vitro* mating-factor-dependent GTPase activity. Further analysis of the suppressor mutations will be useful to elucidate the function of GP1α in the signal transducing pathway.

6.3.2 *Cyclic AMP pathway at early G1 phase*
The early G1 stage is blocked by nutrient limitation. Genetic and biochemical studies have shown that cyclic AMP plays an important role at this stage (Matsumoto *et al.*, 1985). The regulatory role of cyclic AMP in yeast has been studied by isolation of mutants defective in adenylate cyclase and cyclic AMP-dependent protein kinase (Matsumoto *et al.*, 1982). The *cyr1* mutants defective in adenylate cyclase activity were arrested at the early G1 phase of the cell cycle

in the absence of cyclic AMP (Matsumoto *et al.*, 1983, 1984). Cyclic AMP exerts its effect by binding regulatory subunits (*BCY1* gene product) of cyclic AMP-dependent protein kinase, thereby freeing active catalytic subunits (Matsumoto *et al.*, 1982; Johnson *et al.*, 1987; Toda *et al.*, 1987; Yamano *et al.*, 1987), *bcy1* mutants have decreased levels of these regulatory subunits, thereby allowing the protein kinase to function in the absence of cyclic AMP (Matsumoto *et al.*, 1982). Thus *bcy1* can suppress the growth defect of *cyr1*. In addition, the early G1 arrest caused by nutritional limitation was overcome by *bcy1* (Matsumoto *et al.*, 1983).

The signal transduction system regulating adenylate cyclase activity in mammalian cells involves two G proteins, Gs and Gi (Gilman, 1987). Binding of ligands to a receptor results in stimulation or inhibition of adenylate cyclase activity. The adenylate cyclase system of yeast consists of at least two protein components, the catalytic and regulatory subunits, and is regulated by guanine nucleotides in the presence of magnesium ions (Casperson *et al.*, 1983). It has been shown that yeast adenylate cyclase is regulated by the *RAS* proteins (Broek *et al.*, 1985; Toda *et al.*, 1985). Compared to the wild-type strain, intracellular cyclic AMP levels were slightly low in *ras1*, significantly depressed in *ras2*, and almost undetectable in *ras1 ras2 bcy1* cells (Table 6.1). A *ras2* allele with a missense mutation designated $RAS2^{val19}$ has been isolated by Kataoka *et al.*, (1984). This replacement of glycine at position 19 of the *RAS2* protein is equivalent to the change at amino acid position 12 of the mammalian *ras* p21 that results in mammalian cell transformation. Intracellular cyclic AMP levels were significantly elevated in $RAS2^{val19}$. Membrane fractions from wild-type cells had high adenylate cyclase activity when assayed in the presence of Mn^{2+} but low activity in Mg^{2+} (Table 6.1). Adenylate

Table 6.1 Adenylate cyclase activity and cyclic AMP levels of yeast *RAS* mutants (Matsumoto *et al.*, 1985; Toda *et al.*, 1985).

Genotype	cAMP level[a]	Adenylate cyclase[b]		
		Mn^{2+}	Mg^{2+}	Mg^{2+}, Gpp (NH)p
RAS1 RAS2	2.4	49.2	3.7	14.3
ras1 RAS2	1.8	62.0	1.2	5.9
RAS1 ras2	0.6	42.1	1.0	0.8
ras1 ras2 bcy1	0.1	51.0	1.1	0.7
$RAS1\ RAS2^{val19}$	8.3	51.9	14.2	16.1
RAS1 RAS2 cyr1	—	0.3	0.1	0.2
ras1 ras2 bcy1				
+	—	29.5	2.3	23.5
RAS1 RAS2 cyr1				

[a]Cyclic AMP levels are expressed as pmol per mg protein.
[b]Membranes from the indicated strains were prepared and adenylate cyclase was assayed in the presence of 2.5 mM Mn^{2+}, or 2.5 mM Mg^{2+}, or 2.5 mM Mg^{2+} and 10 μM of the nonhydrolysable GTP analogue Gpp (NH)p. Adenylate cyclase activity is expressed in units of pmol of cyclic AMP generated per mg of membrane protein per minute.

cyclase activity assayed with Mg^{2+} was stimulated significantly by the addition of GTP. Membranes from *ras1 ras2 bcy1* lacked GTP-stimulated adenylate cyclase activity present in membranes from wild-type cells, and membranes from the $RAS2^{va119}$ strain had elevated levels of an apparently GTP-independent adenylate cyclase activity (Table 6.1). Mixing membranes from *ras1 ras2 bcy1* with membranes from *cyr1* reconstituted a GTP-dependent adenylate cyclase (Table 6.1). Broek *et al.* (1985) have reconstituted the system *in vitro* by adding purified RAS protein to membrane fractions devoid of RAS protein. The adenylate cyclase activity was dependent on the presence of exogenously added RAS protein. The simplest interpretation of these results is that the RAS protein directly stimulates adenylate cyclase in a manner analogous to the stimulation of mammalian adenylate cyclase by Gs protein (Gilman, 1987).

Based on analogy with the G protein, RAS protein may act as a transducer to convey extracellular signals to an intracellular amplifier pathway. The presence of glucose and its derivatives results in a rapid increase in intracellular cyclic AMP in yeast in a manner similar to the effect of various hormones and growth factors on mammalian cells (Eraso and Gancedo, 1985). We tested the stimulation of glucose-mediated cyclic AMP formation in *ras* mutants. In glucose-starved wild-type cells, glucose stimulated cyclic AMP formation transiently (Figure 6.10*A*). In *ras1 ras2 bcy1* cells, glucose slightly increased cyclic AMP formation (K. Kaibuchi, unpublished data). This suggests that adenylate cyclase activity depends on more than just *CYR1* and *RAS* gene products. We have found that the yeast G protein, GP2α, is involved in the regulation of cyclic AMP levels (Nakafuku *et al.*, 1988).

To examine the possibility that *GPA2* may be involved in the regulation of cyclic AMP levels, we have studied the effect of *GPA2* on cyclic AMP formation. The *gpa2* disruption mutant is viable and did not affect glucose-induced cyclic AMP formation (Figure 6.10*B*). On the other hand, a multicopy plasmid carrying *GPA2* (YEp*GPA2*) remarkably increased the level of glucose-induced synthesis of cyclic AMP (Figure 6.10*C*) and suppressed the growth defect in a temperature-sensitive *ras2* mutant (*ras2-ts*) at high temperature. In the *ras2-ts* mutant, glucose did not induce any cyclic AMP formation at the restrictive temperature (Figure 6.10*D*). YEp*GPA2* restored glucose-induced cyclic AMP formation in the *ras2-ts* mutant at high temperature (Figure 6.10*D*). These results indicate that *GPA2*, in addition to *RAS1* and *RAS2*, regulates the level of cyclic AMP in yeast. Since the level of cyclic AMP is determined by adenylate cyclase and phosphodiesterase activities, *GPA2* on multicopy plasmids may directly either activate adenylate cyclase or inhibit phosphodiesterase. If the function of *GPA2* is to regulate adenylate cyclase, the non-lethal phenotype of the *GPA2* null mutant suggests that an additional G protein may be involved in the activation of adenylate cyclase in yeast. On the other hand, the lack of lethality of the *GPA2* null mutant is consistent with the idea that *GPA2* maintains the cyclic AMP level

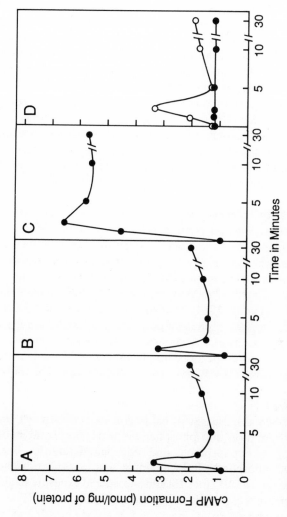

Figure 6.10 Kinetics of the glucose-induced cyclic AMP formation. Yeast cells were stimulated with 25 nM glucose and incubated at 30°C (*A*, *B* and *C*) or 37°C (*D*). (*A*) Wild-type strain; (*B*) *GPA2*-disrupted strain; (*C*) wild-type strain with YEp*GPA2*; (*D*) *ras2*^{ts} strain without (●) and with (○) YEp*GPA2* (Nakafuku *et al.*, 1988).

by inhibiting phosphodiesterase activity which is not essential for growth (Nikawa et al., 1987). Alternatively, the effect of GPA2 could be indirect and thus, it may not be normally involved in the regulation of cyclic AMP level. In mammalian cells, hormonal stimuli dissociate β- and γ-subunits from the α-subunit of G protein. Excess $\beta\gamma$ is inhibitory to the activation of adenylate cyclase by the GTP bound form of the Gsα subunit (Katada et al., 1984). Likewise, GP2α when expressed at a high level may form a complex with free β- and γ-subunits which are otherwise inhibitory to other G protein(s) involved in cyclic AMP formation although the occurrence of β- and γ-subunits in yeast is not clear at present.

In the previous section, we presented the possibility that GP1α interacts with the mating factor receptors (STE2 and STE3 products) in a manner analogous to mammalian G proteins interactions with β-adrenergic receptors. Therefore, it may be reasonable to assume that GP2α also interacts with a receptor molecule in the cytoplasmic membrane of yeast cells to transmit a signal to an amplifier molecule.

6.3.3 Inositol phospholipid turnover pathway

Growth factors, such as platelet-derived growth factor and fibroblast growth factor, activate phospholipase C to induce hydrolysis of membrane inositol phospholipids (Habenicht et al., 1981). The initial products of this reaction are diacylglycerol and inositol triphosphate (InsP$_3$). Diacylglycerol remains in the membrane and serves as a second messenger for activation of protein kinase C (Nishizuka, 1984; Takai et al., 1984). Diacylglycerol is then metabolized to inositol phospholipids by way of phosphatidic acid and CDP-diacylglycerol. Water-soluble InsP$_3$ also acts as a second messenger for Ca^{2+} mobilization from intracellular Ca^{2+} stores (Berridge and Irvine, 1984). These two second messengers, as well as cyclic AMP, may play critical roles in the transition from the G0/G1 phase of the cell cycle in mammalian cells (Rozengurt, 1982; Kaibuchi et al., 1986a).

Kaibuchi et al. (1986b) demonstrated that in glucose-starved yeast glucose stimulated ^{32}P incorporation into phosphatidic acid, phosphatidylinositol, phosphatidylinositol monophosphate, and phosphatidylinositol bisphosphate, but not into phosphatidylethanolamine and phosphatidylcholine. Preincubation of yeast with [^{3}H] inositol and subsequent exposure to glucose resulted in rapid formation of [^{3}H] inositol monophosphate and [^{3}H] inositol triphosphate, presumably derived from phosphatidylinositol and phosphatidylinositol bisphosphate. Under similar conditions, glucose elicited both efflux and influx of Ca^{2+} in yeast. These results suggested that glucose stimulates phospholipase C activation to hydrolyse inositol phospholipids in a manner similar to that of a variety of hormones and growth factors in mammalian cells. Glucose has been shown to stimulate inositol phospholipid turnover in β-cells of pancreatic islets (Clements et al., 1981). Non-metabolizable derivatives of glucose such as 2-deoxyglucose and 6-deoxyglucose have no effect on inositol phospholipid turnover in either β-cells or yeast. In contrast, 2-deoxyglucose and 6-deoxyglucose have been shown to stimulate cyclic AMP formation in yeast cells (Eraso and Gancedo, 1985). It is, therefore, conceivable that glucose itself is responsible for adenylate cyclase activation, and glucose metabolites result in phospholipase C activation. In mammalian cells, inositol phospholipid turnover is linked to activation of protein kinase C and Ca^{2+} mobilization (Berridge and Irvine 1984; Takai et al., 1984). $InsP_3$ serves as a second messenger for Ca^{2+} translocation in various eukaryotic cells. However, protein kinase C activity has not been shown in yeast and it is not known whether $InsP_3$ induces Ca^{2+} mobilization in yeast cells. Identification of protein kinase C and its substrate in yeast and the elucidation of the effect of $InsP_3$ on Ca^{2+} mobilization are necessary for a better understanding of the role of inositol phospholipid turnover in cell proliferation.

To test the hypothesis that RAS proteins modulate inositol phospholipid turnover as well as adenylate cyclase activity, Kaibuchi et al., (1986b) determined ^{32}P incorporation into inositol phospholipids and formation of [^{3}H] inositol phosphates in yeast carrying mutant RAS genes (Table 6.2). It was found that ras mutations led to increased inositol phospholipid turnover. Disruption of the RAS1 gene greatly enhanced inositol phospholipid turnover while disruption of the RAS2 gene was less effective. Strains lacking both RAS genes or containing the $RAS2^{val19}$ also greatly induced inositol phospholipid turnover. These results suggest that RAS1 protein may play a major inhibitory role in glucose-induced inositol phospholipid turnover and RAS2 protein may be less important. However, the $RAS2^{val19}$ mutation activated inositol phospholipid turnover. Although we do not know the mechanism by which

Table 6.2 ^{32}P incorporation into PtdIns and formation of [^{3}H] InsP3 in yeast RAS mutants (Kaibuchi et al., 1986b).

Genotype	Addition	^{32}P incorporation into PtdIns cpm	[^{3}H] PtdIns cpm	[^{3}H] InsP3 formation cpm
RAS1 RAS2	None	1 050	481 960	140
	Glucose	7 650	512 160	360
ras1 RAS2	None	1 080	481 630	150
	Glucose	16 320	522 120	620
RAS1 ras2	None	1 010	533 190	150
	Glucose	8 190	566 760	420
ras1 ras2 bcy1	None	2 250	557 630	290
	Glucose	32 670	587 310	1160
RAS1 RAS2 bcy1	None	1 110	461 120	130
	Glucose	6 370	481 970	330
RAS1 RAS2^{val19}	None	1 370	541 670	250
	Glucose	25 270	571 720	920

PtdIns, phosphatidylinositol monophosphate; InsP3, inositol triphosphate.

RAS proteins modulate inositol phospholipid turnover, a conformational change may convert the RAS2 protein into an activating molecule. It can be speculated that $RAS2^{val19}$ cells tend to continue growth without entering a G0/G1 arrest because the level of inositol phospholipid turnover and cyclic AMP are higher than in wild-type cells. Since Ha- and Ki-ras proteins are homologous to yeast RAS proteins (Defeo-Jones et al., 1983; Powers et al., 1984), it is possible that in mammalian cells ras proteins affect inositol phospholipid turnover as well as cyclic AMP formation, leading to uncontrollable cell growth. Recently, Fleischman et al. (1986) suggested that mutated ras may directly stimulate phospholipase C. Furthermore, expression of the N-ras gene in NIH 3T3 cells leads to the coupling of certain growth factor receptors to inositol phosphate production (Wakelam et al., 1986). However, it was reported that the transforming potential of mutated ras does not result from persistent activation of phospholipase C (Seuwen et al., 1988). Benjamin et al. (1987) and Parries et al. (1987) suggested that mutated ras does not stimulate phospholipase C but reduces growth factor-stimulated phospholipase activity. Further studies are needed to clarify the involvement of ras proteins in inositol phospholipid turnover.

Several lines of evidence indicate that agonist-induced activation of both phospholipase C and adenylate cyclase is regulated by G proteins such as Gs, Gi, and Go in mammalian cells (Nakamura and Ui, 1985: Smith et al., 1985; Gilman 1987). Glucose stimulates inositol phospholipid turnover in the absence of RAS proteins as shown in the ras1 ras2 bcy1 strain (Table 6.2). It is conceivable that GTP-binding proteins other than RAS proteins may be responsible for phospholipase C activation in yeast, and that inositol phospholipid turnover is dually regulated by RAS proteins and these putative GTP-binding proteins.

Acknowledgements

We thank Charles Brenner, Hiroshi Itoh, Atsushi Miyajima, Ikuko Miyajima, Shun Nakamura, Naoki Nakayama, and Tomoko Obara for their collaboration.

References

Arai, K., Clark, B.F.C., Duffy, L., Jones, M.D., Kaziro, Y., Laursen, R.A., L'Italien, J., Miller, D.L., Nagarkatti, S., Nakamura, S., Nielsen, K.M., Petersen, T.E., Takahashi, K. and Wade. M. (1980) Primary structure of elongation factor Tu from Escherichia coli. Proc. Natl. Acad. Sci. USA 77: 1326.
Bender, A., and Sprague, G.F. Jr., (1986) Yeast peptide pheromone, a-factor and α-factor, activate a common response mechanism in their target cells. Cell 47: 929.
Benjamin, C. W., Tarpley, W.G., and Gorman. R.R. (1987) Loss of platelet-derived growth factor-stimulated phospholipase activity in NIH-3T3 cells expressing the EJ-ras oncogene. Proc. Natl. Acad. Sci. USA 84: 546.
Berridge, M.J., and Irvine, R.F. (1984) Inositol triphosphate, a novel second messenger in cellular signal transduction. Nature 312: 315.
Bourne, H.R. (1986) One molecular machine can transduce diverse signals. Nature 321: 814.
Bray, P., Carter, A., Simons, C., Guo, V., Puckett, C., Kamholz, J., Spiegel, A. and Nirenberg, M.

(1986) Human cDNA for four species of Gαs signal transduction protein. *Proc. Natl. Acad. Sci. USA* **83**: 8893.

Bray, P., Carter, A., Guo, V., Puckett, C., Kamholz, J., Spiegel, A. and Nirenberg, M. (1987) Human cDNA clones for an α subunit of Gi signal transduction protein. *Proc. Natl. Acad. Sci. USA* **84**: 5115.

Broek, D., Samiy, N., Fasano, O., Fujiyama, A., Tamanoi, F., Northup, J. and Wigler, M. (1985) Differential activation of yeast adenylate cyclase by wild-type and mutant *RAS* proteins. *Cell* **41**: 763.

Burkholder, A.C. and Hartwell, L.H. (1985) The yeast α-factor receptor: structural properties deduced from the sequence of the *STE2* gene. *Nucleic Acids Res.* **13**: 8463.

Casperson, G.F., Walker, N., Brasier, A.R. and Bourne, H.R. (1983) A guanine nucleotide-sensitive adenylate cyclase in the yeast *Saccharomyces cerevisiae*. *J. Biol. Chem.* **258**: 7911.

Clements, R.S., Evans, M.H. Jr., and Pace, C.S. (1981) Substrate requirements for the phosphoinositide response in rat pancreatic islets. *Biochim. Biophys. Acta* **674**: 1.

Defeo-Jones, D., Scolnick, E.M., Koller, R. and Dhar, R. (1983) *ras*-related gene sequences identified and isolated from *Saccharomyces cerevisiae*. *Nature* **306**: 707.

Defeo-Jones, D., Tatchell, K., Robinson, L.C., Sigal, I.S., Vass, W.C., Lowy, D.R. and Scolnick, E.M. (1985) Mammalian and yeast *ras* gene products: biological function in their heterologous systems. *Science* **228**: 179.

Dietzel, C. and Kurjan, J. (1987) The yeast *SCG1* gene: a Gα-like protein implicated in the *a*- and α-factor response pathway. *Cell* **50**: 1001.

Dixon, R.A.F., Kobilka, B.K., Strader, D., Benovic, J.F., Dohlman, H.G., Frielle, T., Bolonowski, M., Bennett, C.D., Rands, E., Diehl, R.E., Mumford, R.A., Slater, E.E., Sigal, I.S., Caron, M.G., Lefkowitz, R.J. and Strader, C.D. (1986) Cloning of the gene and cDNA for mammalian β-adrenergic receptor and homology with rhodopsin. *Nature* **321**: 75.

Eraso, P. and Gancedo, J.M. (1985) Use of glucose analogues to study the mechanism of glucose-mediated cAMP increase in yeast. *FEBS Lett.* **191**: 51.

Fleischman, L. F., Chahwala, S. B. and Cantley, L. (1986) *Ras*-transformed cells: altered levels of phosphatidylinositol-4, 5-bisphosphate and catabolites. *Science* **231**: 407.

Fong, H.K., Hurley, J.B., Hopkins, R.S., Lye, R.M., Johnson, M.S., Doolittle, R.F. and Simon, M.I. (1986) Repetitive segmental structure of the transducin β subunit: homology with the *CDC4* gene and identification of related mRNAs. *Proc. Natl. Acad. Sci. USA* **83**: 2162.

Fukui, Y. and Kaziro, Y. (1985) Molecular cloning and sequence analysis of a *ras* gene from *Schizosaccharomyces pombe*. *EMBO J.* **4**: 687.

Gallwitz, D., Donath, C. and Sander, C. (1983) A yeast gene encoding a protein homologous to the human c-has/bas proto-oncogene product. *Nature* **306**: 704.

Gibbs, J.B., Sigal, I.S., Poe, M. and Scolnick, E.M. (1984) Intrinsic GTPase activity distinguishes normal and oncogenic *ras* p21 molecules. *Proc. Natl. Acad. Sci. USA* **81**: 5704.

Gilman, A.G. (1987) G proteins: transducers of receptor-generated signals. *Ann. Rev. Biochem.* **56**: 615.

Habenicht, A.J.R., Glomset, J.A., King, W.C., Nist, C. Mitchell, C.D. and Ross, R. (1981) Early changes in phosphatidylinositol and arachidonic acid metabolism in quiescent Swiss 3T3 cells stimulated to divide by platelet-derived growth factor. *J. Biol. Chem.* **256**: 12329.

Hagen, D.C., McCaffrey, G. and Sprague, G.F. Jr., (1986) Evidence that yeast *STE3* gene encodes a receptor for the peptide pheromone *a* factor: gene sequence and implication for the structure of the presumed receptor. *Proc. Natl. Acad. Sci. USA* **82**: 1418.

Holz, G.G., Rane, S.G. and Dunlap, K. (1986) GTP-binding proteins mediate transmitter inhibition of voltage-dependent calcium channels. *Nature* **319**: 670.

Itoh, H., Kozasa, T., Nagata, S., Nakamura, S., Katada, T., Ui, M., Iwai, S., Otsuka, E., Kawasaki, H., Suzuki, K. and Kaziro, Y. (1986) Molecular cloning and sequence determination of cDNAs for α subunits of the guanine nucleotide-binding proteins Gs, Gi, and Go from rat brain. *Proc. Natl. Acad. Sci. USA* **83**: 3776.

Jenness, D.D., Burkholder, A.C. and Hartwell, L.H. (1983) Binding of α-factor pheromone to yeast *a* cells: chemical and genetic evidence for an α-factor receptor. *Cell* **35**: 521.

Jones, D.T. and Reed, R.R. (1987) Molecular cloning of five GTP-binding protein cDNA species from rat olfactory neuroepithelium. *J. Biol. Chem.* **202**: 14241.

Johnson, K.E., Cameron, S., Toda, T., Wigler, M. and Zoller, M.J. (1987) Expression in *Escherichia coli* of *BCY1*, the regulatory subunit of cyclic AMP-dependent protein kinase from *Saccharomyces cerevisiae*. *J. Biol. Chem.* **262**: 8636.

Jurnak, F. (1985) Structure of the GDP domain of EF-Tu and location of the amino acids homologous to ras oncogene proteins. Science 230: 32.

Kaibuchi, K., Tsuda, T., Kikuchi, A., Tanimoto, T., Yamashita, T. and Takai, Y. (1986a) Possible involvement of protein kinase C and calcium ion in growth factor-induced expression of c-myc oncogene in Swiss 3T3 fibroblasts. J. Biol. Chem. 261: 1187.

Kaibuchi, K., Miyajima, A., Arai, K. and Matsumoto, K. (1986b) Possible involvement of RAS-encoded proteins in glucose-induced inositol phospholipid turnover in Saccharomyces cerevisiae. Proc. Natl. Acad. Sci. USA 83: 8172.

Katada, T., Bokoch, G.M., Northup, J.K., Ui, M. and Gilman, A.G. (1984) The inhibitory guanine nucleotide-binding regulatory component of adenylate cyclase. J. Biol. Chem. 259: 3568.

Kataoka, T., Powers, S., McGill, C., Fasano, O., Strathern, J., Broach, J. and Wigler, M. (1984) Genetic analysis of yeast RAS1 and RAS2 genes. Cell 37: 437.

Kataoka, T., Powers, S., Cameron, S., Fasano, O., Goldfarb, M., Broach, J. and Wigler, M. (1985) Functional homology of mammalian and yeast RAS genes. Cell 40: 19.

Kaziro, Y. (1978) The role of guanosine 5'-triphosphate in polypeptide chain elongation. Biochim. Biophys. Acta 505: 95.

Lochrie, M.A., Hurley, J.B. and Simon, M.I. (1985) Sequence of the alpha subunit of photoreceptor G protein: homologies between transducin, ras, and elongation factors. Science 228: 96.

Madaule, P. and Axel, R. (1985) A novel ras-related gene family. Cell 41: 31.

Madaule, P., Axel, R. and Meyers, A. (1987) Characterization of two members of the rho gene family from the yeast Saccharomyces cerevisiae. Proc. Natl. Acad. Sci. USA 84: 779.

Mattera, R., Codina, J., Crozat, A., Kidd, V., Woo, S.L.C. and Birnbaumer, L. (1986) Identification by molecular cloning of two forms of the α-subunit of the human liver stimulatory (Gs) regulatory component of adenylate cyclase. FEBS Lett. 206: 36.

Matsumoto, K., Uno, I., Oshima, Y. and Ishikawa. T. (1982) Isolation and characterization of yeast mutants deficient in adenylate cyclase and cAMP-dependent protein kinase. Proc. Natl. Acad. Sci. USA 79: 2355.

Matsumoto, K., Uno, I. and Ishikawa, T. (1983) Control of cell division in Saccharomyces cerevisiae mutants defective in adenylate cyclase and cAMP-dependent protein kinase. Exp. Cell. Res. 146: 151.

Matsumoto, K., Uno, I. and Ishikawa, T. (1984) Identification of the structural gene and nonsense alleles for adenylate cyclase in Saccharomyces cerevisiae. J. Bacteriol. 157: 277.

Matsumoto, K., Uno, I. and Ishikawa, T. (1985) Genetic analysis of the role of cAMP in yeast. Yeast 1: 15.

Medynski, D.C., Sullivan, K., Smith, D., Van Dup, C., Chang, F.-H., Fung, B.-K., Seeburg, P.H. and Bourne, H.R. (1985) Amino acid sequence of the α subunit of transducin deduced from the cDNAsequence. Proc. Natl. Acad. Sci. USA 82: 43411.

Meurs, K.P.V., Angus, C.W., Lavu, S., Kung, H.-F., Czarnecki, S.K., Moss, J. and Vaughan, M. (1987) Deduced amino acid sequence of bovine retinal Goα: similarities to other guanine nucleotide-binding proteins. Proc. Natl. Acad. Sci. USA 84: 3107.

Miyajima, I., Nakafuku, M., Nakayama, N., Brenner, C., Miyajima, A., Kaibuchi, K., Arai, K., Kaziro, Y. and Matsumoto, K. (1987) GPA1, a haploid-specific essential gene, encodes a yeast homolog of mammalian G protein which may be involved in mating factor signal tranduction. Cell 50: 1011.

Nakafuku, M., Itoh, H., Nakamura, S. and Kaziro. Y. (1987) Occurrence in Saccharomyces cerevisiae of a gene homologous to the cDNA coding for the α subunit of mammalian G proteins. Proc. Natl Acad. Sci. USA 84: 2140.

Nakafuku, M., Obara, T., Kaibuchi, K., Miyajima, I., Miyajima, A., Itoh, H., Nakamura, S., Arai, K., Matsumoto, K. and Kaziro, Y. (1988) Isolation of a second G protein homologous gene (GPA2) from Saccharomyces cerevisiae and studies on its possible functions. Proc. Natl. Acad. Sci. USA 85: 1374.

Nakamura, T. and Ui, M. (1985) Simultaneous inhibitors of inositol phospholipid breakdown, arachidonic acid release, and histamine secretion in mast cells by islet-activating protein, pertussis toxin. J. Biol. Chem. 260: 3584.

Nakayama, N., Miyajima, A. and Arai, K. (1985) Nucleotide sequence of STE2 and STE3, cell type-specific sterile genes from Saccharomyces cerevisiae. EMBO J. 4: 2643.

Nakayama, N., Miyajima, A. and Arai, K. (1987) Common signal transduction system shared by STE2 and STE3 in haploid cells of Saccharomyces cerevisiae: autocrine cell-cycle arrest results from forced expression of STE2. EMBO J. 6: 249.

Nathans, J. and Hogness, D.S. (1984) Isolation and nucleotide sequence of the gene encoding human rhodopsin. Proc. Natl. Acad. Sci. USA 81: 4851.

Nikawa, J., Sass, P. and Wigler, M. (1987) Cloning and characterization of the low-affinity cyclic AMP phosphodiesterase gene of Saccharomyces cerevisiae. Mol. Cell. Biol. 7: 3629.

Nishizuka, Y. (1984) The role of protein kinase C in cell surface signal transduction and tumor promotion. Nature 308: 693.

Nukada, T., Tanabe, T., Takahashi, H., Noda, M., Hirose, T., Inayama, S. and Numa, S. (1986a) Primary structure of the α-subunit of bovine adenylate cyclase-stimulating G-protein deduced from the cDNA sequence. FEBS Lett. 195: 220.

Nukada, T., Tanabe, T., Takahashi, H., Noda, M., Haga, K., Haga, T., Ichiyama, A., Kangawa, K., Hiranaga, M., Matsuo, H. and Numa, S. (1986b) Primary structure of the α-subunit of bovine adenylate cyclase-inhibiting G-protein deduced from the cDNA sequence. FEBS Lett. 197: 305.

Ohta, H., Okajima, F. and Ui, M. (1985) Inhibition by islet-activating protein of a chemotactic peptide-induced early breakdown of inositol phospholipids and Ca^{2+} mobilization in guinea pig neutrophils. J. Biol. Chem. 260: 15771.

Parries, G., Hoebel, R. and Racker, E. (1987) Opposing effects of a ras oncogene on growth factor-stimulated phosphoinositide hydrolysis: desensitization to platelet-derived growth factor and enhanced sensitivity to bradykinin. Proc. Natl. Acad. Sci. USA 84: 2648.

Pfaffinger, P.J., Martin, J.M., Hunter, D.D., Nathanson, N.M. and Hille, B. (1985) GTP-binding proteins couple cardiac muscarnic receptors to a K channel. Nature 317: 536.

Powers, S., Kataoka, T., Fasano, O., Goldfarb, M., Strathern, J., Broach, J. and Wigler, M. (1984) Genes in S. cerevisiae encoding proteins with domains homologous to the mammalian ras proteins. Cell 36: 607.

Reed, S.I. (1980) The isolation of S. cerevisiae mutants defective in the start event of cell division. Genetics 95: 561.

Reymond, C.D., Gomer, R.H., Mehdy, M.C. and Firtel, R.A. (1984) Developmental regulation of a Dictyosterilum gene encoding a protein homologous to mammalian ras protein. Cell 39: 141.

Robishaw, J.D., Russell, D.W., Harris, B.A., Smigel, M.D. and Gilman, A.G. (1986) Deduced primary structure of the α subunit of the GTP-binding stimulatory protein of adenylate cyclase. Proc. Natl. Acad. Sci. USA 83: 1251.

Rozengurt, E. (1982) Synergistic stimulation of DNA synthesis by cyclic AMP derivatives and growth factors in mouse 3T3 cells. J. Cell. Physiol. 112: 243.

Salminen, A. and Novick, P.J. (1987) A ras-like protein is required for a post-golgi event in yeast secretion. Cell 49: 527.

Schmitt, H.D., Wagner, P., Pfaff, E. and Gallwitz, D. (1986) The ras-related YPT1 gene product in yeast: a GTP-binding protein that might be involved in microtubule organization. Cell 47: 401.

Seeburg, P.H., Colby, W.W., Capon, D.J., Goeddel, D.V. and Levinson, A.D. (1984) Biological properties of human cHa-ras1 genes mutated at codon 12. Nature 312: 71.

Segev, N. and Botstein, D. (1987) The ras-like yeast YPT1 gene is itself essential for growth, sporulation and starvation response. Mol. Cell. Biol. 7: 2367.

Seuwen, K., Lagarde, A. and Pouyssegur, J. (1988) Deregulation of hamster fibroblast proliferation by mutated ras oncogene is not mediated by constitutive activation of phosphoinositide-specific phospholipase C. EMBO J. 7: 161.

Shilo, B. and Weinberg, R. (1981) DNA sequences homologous to vertebrate oncogenes are conserved in Drosophila melanogaster. Proc. Natl. Acad. Sci. USA 78: 6789.

Smith, C.D., Lane, B.C. Kusaka, I., Verghese, M.W. and Snyderman, R. (1985) Chemoattractant receptor-induced hydrolysis of phosphatidylinositol 4, 5-bis-phosphate in human polymor-phonuclear leukocyte members. J. Biol. Chem. 260: 5875.

Sprague, G.F., Jr., Blair, L.C. and Thorner, J. (1983) Cell interactions and regulation of cell type in the yeast. Ann. Rev. Microbiol. 37: 623.

Sternweis, P.C. and Robishaw, J.D. (1984) Isolation of two proteins with high affinity for guanine nucleotides from membranes of bovine brain. J. Biol. Chem. 259: 13806.

Stryer, L. and Bourne, H.R. (1986) G proteins: a family of signal transducers. Ann. Rev. Cell Biol. 2: 391.

Suki, W.N., Abramowitz, J., Mattera, R., Codina, J. and Birnbaumer, L. (1987) The human genome encodes at least three non-allelic G proteins with αi-type subunits. FEBS Lett. 220: 187.

Sullivan, K.A., Liao, Y.-C., Alborzi, A., Beiderman, B., Chang, F.-H., Masters, S.B., Levinson, A.D. and Bourne, H.R. (1986) Inhibitory and stimulatory G proteins of adenylate cyclase: cDNA and amino acid sequences of the α chains. Proc. Natl. Acad. Sci. USA 83: 6687.

Takai, Y., Kikkawa, U., Kaibuchi, K. and Nishizuka, Y. (1984) Membrane phospholipid metabolism and signal transduction for protein phosphorylation. *Adv. Cyclic Nucleotide Protein Phosphorylation Res.* **18**: 119.

Tanabe, T., Nukada, T., Nishikawa, Y., Sugimoto, K., Suzuki, H., Takahashi, H., Noda, M., Haga, T., Ichiyama, A., Kangawa, K., Minamino, N., Matsuo, H. and Numa. S. (1985) Primary structure of the α-subunit of transducin and its relationship to *ras* proteins. *Nature* **315**: 242.

Tatchell, K., Chaleff, D.T., Defeo-Jones, D. and Scolnick, E. (1984) Requirement of either of a pair of *ras*-related genes of *Saccharomyces cerevisiae* for spore viability. *Nature* **309**: 523.

Toda, T., Uno, I., Ishikawa, T., Powers, S., Kataoka, T., Broek, D., Cameron, S., Broach, J., Matsumoto, K. and Wigler, M. (1985) In yeast, *RAS* proteins are controlling elements of adenylate cyclase. *Cell* **40**: 27.

Toda, T., Cameron, S., Sass, P., Zoller, M., Scott, J.D., McMullen, B., Hurwitz, M., Krebs, E.G. and Wigler. M. (1987) Cloning and characterization of *BCY1*, a locus encoding a regulatory subunit of the cyclic AMP-dependent protein kinase in *Saccharomyces cerevisiae*. *Mol. Cell. Biol.* **7**: 1371.

Yamano, S., Tanaka, K., Matsumoto, K. and Toh-e. A. (1987) Mutant regulatory subunit of 3′, 5′-cAMP-dependent protein kinase of yeast *Saccharomyces cerevisiae*. *Mol. Gen. Genet.* **210**: 413.

Yatsunami, K. and Khorana, H.G. (1985) GTPase of bovine rod outer segments: the amino acid sequence of the α subunit as derived from the cDNA sequence. *Proc. Natl. Acad. Sci. USA* **82**: 4316.

Wagner, P., Molenaar, C.M.T., Ranh, A.J.G., Brokel, R., Schmitt, H.D. and Gallwitz, D. (1987) Biochemical properties of the *ras*-related *YPT1* protein in yeast: a mutational analysis. *EMBO J.* **6**: 2373.

Wakelam, M.J.O., Davies, S.A., Houslay, M.D., McKay, I., Marshall, C.J. and Hall, A. (1986) Normal p21[N-ras] couples bombesin and other growth factor receptors to inositol phosphate production. *Nature* **323**: 173.

7 Controlling elements in the cell division cycle of *Schizosaccharomyces pombe*

TATSUYA HIRANO and MITSUHIRO YANAGIDA

7.1 Introduction

The basic mechanisms of the eukaryotic cell cycle may be the same among largely different organisms. Selection of a model organism is therefore essential. The fission yeast *Schizosaccharomyces pombe* has several advantages for analysis of the cell cycle. It has a small genome (approximately 14 000-kb in three chromosomes), is stable in haploid, grows in synthetic media and divides by fission with a short generation time of 130 minutes at 30°C in rich medium. There are discrete G1, S, G2 and M phases in the cell cycle of *S. pombe*. The genetics are efficient and convenient due to the isogenic strains used, and 460 markers are known to date (Kohli, 1987). Availability of the cloning vehicles of *S. pombe* genes by transformation (Beach and Nurse, 1981) also enables isolation of a number of interesting genes which are implicated in the cell cycle. We shall describe here recent developments in the study of controlling elements in the cell cycle of *S. pombe*. Readers may refer to previous reveiws for genetics, physiology and molecular biology of the cell cycle of this yeast (Gutz *et al.*, 1974; Mitchison, 1970; Nurse, 1985; Yanagida *et al.*, 1986).

7.2 Structural elements in the cell cycle control

7.2.1 *Increase in cell length*
The cells of *Schizosaccharomyces pombe* are round-ended cylinders of nearly constant diameter. They grow by length extension alone, and divide by fission (Figure 7.1). The growth in length occurs during the first three-quarters of the cell cycle after division. During the last quarter of the cycle, the length remains constant, while a septum forms across the middle of the cell. Mitosis occurs at the beginning of the constant stage. The G1 phase is very short in the wild-type cell cycle. DNA is replicated at the time of cell division, so that the G2 phase occupies the remainder (about 70%) of the cell cycle, in which the nucleus has a 2C quantity of DNA (Mitchison, 1970).

We use Mitchison's definition for the progress of cell cycle (0 for the birth of the new cell and 1 for the end of cell division). By this definition, mitosis occurs at 0.75 in wild-type cells at 30°C. At the beginning of the cell cycle (in the early G2 phase), growth is restricted to the old end of the cell, which existed in the

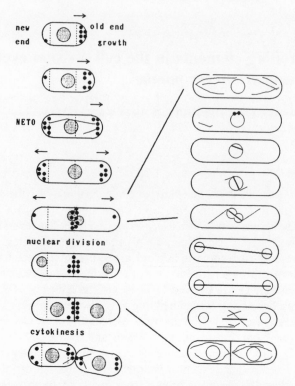

Figure 7.1 Cytoskeletal alterations during the cell cycle of *S. pombe*. The arrows indicate the direction of cell elongation. In the series on the left, broken lines represent division scars, and dots (also thin filaments at NETO and cytokinesis) show the distributions of actin. The shaded circles indicate nuclei. Microtubules and spindles in mitotic cells are shown respectively by thin filaments and thick lines in the series on the right. (See Marks *et al.*, 1986; Hagan and Hyams, 1988).

previous cycle (Figure 7.1). The new end then starts to grow at a point in the G2 phase called NETO ('new end take-off'). Fluorescence microscopy has been used on cells stained with Calcofluor to study NETO in wild-type and mutant cells which undergo division at different sizes (Mitchison and Nurse, 1985). The results showed that NETO is under two controls. First, cells must reach a critical length of 9.0–9.5-μm. Second, they must have passed the 0.3– 0.35 period in the cell cycle. If either of these two requirements is not met, NETO cannot take place. In cells that divide at a small size, NETO is delayed until cells attain the critical length, and in large dividing cells NETO is dependent upon the completion of the event in early G2.

Time-lapse photomicrography was used to follow the rate of length growth in single cells (Mitchison and Nurse, 1985). The total length curve shows a pattern of two linear segments in the first three-quarters of the cycle with a rate-change point (RCP). In the last quarter of the cycle, length growth stops and the septum appears. In wild-type cells, the RCP is coincident with NETO, where the rate of total increase in length increases by 35%. However, the increase is not due simply to the start of new-end growth, since old-end growth slows down at the RCP in many cases.

7.2.2 Nuclear structural alteration

The nucleus of S. pombe consists of two hemispherical regions: the DNA-rich chromosomal (chromatin) region and the non-chromosomal, RNA-rich region that includes the nucleolus (McCully and Robinow, 1971; Toda et al., 1981; Yanagida et al., 1986). In thin-section electron microscopy the electron dense region is RNA-rich, whereas chromosomes are rarely visualized in the other hemisphere, probably because the staining is weak, and the extent of condensation (Umesono et al., 1983) may not be sufficient. Chromosomes can more easily be observed by fluorescence microscopy (see below). The nuclear envelope remains throughout the nuclear division of S. pombe, as found in most fungi.

The spindle pole body (SPB) duplicates in early mitosis (McCully and Robinow, 1971; Tanaka and Kanbe, 1986), in contrast to the duplication of SPB in the late G1 phase of Saccharomyces cerevisiae. In the S. pombe interphase, SPBs locate on the outer membrane of the nuclear envelope, while in mitosis they are within the nuclear envelope. Pole-to-pole microtubules are clearly visible in specimens prepared by freeze-substitution. Tanaka and Kanbe (1986) suggested that kinetochore microtubules radiate from the poles, although an earlier study (McCully and Robinow, 1971) had reported that the chromosomes are directly associated with the pole bodies. The results of immunofluorescence microscopy using anti-tubulin antibody (Hiraoka et al., 1984; Hagan and Hyams, 1988) are consistent with the presence of kinetochore microtubules, but direct evidence is needed for the establishment for kinetochore microtubules.

Toda et al. (1981) investigated the sequential alterations in the nuclear chromatin region (defined as the nuclear area stained with a DNA-specific probe: DAPI) during mitosis by video-connected fluorescence microscopy. The interphase hemisphere of the nuclear chromatin region changes into a condensed ellipsoid, then segregates into a U-shaped intermediate structure, and separates into two smaller hemispheres which move to opposite ends of the cell. The condensation of chromosomes is particularly obvious in mitotically arrested cells. Individual chromosomes are observed in the cells of temperature-sensitive (ts) cdc13 (Nasmyth and Nurse, 1981) and cold-sensitive (cs) nda3 (Umesono et al., 1983b; Hiraoka et al., 1984) incubated at restrictive temperatures, which are arrested in the middle of mitosis. Furthermore, the wild-type cells treated with thiabendazole, an anti-microtubule agent, show chromosome-like bodies. The number of chromosomes (three) is in agreement with that of genetic linkage groups (Kohli et al., 1977) and the electrophoretic karyotype demonstrated by the pulsed field gel electrophoresis (Smith et al., 1987).

7.2.3 Microtubule distribution during the cell cycle

The changes in microtubule organization which occur through the mitotic cell cycle are shown by indirect immunofluorescence microscopy using antibody against tubulin (Hagan and Hyams, 1988; Hiraoka et al., 1984; see also Figure 7.1).

In interphase, cytoplasmic microtubules are aligned along the long axis of the cell. Many of them run between the cell tips and appear to be independent of SPBs. These may be involved in positioning the nucleus at the cell equator and in the establishment of cell polarity.

At the onset of mitosis, the interphase microtubule array disappears and an intranuclear spindle appears, extending between the duplicated SPBs. The short spindle has a greater staining intensity at the ends than the middle, suggesting the existence of two sets of microtubules: pole-to-pole and kinetochrome. When the spindle completely spans the nucleus, such differential staining is lost, and astral microtubules emanating from the cytoplasmic face of SPBs appear. The spindle elongates until the daughter nuclei reach each of the cell ends. At the end of anaphase, the spindle disappears and two new microtubule organizing centers (MTOCs), which re-establish the interphase array, appear at the cell equator. It should be noted that the structural rearrangement of microtubules occurs at two distinct points in the mitotic cell cycle: at the G2/M boundary and at mitotic telophase. Each of these seems to be under the control of distinct MTOCs.

Thus the spindle dynamics of *S. pombe* are similar to those of higher eukaryotic cells; in *S. cerevisiae*, the spindle is present during most of the cell cycle so that structural transition at the onset of mitosis is not clear (Byers and Goestch, 1975; Kilmartin and Adams, 1984).

7.2.4 Actin distribution during the cell cycle

The distribution of F-actin was investigated by fluorescence microscopy using rhodamine-conjugated phalloidin (Marks and Hyams, 1985; Marks et al., 1986). In newly divided cells growth occurs intially only at the old end that existed prior to septation. At NETO, bipolar growth is initiated. This is maintained until the onset of mitosis, at which point growth ceases and cytokinesis and cell separation occur. The distribution of actin coincides precisely with these features of cell growth (Figure 7.1). In newly divided cells actin is localized at the single growing (old) end, while initiation of bipolar growth coincides with the appearance of actin at both ends of the cell. The actin filaments are temporarily observed at this NETO transition. As cells stop growing and enter mitosis, actin disappears from the poles and reappears as a ring at the cell equator where it overlies the dividing nucleus. The position of the actin ring anticipates the deposition of septum which stains intensely with Calcofluor. At the completion of septation, actin leaves the cell equator by the breakdown of the ring and moves to the opposite end of each daughter cell to initiate new cell growth. This transition is accompanied by the transient appearance of actin filaments.

7.3 Genetic elements in cell cycle control

7.3.1 Identification of cell cycle genes
Three different approaches have been employed to identify the cell cycle genes and their functions in *S. pombe*.

(i) *Isolation of conditional lethal mutants which arrest at specific stages of the cell cycle.* A number of temperature-sensitive *cdc* (cell division cycle) mutants have been isolated and characterized (Nurse *et al.*, 1976; Nasmyth and Nurse, 1981). These mutations define 26 unlinked genes, which are involved in DNA synthesis, nuclear division, or cell plate formation. Toda *et al.* (1983) screened a collection of cold-sensitive mutants and identifed 12 *nda* (nuclear division arrest) mutants which are defective in nuclear division. By examining different cytological phenotypes, novel classes of cell cycle mutants which were not identified by the previous screening for temperature-sensitive mutations have also been isolated (Hirano *et al.*, 1986; Hirano *et al.*, 1988; Ohkura *et al.*, 1988). These cell cycle genes for which conditional lethal mutations are available can be cloned by complementation. The deduced amino acid sequence from the cloned gene may provide information about its gene function, and allows antisera to be prepared against either the fused protein expressed in bacterial cells or short synthetic oligopeptides. The obtained antisera may be a powerful tool to investigate the biochemical function and/or cellular localization of the gene products.

(ii) *Isolation of mutants defective in a known gene product by assay screening.* Uemura and Yanagida (1984) have successfully isolated type I and II topoisomerase mutants by assaying cell extracts from mutagenized cells.

(iii) *Cloning of the genes possibly involved in cell cycle control by using heterologous gene probes.* The genes encoding highly conserved proteins, such as histones and calmodulin (Matsumoto and Yanagida, 1985; Takeda and Yamamoto, 1987), can be isolated by their DNA sequence homology to heterologous probes. Site-directed mutagenesis and gene replacement techniques make it possible to construct mutants and characterize their defective phenotypes in the cell cycle.

7.3.2 G1 phase
'Start' is the point in G1 at which cells become committed to the mitotic cell cycle (Hartwell *et al.*, 1974). Cells which have completed this point are committed to the mitotic cycle in progress and are unable to undergo an alternative developmental pathway, such as conjugation and meiosis. Completion of 'start' is also the major rate-limiting step of the cell cycle which regulates the rate of cell division. In fission yeast, two genes, $cdc2^+$ and $cdc10^+$, have been identified whose functions are required for 'start' (Nurse and Bissett, 1981); these two mutants have the ability to conjugate after arrest at the restrictive temperature. The $cdc2^+$ gene function also participates in another

major cell cycle control which determines the timing of mitosis (Nurse *et al.*, 1976; Nurse and Bissett, 1981). The $cdc2^+$ gene product appears to have a central role in the regulation of the cell cycle (see below).

The $cdc10^+$ gene has been cloned by rescue of mutant function (Aves *et al.*, 1985). The transcriptional level of the $cdc10^+$ gene was investigated and found to be constant during the mitotic cell cycle. Moreover, no change could be detected during a shift from exponential growth to stationary phase. This suggests that commitment to the cell cycle is not controlled by regulation of $cdc10^+$ transcript level. Nucleotide sequence determination has shown that the $cdc10^+$ gene encodes a putative 85 kD protein (Aves *et al.*, 1985). Recently, the *SW16* protein of *Saccharomyces cerevisiae* has been shown to have a significant homology to the $cdc10^+$ product (Breeden and Nasmyth, 1987); 138 of the 420 C-terminal amino acids of the $cdc10^+$ protein were identical to those of *SW16* when nine gaps were introduced. *SW16* has been identified as a *trans*-acting activator of the HO gene which encodes the endonuclease that initiates the mating-type switching. HO transcription is dependent on 'start', suggesting that the two yeasts may execute this crucial control step by common regulatory mechanisms.

The other two genes, $cdc20^+$ and $cdc22^+$, have their functions in G1 after 'start', since these mutants cannot conjugate from their arrest points (Nurse and Bissett, 1981). They are likely to be required for the initiation of DNA replication (Nurse and Nasmyth, 1981); the $cdc20$ and $cdc22$ mutants immediately blocked DNA synthesis upon shift to the restrictive temperature (36°C), but were able to undergo one round of cell division at this temperature after release from a block by hydroxyurea, an inhibitor of DNA synthesis. Their gene functions therefore seem to be completed before the block point of hydroxyurea. The $cdc22^+$ gene has been cloned and shown to encode a transcript of 3.3 kb (Gordon and Fantes, 1986). The level of $cdc22^+$ transcript during the cell cycle was investigated in synchronous culture and found to be cell-cycle regulated, with a maximum level during the late G1/S phase. With the exception of the histone genes (Aves *et al.*, 1985; Matsumoto *et al.*, 1987), this gene is the only reported case of cell-cycle regulated expression in *S. pombe*.

7.3.3 S phase

Some *cdc* mutants which have defects in the S phase have been reported (Nasmyth and Nurse, 1981). The position of a *cdc* gene product function with respect to the point in the cell cycle blocked by hydroxyurea was investigated by reciprocal shift experiments. The *cdc17* and *cdc24* mutants undergo a complete round of DNA synthesis at restrictive temperature (36°C) but the genes cannot complete their functions when the cells are arrested by hydroxyurea before temperature shift-up. The DNA made at 36°C must therefore be defective in some way even though there is bulk DNA synthesis. *cdc23* is defective in bulk DNA synthesis, and *cdc21* may have a partial defect in the initiation of DNA replication. Little is known about the molecular functions of these gene products with the exception of $cdc17^+$.

Nasmyth has reported that the $cdc17$ mutant at the restrictive temperature accumulated nascent DNA in the form of Okazaki fragments, was sensitive to ultraviolet irradiation, and had a deficiency in the DNA ligase activity of crude extract (Nasmyth, 1977). Furthermore, DNA ligase partially purified from the mutants was found to be thermolabile, strongly suggesting that the $cdc17^+$ gene encodes the structural gene for DNA ligase (Nasmyth, 1979). Recently, the $cdc17^+$ gene has been cloned by complementation (Johnston et al., 1986) and sequenced (Barker et al., 1987). The deduced amino acid sequence predicted a protein of 86 kD, and an overall 53% homology with DNA ligase of Saccharomyces cerevisiae which is encoded by the CDC9 gene. A short stretch containing the putative ATP binding site is also conserved in T4 and T7 DNA ligases.

Histone genes of S. pombe have been cloned by hybridization and their nucleotide sequence has been determined (Matsumoto and Yanagida, 1985). The genome of S. pombe has a single, isolated H2A, a pair of H2A-H2B, and three pairs of H3-H4. This gene organization is distinct from that of S. cerevisiae which consists of two sets each of the paired H2A–H2B and H3–H4 (Hereford et al., 1979; Smith and Andresson, 1983). A 17-bp highly homologous sequence (AACCCT box) is found in the intergene spacer sequences or in the 5' upstream region of all the histone genes. This sequence may play an important role in the regulation of histone gene expression.

In many types of eukaryotic cells, it is known that there is an increase in histone transcription level at the G1/S boundary which decreases towards the end of S phase. In S. pombe, the level of histone H2A transcript was shown to vary periodically during the cell cycle with peaks at the time of S phase (Aves et al., 1985). Recently, the pattern of histone H2B transcript accumulation has been monitored in temperature-sensitive mutants which block at different times during the cell cycle, at 'start' in G1, at G1 after 'start', and during S phase (Matsumoto et al., 1987). The blocking of cells before 'start' using $cdc10$ prevents an increase in H2B transcript level, indicating that completion of 'start' is required for the increase. The blocking of cells in late G1 using $cdc20$ and $cdc22$ allows levels to rise, suggesting that the increase in transcripts can be uncoupled from DNA synthesis since there is no increase in DNA content with these mutants. When the completion of DNA replication is blocked in late S phase by the $cdc17$ and $cdc24$ mutants, histone H2B transcripts increase normally but remain at a high level. Even though there has been a doubling in DNA content, the mechanism leading to a normal fall in transcript level at the end of S phase is not activated in these mutants.

7.3.4 Late G2 and initiation of mitosis

The control regulating the initiation of mitosis has been investigated using mutants altered in the cell size at which they undergo mitosis. The wee

Table 7.1 The cell cycle genes of *Schizosaccharomyces pombe*.

Gene	Defect	Mutation	Mutant phenotype/ gene product	Map position	Reference
cdc1	G2/M	ts		I R	1
cdc2	start, G2/M	ts, wee	Required at two points in the cell cycle; encodes protein kinase p34	II L	1, 4, 7, 8, 9, 10, 11, 12, 13
cdc6	G2	ts		II	1
cdc10	start		Encodes an 85-kD protein	II R	1, 22, 23
cdc13	M	ts	Suppresses *cdc2*	II	1, 2, 23
cdc17	S	ts	Encodes DNA ligase	—	2, 25, 26, 27, 28
cdc18	S/M?	ts		II R	2
cdc19	S/M?	ts		—	2
cdc20	G1	ts		—	2
cdc21	S	ts		—	2
cdc22	G1	ts	Periodic transcription	—	2, 24
cdc23	S	ts		—	2
cdc24	S	ts		—	2
cdc25	G2/M	ts	Mitotic inducer; encodes a 67-kD protein	I L	6, 14
cdc27	G2/M	ts		—	2
cdc28	G2/M	ts		—	2
wee1	G2/M	wee	Mitotic inhibitor; encodes protein kinase	III R	3, 4, 5, 15
nim1	—	—	Suppresses *cdc25*; negative regulator of *wee1*; encodes protein kinase	—	16
suc1	—	—	Suppresses *cdc2*; regulatory component of *cdc2*; encodes a 13-kD protein	—	17, 18, 19, 20
nda1	M	cs		II L	29, 34
nda2	M	cs	Supersensitive to TBZ; encodes α1-tubulin	II R	29, 30, 31, 32, 34
atb2	—	—	Encodes α2-tubulin	II L	32, 34
nda3	M	cs	Supersensitive or resistant to TBZ; encodes β-tubulin	II R	29, 30, 33, 34
nda4	M	cs		I R	29, 34
nda5–12	M	cs		—	29
ben4	M	cs	Resistant to TBZ	I	35
top1	—	—	Encodes topoisomerase I	II R	34, 36, 40
top2	M	ts, cs	Encodes topoisomerase II	II R	34, 36, 37, 38, 39
cut1	M	ts	Blocks nuclear division but not cytokinesis	III R	34, 41
cut2	M	ts	Blocks nuclear division but not cytokinesis	II R	34, 41
cut3–9	M	ts	Blocks nuclear division but not cytokinesis	—	34, 41
nuc1	—	ts	Alters nuclear structure; encodes the large subunit of RNA polymerase I	II R	34, 41
nuc2	M	ts	Blocks spindle elongation; encodes a nuclear scaffold-like protein p67	I R	34, 42

(*Contd.*)

Table 7.1 (*Contd.*)

Gene	Defect	Mutation	Mutant phenotype/ gene product	Map position	Reference
dis1–3	M	cs	Defective in sister chromatid separation; caffeine supersensitive	—	43
cam1	—	null	Lethal; encodes calmodulin	—	44
ras1	—	null	Defective in conjugation; encodes a *ras* homolog	—	45, 46

(1) Nurse et al. (1976); (2) Nasmyth and Nurse (1981); (3) Nurse (1975); (4) Thuriaux et al. (1978); (5) Nurse and Thuriaux (1980); (6) Fantes (1979); (7) Nurse and Bissett (1981); (8) Beach et al. (1982); (9) Hindley and Phear (1984); (10) Booher and Beach (1986); (11) Simanis and Nurse (1986); (12) Lee and Nurse (1987); (13) Draetta et al., (1987); (14) Russell and Nurse (1986); (15) Russell and Nurse (1987a); (16) Russell and Nurse (1987b); (17) Hayles et al., (1986a); (18) Hayles et al. (1986b); (19) Hindley et al. (1987); (20) Brizuela et al. (1987); (21) Booher and Beach (1987); (22) Aves et al. (1985); (23) Breeden and Nasmyth (1987); (24) Gordon and Fantes (1986); (25) Nasmyth (1977); (26) Nasmyth (1979); (27) Johnston et al., (1986); (28) Barker et al. (1987); (29) Toda et al. (1983); (30) Umesono et al. (1983a); (31) Toda et al. (1984); (32) Adachi et al. (1986); (33) Hiraoka et al. (1984); (34) Yanagida et al. (1986); (35) Roy and Fantes (1982); (36) Uemura and Yanagida (1984); (37) Uemura and Yanagida (1986); (38) Uemura et al. (1986); (39) Uemura et al. (1987a); (40) Uemura et al. (1987b); (41) Hirano et al. (1986); (42) Hirano et al. (1988); (43) Ohkura et al. (1988); (44) Takeda and Yamamoto (1987); (45) Fukui and Kaziro (1985); (46) Fukui et al., (1986).

mutants, which initiate mitosis and cell division at reduced cell size, were found to map at two loci, *wee1* and *cdc2* (Nurse, 1975; Thuriaux *et al.*, 1978). Genetic evidence indicates that the *wee1⁺* gene product acts as an inhibitor in mitotic control (Nurse and Thuriaux, 1980). There are two classes of *cdc2* mutations: one class is temperature-sensitive lethal, recessive mutations which block in late G2 before mitosis (Nurse *et al.*, 1976); the second class is dominant *wee* mutations, called *cdc2w* (Thuriaux *et al.*, 1978). This indicates that the *cdc2⁺* gene product may act as an inducer of mitosis which, when modified into an activated form by *cdc2w* mutations, advances cells into mitosis at a reduced size. A third gene, *cdc25⁺*, may also be involved in the mitotic control. Temperature-sensitive lethal phenotype of *cdc25* which block in late G2 before mitosis (Nurse *et al.*, 1976) are completely suppressed by *wee1* mutations, suggesting that *cdc25⁺* may in some way interact with *wee1⁺* (Fantes, 1979).

The *cdc2⁺* gene was cloned by complementation of a temperature-sensitive lethal *cdc2* mutant (Beach *et al.*, 1982). The *S. cerevisiae CDC28* gene was also found to have the ability to complement *cdc2*, suggesting that the two genes have a homologous function in cell cycle control. The *CDC28* gene of *S. cerevisiae* like the *cdc2⁺* gene (Nurse and Bissett, 1981) has two points of function during the mitotic cell cycle, one in G1 at 'start' and one in late G2 before mitosis (Piggott *et al.*, 1982). The predicted proteins encoded by the *cdc2⁺* and *CDC28* genes share 62% overall amino acid homology (Hindley and Phear, 1984). They are of approximately 34-kD and show significant homology with vertebrate protein kinases, such as bovine cAMP-dependent protein kinase and the *src* family of viral oncogenes, particularly in the ATP binding site and the putative phosphorylation site. Site-directed mutagenesis of the $cdc2^+$ gene has supported the idea that the predicted protein kinase activity is essential for the function of $cdc2$ (Booher and Beach, 1986): introduction of mutations within the putative ATP binding site and in the region of the phosphorylation site results in loss of ability to complement $cdc2$ mutant.

Simanis and Nurse (1986) have prepared antibodies against the $cdc2^+$ protein using synthetic peptides, and demonstrated that the $cdc2^+$ gene product is a 34-kD phosphoprotein with protein kinase activity. They observed no changes in the protein level or phosphorylation state during the mitotic cell cycle. However, when cells are arrested in G1 by nitrogen starvation, the protein becomes dephosphorylated and loses protein kinase activity. On readdition of nutrients, the protein gains protein kinase activity and becomes phosphorylated, suggesting that entry into the mitotic cycle could be regulated by the modulation of $cdc2^+$ protein kinase activity, possibly by phosphorylation.

Recently, human homologue of the $cdc2^+$ gene has been cloned by expressing a human cDNA library in S. pombe and selecting those clones which can complement a mutation of the $cdc2$ gene (Lee and Nurse, 1987). The human $cdc2^+$ gene encodes a protein of the same molecular weight as S. pombe $cdc2^+$ with a 63% identity of amino acid sequence. Another approach to identification of a human homologue of $cdc2^+$ has been reported by Draetta et al. (1987). Monoclonal antibodies were raised against the $cdc2^+$ protein expressed in E. coli and screened for those which recognize both the $cdc2^+$ and S. cerevisiae CDC28 products. The antibodies obtained detected a 34-kD protein in Hela cells. Peptide mapping experiments revealed that human p34 was structurally related to the two yeast proteins. Further, this 34-kD protein was shown to share two biochemical characteristics with the S. pombe p34: first, an immune complex containing human p34 has protein kinase activity, and, second, human p34 is physically associated with p13, the human homologue of the $suc1^+$ gene product of S. pombe (described later). This evidence indicates that elements of cell cycle control are probably conserved between yeast and humans.

The $cdc25^+$ gene has been cloned by complementation of a $cdc25$ mutant (Russell and Nurse, 1986) and has been shown to encode a putative 67-kD protein, which has no homology with any sequence in computer databases. Increased expression of $cdc25^+$ has been shown to result in the initiation of mitosis at a reduced cell size, indicating that $cdc25^+$ functions as a dosage-dependent inducer of mitosis. A null allele of $cdc25$, as well as temperature-sensitive alleles, is suppressed by a $wee1$ mutation. The wee phenotypes caused by increased expression of $cdc25^+$ and inactivation of $wee1$ by mutation are additive and in combination are lethal. These results suggest that the $cdc25^+$ and $wee1^+$ gene functions are counteractive and independent in their regulation of the initiation of mitosis (Figure 7.2).

The lethal phenotype, caused by overexpression of $cdc25^+$ and the $wee1$ mutations, was used to clone the $wee1^+$ gene by complementation (Russell and Nurse, 1987a). The $wee1^+$ gene encodes a 112-kD protein which contains the consensus sequences of protein kinases. When $wee1^+$ expression is increased, mitosis is delayed until cells grow to a greater size, demonstrating that the $wee1^+$ gene product functions as a dosage-dependent inhibitor of mitosis. Genetic evidence indicates that there is allele-specific interaction between $cdc2w$ alleles and $wee1$ (or $cdc25$) mutations; the $cdc2-1w$ product is largely insensitive to changes of $wee1^+$ expression but still require $cdc25^+$ function, whereas the $cdc2-3w$ product is responsive to $wee1^+$ function but is independent of $cdc25^+$ function. These results suggest that the $cdc2^+$ function is rate-limiting for the initiation of mitosis and is controlled by $cdc25^+$ and $wee1^+$.

The $nim1^+$ gene was identified by its ability to suppress a $cdc25$ mutation when present on a multicopy plasmid (Russell and Nurse, 1987b). The predicted $nim1^+$ gene product is a 50-kD protein which has strong homology to protein kinases. Manipulation of the cloned $nim1^+$ gene demonstrates that $nim1^+$ is a dosage-dependent inducer of mitosis. Increased $nim1^+$ expression suppresses a null allele of $cdc25$, but not $cdc2$ mutations. Deletion or overexpression of $nim1^+$ does not affect the timing of mitosis in $wee1$ mutant, suggesting that the $nim1^+$ gene product functions as a mitotic inducer by blocking mitotic inhibition by $wee1^+$.

A model of the regulatory relationships of these gene product is shown in Figure 7.2 (Russell and Nurse, 1987b). In summary, (i) $cdc2^+$ mitotic function is rate-limiting for the initiation of mitosis; (ii) $cdc2^+$ function is stimulated by $cdc25^+$ and inhibited by $wee1^+$; (iii) $wee1^+$ activity is negatively regulated by $nim1^+$. Genetic evidence is consistent with this model, while there is no biochemical evidence to indicate direct interaction between the products of these genes. This regulatory network controlling the initiation of mitosis is

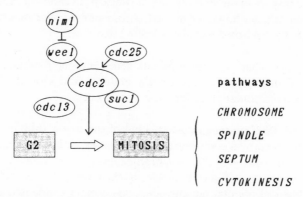

Figure 7.2 Genes required for the initiation of mitosis. Four mitotic pathways that become independent under certain circumstances (see text) take place only after the completion of the $cdc2^+$ gene functions.

likely to involve protein phosphorylation since three ($cdc2^+$, $wee1^+$ and $niml^+$) of the four elements have been identified as protein kinases.

The $sucl^+$ gene was first identified as a DNA sequence which, when present on a multicopy vector, suppresses certain alleles of $cdc2$ mutation (Hayles et al., 1986a). Subsequently, two extragenic suppressor mutations of $cdc2$ mutants were mapped on the $sucl$ locus (Hayles et al., 1986b). A null allele of the $sucl^+$ gene causes cell cycle arrest (Hayles et al., 1986b; Hindley et al., 1987). Suppression of $cdc2$ by mutations or overexpression of the $sucl^+$ gene is highly allele-specific, suggesting direct interaction between the $cdc2^+$ and $sucl^+$ gene product. The nucleotide sequence of $sucl^+$ has been determined and shown to encode a 13-kD protein (p13), which has no homology with any known proteins (Hindley et al., 1987). Anti-p13 antiserum has been prepared and used to identify p13 in S. pombe crude extract (Brizuela et al., 1987). p13 is found in a complex with the $cdc2^+$ gene product p34. The stability of the complex is reduced in strains carrying temperature-sensitive alleles of $cdc2$ that are suppressible by overexpression of $sucl^+$. These results suggest that the $sucl^+$ gene product acts as a regulatory component of the $cdc2^+$ gene product. However, there is no evidence that p13 is a substrate of p34 protein kinase.

The $cdc13^+$ gene is one of the cdc genes required for the progression of mitosis (Nurse et al., 1976; Nasmyth and Nurse, 1981). Recently, an extragenic suppressor mutation of a cold-sensitive allele of $cdc2$ has been shown to be an allele of $cdc13$ (Booher and Beach, 1987). Genetic analyses demonstrate that there are a variety of allele-specific interactions between $cdc2^+$ and $cdc13^+$. The $cdc13^+$ gene product might be a G2-specific substrate of the $cdc2^+$ protein kinase.

7.3.5 Genes required for mitosis
In contrast to the genes required for the late G2 or the initiation of mitosis described above, mutations in the genes required for mitosis show certain terminal phenotypes in the steps during the nuclear division (an exception is a $cdc13$ mutant which shows condensed chromosomes). We describe below genes required for mitosis and also some related genes.

(i) *Tubulin genes.* There are two α-tubulin genes and one β-tubulin gene in S. pombe. Detailed analyses using tubulin mutants and cloned genes have revealed different cellular roles for these tubulin genes.

(ii) $nda2^+$ and $atb2^+$: α-tubulin genes. One of the two α-tubulin genes was originally identified by a cold-sensitive mutation $nda2$ that is impaired in the nuclear division (Toda et al., 1983). Chromosome condensation takes place at restrictive temperature but the block in chromosome separation is not uniform in the cell population; chromosomes remain condensed in roughly half of the cells, while in the remainder chromosomes are separated by a short distance. Anaphase A but not B might have taken place in the latter cells; this hypothesis is consistent with the fact that functional tubulin molecules consisting of $\alpha 2$ and β (see below) exist in the mutant cells. The nda2 mutant shows pleiotropic phenotypes at a restrictive temperature; the nucleus is displaced from the centre of the cell, normal mitotic spindle is not observed, and the duplicated SPBs are abnormally located on the constricted nuclear envelope. Subsequently, nda2 was found to be one of the two major loci that confer supersensitivity to thiabendazole (TBZ) and related benomyl compounds known as tubulin inhibitors (Umesono et al., 1983a). These observations suggest that the $nda2^+$ gene product is involved in the nuclear and cytoplasmic microtubular organization.

Two independent genomic sequences were cloned by complementation of a cold-sensitive nda2 mutation (Toda et al., 1984). Predicted amino acid sequences showed that both encode α-tubulins and share 85% overall homology. The homology to porcine α-tubulin was 76% in both cases. By chromosomal integration, one of them ($\alpha 1$-tubulin gene) was found to be derived from the nda2 locus while the other ($\alpha 2$-tubulin gene: designated atb2) was from a locus not linked to nda2 (atb2 was recently mapped in the short arm of chromosome II distal to mei3; Y. Chikashige, unpublished result). Gene disruption experiments showed that the $\alpha 1$-tubulin gene is essential whereas the $\alpha 2$ gene is dispensable (Adachi et al., 1986). Although the locus encoding $\alpha 2$-tubulin is mutationally silent, the following evidence indicates that $\alpha 2$-tubulin is functional (i) the $\alpha 2$ transcripts are detected; (ii) $\alpha 2$-tubulin polypeptides, which are not found in the $\alpha 2$-disrupted cells, can be detected in the wild-type cells: (iii) increased expression of the $\alpha 2$ gene, when present on a multicopy plasmid, suppresses the $\alpha 1$ mutation; (iv) the $\alpha 2$-disrupted cells show an increased sensitivity to TBZ; (v) $\alpha 2$ disruption, when combined with cs $\alpha 1$ mutation, results in a semilethal phenotype. Northern blot analyses showed that the transcriptional regulation of the $\alpha 1$ gene is quite different from that of the $\alpha 2$ genes: $\alpha 1$ expression is modulative whereas $\alpha 2$ is constitutive. The essentiality of the $\alpha 1$ gene and dispensability of the $\alpha 2$ gene seem to depend on their regulatory sequences instead of the coding sequences.

(iii) $nda3^+$: β-tubulin gene. Cells of a cold-sensitive nda3 mutant show phenotypes similar but not identical to nda2. Three condensed chromosomes were observed (Toda et al., 1983; Umesono et al., 1983b). Genetic analysis indicated that nda3 is closely linked to the ben1 locus which confers resistance to benzimidazole compounds (Yamamoto, 1980). The identity of ben1 to nda3 was suggested by the isolation of an allele of nda3 that simultaneously shows cold-sensitive and drug-resistant phenotypes. Furthermore, four of the eleven supersensitive mutations to TBZ are mapped at nda3 (Umesono et al., 1983a). These results strongly suggest that the $nda3^+$ gene product is related to microtubular functions. Direct evidence that the $nda3^+$ gene encodes β-tubulin was shown by cloning of the gene and determination of the nucleotide sequence (Hiraoka et al., 1984); the predicted amino acid sequence contains

448 residues and is 75% homologous to chicken β-tubulin. Southern hybridization indicated that the genome of S. *pombe* has a single copy of β-tubulin gene.

The cells of a cold-sensitive *nda3* mutation are arrested at the restrictive temperature in a highly synchronous manner at a step similar to mitotic prometaphase (Hiraoka *et al.*, 1984). The nuclear chromatin region condenses and segregates into three chromosomal bodies. No mitotic spindle is observed by immunofluorescence microscopy. A rapid, and reversible reactivation of the cs β-tubulin was demonstrated by temperature shift-up experiments to the permissive temperature; six minutes after the shift, the spindle appeared and became elongated. The chromosomes were separated at a constant speed (relative velocity 1 μm/min), and the spindle disappeared after the chromosomes reached opposite ends of the cell. Thus, high synchrony obtained by the shift-up of *nda3* may provide a convenient system by which to investigate spindle dynamics in mitosis.

(iv) *nda1*$^+$, *nda4*$^+$ and *ben4*$^+$ *genes*. These three genes appear to be required for the nuclear division of S. *pombe* (Toda *et al.*, 1983; Roy and Fantes, 1982; Yanagida *et al.*, 1986), but their gene products have not been characterized.

(v) *DNA topoisomerase genes*. DNA topoisomerases are the enzymes that control the topological states of DNA by transient breakage and subsequent rejoining of DNA strands. The enzymes are classified in two types: type I topoisomerases (topo I) which break and rejoin one of the two DNA strands, and type II topoisomerases (topo II) which break and rejoin a double-strand.

Mutants defective in DNA topoisomerases were isolated by screening individual extracts of mutagenized cells (Uemura and Yanagida, 1984). Type I topoisomerase mutants (*top1*), which had heat-sensitive topo I activity, were found to be viable, suggesting that topo I is dispensable for growth. A possibility that residual topo I activities in these mutants might be sufficient for growth was rejected by construction of a null allele of *top1* which was also viable, although its generation time was 20% longer than that of wild-type (Uemura *et al.*, 1987b). In extract of *top1*, Mg^{2+}- and ATP-dependent topo II activity could be detected. Three temperature-sensitive mutants, which had heat-sensitive topo II enzymes, were isolated and mapped to the same (*top2*) locus (Uemura and Yanagida, 1984).

The *top2* cells produce abnormal chromosomes at the time of mitosis under the restrictive condition (Uemura and Yanagida, 1986). The chromosomes are transiently extended into filamentous structures along with the elongating mitotic spindle but are not separated. The undivided nuclei are then cut across by the septum and broken by cytokinesis. A primary defect in *top2* appears to be the formation of aberrant mitotic chromosomes inseparable by the force generated by the spindle apparatus.

Single *top2* mutants are not appropriate for examining the blocked stages in chromosome separation because the mutant phenotypes are obscured by other mitotic events, including spindle dynamics and cytokinesis. For this reason, a cold-sensitive allele of *top2* was isolated, and a double *cs top-cs nda3* mutant was constructed, whereby the *nda3* mutation was expected to block spindle formation and cytokinesis (Uemura *et al.*, 1987). The double mutant was arrested at an earlier stage than the single *nda3* mutant, and showed long, entangled chromosomes at restrictive temperature. Upon the shift to permissive temperature, the chromosomes condensed into rod-like chromosomes, which are seen in the single *nda3* mutant, and then separated. Thus, inactivation of topo II in the absence of the spindle leads to the production of extended chromosomes, while reactivation of topo II leads to shortened chromosomes. This suggests that topo II is required for chromosome condensation. Experiments with *ts top2-cs nda3* cells showed that topo II is also required for chromosome separation in mitotic anaphase: inactivation of topo II and activation of β-tubulin allows normal spindle formation but results in 'streaked' chromosomes. Chromosome separation is completely blocked by prior inactivation of *ts* topo II at 36°C in glucose-deficient medium.

The phenotype of the *top1-top2* double mutant is strikingly different from that of the single *top2* mutant (Uemura and Yanagida, 1984). The double mutant cells incubated at the restrictive temperature are rapidly arrested at stages throughout the cell cycle, producing an altered nuclear chromatin region which looks like a ring or hollow structure. To explain the different phenotypes among *top1*, *top2* and *top1-top2*, Uemura and Yanagida (1984) postulated that topo II has two functions *in vivo*. One is to control, throughout the cell cycle, the superhelical density on chromatin by relaxing negative or positive supercoils; this function is redundant because topo I can play a similar role. The second function is the requirement of topo II for chromosome condensation and separation in mitosis through unknotting and decatenating the chromosomal DNA. Because the lack of the topo I enzyme can be complemented by topo II, *top1* mutants are viable. The defect of the topo II enzyme would block chromosome condensation or separation, because events in other parts of the cell cycle can be advanced with the complementing activity of the topo I enzyme. The defect in controlling the superhelical density can only be realized in the *top1-top2* mutants that cause a rapid arrest, irrespective of cell cycle stage.

Further experiments with a null allele of *top1* suggested that the total level of the relaxing activity by the topo I and topo II enzymes determines the growth rates of cells (Uemura *et al.*, 1987b); the *top1*-disrupted cells are viable but, when combined with *top2* mutations, grow poorly or become lethal. The *top1*⁺ gene appears to become essential when the relaxing activity of topo II enzyme is not abundant.

The *top1*⁺ and *top2*⁺ genes have been cloned by hybridization and transformation (Uemura and Yanagida, 1986; Uemura *et al.*, 1987b). The *S. pombe top1*⁺ gene encodes a 94-kD protein that shared 47% homology with the

S. cerevisiae TOP1 product. The *S. pombe top2*$^+$ gene can complement the *S. cerevisiae top2* mutant as well as the *S. pombe top2* mutants, although the *S. pombe* and *S. cerevisiae* sequences share only 49% overall homology (Uemura *et al.*, 1986). A weak but significant homology to the sequence of bacterial DNA gyrase subunits is found in yeast *top2*$^+$ sequences. The amino half of *top2*$^+$ is homologous to the ATP-binding *gyrB* subunit, and the central-to-latter part of *top2*$^+$ is homologous to the amino domain of the catalytic *gyrA* subunit, suggesting a possible evolutionary consequence of gene fusion of the bacterial DNA gyrase subunits into the eukaryotic topo II gene.

(vi) *cut1*$^+$, *cut2*$^+$ *and other cut genes.* Uncoordinated mitosis, that is, the occurrence of spindle dynamics and cytokinesis in the absence of chromosome separation, is revealed in *top2* mutants. In all previously isolated *ts cdc* and *cs nda* mutants, the arrest of nuclear division is always accompanied by the blocking of cytokinesis (Nurse *et al.*, 1976; Toda *et al.*, 1983). The phenotype of *top2*, however, is not exceptional among the mutants defective in nuclear division; a number of *ts* mutants showing cytological phenotypes similar to *top2* were isolated and classified into nine complementation groups (Hirano *et al.*, 1986). The phenotypes of *cut* mutants are similar to that of *top2* but differ in some points. First, the cytological phenotype of *cut* mutants is independent of *top1*. Second, *cut1* and *cut2* mutants show aberrant forms of chromosome separation differing from that of *top2*. In contrast to the partially extended chromosomes in *top2*, an archery bow-like structure consisting of a thin straight filament and thick curved region is observed in *cut1* and *cut2*. The *cut* gene product might be directly involved in chromosome organization. The *cut1*$^+$ gene was cloned by transformation and shown to complement *cut2* as well as *cut1*, revealing a functional relationship between the two genes. Extracts of the *cut1* and *cut2* mutants contained the normal topo II activity.

(vii) *nuc1*$^+$. A *ts* mutant *nuc1* was originally isolated as a mutant which shows a cytological phenotype similar to the double mutant *top1-top2* (Hirano *et al.*, 1986). Cells of the *nuc1* mutant, which have a normal level of topo I and topo II activity, show a strikingly altered nuclear chromation structure; the hemispherical nuclear chromatin region changes to a ring or hollow-like structure at the restrictive temperature. Predicted amino acid sequence suggests that the *nuc1*$^+$ gene may encode the largest subunit of RNA polymerase I which is known to transcribe ribosomal RNA genes (Hirano *et al.*, unpublished). Consistently, the *nuc1*$^+$ gene products are found to be located in the nucleolar region as determined by immunofluorescence microscopy using antisera raised against a *lacZ-nuc1*$^+$ fusion protein. Reduced rRNA synthesis by inactivation of RNA polymerase I might cause nucleolar disorganization and concomitant alteration in the nuclear structure.

(viii) *nuc2*$^+$. A *ts* mutant *nuc2* blocks mitosis specifically at a step in anaphase spindle elongation (Hirano *et al.*, 1988). The phenotype of this mutant is similar to that of tubulin mutants in that the chromosomes condense and the nucleus becomes displaced. However, in contrast to tubulin mutants which lack the spindle, a short spindle forms at the restrictive temperature. The spindle runs through a metaphase-plate-like structure consisting of three chromosomes. It is surprising that the spindle and chromosomes show such a regular ordered structure reminiscent of the metaphase plate in higher eukaryotes. If the metaphase plate formation in higher eukaryotes were required for concerted sister chromatid separation, then similar events might take place in *S. pombe*.

The *nuc2*$^+$ gene has been cloned and sequenced. Its coding region predicts a 665-residue internally-repeating protein (MW 76000). The *nuc2*$^+$ gene product (designated p67; MW 67000) was identified by immunoblots using antisera raised against the *lacZ-nuc2*$^+$ fused protein, and found to be enriched in an insoluble nuclear structure like a nuclear scaffold. In *nuc2* mutant cells, however, a soluble p76, perhaps an unprocessed precursor, accumulated in addition to insoluble p67. Three possible roles of the *nuc2*$^+$ gene product have been noted (Hirano *et al.*, 1988). First, p67 is a nuclear scaffold-like structure directly or indirectly interacting with microtubules or the spindle apparatus. The *nuc2*$^+$ gene product is not required for the initiation of spindle assembly but becomes essential for its elongation. Secondly, p67 may interact with a chromosomal domain such as the centromere. The short spindle cannot elongate, possibly due to the failure in the shortening of kinetochore microtubules during anaphase A. Certain kinetochore proteins behave as nuclear scaffold proteins (Earnshaw *et al.*, 1984). Thirdly, p67 might be involved in the elongation of nuclear membranes, although p67 does not behave as a membrane-bound protein. The role of the *nuc2*$^+$ gene may be to interconnect nuclear and cytoskeletal functions in the chromosome separation.

(ix) *dis1*$^+$, *dis2*$^+$ and *dis3*$^+$. Novel classes of cold-sensitive mutants that block mitotic chromosome separation were isolated and mapped into the three loci, *dis1*, *dis2* and *dis3* (Ohkura *et al.*, 1988). Their defective phenotypes at restrictive temperature are similar; the chromosomes condense and anomalously move to the cell ends in the absence of disjoining so that they are unequally distributed at the two cell ends. The cells become lethal during mitosis. The spindle forms and elongates although the initial spindle structure and the terminal spindle degradation differ from those of the wild type. The *dis*$^+$ gene products are not essential for chromosome condensation and spindle elongation but are required for the sister chromatid separation in mitosis. Interestingly, all the *dis* mutants were found to be supersensitive to caffeine at permissive temperature. The cytological phenotype of the *dis1* mutant produced by caffeine is similar to that produced at restrictive temperature. The *dis* gene function might be controlled by intracellular cAMP concentration since caffeine is known to be an inhibitor of cAMP phosphodiesterase.

(x) *cam1$^+$; calmodulin gene.* Calmodulin, the ubiquitous intracellular Ca^{2+}-receptor, mediates a variety of the Ca^{2+} regulated events in eukaryotic cells. The gene encoding calmodulin of *S. pombe* has been cloned by hybridization using synthetic oligonucleotide probes corresponding to the sequence of *Tetrahymena* calmodulin (Takeda and Yamamoto, 1987). The predicted amino acid sequence contains 149 residues, and shares 74% and 56% homology, respectively, with bovine and *S. cerevisiae* calmodulin. Gene disruption experiments revealed that the calmodulin gene (*cam1$^+$*) has an essential function in vegetative growth.

(xi) *Actin gene.* The actin gene of *S. pombe* has been cloned by using the *S. cerevisiae* actin gene as a hybridization probe (Mertins and Gallwitz, 1987). Interestingly, the amino acid sequence deduced from the nucleotide sequence showed that the *S. pombe* actin is more closely related to the mammalian γ-actin than to the actin of *S. cerevisiae.*

(xii) *ras1$^+$.* The genome of *S. pombe* has a single gene (designated *ras1$^+$*) which is homologous to mammalian *ras* genes (Fukui and Kaziro, 1985). Predicted *ras1* product consists of 219 amino acids and is much closer in size to mammalian *ras* than *S. cerevisiae RAS1* and *RAS2.* Gene disruption showed that *S. pombe ras1$^+$* is required for conjugation but not for vegetative growth (Fukui *et al.*, 1986). Furthermore, *ras1$^+$* of *S. pombe* appears to have no effect on adenylate cyclase activity, in contrast to *RAS1* and *RAS2* of *S. cerevisiae* which have essential functions for cell proliferation through modulation of the adenylate cyclase activity (Toda *et al.*, 1985). The function *in vivo* of ras homologue seems to be quite different between the two yeasts.

7.4 Mitotic pathways

Among the genes required for the initiation of mitosis (Figure 7.2), *cdc2$^+$* plays a central role. No mitotic events, not even initial ones such as the dissolution of cytoplasmic microtubules, take place in the absence of *cdc2$^+$*. Much is known of the controlling elements of *cdc2$^+$* which include *wee1$^+$*, *cdc25*, *suc1$^+$*, *nim1$^+$* and *cdc13$^+$* (Simanis and Nurse, 1986; Russell and Nurse, 1986, 1987a, b; Hayles *et al.*, 1986b; Hindley *et al.*, 1987; Booher and Beach, 1987). However, a large gap exists between molecular understanding of the *cdc2$^+$* gene function and the actual mechanism to trigger mitosis. An important question, as yet unanswered, is the role of the target protein(s) of *cdc2$^+$* kinase in mitosis. More steps may still be required to initiate mitosis after the completion of the *cdc2$^+$* gene function.

A number of mutants including *nda3* and *nuc2* that block nuclear division also block the ensuing cytokinesis (Figure 7.3). We found uncoordinated mitosis, however, which is the occurrence of spindle dynamics and cytokinesis in the absence of nuclear division, in many other mutants such as *top2, cut1–9*

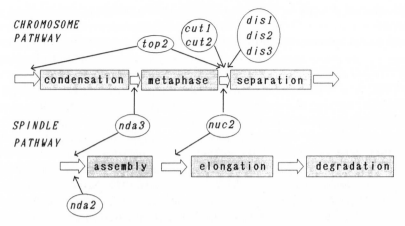

Figure 7.3 Steps and genes required for chromosome and spindle pathways during mitosis (see text). Known gene products are DNA topoisomerase II ($top2^+$), α1-tubulin ($nda2^+$), β-tubulin ($nda3^+$) and insoluble nuclear protein ($nuc2^+$).

and $dis1-3$ (Figure 7.3; Uemura and Yanagida 1986; Hirano et al., 1986; Ohkura et al., 1988). The uncoupling of the chromosome pathway from the spindle (and cytokinesis) pathway must be a general phenomenon, since mutations in more than 13 genes cause such a phenotype. These gene products may become essential after the cells have passed the stage that commits them to spindle dynamics and cytokinesis. Other examples of uncoordinated cell cycle events known in S. pombe are that (i) blocking of cytokinesis causes multiseptated cells, and (ii) blocking of septum formation causes multinucleated cells (Nurse et al., 1976): the cycles of nuclear division can occur in the absence of septum formation and cell separation. Studies on $nda2$ mutants show uncoupling between chromosome separation and spindle dynamics (Toda et al., 1983 and unpublished results); chromosomes appear to separate for a short distance although no normal spindle is present. Thus in S. pombe there are several independent mitotic pathways. Their independent natures are revealed only in mutant cells. In wild-type mitosis, they occur temporarily in succession, but the mechanism of coordinated mitosis is unknown.

A hypothesis was presented that sister chromatid separation is coupled to the transient decrease of intracellular cyclic AMP (cAMP) concentration in Schizosaccharomyces pombe mitosis (Ohkura et al., 1988). This hypothesis could explain how multiple chromosomes are disjoined concertedly in mitotic anaphase A so that the error due to non-disjunction or loss is minimized. A striking phenotype is that $nuc2$ and all the dis mutants are supersensitive to caffeine at permissive temperature (Ohkura et al., 1988; Hirano T., unpublished). Furthermore, the arrested phenotype was basically the same between conditions at restrictive temperature or in the presence of caffeine at permissive temperature. Thus, the $nuc2$ and dis gene expressions

J

are affected directly or indirectly by caffeine. The effect of caffeine may be pleiotropic but it is a known inhibitor of cAMP-dependent phosphodiesterase and would increase the intracellular concentration of cAMP. Consistently, such an increase occurs in the wild-type *S. pombe*. with the addition of caffeine (T. Hirano *et al.*, unpublished). The *dis* mutants seem to be more sensitive to cAMP than wild-type cells (unpublished result). In addition, multicopy plasmids carrying *S. cerevisiae* phosphodiesterase (PDE) genes (Sass *et al.*, 1985) complemented *dis* mutations. These results suggest that the rise of intracellular cAMP concentration by caffeine might effectively inactivate the mutant gene products at the permissive temperature. Alternatively, the inactivation of the gene products might cause the rise of intracellular concentration of cAMP. It is well established that cAMP greatly influences the cell cycle pathway in *S. cerevisiae*, especially at the G1 phase (Matsumoto *et al.*, 1982), but the notion that cAMP is also involved in sister chromatid separation is novel. It has been known that in *S. pombe* the expression of certain meiotic genes is affected by cAMP or caffeine (Beach *et al.*, 1985). Mitotic pathways involving the *nuc2* and the three *dis* gene functions which are required for the sister chromatid separation appear to be directly or indirectly related to the intracellular concentration of cAMP.

The cAMP concentration in the synchronized *S. pombe* wild-type cells has been measured and preliminary results show that the levels of the cAMP decreases temporally in mitosis (T. Hirano *et al.*, unpublished). Timing for the minimal level of cAMP coincided with the maximal frequency of mitotic cells containing separating chromosomes. Further investigation is required to determine the levels of cAMP in the mutant cells, but it appears that sister chromatid separation and transient decrease of cAMP in mitosis are coupled. Ohkura *et al* (1988), in addition to the $dis2^+$ and $dis3^+$ genes, cloned multicopy extragenic suppressor sequences which complement *dis1* and *dis2* mutations (Ohkura *et al.*, 1988). Predicted polypeptide of one of suppressor DNAs ($sds22^+$) is surprisingly similar to the non-catalytic domain of adenylate cyclase (H. Ohkura *et al.*, unpublished). A complex regulatory system may exist for the execution of the dis^+ gene functions. We speculate that, although a large number of gene products must be required for mitotic pathways, a relatively small number of the factors, such as cAMP or calcium ion, act *in trans* for the concerted and temporal order of these pathways.

References

Adachi, Y., Toda, T., Niwa, O. and Yanagida, M. (1986) Differential expression of essential and nonessential α-tubulin genes in *Schizosaccharomyces pombe*. *Mol. Cell. Biol.* **6**: 2168.

Aves, S.J., Durkacz, B.W., Carr, A. and Nurse, P. (1985) Cloning, sequencing and transcriptional control of the *Schizosaccharomyces pombe cdc10* 'start' gene. *EMBO J.* **4**: 457.

Barker, D.G., White, J.H.M. and Johnston, L.H. (1987) Molecular characterization of the DNA ligase gene, *CDC17*, from the fission yeast *Schizosaccharomyces pombe*. *Eur. J. Biochem.* **162**: 659.

Beach, D. and Nurse, P. (1981) High frequency transformation of the fission yeast *Schizosaccharomyces pombe*. *Nature* **290**: 140.

Beach, D., Durkacz, B. and Nurse, P. (1982) Functionally homologous cell cycle genes in budding and fission yeast. *Nature* **300**: 706.

Beach, D., Rodgers, L. and Gould, J. (1985) *RAN1*[+] controls the transition from mitotic division to meiosis in fission yeast. *Curr. Genet.* **10**: 297.

Booher, R. and Beach, D. (1986) Site-specific mutagenesis of *cdc2*[+], a cell cycle control gene of the fission yeast *Schizosaccharomyces pombe*. *Mol. Cell. Biol.* **6**: 3523.

Booher, R. and Beach, D. (1987) Interaction between *cdc13*[+] and *cdc2*[+] in the control of mitosis in fission yeast; dissociation of the G1 and G2 roles of the *cdc2*[+] protein kinase. *EMBO J.* **6**: 3441.

Breeden, L. and Nasmyth, K. (1987) Similarity between cell-cycle genes of budding yeast and fission yeast and the *Notch* gene of *Drosophila. Nature* **329**: 651.

Brizuela, L., Draetta, G. and Beach, D. (1987) p13[suc1] acts in the fission yeast cell division cycle as a component of the p34[cdc2] protein kinase. *EMBO J.* **6**: 3507.

Byers, B. and Goetsch, L. (1975) Behavior of spindles and spindle plaques in the cell cycle and conjugation of *Saccharomyces cerevisiae. J. Bacteriol.* **124**: 511.

Draetta, G., Brizuela, B., Potashkin, J. and Beach, D. (1987) Identification of p34 and p13, human homologs of the cell cycle regulators of fission yeast encoded by *cdc2*[+] and *suc1*[+]. *Cell* **50**: 319.

Earnshaw, W.C., Halligan, N., Cooke, C. and Rothfield, N. (1984) The kinetochore is part of the metaphase chromosome scaffold. *J. Cell. Biol.* **98**: 352.

Fantes, P. (1979) Epistatic gene interactions in the control of division in fission yeast. *Nature* **279**: 428.

Fukui, Y. and Kaziro, Y. (1985) Molecular cloning and sequence analysis of a *ras* gene from *Schizosaccharomyces pombe. EMBO J.* **4**: 687.

Fukui, Y., Kozasa, T., Kaziro, Y., Takeda, T. and Yamamoto, M. (1986) Role of a *ras* homolog in the life cycle of *Schizosaccharomyces pombe. Cell* **44**: 329.

Gordon, C.B. and Fantes, P.A. (1986) The *cdc22* gene of *Schizosaccharomyces pombe* encodes a cell cycle-regulated transcript. *EMBO J.* **5**: 2981.

Gutz, H., Heslot, H., Leupold, U. and Loprieno, N. (1974) *Schizosaccharomyces pombe.* In *Handbook of Genetics 1*, ed. King, R.C., Plenum, New York, p. 395.

Hagan, I.M. and Hyams, J.S. (1988) The use of cell division cycle mutants to investigate the control of microtubule distribution in the fission yeast *Schizosaccharomyces pombe. J. Cell. Sci.* **89**: 343.

Hartwell, L.H., Colotti, J., Pringle, J.R. and Reid, B.J. (1974) Genetic control of the cell division cycle in yeast: a model. *Science* **183**: 46.

Hayles, J., Beach, D., Durkacz, B. and Nurse, P. (1986a) The fission yeast cell cycle control gene *cdc2*; isolation of a sequence *Suc1* that suppresses *cdc2* mutant function. *Mol. Gen. Genet.* **202**: 291.

Hayles, J., Aves, S. and Nurse, P. (1986b) *suc1* is an essential gene involved in both the cell cycle and growth in fission yeast. *EMBO J.* **5**: 3373.

Hereford, L.M., Fahrner, K., Woolford, J. Jr., Rosbash, M. and Kaback, D.B. (1979) Isolation of yeast genes H2A and H2B. *Cell* **18**: 1261.

Hindley, J. and Phear, G.A. (1984) Sequence of the cell division gene *CDC2* from *Schizosaccharomyces pombe*; patterns of splicing and homology to protein kinase. *Gene* **31**: 129.

Hindley, J., Phear, G., Stein, M. and Beach, D. (1987) *suc1*[+] encodes a predicted 13-kilodalton protein that is essential for cell viability and is directly involved in the division cycle of *Schizosaccharomyces pombe. Mol. Cell. Biol.* **7**: 504.

Hirano, T., Funahashi, S., Uemura, T. and Yanagida, M. (1986) Isolation and characterization of *Schizosaccharomyces pombe cut* mutants that block nuclear division but not cytokinesis. *EMBO J.* **5**: 2973.

Hirano, T., Hiraoka, Y. and Yanagida, M. (1988) A temperature-sensitive mutation of the *Schizosaccharomyces pombe* gene *nuc2*[+] that encodes a nuclear scaffold-like protein blocks spindle elongation in mitotic anaphase. *J. Cell. Biol.* **106**: 1171.

Hiraoka, Y., Toda, T. and Yanagida, M. (1984) The *NDA3* gene of fission yeast encodes β-tubulin: a cold-sensitive *nda3* mutation reversibly blocks spindle formation and chromosome movement in mitosis. *Cell* **39**: 349.

Johnston, L.H., Barker, D.G. and Nurse, P. (1986) Cloning and characterization of the *Schizosaccharomyces pombe* DNA ligase gene *CDC17. Gene* **41**: 321.

Kilmartin, J.V. and Adams, A.E.M. (1984) Structural rearrangements of tubulin and actin during the cell cycle of the yeast *Saccharomyces. J. Cell. Biol.* **98**: 922.

Kohli, J., Hottinger, H., Munz, P., Strauss, A. and Thuriaux, P. (1977) Genetic mapping in *Schizosaccharomyces pombe* by mitotic and meiotic analysis and induced haploidization. *Genetics* **87**: 471.

Kohli, J. (1987) Genetic nomenclature and gene list of the fission yeast *Schizosaccharomyces pombe*. *Curr. Genet.* **11**: 575.

Lee, M.G. and Nurse, P. (1987) Complementation used to clone a human homolog of the fission yeast cell cycle control gene *cdc2*⁺. *Nature* **327**: 31.

Marks, J. and Hyams, J. (1985) Localization of F-actin through the cell division cycle of *Schizosaccharomyces pombe*. *Eur. J. Cell. Biol.* **39**: 27.

Marks, J., Hagan, I. and Hyams, J.S. (1986) Growth polarity and cytokinesis in fission yeast: the role of the cytoskeleton. *J. Cell. Sci. Suppl.* **5**: 229.

Matsumoto, S. and Yanagida, M. (1985) Histone gene organization of fission yeast: a common upstream sequence. *EMBO J.* **4**: 3531.

Matsumoto, S., Yanagida, M. and Nurse, P. (1987) Histone transcription in cell cycle mutants of fission yeast. *EMBO J.* **6**: 1093.

May, J.W. and Mitchison, J.M. (1986) Length growth in fission yeast measured by two novel techniques. *Nature* **322**: 752.

McCully, E.K. and Robinow, C.F. (1971) Mitosis in the fission yeast *Schizosaccharomyces pombe*: A comparative study with light and electron microscopy. *J. Cell. Sci.* **9**: 475.

Mertins, P. and Gallwitz, D. (1987) A single intronless actin gene in the fission yeast *Schizosaccharomyces pombe*: nucleotide sequence and transcripts found in homologous and heterologous yeast. *Nucleic Acids Res.* **15**: 7376.

Mitchison, J.M. (1970) Physiological and cytological methods for *Schizosaccharomyces pombe*. In *Methods in Cell Physiology*, ed. D.M. Prescott, Vol. 4, Academic Press, New York, 131.

Mitchison, J.M. and Nurse, P. (1985) Growth in cell length in the fission yeast *Schizosaccharomyces pombe*. *J. Cell. Sci.* **75**: 357.

Nasmyth, K. (1977) Temperature-sensitive lethal mutants in the structural gene for DNA ligase in the fission yeast *Schizosaccharomyces pombe*. *Cell* **12**: 1109.

Nasmyth, K. (1979) Genetic and enzymatic characterization of conditional lethal mutants of the yeast *Schizosaccharomyces pombe* with a temperature-sensitive DNA ligase. *J. Mol. Biol.* **130**: 273.

Nasmyth, K. and Nurse, P. (1981) Cell division cycle mutants altered in DNA replication and mitosis in the fission yeast *Schizosaccharomyces pombe*. *Mol. Gen. Genet.* **182**: 119.

Nurse, P. (1975) Genetic control of cell size at cell division in yeast. *Nature* **256**: 547.

Nurse, P., Thuriaux, P. and Nasmyth, K. (1976) Genetic control of the cell division cycle in the fission yeast. *Schizosaccharomycses pombe*. *Mol. Gen. Genet.* **146**: 167.

Nurse, P. and Thuriaux, P. (1980) Regulatory genes controlling mitosis in the fission yeast *Schizosaccharomyces pombe*. *Genetics* **96**: 627.

Nurse, P. and Bissett, Y. (1981) Gene required in G1 for commitment to cell cycle and in G2 for control of mitosis in fission yeast. *Nature* **292**: 558.

Nurse, P. (1985) Cell cycle control genes in yeast. *Trends Genet.* **1**: 51.

Ohkura, H., Adachi, Y., Kinoshita, N., Niwa, O., Toda, T. and Yanagida, M. (1988) Cold-sensitive and caffeine-sensitive mutants of the *Schizosaccharomyces pombe dis* genes implicated in sister chromatid separation during mitosis. *EMBO J.* **7**: 1465.

Piggott, J., Rai, R. and Carter, B. (1982) A bifunctional gene product involved in two phases of the yeast cell cycle. *Nature* **298**: 391.

Roy, D. and Fantes, P. (1982) Benomyl resistant mutants of *Schizosaccharomyces pombe* cold-sensitive for mitosis. *Curr. Genet.* **6**: 195.

Russell, P.R. and Nurse, P. (1986) *cdc25*⁺ functions as an inducer of mitotic control of fission yeast. *Cell* **45**: 145.

Russell, P. and Nurse, P. (1987a) Negative regulation of mitosis by *wee1*⁺, a gene encoding a protein kinase homolog. *Cell* **49**: 559.

Russell, P.R. and Nurse, P. (1987b) The mitotic inducer *nim1*⁺ functions in a regulatory network of protein kinase homologs controlling the initiation of mitosis. *Cell* **49**: 569.

Sass, P., Field, J., Nikawa, J., Toda, T. and Wigler, M. (1986) Cloning and characterization of the high affinity cAMP phosphodiesterase of *S. cerevisiae*. *Proc. Natl. Acad. Sci. USA* **83**: 9303.

Simanis, V. and Nurse, P. (1986) The cell cycle control gene *cdc2*⁺ of fission yeast encodes a protein kinase potentially regulated by phosphorylation. *Cell* **45**: 261.

Smith, M.M. and Andresson, O.S. (1983) DNA sequences of yeast H3 and H4 histone genes from two non-allelic gene sets encode identical H3 and H4 proteins. *J. Mol. Biol.* **169**: 669.

Smith, C.L., Matsumoto, T., Niwa, O., Klco, S., Fan, J.-B., Yanagida, M. and Cantor, C. (1987) An electrophoretic karyotype for *Schizosaccharomyces pombe* by pulse field gel electrophoresis. *Nucleic Acids Res.* **15**: 4481.

Takeda, T. and Yamamoto, M. (1987) Analysis and *in vivo* disruption of the gene coding for calmodulin in *Schizosaccharomyces pombe*. *Proc. Natl. Acad. Sci. USA* **84**: 3580.

Tanaka, K. and Kanbe, T. (1986) Mitosis in the fission yeast *Schizosaccharomyces pombe* as revealed by freeze-substitution electron microscopy. *J. Cell. Sci.* **80**: 253.

Thuriaux, P., Nurse, P. and Carter, B. (1978) Mutants altered in control co-ordinating cell division with cell growth in the fission yeast *Schizosaccharomyces pombe*. *Mol. Gen. Genet.* **161**: 215.

Toda, T., Yamamoto, M. and Yanagida, M. (1981) Sequential alterations in the nuclear chromatin region during mitosis of the fission yeast *Schizosaccharomyces pombe*: video fluorescence microscopy of synchronously growing wild-type and cold-sensitive *cdc* mutants by using a DNA-binding fluorescent probe. *J. Cell. Sci.* **52**: 271.

Toda, T., Umesono, K., Hirata, A. and Yanagida, M. (1983) Cold-sensitive nuclear division arrest mutants of the fission yeast *Schizosaccharomyces pombe*. *J. Mol. Biol.* **168**: 251.

Toda, T., Adachi, Y., Hiraoka, Y. and Yanagida, M. (1984) Identification of the pleiotropic cell cycle gene *NDA2* as one of two different α-tubulin genes in *Schizosaccharomyces pombe*. *Cell* **37**: 233.

Toda, T., Uno, I., Ishikawa, T., Powers, S., Kataoka, T., Broek, D., Cameron, S., Broach, J., Matsumoto, K. and Wigler, M. (1985) In yeast, ras proteins are controlling elements of adenylate cyclase. *Cell* **40**: 27.

Uemura, T. and Yanagida, M. (1984) Isolation of type I and II DNA topoisomerase mutants from fission yeast: single and double mutants show different phenotypes in cell growth and chromatin organization. *EMBO J.* **3**: 1737.

Uemura, T. and Yanagida, M. (1986) Mitotic spindle pulls but fails to separate chromosomes in type II DNA topoisomerase mutants: uncoordinated mitosis. *EMBO J.* **5**: 1003.

Uemura, T. Morikawa, K. and Yanagida, M. (1986) The nucleotide sequence of the fission yeast DNA topoisomerase II gene: structural and functional relationships to other DNA topoisomerases. *EMBO J.* **5**: 2355.

Uemura, T., Ohkura, H., Adachi, Y., Morino, K., Shiozaki, K. and Yanagida, M. (1987*a*) DNA topoisomerase II is required for condensation and separation of mitotic chromosomes in *S. pombe*. *Cell* **50**: 917.

Uemura, T., Morino, K., Uzawa, S., Shiozaki, K. and Yanagida, M. (1987*b*) Cloning and sequencing of *Schizosaccharomyces pombe* DNA topoisomerase I gene, and effect of gene disruption. *Nucleic Acid Res.* **15**: 9727.

Umesono, K., Toda, T., Hayashi, S. and Yanagida, M. (1983*a*) Two cell division cycle gene *NDA2* and *NDA3* of the fission yeast *Schizosaccharomyces pombe* control microtubular organization and sensitivity to anti-mitotic benzimidazole compounds. *J. Mol. Biol.* **168**: 271.

Umesono, K., Hiraoka, Y., Toda, T. and Yanagida, M. (1983*b*) Visualization of chromosomes in mitotically arrested cells of the fission yeast *Schizosaccharomyces pombe*. *Curr. Genet.* **7**: 123.

Yamamoto, M. (1980) Genetic analysis of resistant mutants to antimitotic benzimidazole compounds in *Schizosaccharomyces pombe*. *Mol. Gen. Genet.* **180**: 231.

Yanagida, M. and Wang, J.C. (1987) Yeast DNA topoisomerases and their structural genes. In *Nucleic Acid and Molecular Biology*, Vol. 1, eds Eckstein, F. and Lilley, D.M. 196.

Yanagida, M., Hiraoka, Y., Uemura, T., Miyake, S. and Hirano, T. (1986) Control mechanisms of chromosome movement in mitosis of fission yeast. In *Yeast Cell Biology*, ed. Hicks, J., Alan R. Liss, New York, 279.

Yanagida, M. (1987) Yeast tubulin genes. *Microbiol. Sci.* **4**: 115.

Articles by Dunphy *et al.*, Gautier *et al.*, Gocbe and Byen and Solomon *et al.* (all 1988) in *Cell* **54** are also relevant.

8 Molecular mechanisms underlying the expression and maintenance of the type 1 killer system of Saccharomyces cerevisiae

STEPHEN L. STURLEY AND KEITH A. BOSTIAN*

8.1 Introduction

The production of exocellular toxins with wide ranges of killing specificity is a relatively common trait exhibited by numerous fungal genera (Young and Yagiu, 1978). The best characterized of these by far is the dsRNA mycoviral 'killer' system of *Saccharomyces cerevisiae*. Initially observed by Makower and Bevan in 1963, the yeast killer system has been extensively investigated, yielding a wealth of information concerning the replication and maintenance of dsRNA genomes and host–nucleocytoplasmic interactions, while serving as an eminent paradigm for protein maturation and secretion in yeast. More recently, the killer system has provided insight into potential mechanisms involved in the dissemination of mycoviruses (El-Sherbeini and Bostian, 1987; Sturley *et al.*, 1988*a*) and intracellular protein trafficking processes that apparently direct the expression of immunity to killer toxin (Sturley *et al.*, 1988*c*).

The type 1 killer or K1 mycoviral system, like its mammalian reoviral counterpart, has a segmented genome consisting minimally of two inter-related dsRNA genomes: a 4.7-kilobasepair (kb) L_1-dsRNA, present in the cell cytoplasm and encapsidated in a 40-nm icosahedral virus-like particle (ScV-L_1 VLPs), and a similarly encapsidated 1.9-kb M_1-dsRNA (Hopper *et al.*, 1977; Bostian *et al.*, 1980). The L_1 plasmid provides encapsidation, replication and maintenance functions for both viruses, whereas the M_1 genome directs the production of a dimeric, exocellular protein toxin and specific immunity to this toxin. Mature, secreted mycotoxin consists of two disulphide-linked, non-glycosylated subunits of 11.4 and 9.0 kilodaltons (kD), denoted α and β, respectively. Toxin-producing cells kill sensitive strains by disrupting cytoplasmic membrane function (Bussey, 1981), although the exact mechanisms of killing and immunity remain to be unequivocally established.

In addition to the *cis* encoded properties of the dsRNA molecules, the killer system is dependent upon numerous complex interactions with *trans*-acting factors provided by the host (summarized in Table 8.1). These factors range

---

*To whom correspondence should be addressed.

Table 8.1 Some of the *trans*-acting factors mediating gene expression in the *S. cerevisiae* killer system

Gene	Product	Function	Reference
KEX1	Carboxypeptidase	Protoxin processing/ activity	1
KEX2	Endopeptidase	Protoxin processing	2
SEC1, 7, 18	?	Toxin secretion	3
SEC53	?	Toxin secretion	4
SK15	Exocellular peptidase	Toxin stability	5
MAK1	Topoisomerase I	M-dsRNA maintenance	6
MAK8	Ribosomal protein L3	M-dsRNA maintenance	7
MAK2, 4–7, 9, 11–30	?	M-dsRNA maintenance	8
MAK3, 10, 31, 32	?	M and L-dsRNA maintenance	8
SK12–4, 6–8	?	dsRNA maintenance	8
KRE1, 2, 5	Cell wall glucan components	Toxin binding	9
KRE3	Plasma membrane components	Toxin binding	9
REX	Endopeptidase?	Protoxin maturation, immunity expression	10
VPL3, 6	Vacuolar protein localization	Immunity expression	11
END1	Pinocytosis	Immunity expression	12

Literature citations are (1) Dmochowska *et al.* (1987); Bussey *et al.* (1983); (2) Julius *et al.* (1984); Bussey *et al.* (1983); (3) Bussey *et al.* (1983); (4) Lolle and Bussey (1986); (5) Bussey *et al.* (1983); (6) Thrash *et al.* (1984); (7) Wickner *et al.* (1982); (8) Wickner (1986); (9) Al-Aidroos and Bussey (1978); Bussey (1973); Hutchins and Bussey (1983); (10) Wickner and Leibowitz (1976); Sturley *et al.* (1988*b*); (11) Rothman and Stevens, (1986); Sturley *et al.* (1988*c*); (12) Chvatchko *et al.* (1986); Sturley *et al.* (1988*c*).

from elements mediating the maintenance and copy number of the dsRNA plasmids (e.g. at least 32 maintenance of killer (*MAK*) genes, and 7 superkiller (*SKI*) genes), to factors mediating maturation of the primary translation product of $M_1(M_1$-P1 or preprotoxin), such as those encoded by the *SEC* (secretion) and *KEX* (killer expression) genes. In addition, numerous host-encoded factors are involved in the process of immunity, ranging from gene products involved in host-mediated resistance to toxin, encoded by the *KRE* (killer resistance) genes, to gene products that participate in the process of viral mediated immunity to toxin, such as *VPL* (vacuolar protein localization), *END* (endocytosis), and *REX* (resistance expression).

The relatively recent creation of partial or complete cDNA copies of the major extra-chromosomal elements of the killer system (Bostian *et al.*, 1984; Lolle *et al.*, 1984; Bruenn *et al.*, 1985) has set the foundation for their detailed molecular analysis. Previous reviews have admirably covered the extensive advances made in elucidating the genetic (for review, see Wickner, 1986), and biochemical features of this system (reviewed by Bussey, 1981; Tipper and Bostian, 1984). This review will address the recent advances made in understanding the molecular principles underlying propagation of the virus and its phenotypic expression. The intriguing property of the translation product of the preprotoxin gene to endow the two distinct but intricately related viral phenotypes, production of killer toxin and self-immunity to toxin, will be a primary focus for discussion.

8.2 Genomic structure, maintenance and transmission of the mycovirus

8.2.1 The M_1 genome

The 1.9-kb M_1-dsRNA genome encodes the preprotoxin gene, which confers both the killing and immunity phenotypes to host yeast strains. This extrachromosomal element is separated by an adenine-uracil-rich denaturation 'bubble' into two regions of approximately 1.0 and 0.6 kb (Figure 8.1; Fried and Fink, 1978; Bostian *et al.*, 1983*a*; Hannig *et al.*, 1986). The denaturation 'bubble' is between 130 and 200-bp long. The 1.0-kb region contains the toxin encoding preprotoxin gene (Bostian *et al.*, 1980*a*, 1984), while the 0.6-kb heteroduplex region and the denaturation 'bubble' possess no substantial open reading frames (ORFs) (Hanning *et al.*, 1986), and are probably involved in maintenance of the virus and its encapsidation by products of the L genome.

Naturally occurring defective interfering particles are deletion derivatives of the full length M genome that lack the majority of the preprotoxin coding region. The remaining sequences are sufficient for the maintenance of M_1-dsRNA. In particular, the presence of five repeats of a consensus sequence, (T/C)GCGATTCGC(A/G) in the 3' portion of M_1 and M_2 (encoding the type 2 killer toxin), may play a significant role in these mechanisms (Lee *et al.*, 1986). There are additional short regions of homology between M_1 and M_2-dsRNAs. The longest of these has the sequence 3'-ACACACGAGU-5', and is situated 105-bp from the 3' terminus of both molecules. These domains could act as recognition sequences for the *MAK* maintenance functions provided by the host genome, since they are not found in the sequenced portions of L_1 (Hannig and Leibowitz, 1985). Plasmid mutations (e.g. [Kil-b]; Tyagi *et al.*, 1984; Wickner, 1976) that alter the patterns of maintenance of the M genome, may also map to some of the non-coding M_1 regions.

We constructed a DNA version of the 1.0-kb preprotoxin encoding segment of the M_1-dsRNA by cDNA cloning (Bostian *et al.*, 1984). Comparison of the N-terminal sequences of the secreted toxin subunits with the nucleotide sequence of the cDNA clone located both of them within a 316 amino-acid ORF (preprotoxin or M_1-P1) with the linear domain structure shown in Figure 8.1 (Bostian *et al.*, 1984; Skipper *et al.*, 1984). Amino and carboxy terminal amino-acid sequence analysis of secreted toxin indicated that the 103 residue α-subunit commences within preprotoxin at residue 45 (P2) and ends at residue 147 (P3), with a predicted molecular mass of 11.4-kD (Zhu *et al.*, 1987). The α-domain is preceded by a 44-amino-acid N-terminal segment called δ which has many characteristics of a typical secretion signal peptide

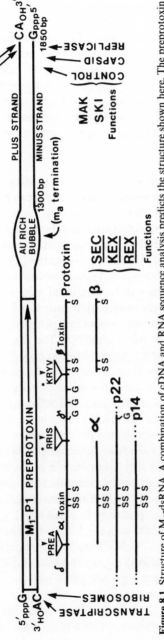

Figure 8.1 Structure of M_1-dsRNA. A combination of cDNA and RNA sequence analysis predicts the structure shown here. The preprotoxin open reading frame, M_1-P1, is situated at the 5' end of the molecule and encodes a protein with the linear domain structure, δ-α-γ-β. Proteolytic cleavage mediated by the $KEX2$ endopeptidase (▼) and $KEX1$ carboxypeptidase (*), glycosylation (G) and potential disulphide bond formation at cysteine residues (S) produces the α, β, p22 and p14 moieties from preprotoxin. The termini of p14 and p22 are not precisely known and are therefore indicated with dotted lines. The positions of transcriptional ('transcriptase') and translational initiation ('ribosomes') are also indicated. In addition, the host mediated functions (MAK and SKI) involved in encapsidation and replication and their *cis* encoded recognition sequences (capsid and replicase respectively) are shown.

(Bostian et al., 1983b; Lolle and Bussey, 1986). The α-subunit and the C-terminal domain of preprotoxin, β (83 amino-acids), are separated by a segment of 86 amino acids called γ, carrying all three N-glycosyl substituents of protoxin. Processed γ has never been detected in vivo. α and β possess the only cysteine residues (3 in each domain) in preprotoxin which form the disulphide bridges presumably essential for the activity of toxin. The preprotoxin cDNA, lacking most of δ, confers both the killing and immunity phenotypes when expressed from the promoter and secretion leader of the repressible acid phosphatase structural gene (PHO5) in an M_1-dsRNA-free yeast strain (Hanes et al., 1986). This was the first indication that a single open reading frame is sufficient for the expression of both phenotypes.

8.2.2 The L_1 genome

The complete nucleotide sequence of the capsid encoding 4.7-kb L_1-dsRNA genome is currently being determined in several laboratories studying yeast viral replication (J. Bruenn; R. Wickner, personal communication). Some generalizations concerning its structure are already possible, based upon the nucleotide sequence of partial cDNA clones derived from dsRNA transcripts (Bobek et al., 1982; Bruenn et al., 1985). Approximately 2.5-kb from the 5' end of the plus strand of L is a 170-bp inverted repeat separated by 328-bp of non-repetitive sequence. The presence of translational stop codons in all three reading frames within this repeat implies that only the 5' 2.5-kb segment is utilized as coding sequence, since yeast ribosomes primarily translate only the first ORF commencing at the 5' end of a mRNA. The size of the ORF (about 2.4-kb) approximates the predicted size required to encode the 88-kD capsid polypeptide (Hopper et al., 1977). Interestingly, translation could reinitiate at an AUG initiation codon within the 3' branch of the inverted repeat. More complicated processes leading to the translational activation of the 3' region of L, such as differential splicing of the mRNA, cannot be ruled out.

Several laboratories have demonstrated the coexistence of multiple species of L in a variety of strains of S. cerevisiae (Sommer and Wickner, 1982; Field et al., 1982; El-Sherbeini et al., 1983). Examination of the in vivo and in vitro translation products from these strains has identified four unique species of L-sized dsRNA; L_{1A}, L_{1BC}, L_{2A}, and L_{2BC} (El-Sherbeini et al., 1984). L_{1A} encodes VL_{1A}-P_1, the 88-kD protein encapsidating L_{1A}-dsRNA. L_{1BC} is contained in VLPs with a unique 82-kD capsid protein, VL_{1BC}-P1, encoded by L_{1BC}. In strains containing both L_{1A} and L_{1BC}, M_1 is encapsidated only in VL_{1A}-P1. L_{2A} and L_{2BC} encode the capsid proteins VL_{2A}-P1 (84-kD), and VL_{2BC}-P1 (82-kD). Although L_{1A} and L_{2A} share some sequence homology, they encode distinctly different capsid proteins and have dramatically different S. aureus V8 protease cleavage profiles (El-Sherbeini et al., 1984). It will be interesting to see if they also encode distinct RNA polymerases. Certain variants of L-dsRNA have been correlated with an ability to exclude or maintain M_2 (expression of the type 2 killer phenotype) and M_1-dsRNAs in different genetic backgrounds.

For example, the [HOK] phenotype, characterized by its ability to maintain replication defective mutant forms of M-dsRNA, has been correlated by genetic and biochemical experiments with L_{1A} (Sommer and Wickner, 1982; El-Sherbeini et al., 1983). Similarly, another form of L_{1A}-dsRNA correlates with the [EXL] encoded ability to exclude M_2-dsRNA in crosses involving these dsRNA species (Sommer and Wickner, 1982).

8.2.3 Maintenance and expression of the dsRNA genomes

M_1-dsRNA is found only in cells possessing L-dsRNA, which confers no recognizable phenotype other than support of the M genome. The identification of the 88-kD capsid protein VL_{1A}-P1 (ScV-P_1) as the major gene product of L_1, and the demonstration that VL_{1A}-P1 packages both L_1 and M_1-dsRNA, explains part of this dependency relationship (Bostian et al., 1980; El-Sherbeini et al., 1984). It is also likely that L provides a part or all of the polymerase and replicase activity involved in the propagation of both dsRNA plasmids. The genomic location of the RNA polymerase has not yet been established, although likely candidates are the MAK chromosomal genes or the L plasmid itself. There is clearly sufficient information in L to encode both capsid and polymerase genes in one di-cistronic unit. The isolation of a DNA independent RNA-polymerase from dsRNA-containing virions and characterization of its activity in vitro has indicated that both replication and transcription of the killer virions are conservative and may be closely related (Williams and Leibowitz, 1987). Transcriptional and replicative RNA polymerase activities have not been distinguished as separate entities, although stationary phase cells have a preponderance of the former activity. The initial production of a full length plus strand message (m), and its eventual release from the VLP, could function both in translation and as a template for the synthesis of the minus strand in replication (Leibowitz et al., 1988). Esteban and Wickner (1986) have proposed a 'head-full' hypothesis for the replication of L and M dsRNAs, based upon the initial retention of the plus strand within the virion and production of the minus strand, at which point all subsequent transcripts are exuded from the capsid due to lack of room. Two copies of the M genome (approximately half the size of L) are packaged into the capsid, although replication intermediates containing a single plasmid or a single plus strand can be detected. Replication therefore appears to precede transcription, although the enzymes mediating these processes are conceivably identical. The partial length m_a transcripts are probably the result of transcriptional 'pausing', and presumably play no direct role in virus replication.

The degree to which the dsRNA plasmids of the killer systems interact with the host is probably best exemplified by the action of the MAK and SKI genes. It is interesting to note that only 4 MAK loci, 3, 8, 31, and 32 (the latter pair linked as the PET18 complex) are required for maintenance of L although all are required for M_1 (Wickner and Toh-e, 1982). Many of the MAK genes have been isolated and in some cases sequenced (reviewed by Wickner, 1986). For example, $MAK 8 (TCM1)$ encodes ribosomal protein L3 (Wickner *et al.*, 1982) and the $MAK1$ ($TOP1$) gene product is DNA topoisomerase 1 (Thrash *et al.*, 1984). It is not immediately obvious how any of the MAK genes exert an effect on the killer system. They are clearly involved in other cellular processes, since many are required for vegetative cell growth at any temperature (e.g. mutations such as $mak13$ or $mak15$; Wickner and Leibowitz, 1979), while other mutations (Ridley *et al.*, 1984) lead to inviability at elevated ($mak16.1$) or reduced ($mak6$) temperatures.

The SKI gene products repress replication of M and L dsRNAs. Mutations at all of these loci except $SKI5$ (which encodes an extracellular protease) and $SKI1$, lead to an elevation of the dsRNA copy number and thus a super-killer phenotype. In many cases, ski mutations suppress certain mak alleles as well as rendering M_1 independent of [HOK], the attribute of L_A-dsRNA that mediates M_1 replication. In addition, elevated copy numbers of the M_1 genome cause a cold-sensitive (8°C) phenotype which is not a result of over-expression of the preprotoxin, ORF, since deletion of this region (in S3 defective interfering particles) does not relieve the cold-sensitive defect (Wickner, 1986). Clearly, then, the complex interaction of the MAK and SKI gene products with the *cis* elements of the dsRNA molecule has evolved into a homeostatic mechanism that protects the cell from over-replication of the killer genomes.

8.2.4 Dissemination of the virus

Vertical transmission during cell fusion is the accepted and probably most common mechanism of mycoviral transfer in nature and in the laboratory. It has generally been believed that mycoviruses, including the yeast killer virus, are incapable of extracellular infection, although protoplast uptake of VLPs has been reported in other fungal organisms (Stanway and Buck, 1984). Recently, however, evidence for the infectivity of the killer mycovirus has accumulated. El-Sherbeini and Bostian (1987) have developed infection procedures by modification of established protocols for DNA transformation and VLP preparation, and used them to successfully 'infect' yeast cells with the VLPs of the killer system. It appears that any process leading to permeabiliz-ation of the yeast cell wall renders yeasts of several different genera competent for uptake of the VLPs. The relevance of this discovery to the natural dissemination of the virus remains to be established, although its technical potential for researchers in the field is obvious. Natural conditions that could lead to such permeabilization have not been determined or tested, although numerous such environments can be envisaged. Of particular interest is the possibility that the action of sublethal concentrations of killer toxin (acting as an ionophore) on a sensitive yeast population could render a finite number of cells competent for uptake of the dsRNA viruses, and thus provide a horizontal route of transmission. The demonstration that mating yeast cells are competent for viral uptake is another intriguing observation, suggesting

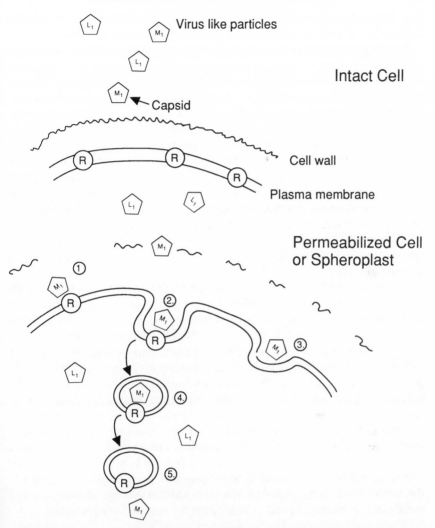

Figure 8.2 The demonstration of killer virus uptake and possible mechanisms of transfer. The upper panel depicts the experimental conditions used by El-Sherbeini and Bostian (1987) to detect mycoviral infection. An L-dsRNA, containing non-killer strain of *S. cerevisiae* competent for propagation of ScV-M_1, was incubated with preparations of M_1 and L_1 virus-like particles (VLPs) in roughly equimolar concentrations. The acquisition of the killer phenotype was used to score a successful infection. Due to the barrier of the yeast cell wall, intact cells were not detectably sensitive to infection by the killer mycoviruses. However, if the cell wall was permeabilized *in vitro* (i.e., by transformation), or *in vivo* during conjugation, the virus-like particles were taken up. Shown in the lower panel are two likely mechanisms for the uptake of ScV-M_1 VLPs. Uptake (1) could occur via interaction with a receptor (R) followed by its passage into clathrin coated vesicles (2). Endocytosis of these clustered pits could then result in passage of the virus-receptor complex into an endosome-like vesicle (4) and ultimately to the vacuole. At this point, a reduction in pH would result in the disassociation of receptor and virus and liberation of M_1-dsRNA or ScV-M_1 VLPs into the cytoplasm (5). Alternatively, the process may be more passive and receptor-independent (3).

that horizontal transmission via infection could be a plausible and relevant mechanism of viral dissemination (El-Sherbeini and Bostian, 1987).

The experiment demonstrating viral uptake is illustrated in Figure 8.2, along with possible models for the infection process. The exact mechanism remains to be determined. Fluid phase endocytosis is the most probable route of entry, given the dependence of infection on incubation time and temperature, and VLP concentration (El-Sherbeini and Bostian, 1987). Alternatively, viral entry could take place through a receptor-mediated process. *S. cerevisiae* has been shown to endocytose Vesicular Stomatitis Virus (Makarow, 1986), and it could be interesting to determine if the same mechanisms are involved in mycoviral uptake. The immense variability in both strain susceptibility and viral infectivity indicates a complex, possibly multigenic component to the process. Furthermore, the demonstration that distantly related fungal genera such as *Yarrowia lipolytica* can be infected with components of the *S. cerevisiae* killer system (Sturley *et al.*, 1988a; El-Sherbeini *et al.*, 1988) may indicate that the mechanisms of uptake are related in these organisms and possibly universal in nature.

A wide variety of mycoviral systems are known to exist and studying the relationships and properties of these viruses will be greatly facilitated by the infection procedure. The development of transformation protocols for fungi such as *Ustilago maydis* (Wang *et al.*, 1988) and *Schizosaccharomyces pombe* (Beach and Nurse, 1981) provides a unique opportunity to study the interactions of totally unrelated viruses in a variety of hosts, as well as the general evolution of mycoviruses. In addition, the expression of the M_1-dsRNA preprotoxin gene in non-*Saccharomyces* fungal hosts as well as expression of other killer toxins in *S. cerevisiae* may help define host contributions to the mechanisms of toxin secretion and the expression of immunity. The mycoviral infection procedures are superior to the techniques of cell mating or fusion to attain mycoviral transfer between strains. For the first time, kinetic properties of viral propagation particularly with regard to the factors controlling amplification, copy number control, interference and exclusion can be determined in homogeneous genetic backgrounds.

8.3 Processing and secretion of toxin

The eukaryotic processes of co- and post-translational modification leading to secretion of bioactive peptides are probably best exemplified by the maturation of killer toxin and α-factor mating pheromone in yeast. Nearly all of the enzymes which mediate the biosynthetic pathways have been unambiguously identified, and characterized genetically and/or biochemically. The killer system has an added level of complexity in that the toxin precursor confers both a killing and an immunity phenotype. It is likely that certain aspects of the processing events mediate the separate expression of both traits, and a

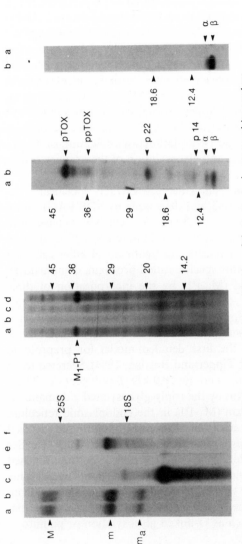

Figure 8.3 Transcription, translation and covalent modification of the preprotoxin gene and its product.

Panel A: *Northern hybridization of RNA from strains containing M₁-dsRNA or PTX1 using a preprotoxin specific radiolabelled probe.* Total yeast nucleic acid from strains K12-1 containing M₁-dsRNA, lanes (*a*), (*b*); and GG100-14D transformed with *PTX1*, lanes (*c*), (*d*); or *PTX1-316*, lanes (*e*), lanes (*f*); was extracted from cells grown on low-Pi medium, lanes (*a*), (*c*), and (*e*); and high-Pi medium, lanes (*b*), (*d*) and (*f*). Molecular mass markers were the 25 S and 18 S ribosomal RNAs (1.5 and 3.0 kb respectively). *M* is the undenatured M₁-dsRNA plasmid present in strain K12-1 and m and m_a are the 1.9- and 1.2-kb transcripts derived from *M*. Strains expressing *PTX1*, lane (*c*), produce a stable 1.0-kb transcript, whereas the *PTX1-316* allele, lane (*e*), directs the production of a least 2 transcripts approximately 2.1 and 2.9 kb in length.

Panel B: *In-vitro translation of preprotoxin mRNA*. Total RNA was translated in a wheat-germ system using [³H]-leucine as label, the products precipitated with anti-toxin IgG and analyzed by SDS-PAGE and autoradiography. RNA was from strains K12-1, lane (*a*); GG100-14D transformed with YEp24 grown on low-Pi medium, lane (*b*); or *PTX1* grown on high-Pi, lane (*c*) and low-Pi medium, lane (*d*) respectively. M₁-P1 is the 35-kD primary translation product of denatured M₁-dsRNA, m, m_a, or *PTX1* transcripts.

Panels C and D. *In-vivo-synthesized proteins from PTX1 transformants.* Proteins were fractionated by SDS-polyacrylamide electrophoresis, transferred to nitrocellulose and incubated with anti-toxin IgG. Antigen-antibody complexes were detected using ¹²⁵I-protein A. Membrane-bound intracellular proteins (panel C, or concentrated secreted proteins (panel D), from strain GG100-14D transformed with *PTX1* grown in high-Pi; lane (*a*), or low-Pi; lane (*b*) medium are shown. The species protoxin (pTOX), preprotoxin (ppTOX), p22, p14, α and β are described in the text. None of these species appears in protein extracts of YEp24 transformants of GG100-14D grown in low-Pi medium (data not shown).

thorough understanding of the mechanisms involved may have implications for preprohormone processing in higher eukaryotes.

8.3.1 Biochemical studies of preprotoxin synthesis and maturation

The primary in vitro translation product of denatured M_1-dsRNA (or the m and m_a in vivo killer transcripts derived from M_1; Figure 8.3A) is a 35-kD polypeptide, M_1-P1, that is immunoreactive with antisera raised against purified toxin (Figure 8.3B). M_1-P1 possesses the majority of Staphylococcus aureus protease-generated peptide components of secreted toxin (Bostian et al., 1980a) and therefore can be considered a preprotoxin, the intracellular precursor to toxin. In vitro translation of denatured M_1-dsRNA, or its m or m_a transcripts, in a dog pancreas microsomal membrane system yielded two species of 32.4 and 43 kD (Bostian et al., 1983b). The appearance of these species is consistent with co- and post-translational modification of M_1-P1 involving both proteolytic cleavage and glycosylation of the primary translation product.

The only radiolabelled protein from killer cell extracts specifically immunoprecipitated with anti-toxin IgG and detectable by SDS-PAGE autoradiography was a membrane-bound, 43-kD species, which was chased into exocellular toxin with an approximate half-life of 25 minutes at 30°C (Bostian et al., 1983a; Bussey et al., 1983). Tunicamycin treatment of killer cells and endoglycosidase H treatment of this species, called protoxin, demonstrated that it differed from preprotoxin (M_1-P1) by the incorporation of three mannose-based $(GlcNAc)_2(Man)_9$-$(Glu)_3$ glycosylation units (Bostian et al., 1983a; Bussey et al., 1983). This information together with the identification of the termini of the secreted toxin subunits and the elucidation of the δ-α-γ-β domains of preprotoxin led to the first detailed model for preprotoxin maturation (Bostian et al., 1983b; Tipper and Bostian, 1984). Secreted toxin consists of the 11.4-kD α-subunit and the 9.0-kD β-polypeptide. These molecules are separated in protoxin by the triply-glycosylated γ domain. N-linked glycosylation of preprotoxin (M_1-P1) in the endoplasmic reticulum occurs solely in the γ domain by addition of three mannose-based glycosyl subunits, presumably at the three asparagine residues found in Asn-X-Thr/Ser glycosylation consensus sites within the region. In addition to the proposed leader cleavage within the δ-domain, at least three proteolytic cleavage events are required for the release of the termini of α and the N-terminus of β from protoxin (Figure 8.1). The presence or relevance of other secondary modifications to protoxin maturation such as O-linked glycosylation or palmitoylation have not been investigated.

8.3.2 Creation of PTX1, a repressible cDNA copy of M_1

To further investigate the maturation and functions of preprotoxin, we fused a full-length DNA copy of the preprotoxin ORF to the PHO5 promoter (Sturley

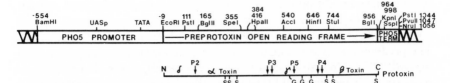

Figure 8.4 The *PTX1* allele; a repressible cDNA copy of the preprotoxin ORF. The structure of the *PHO5*-preprotoxin cDNA gene and its eventual product, protoxin, is shown with characteristic restriction endonuclease cleavage sites, *PHO5* upstream activation site (UAS), and transcription termination sequences (TERM). *KEX2* endoproteolytic cleavage (single arrow) and positions of carboxypeptidase 'trimming' (double arrow) at P3 and P4 are also shown. P2 and P3 demarcate the N- and C-termini of α, whereas P4 indicates the N-terminus of β. P5 is the proposed C-terminus of the p22a molecule. Sites of glycosylation (G) and potential disulphide bond formation at cysteine residues (S) are shown.

et al., 1986*a*; Sturley *et al.*, 1988*a*). The resulting chimera was denoted *PTX1-316* (*PHO5-TOX*) and transferred to multiple copy vectors. The 'allelic' designation, 316, refers to the position of the last amino acid prior to translation termination. This nomenclature is also used for mutations in the *PHO5*-preprotoxin gene where the number indicates the position of the first alteration from the wild-type sequence (see below). The complete *PTX1* gene (Figure 8.4) was constructed by the fusion of a synthetic 120-base oligonucleotide, containing the *PHO5* transcription terminator, immediately downstream of the preprotoxin translational termination codon in *PTX1-316*.

Plasmids bearing these alleles were transformed into cells of an M_1-dsRNA-free, non-killer, yeast strain, where they displayed phosphate regulated killing and immunity phenotypes. Levels of phenotypic expression were higher than those obtained with the same strain infected with M_1-dsRNA. Killing and immunity phenotypes expressed by *PTX1* and *PTX1-316* were also considerably stronger than the levels expressed by pSH-GB1, in which the majority of δ was replaced with the *PHO5* secretion leader (Hanes *et al.*, 1986). Northern blot hybridization analysis of total RNA from these transformants confirmed the phosphate regulated expression of the *PTX1-316* gene (Sturley *et al.*, 1986*a*). Similar analysis of RNA transcripts derived from phosphate derepressed cells carrying the complete *PTX1* gene demonstrated the presence of high levels of a stable 1 kb transcript (Figure 8.3*A*). *In vitro* translation of total RNA from phosphate derepressed cells harbouring *PTX1-316* produced a 35-kD protein, immunoprecipitable with anti-toxin IgG, with the same mobility as M_1-P1, the immunoprecipitable translation product of wild-type M_1-dsRNA transcripts (Figure 8.3*B*).

8.3.3 *Intermediates and end-products of protoxin maturation*

Examination of preprotoxin processing intermediates in derepressed *PTX1* transformants by immunoblotting cell extracts with anti-toxin IgG has provided new insight into the maturation of protoxin (Sturley *et al.*, 1986*a*). The *PTX1-316* gene directs the secretion of α and β toxin subunits at high levels in cells grown under conditions of phosphate limitation (Figure 8.3*D*). The predominant toxin-related protein in membrane fractions of the dere-pressed *PTX1-316* transformants is a 43-kD species, identical in mobility to protoxin extracted from wild-type killer cells (data not shown). In addition, a less abundant 35-kD preprotoxin species is discernible in the same prepar-ations. This species is presumably equivalent to M_1-P1. Two minor protein species with molecular weights of 38 and 40 kD are glycosylation intermed-iates, containing one and two core units respectively, as indicated by their change in mobility upon hydrolysis with endoglycosidase H (data not shown).

In addition to the biosynthetic intermediates, four other intracellular protein species are consistently present in extracts of *PTX1* transformants. Two of these species, of molecular weight 11.4 and 9.0 kD, represent intracellular forms of the secreted α- and β-toxin subunits, or toxin eluted from cell wall fragments present in the membrane preparation. In these extracts, α is less abundant than β. The two other previously unseen antigenic species, termed p22 and p14 based on their respective mobilities, are also detectable (Figure 8.3*C*). The p14 species has the mobility of an uncleaved δ-α complex. The size and degree of glycosylation of p22, however, infers cleavage of protoxin at a previously undiscovered processing site within the γ domain, indicated as P5 in Figure 8.4. The possible functions of these proteins are discussed below.

In subsequent experiments involving endoglycosidase H hydrolysis, p22 has in fact migrated as a doublet, consisting of the two portions of protoxin defined by cleavage at site P5 as shown in Figure 8.6. p22a, the N-terminal segment prior to P5, contains one glycosyl unit, and p22b, the C-terminal segment downstream from P5, contains 2 glycosyl units (Sturley *et al.*, 1988*b*). The p22 doublet also reacts with an anti-γIgG, raised against a β-galactosidase-γ protein fusion (S. Sturley, J. LeVitre and K. Bostian, unpublished observ-ations). The proportion of p22a to p22b is not clear from existing data. The p22b species could act as an additional source of the β-domain and may explain the abundance of this domain relative to α observed in intracellular protein preparations. The two p22 species have subsequently been detected in wild-type, M_1-dsRNA killer cell extracts, indicating they are not artefacts of the *PHO5*-regulated expression of *PTX1-316*.

The δ-domain of protoxin has many of the features of a conventional secretion signal peptide, including a hydrophobic membrane spanning peptide sequence between residues 12 and 27 (see below, Figure 8.7; Bostian *et al.*, 1984) and a potential leader endopeptidase cleavage site (Perlman and Halvorson, 1983) at residue 26 which appears to be cleaved *in vitro* in the presence of dog pancreas membranes (Bostian *et al.*, 1983*a*). However, the evidence for internal cleavage of δ *in vivo* is somewhat conflicting. For example, the effects of the serine protease inhibitor tosyl-L-phenylalanylchloromethylketone

(TPCK) on preprotoxin maturation suggest that cleavage may not occur, at least until after glycosylation (Hanes *et al.*, 1986; Sturley *et al.*, 1986*b*). Alternatively, Lolle and Bussey (1986) have demonstrated that *sec53* mutant strains (defective in early post-translational stages of passage through the ER) accumulate a preprotoxin molecule with a mobility suggesting leader cleavage prior to glycosylation. If δ does persist to a later stage of maturation, it may serve as a membrane anchor to facilitate correct or differential endopeptidase processing in the golgi or secretory vesicles. Definitive evidence for or against co-translational signal cleavage awaits N-terminal sequencing of protoxin.

8.3.4 Factors that mediate proteolytic processing of protoxin

A detailed mutagenic analysis of the killer system performed by Wickner and Leibowitz in 1976 led to the identification of several *trans*-acting factors that mediate expression of killing and immunity and secretion of toxin. Since then, the maturation of preprotoxin in wild-type killer cells has been shown to require functional *SEC* (Novick *et al.*, 1980), *KEX* and possibly *REX1* alleles as well as the action of enzymes sensitive to the chymotrypsin inhibitor, TPCK (Tipper and Bostian, 1984; Hanes *et al.*, 1986). Lesions in these genes result in a failure to correctly process the toxin precursor and lead to an intracellular accumulation of protoxin. Mutations in the *SEC* genes define various points in the yeast secretory pathway, and the effects of various *sec* mutants (*sec1, 7, 18* and *53*) on preprotoxin maturation indicate that it is classically processed via the ER, golgi, and secretory vesicles (Bussey *et al.*, 1983). Although *kex1* and *kex2* mutants are unable to secrete active toxin and kill sensitive yeasts, they retain immunity to exogenously applied toxin (Wickner and Leibowitz, 1976). Cleavage of protoxin by these gene products is therefore not essential for expression of immunity. However, a third *trans*-acting factor, the *REX1* gene product, is required for immunity. Mutations at this locus render cells sensitive to the toxin they produce. *REX1* has therefore been implicated as a mediator of immunity via the liberation of the immunity determinant from protoxin (Sturley *et al.*, 1988*a*, *b*).

A further clarification of the processing events involved in preprotoxin maturation comes from the isolation and DNA sequence analysis of the *KEX1* (Dmochowska *et al.*, 1987) and *KEX2* (Julius *et al.*, 1984) genes. As stated previously, lesions in both genes were originally identified by the loss of toxin activity. Subsequently they were shown to be required for the maturation of the mating pheromone, α-factor (Leibowitz and Wickner, 1976; Dmochowska *et al.*, 1987). Neither gene product is essential for cell viability. Deletion of the *KEX2* gene, however, confers a cold-sensitive growth phenotype (R. Fuller, personal communication).

The *KEX2* gene was isolated by complementing the defect in α-factor activity (Julius *et al.*, 1984). The gene encodes a calcium-dependent, neutral serine protease with a marked degree of substrate specificity towards paired basic residues. Purified *KEX2* endopeptidase readily cleaves small peptides *in vitro* at positions following Lys-Arg or Arg-Arg residues.

There are four potential dibasic *KEX2* cleavage sites within prepro-α-factor and three within preprotoxin. All four of the sites in prepro-α-factor are recognized by the *KEX2* enzyme, and four repeats of an immature α-factor polypeptide are released (Julius *et al.*, 1984). The situation for preprotoxin, however, is less clear. During maturation of protoxin, recognition of two of the predicted sites leads to cleavage at the N-terminus of the β-domain (P4) and at a site (P3) near the C-terminus of the α-domain. Cleavage at the third potential *KEX2* site, located at residue 188 within the γ domain (P5), would result in the liberation of molecules the size of p22a and p22b. An additional sequence surrounding the N-terminus of α, ProArgGluAla (P2), bears some resemblance to that found in prepro-α-factor (a pair of basic residues followed by GluAla) and may also be a substrate for the *KEX2* endoprotease.

A similar approach led to the isolation of the *KEX1* gene and the demonstration that it is probably the carboxypeptidase required for removal of the dibasic, C-terminal residues left in prepro-α-factor and preprotoxin following *KEX2* cleavage (Dmochowska *et al.*, 1987). DNA sequence analysis of the *KEX1* gene revealed regions with homology to the yeast vacuolar serine protease, carboxypeptidase Y (CPY). This new information refutes the earlier notion (Tipper and Bostian, 1984) that *KEX1* is responsible for cleavage at the δ-α-junction (P2), which is in fact supported by recent data demonstrating that a *PHO5* preprotoxin gene fusion lacking δ through replacement with the acid phosphatase signal sequence, is still dependent upon *KEX1* for production of active toxin.

The comparison of *KEX1* and carboxypeptidase Y also indicates the conservation of both the active site serine residue and the histidine residues thought to be responsible for TPCK inhibition of CPY (Dmochowska *et al.*, 1987), suggesting that the *KEX1* enzyme may also be inhibited by TPCK. *KEX1* therefore may be the TPCK-sensitive enzyme detected in earlier studies on preprotoxin maturation (Bussey *et al.*, 1983). The compelling model posited for *KEX2* and *KEX1* action implies an epistatic relationship between the two genes, with *KEX2* cleavage preceding *KEX1*. This is the opposite of that concluded from the protoxin stability studies performed previously (Bussey *et al.*, 1983).

Studies of prepro-α-factor maturation are equally consistent with the model for *KEX1* action (Dmochowska *et al.*, 1987). The *MFα1* gene with the C-terminal α-factor repeat unit deleted becomes completely dependent upon the removal of the LysArg C-terminal residues from the remaining repeats for production of active α-factor. The process is *KEX1*-dependent. Furthermore, immature and inactive α-factor peptides from *kex1* mutants bearing the truncated prepro-α-factor allele can be activated *in vitro* by the action of carboxypeptidase B.

Closer examination of the nucleotide sequence of the *KEX* genes indicates the presence of apparent signal leader sequences, glycosylation consensus sequences and *trans*-membrane domains (TMD) consistent with the entry and retention of both proteins in the yeast secretory pathway. Deletion of the TMD of *KEX2* results in the secretion of the enzyme and presumably a decrease in the efficiency of killer and α-factor maturation (R. Fuller and J. Thorner, personal communication). A similar phenotype is observed in cells with a disrupted copy of the gene for clathrin heavy chain (*CHC1*), indicating clathrin to be important in the retention and compartmentalization of such enzymes (G. Payne, personal communication).

8.3.5 Expression of PTX1 in processing deficient backgrounds

We extended our investigations into preprotoxin maturation by analysing the toxin-related proteins extracted from killer strains harbouring wild-type and mutant alleles of genes involved in the processing pathway (Figure 8.5). In one series of experiments, null mutant strains bearing *in vitro*-generated deletions of the *KEX2* gene were constructed and analysed following transformation with the *PTX1* allele (Sturley *et al.*, 1988a,b). While the expression of immunity was unaffected in these strains, killing and secretion of active toxin were reduced only to low levels. Another endopeptidase must therefore be able

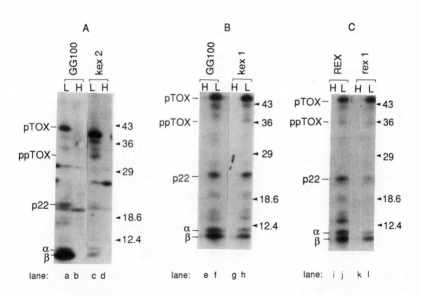

Figure 8.5 Proteolytic processing deficiencies in *kex* and *rex* mutants. GG100-14D (wild-type) and congenic *kex2*, *kex1* and *rex1* derivatives were transformed with the *PTX1* allele and grown on high (H) or low (L) phosphate medium. Detergent-soluble cell extracts were prepared and analysed by immunoblotting using anti-toxin IgG and ¹²⁵I-protein A. *REX* is a spontaneous, immune revertant of the *rex1* mutant strain shown here. pTOX, ppTOX, p22, α and β are as described in Figure 8.3 and in the text.

to substitute inefficiently for *KEX2*. These strains displayed a marked intracellular accumulation of protoxin, with detectable but very low levels of p22, α and β. All species had apparently normal mobilities on SDS-PAGE gels. While *KEX2* probably directs the primary steps in proteolysis of protoxin, there must be other enzymes which have similar specificity but reduced activity or abundance, or are incorrectly localized. The reduction in levels of p22 in the null mutant strains suggests that *KEX2* may cleave at residue 188 to generate p22a and p22b, or alternatively, that *KEX2* cleavage elsewhere in the molecule may be a prerequisite for formation of the two species.

A similar analysis of *kex1.1* mutant strains indicates that although they display reduced levels of killer activity, they secrete wild-type levels of apparently normal-sized toxin (Sturley *et al.*, 1988*a,b*). The intracellular protein profile of this mutant is also apparently wild-type, with a possible accumulation of p22, α and β. Defects in the *KEX1* gene therefore appear to

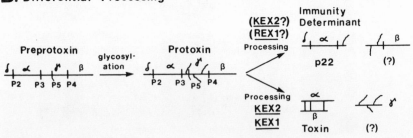

Figure 8.6 A model for preprotoxin maturation. Two models are proposed for the generation of an immunity determinant from preprotoxin. Model A (excess precursor) proposes that protoxin, possibly modified by *REX1*, serves as the immunity determinant, with the subsequent processing of excess protoxin to produce active α and β. Model B (differential processing) predicts two fates for an intracellular pool of protoxin: *REX1* and/or *KEX2* mediated cleavage to give p22a and p22b, and an alternate series of proteolytic events mediated by the *KEX* gene products to produce the secreted α and β. P2-P5 are potential processing sites using the nomenclature of Sturley *et al.* (1986). Glycosylation is denoted by the appendages to the γ domain. Hypothetical disulphide linkages between α and β are indicated by the vertical lines.

have little effect on secretion of toxin but a marked debilitation of toxin activity. Removal of the two C-terminal amino acids by the $KEX1$ product, although not detectable on SDS gels, is thus apparently essential for the killing activity of the α-β dimer.

Finally, $rex1$ mutants were analysed for their profile of preprotoxin processing intermediates and secreted proteins (Sturley *et al.*, 1988*a*, *b*). While the p22 species were present in parental, wild-type $PTX1$ transformants and parental M_1-dsRNA containing strains, they were absent from $rex1$ mutant cell extracts, despite the fact that protoxin accumulates at high levels, as do the intracellular forms of α and β. The data from analysis of mutants at all three loci suggest a model for the production of p22, α and β as shown in Figure 8.6. The model invokes differential processing of a common pool of protoxin into p22 (mediated by $REX1$) and separate maturation of the precursor into α and β under the influence of the $KEX1$ and $KEX2$ peptidases. There are potential endoprotease cleavage sites at residue 188 ($KEX2$-like, LysArgSerAsp) and at residue 180 (AlaAsnAla with cleavage after the second alanine residue). Cleavage at either position would produce the 22-kD doublet, with the physical properties of p22a and p22b. The implications of such a model with regard to the expression of immunity are discussed below.

8.3.6 *Preprotoxin, prepro-α-factor and higher eukaryotic preprohormone maturation*

It is interesting to note the marked similarities between the maturation of preprotoxin and the processing of mammalian hormones such as insulin (Chan *et al.*, 1979). The primary translation products of the preprotoxin and insulin genes are both produced in the ER, where signal peptide removal probably precedes glycosylation. In the golgi or secretory vesicles, cleavage at the N- and C-termini by proteases with specificity towards dibasic residues ($KEX2$-type activity) followed by carboxypeptidase trimming of the termini ($KEX1$-like activity) ultimately results in the secretion of an active polypeptide dimer covalently joined by disulphide bridges between cysteine residues. In the case of insulin, the C-peptide is directly analogous to the γ-domain of preprotoxin. If the model for differential processing of preprotoxin represents the actual processing pathway for the molecule, then the similarities between toxin production and generation of mammalian opiate peptides is even more striking. For example, the corticotropin-β-lipotropin precursor is cleaved at pairs of dibasic residues to yield several different hormones (Nakanishi *et al.*, 1979). Some of the cleavages occur within the domains of other active hormones, analogous to the α-γ site within p22 at residue 147. Furthermore, expression of $KEX2$ in mammalian tissue culture cells mediates the cleavage of proopiomelanocortin to generate β-endorphin (R. Fuller and J. Thorner, personal communication). It appears therefore that these processes are quite ubiquitous in nature.

The maturation of prepro-α-factor is also mediated by very similar mechanisms. The process diverges from this general model of precursor processing in that the action of a type IV dipeptidyl aminopeptidase specifically removes -X-Ala repeats from the N-terminus of α-factor following *KEX2* cleavage (Julius *et al.*, 1983). Strains bearing mutations in the *STE13* gene lack this activity and therefore secrete inactive α-factor. A similar sequence (GluAlaPro) exists in protoxin at the N-terminus of the α domain (Bostian *et al.*, 1984). However, since these residues persist in secreted toxin, they are presumably not *STE13* substrates. This may be due to the resistance to cleavage of peptide bonds involving proline residues. Alternatively, N-terminal dipeptidyl removal could follow signal peptide cleavage within δ (at P1) to ultimately liberate the α-domain. This activity would not be provided by *STE13*, since strains bearing null mutations in this gene exhibit no defects in protoxin maturation, or in killing or immunity phenotypes (Sturley *et al.*, 1988*b*). Similarly, defects in genes involved in *a*-factor maturation have no detectable effect on toxin or immunity production, although deficiencies in vacuolar degradative functions (e.g. *pep4* mutants; Woolford *et al.*, 1986) do lead to increased stability of protoxin and to higher levels of toxin secretion (Sturley *et al.*, 1988*b*).

8.4 Mutagenesis of preprotoxin genes

The intractability of the multiple-copy dsRNA system to mutational analysis has rendered assignment of structure–function relationships within the viral genome a difficult task. Furthermore, due to the paucity of naturally occurring or artificially induced killer plasmid mutations (Bussey *et al.*, 1982), relatively few novel phenotypes have arisen. Recombination within each RNA genomic segment has not been observed, making a genetic analysis of the *cis*-encoded functions of M_1-dsRNA virtually impossible. To overcome this, several groups have utilized cDNA copies of the preprotoxin for use in mutational analysis (Hanes *et al.*, 1986; Lolle *et al.*, 1984; Boone *et al.*, 1986; Sturley *et al.*, 1986). Our approach of regulating the expression of wild-type and *in vitro* generated mutants of the *PTX1* allele (see Figure 8.4 above; Sturley *et al.*, 1986*a*) avoids the further problem of selectional disadvantage conferred by mutations in the preprotoxin gene that disrupt immunity without an effect on killing.

A series of mutations were generated, primarily by linker insertion and site-directed mutagenesis at specific positions within the coding sequence of *PTX1-316* (Sturley *et al.*, 1986*a*). Most mutations occurred at locations with significant predicted secondary structures (Figure 8.7). All were verified by DNA sequence analysis.

The phenotypes of the mutants were studied under derepressed conditions in a non-killer yeast strain. RNA blot hybridizations confirmed the existence of the expected, regulated transcripts in all of the mutants. Protein products of the correct mobility on SDS-PAGE gels were confirmed by *in vitro* translation

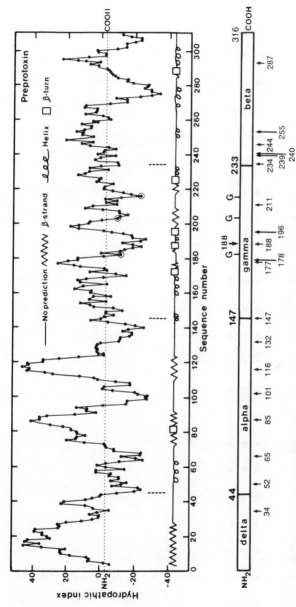

Figure 8.7 Predicted secondary structure and hydropathy analysis of type 1 yeast killer preprotoxin. Secondary structures and hydropathicity were predicted by the methods of Chou and Fasman (1974) and Kyte and Doolittle (1982), respectively. Circled residues are the predicted points (Asn-X-Thr/Ser) of glycosylation in protoxin. Also shown, in the lower panel, is the domain structure of the type 1 killer protoxin. Numbers indicate the amino-acid residues of protoxin, with potential processing residues (44, 147, 188 and 233) and glycosylation sites (G) as shown. The positions of the preprotoxin mutations discussed in the text (Figure 8.8) are also indicated by arrows and residue number.

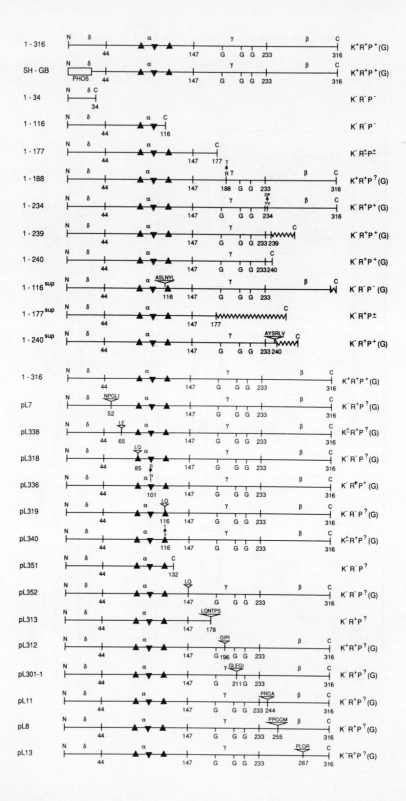

of *in vivo* transcripts and by immunoblotting of *in vivo* synthesized proteins. The gene products of the mutant *PTX1* alleles and their phosphate regulated killer ($K^{+/-}$) and immunity ($R^{+/-}$) phenotypes are shown in Figure 8.8. In addition to assaying killer toxin against intact cells, the sensitivity of spheroplasts ($P^{+/-}$) to the secreted proteins from these strains were also tested. Similar analyses were performed by Boone *et al.* (1986). Their data are also shown in Figure 8.8.

8.4.1 *The 'active site' of α is required for the expression of killing and immunity*

The immunity conferred by pSH-GB eliminates a role for δ in immunity determination. This is confirmed by the inability of the *PTX1-34* allele (in which translation terminates close to the δ-α boundary) to confer immunity. Similarly, it is apparent from the immune phenotypes conferred by alleles with disrupted (*PTX1-234*, pL11, pL8, and pL13) or incomplete β domains (*PTX1-239, 240* and *240^{sup}*) that this region is also non-essential for the expression of immunity. However, since *PTX1-116, 116^{sup}* (termination and disruption of the second hydrophobic region of α, respectively), pL318 and 319 (insertional disruptions of the first and second hydrophobic regions of α, respectively) do not express immunity or killing, this region of protoxin is inherently involved in both processes and may define an 'active site' of α. Alternatively, the perturbations in secondary structure induced by these mutations may alter the context of an active site located elsewhere in α, so that it is no longer presented to the plasma membrane receptor in an active conformation. Mutations flanking this region of α (pL7, 338; D. Tipper, personal communication) have a more deleterious effect on toxin activity than on immunity expression. The partial immunity conferred by the *PTX1-177* and *177^{sup}* alleles (defective in the N-terminal portion γ) and complete immunity exhibited by pL312, pL313 and pL301-1 (insertional mutations within γ) defines the region upstream of residue 177 as the minimum determinant for immunity, negating a role for the C-terminal portion of γ in this process.

A further rebuttal for a role of γ in immunity was obtained by expression of this γ domain in the absence of the other protoxin regions by fusion of γ to the

---

Figure 8.8 Mutagenesis of *PTX1*. A series of mutations was introduced into the preprotoxin ORF and expressed in M_1-dsRNA free strains, as described by Sturley *et al.* (1986) for the *PTX1* series (upper diagram) and Boone *et al.* (1986) for the pL series (lower diagram). *PTX1-316* is the 316-amino-acid, wild-type *PTX1* allele possessing the δ-α-γ-β structure with predicted points of glycosylation (G), and proteolytic cleavage (44, 147, 233), as indicated. The proposed active region of α is indicated by the upright (hydrophobic) and inverted (hydrophilic) triangles. The position of each mutation is indicated by the number of the first residue which differs from the wild-type sequence. Its nature, i.e. insertion or substitution, is denoted by the single letter amino-acid code. Premature truncations were also created. Fusion proteins created in some of the mutations are indicated by the wavy line. The phenotypes conferred (killing of sensitive strains, *K*; immunity to exogenous toxin, *R*; and killing of sensitive spheroplasts, *P*); were analysed under the appropriate conditions as described by the authors. *G* denotes the observation of a glycosylated precursor protein.

PHO5 promoter-leader construct used in pSH-GB1. No discernible phenotype resulted from expression of this construct in a non-killer yeast strain. The production of truncated forms of α (pL351) or changes in protoxin immediately following the C-terminus of mature α (pL352) are also unable to confer immunity. In the latter case, processing of protoxin is probably aberrant and α would not be liberated, possibly implicating a role for mature α in immunity expression. Alternatively, this mutation could implicate molecules with C-terminal extensions of α (e.g. p22) as mediators of immunity. Interestingly, mutations immediately following the C-terminus of α (pL352) which block secretion of active toxin (but not production of glycosylated protoxin) also disrupt immunity, inferring that an intact p22a molecule or fully mature α could be prerequisites for immunity expression.

8.4.2 α *is dependent on* β *for delivery to the plasma membrane*
An early hypothesis for the action of toxin postulated that the α domain interacted lethally with the plasma membrane, but that its localization to this site in whole cells required binding of the β component of the α-β dimer to the cell wall glucan receptor (Tipper and Bostian, 1984). Cells harbouring the *PTX1-239* or *PTX1-240* alleles produce fully glycosylated truncated polypeptides lacking all of the β domain, except the first six residues. Normal processing produced the intracellular α-toxin component, although the secreted form was either labile or inefficiently secreted, since it was not detected in culture supernatants. Mutations elsewhere in β (pL11, 8 and 13) also had a deleterious effect on toxin secretion, indicating that the α-β interaction may be required for efficient secretion. Similarly, a mutation in γ also blocked secretion of toxin although the mutant protoxin was glycosylated normally (pL301-1). Although intact sensitive cells were not killed by cell-free supernatants from derepressed cultures of *PTX1-239-* or *PTX1-240*-bearing strains, spheroplasts of a sensitive strain were killed, indicating a low level of secretion of toxic material. This is consistent with α acting as the lethal toxin component and β bringing α in contact with its site of action at the plasma membrane by binding to the cell wall. Cells harbouring *PTX1-234* accumulated normal levels of a fully glycosylated protoxin of normal size and processed it with secretion of near normal quantities of α- and β-subunits, identified by size and immunogenicity. The *PTX1-234* gene product differs from the wild-type protoxin by the replacement of TyrVal at the N-terminus of β by AspPro. This substitution did not appear to alter processing by *KEX2* or export of mature toxin. The mutant form of toxin was inactive against whole cells, although secreted proteins killed spheroplasts as effectively as wild-type toxin. Moreover, unlike the situation with wild-type killers, very little of the modified β was retained by the cell walls of *PTX1-234* transformants, suggesting that the defect in β function may be binding to the cell wall receptor (S. Sturley, D. Tipper and K. Bostian, unpublished data). A mutation in α (pL336) had a similar phenotype and may define a region of α

that interacts with β. The ability of pL336 culture supernatants to kill spheroplasts but not whole cells may result from the inability of the α-domain encoded by this allele to interact successfully with β, although in situations where β is extraneous, it has a toxic effect on sensitive cells.

8.5 Novel interactions with the host

Despite the vast array of interactions that the killer system displays with the host cell, remarkably few *trans*-acting factors have been implicated in the expression of immunity to killer toxin. There are two probable reasons for this disproportionality. The suicidal nature of such mutations in the presence of active toxin has made isolation of all but leaky representatives virtually impossible. In addition, it is possible that the two phenotypes are so closely inter-related that factors mediating immunity but not killing do not exist or are rare. In a search for mutations of this sort, we have addressed the former problem by utilizing the *PTX1* allele described above (Figure 8.4). Mutagenesis of a M_1-dsRNA-free strain transformed with this allele, followed by the isolation and subsequent phosphate derepression of the mutant population, was performed as described in Figure 8.9 (Sturley *et al.*, 1988c). A series of screening procedures enabled a range of mutant phenotypes to be detected. Mutants displaying resistance to killer toxin in the absence of expression of the *PTX1* gene (i.e. killer-resistant or *kre*-type mutants) were selected by their ability to grow under repressed conditions (i.e. media containing high concentrations of inorganic phosphate) in the presence of active killer toxin. *kre* mutants were surprisingly common, although a full complementation analysis has not yet been performed. From the same mutant enrichment procedure, numerous representatives of mutations in the phosphatase regulatory pathway were detected as constitutive killers. All other mutations were detected after derepression of the *PHO5* promoter. As expected, *kex*-like mutations retaining immunity to toxin but defective in killing were readily detected. Some of these mutants were defective in mating, were concomitantly shown to be unable to secrete α-factor, and thus probably represent *kex2* alleles. In addition, *nks* (non-killer sensitive) mutants that failed to express either phenotype were also detected. Presumably they define events in preprotoxin maturation other than *KEX* or *REX*-mediated proteolytic processing.

   The major emphasis of this study was on the isolation of mutations specifically involved in the mediation of immunity. Contrary to our expectations, the frequency of *rex*-like mutations was relatively high. To date, none of these alleles has been shown to be obviously defective in maturation of protoxin. In a parallel study, we investigated the effect of aberrant vacuolar protein localization and defective endocytosis on expression of killing and immunity (Sturley *et al.*, 1988c). Eight complementation groups of vacuolar protein localization (*vpl*) mutants were originally identified by

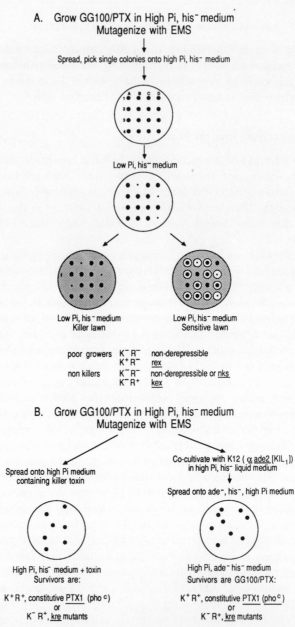

A. Grow GG100/PTX in High Pi, his⁻ medium
Mutagenize with EMS

Spread, pick single colonies onto high Pi, his⁻ medium

Low Pi, his⁻ medium

Low Pi, his⁻ medium
Killer lawn

Low Pi, his⁻ medium
Sensitive lawn

poor growers   K⁻ R⁻   non-derepressible
               K⁺ R⁻   rex
non killers   K⁻ R⁻   non-derepressible or nks
               K⁻ R⁺   kex

B. Grow GG100/PTX in High Pi, his⁻ medium
Mutagenize with EMS

Spread onto high Pi medium
containing killer toxin

Co-cultivate with K12 (α ade2 [KIL₁])
in high Pi, his⁻ liquid medium

Spread onto ade⁻, his⁻, high Pi medium

High Pi, his⁻ medium + toxin
Survivors are:

K⁺ R⁺, constitutive PTX1 (pho ᶜ)
or
K⁻ R⁺, kre mutants

High Pi, ade⁻ his⁻ medium
Survivors are GG100/PTX:

K⁺ R⁺, constitutive PTX1 (phoᶜ)
or
K⁻ R⁺, kre mutants

Figure 8.9 Mutagenesis of *trans*-acting factors in expression of the killer system. An M_1-dsRNA free strain was transformed with the *PTX1* allele and subjected to EMS mutagenesis after growth under repressed conditions (high-Pi medium). As shown in Panel *A*, the single colonies were isolated, derepressed by growth on low-phosphate medium and transferred to low-Pi medium in the presence of a lawn of killer or sensitive cells. *rex* mutants (e.g. colony 2*D*) were detected by their poor growth and killing activity on low-Pi medium. *Kex* mutants (e.g. colony 1*D*) were detected by their normal growth but reduced or absent killing zones. Non-derepressible (nks) mutants (e.g. colony 3*A*), were non-killers and poor growers in the presence of the killer tester strain. Panel *B* shows the enrichment procedure used for isolation of *kre*-type mutants. The mutant collection was either spread onto repressing medium containing active killer toxin or co-cultivated with a constitutive killer followed by auxotrophic selection against the killer tester. Survivors were either constitutive (i.e. phoᶜ) killers or non-killer *kre* mutants.

Rothman and Stevens (1986) by the aberrant secretion of the vacuolar hydrolase, carboxypeptidase Y (CPY). Several representatives of these complementation groups were transformed with $PTX1$ and assayed for killing and immunity phenotypes. Although they secreted virtually wild-type levels of active toxin, all vpl mutants tested were deficient in immunity expression and grew poorly in the presence of active toxin. Interestingly, approximately 30% of the rex mutants isolated by the screening procedures described above were also vpl mutants as judged by the secretion of CPY. The immunity determinant was apparently not degraded due to aberrant localization of vacuolar hydrolases, since vpl $pep4$ double mutants (defective in vacuolar protease activity) expressed poorer levels of immunity than vpl single mutants. Strains characterized as deficient in fluid-phase endocytosis and receptor-mediated uptake of α factor (Chvatchko et al., 1986) were also analysed for defects in killer toxin-associated activities (Sturley et al., 1988c). Introduction of the $PTX1$ allele conferred a wild-type killer phenotype to one of the mutants ($end2$), while a second mutant, defective in the pinocytotic stage of endocytosis ($end1$) was unable to express immunity despite secretion of active killer toxin.

8.6 The mechanism of killing

Killing of sensitive cells by toxin results from a rapidly induced increase in the permeability of the cytoplasmic membrane to protons and the associated leakage of metabolic pools such as ATP from the cell (de la Pena et al., 1981). The action of toxin appears to be a two-stage process (Figure 8.10), initially involving an energy-independent binding of toxin to a 1, 6 β-D-glucan cell wall receptor (Hutchins and Bussey, 1983), probably synthesized by enzymes encoded by the $KRE1$ and $KRE2$ genes. Mutations at either of these loci confer toxin resistance, reduce cell wall binding of toxin and modify the 1, 6 β-D-glucan content of the cell wall. However, $kre1$ and $kre2$ mutant spheroplasts are sensitive to toxin, indicating that the actual killing process occurs at the plasma membrane. Following binding to the cell wall, an energy-dependent process (impaired by metabolic inhibitors such as dinitrophenol or antimycin A; Skipper and Bussey, 1977) transfers toxin to the plasma membrane where, based upon its in vitro effect on lipid bilayers, it is proposed that it forms ion-permeable channels (de la Pena et al., 1981). However, preliminary experiments involving the use of a patch-clamp apparatus (Gustin et al., 1986) to measure ion channels in yeast treated with killer toxin do not confirm the latter observation (M. Gustin, C. Kung, K. Bostian and S. Sturley, unpublished data). Mutations at a third chromosomal locus, $KRE3$, confer toxin resistance to spheroplasts, consistent with the involvement of a plasma membrane located receptor in the killing process (Al-Aidroos and Bussey, 1978). The isolation of additional toxin resistant mutants as described above (Sturley et al., 1988c; S. Sommer, personal communication) has revealed at least two

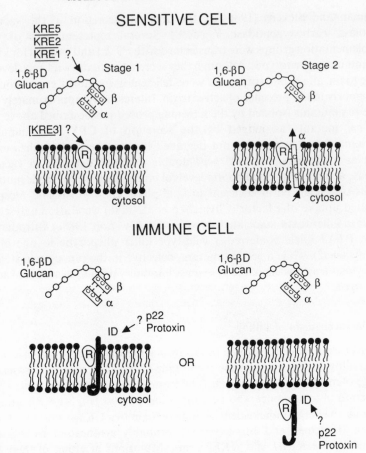

Figure 8.10 Mechanisms of killing and immunity. Killing of a sensitive cell is envisaged as a two-stage process involving initial binding of toxin to 1, 6 β-D-glucan cell wall components (dependent upon the *KRE1*, *KRE2* or *KRE5* gene products) via the β-domain. The α-domain is thus accessible to the *KRE3* plasma membrane receptor (R) where it exerts its lethal effect, leakage of intracellular metabolites (stage 2). Two mechanisms for immunity are represented for an immune cell. Either the immunity determinant (perhaps protoxin or p22) alters or masks the receptor so that it is incapable of interacting with exogenously supplied toxin (the α-domain), or the immunity determinant mediates relocation or removal of the receptor from the plasma membrane so that is no longer available for interaction with α delivered from outside the cell. In the latter model, processes defined by mutations in vacuolar protein localization and endocytosis genes may mediate certain portions of this pathway.

more *KRE* complementation groups, one of which, *KRE5*, has also been implicated in cell wall 1, 6 β-D-glucan synthesis (H. Bussey, personal communication).

The genes responsible for the *KRE1*, *KRE2*, and *KRE5* phenotypes have recently been isolated (H. Bussey, personal communication). All are required for the deposition of 1, 6 β-D-glucan into the yeast cell wall and therefore sensitivity to toxin, but none of the isolated genes is essential for vegetative growth at normal temperatures. The finding that a mutant form of the plasma membrane ATPase (*pma1* mutants) induces partial resistance to toxin in non-killer cells (S. Sturley, J. Haber and K. Bostian, unpublished) also implies a role for membrane potential in sensitivity to toxin and may explain part of the reported energy requirement of transport of toxin to the plasma membrane receptor. Alternatively, the ATPase may itself be the plasma membrane target of killer toxin. In this model *pma* mutants would produce proton pumps that no longer interact with toxin and are therefore resistant to its effect.

Hydropathy analysis of the amino-acid sequence of preprotoxin (Figure 8.7) predicts the α domain to possess two highly hydrophobic regions (residues 72–91 and 112–127) separated by a hydrophilic region containing acidic amino-acids and all of the cysteine residues found in α (Figure 8.5; Bostian *et al.*, 1984). It has been postulated (Tipper and Bostian, 1984), that insertion of this region of α into a membrane by interaction with the putative *KRE3* receptor could create the cation channels responsible for cell death. In confirmation of this model, mutations that disrupt this region render toxin inactive against whole cells or spheroplasts. The observation that mutations in β eliminate the lethality of toxin against intact cells but not spheroplasts implicates this toxin subunit as the ligand for cell wall glucan. These receptor-mediated functions do not require endocytosis to exert their effect on sensitive cells, as shown by the sensitivity of non-killer *end* mutants to toxin.

8.7 Mechanisms of immunity

The exact mechanism of immunity has remained elusive for some time. The most popular models have assumed the action of toxin to be receptor-mediated and have invoked the alteration or masking of the receptor (*KRE3*) by the immunity determinant, rendering it inaccessible to killer toxin (Figure 8.10). Alternatively, it has been proposed (Sturley *et al.*, 1986a, 1988c) that the presently unidentified immunity determinant could tag the receptor for diversion or removal from the plasma membrane. In keeping with such models, the major site of action for immunity appears to be at the plasma membrane since spheroplasts of killer cells are resistant to toxin. Mutant strains (*vpl*) that mislocalize proteins normally destined for the vacuole are concomitantly defective in expression of immunity. This implies that the major mechanism of immunity may be diversion of a receptor/immunity determinant complex to the vacuole, where it is either degraded or altered so that it no longer binds to the α domain. Since pinocytosis mutants (*end1*) are also incapable of expressing immunity, it is possible that portions of the endocytotic pathway are involved in the same mechanism.

The precise component of protoxin that mediates immunity has not been defined. A discrete, intact γ-domain was once proposed to function as the immunity determinant by interacting with the receptor during secretion

K

(Tipper and Bostian, 1984). The mutagenic studies outlined above have ruled out this possibility and instead have shown the active site of α to be essential for immunity expression. The presence of p22 in wild-type killer cell extracts has led to the proposal of two models for immunity involving either differential processing or accumulation of protoxin (see Figure 8.6). Both models postulate that two forms of the 'active site' of α can interact with the hypothetical plasma membrane receptor, through the same residues but with different results. Mature α, combined with β, delivered from outside the cell, kills cells as a consequence of this interaction. In contrast, α, either alone or as part of a larger immunity determinant, would interact with the receptor during secretion, rendering it inaccessible or unrecognizable to exogenous α. Uncleaved protoxin or p22 could confer immunity early in the secretion pathway, before exposure to the processing enzymes presumed to act in a Golgi compartment (Sturley *et al.*, 1986*a*).

The observation that non-immune *PTX1 rex1* transformants have reduced levels of p22 suggests a role for this species as the immunity determinant. The reduction in p22 levels in *kex2* mutants suggests that cleavage at residue 188 is mediated by the *KEX2* endopeptidase. Disruption of the dibasic residues at position 188 of preprotoxin (*PTX1-188*; Figure 8.8) resulted in a reduction in p22 levels, the appearance of a novel toxin-related polypeptide of 25 kD (p25), but no effect on killing or immunity phenotypes (D. Tipper, S. Sturley and K. Bostian, unpublished results). Thus, cleavage at 188 may occur but is not essential for expression of immunity. The presence of other potential endoproteolytic cleavage sites in this region may explain the production of p22a, p25 and expression of immunity and killing in these mutant cells. It is possible that any protein moiety possessing an active α domain (α, p14, p22a or protoxin) could circumvent the lethal interaction of toxin and receptor. Immunity may therefore be a function of the rate of production of immature forms of α (e.g. p14, p22 or protoxin) relative to the formation of an α-β complex. *REX1* mediated cleavage at residue 188, either directly or due to an interaction with the *KEX2* endopeptidase, could favour this equilibrium, possibly by being the initial processing event in protoxin maturation. The production of p22a might thereby preempt further internal processing events (δ-α or α-γ). Alternatively, following the acquisition of immunity, residual p22a could function as a substrate for cleavage between δ-α and α-γ and thereby produce active α polypeptides.

8.8 Conclusions and relevance

We have described here a yeast virus and its associated protein, killer toxin, that has undergone intense examination over the past decade. The advances made in understanding this system have mirrored the general progress made in yeast molecular biology. Concomitantly, it has developed into an archetype for host–viral interactions, preprohormone maturation, protein secretion and membrane biology. To take each in turn, the isolation and molecular characterization of the cis- and trans-acting factors involved in maintenance of the dsRNA plasmids can serve as a model system for comparison with the viruses of higher eukaryotes. Elucidation of the exact requirements for stable maintenance and replication of the killer virus could lead to its exploitation as a broad host range fungal expression system. The production of chimeric dsRNA gene fusions that possess the correct sequences for encapsidation and maintenance by both viral and host mediated proteins may be a tangible prospect. The demonstration of cross-species mycoviral transfer and gene expression in at least one instance (S. cerevisiae to Y. lipolytica and vice versa) demonstrates that dsRNA molecules can be transferred between distantly related fungi. Furthermore, genetic and environmental factors are known that could increase the copy number of the M plasmid to several thousand per cell, making it an attractive vector for heterologous expression in diverse fungi.

The maturation of preprotoxin has provided a sound basis for the development of expression vectors in yeast. The identification of the roles of the KEX gene products in protoxin maturation will obviously effect the strategies used to secrete proteins. Derivatives of the protoxin molecule have been utilized in the secretion of cellulase (Skipper et al., 1985) and β-lactamase from yeast (D. Tipper, personal communication). Further elucidation of the processing pathways involved, particularly with respect to the identification of the immunity determinant, may provide useful parallels to preprohormone processing in higher cells.

The implication of receptor-mediated processes in the expression of killing and immunity phenotypes may be a useful tool for the study of membrane structure and function as well as intracellular vesicular trafficking. The intriguing defect in immunity exhibited by vacuolar protein localization and endocytosis mutants has revealed a hitherto unsuspected aspect of the killer system. In combination with the variety of processes leading to cell wall and plasma membrane deposition and function, possibly defined by the KRE gene products, the mechanism of immunity may provide a useful probe for the study of receptor-mediated endocytosis in yeast. In addition, the mechanism of viral infection may provide an alternate perception of the same processes.

The dual phenotypes of the killer system, production of toxin and immunity, have also been utilized in the wine, beer, and soft drink industries (Ouchi et al., 1979; Young, 1981; T. Young, personal communication) to confer anti-contaminant properties to the beverage concerned. This has been performed both during production (for expression of immunity against contaminating yeasts) and during storage (toxicity against contaminants).

The zymocidal nature of killer toxins in general has generated considerable interest in their suitability as antifungal agents. Unfortunately, their immunogenic properties render them inappropriate for direct use other than as topical agents. However, the mechanism of killing by type 1 toxin may well be universal in that the interaction with the receptor involved is the principal lethal event. The wide range of specificity of action of various toxins could possibly result from minor variations in either the cell wall or plasma membrane receptors. Thus, a more ubiquitous analogue of toxin could be devised that would not be immunogenic. Clearly the isolation and characterization of the plasma membrane receptor will be particularly illuminating in this regard. Furthermore, the unique nature of the yeast cell wall in terms of its glucan composition may result in the development of pathogen specific antifungal agents targeted against the cell wall *KRE* gene products. The variability in susceptibility to toxin has also been exploited to identify *Candida* pathovars using a panel of test killer strains (Polonelli *et al.*, 1983).

Acknowledgements

We thank Cameron Douglas and Stephen Parent for useful comments and discussion of the manuscript, and Howard Bussey, Michael Leibowitz, Donald Tipper, Robert Fuller, Greg Payne and Steve Sommer for communicating information prior to publication. This work was supported in part by Public Health Service Grant GM36441 to K.A.B.

References

Al-Aidroos, K. and Bussey, H. (1978) Chromosomal mutants of *Saccharomyces cerevisiae* affecting the cell wall binding site for killer factor. *Can. J. Microbiol.* 24: 228.

Beach, D., and Nurse, P. (1981) High-frequency transformation of the fission yeast *Schizosaccharomyces pombe*. Nature 290: 140.

Bobek, L.A., Bruenn, J.A., Field, L.J. and Gross, K.W. (1982) Cloning of cDNA to a yeast viral double-stranded RNA and comparison of three viral RNAs. *Gene* 19: 225.

Boone, C., Bussey, H., Greene, D., Thomas, D.Y. and Vernet, T. (1986) Yeast killer toxin: Site-directed mutations implicate the precursor protein as the immunity component. *Cell* 46: 105.

Bostian, K.A., Burn, V.E., Jayachandran, S. and Tipper, D.J. (1983a) Yeast killer dsRNA plasmids are transcribed *in vivo* to produce full and partial-length plus-stranded RNAs. *Nucleic Acids Res.* 11: 1077.

Bostian, K.A., Elliot, Q., Bussey, H., Burn, V., Smith, A. and Tipper, D.J. (1984) Sequence of the preprotoxin dsRNA gene of type 1 killer yeast: multiple processing events produce a two component toxin. *Cell* 36: 741.

Bostian, K.A., Hopper, J.E., Rogers, D.T. and Tipper, D.J. (1980a) Translational analysis of the killer-associated virus-like particle dsRNA genome of *S. cerevisiae*: M-dsRNA encodes toxin. *Cell* 19: 404.

Bostian, K.A., Jayachandran, S. and Tipper, D.J. (1983b) A glycosylated protoxin in killer yeast: Models for its structure and maturation. *Cell* 32: 169.

Bostian, K.A., Sturgeon, J.A. and Tipper, D.J. (1980b) Encapsidation of yeast killer double-stranded ribonucleic acids: dependence of M on L. *J. Bacteriol.* 143: 463.

Bruenn, J.A., Madura, K., Siegal, A., Miner, Z. and Lee, M. (1985) Long internal inverted repeat in a viral dsRNA. *Nucleic Acids Res.* 13: 1575.

Bussey, H. (1981) Physiology of killer factor in yeast. *Adv. Microb. Physiol.* 22: 93.

Bussey, H., Sacks, W., Galley, D. and Saville, D. (1982) Yeast killer plasmid mutations affecting toxin secretion and activity and toxin immunity function. *Mol. Cell. Biol.* 2: 346.

Bussey, H., Saville, D., Greene, D., Tipper, D.J. and Bostian, K.A. (1983) Secretion of yeast killler toxin: processing of the glycosylated precursor. *Mol. Cell. Biol.* 3: 1362.

Bussey, H., Sherman, D. and Somers, J.M. (1973) Action of yeast killer factor: a resistant mutant with sensitive spheroplasts. *J. Bacteriol.* 113: 1193.

Bussey, H., Steinmetz, O. and Saville, D. (1983) Protein secretion in yeast: two chromosomal mutants that oversecrete killer toxin in *S. cerevisiae*. *Curr. Genet.* 7: 449.

Chan, S.J., Patzelt, C., Duguid, J.R., Quinn, P., Labrecque, A., Noyes, B., Keim, P., Heinrikson,

R.L. and Steiner, D.F. (1979) Precursors in the biosynthesis of insulin and other peptide hormones. In *From Gene to Protein*, eds. T.R. Russel, K. Brew, H. Faber and J. Schultz, Academic Press, New York, 361.

Chou, P.Y. and Fasman, G.D. (1978) Prediction of the secondary structure of proteins from their amino acid sequence. *Ann. Rev. Biochem.* **47**: 251.

Chvatchko, Y., Howald, I. and Riezman, H. (1986) Two yeast mutants defective in endocytosis are defective in pheromone response. *Cell* **46**: 355.

de la Pena, P., Barros, F., Gascon, S., Lazo, P.S. and Ramos, S. (1981) The effect of yeast killer toxin on sensitive cells of *Saccharomyces cerevisiae*. *J. Biol. Chem.* **256**: 10420.

Dmochowska, A., Dignard, D., Henning, D., Thomas, D.Y. and Bussey, H. (1987) Yeast *KEX1* gene encodes a putative protease with a carboxypeptidase B-like function involved in killer toxin and α-factor precursor processing. *Cell* **50**: 573.

El-Sherbeini, M., Bevan, E.A. and Mitchell, D.J. (1983) Two biochemically and genetically different forms of L dsRNA of *Saccharomyces cerevisiae* exist: one form, L_2 is correlated with the [HOK] plasmid. *Curr. Genet.* **7**: 63.

El-Sherbeini, M. and Bostian, K.A. (1987) Viruses in fungi: infection of yeast with K1 and K2 killer viruses. *Proc. Natl. Acad. Sci. USA* **84**: 4293.

El-Sherbeini, M., Bostian, K.A., LeVitre, J. and Mitchell, D.J. (1987) Gene-protein assignments within the *Yarrowia lipolytica* dsRNA viral genome. *Current Genetics* **11**: 483.

El-Sherbeini, M., Steriti, J., Ramadan, N. and Bostian, K.A. (1988) Cross-species extracellular transmission of mycoviruses: infection of *Saccharomyces cerevisiae* and *Yarrowia lipolytica* with ScV and Y1V. *J. Virol.* (in press).

El-Sherbeini, M., Tipper, D.J., Mitchell, D.J. and Bostian, K.A. (1984) Virus-like particle capsid proteins encoded by different L double-stranded RNAs of *Saccharomyces cerevisiae*: Their roles in maintenance of M double-stranded killer plasmids. *Mol. Cell. Biol.* **4**: 2818.

Esteban, R. and Wickner, R.B. (1986) Three different M_1 dsRNA containing virus-like particle types in *Saccharomyces cerevisiae*. In vitro M_1 double stranded RNA synthesis. *Mol. Cell. Biol.* **6**: 1552.

Field, L.J., Bobek, L., Brennan, V., Reilly, J.D. and Bruenn, J. (1982) There are at least two yeast viral double-stranded RNAs of the same size: an explanation for viral exclusion. *Cell* **31**: 193.

Fried, H.M. and Fink, G. (1978) Electron microscopic heteroduplex analysis of 'killer' double-stranded RNA species from yeast. *Proc. Natl. Acad. Sci. USA.* **75**: 4224.

Gustin, M.C., Martinac, B., Saimi, Y., Culbertson, M.R. and Kung, C. (1986) Ion channels in yeast. *Science* **233**: 1195.

Hanes, S.D., Burn, V.E., Sturley, S.L., Tipper, D.J. and Bostian K.A. (1986) Expression of a cDNA derived from the yeast killer preprotoxin gene: implications for processing and immunity. *Proc. Natl. Acad. Sci. USA* **83**: 1675.

Hannig, E.M. and Leibowitz, M.J. (1985) Structure and expression of the M_2 genomic segment of a type 2 killer virus of yeast. *Nucleic Acids Res.* **13**: 4379.

Hannig, E.M., Williams, T.L. and Leibowitz, M.J. (1986) The internal polyadenylate tract of yeast killer virus M_1 double-stranded RNA is variable in length. *Virology* **152**: 149.

Hopper, J.E., Bostian, K.A., Rowe, L.B. and Tipper, D.J. (1977) Translation of the L-species dsRNA genome of the killer-associated virus-like particles of *Saccharomyces cerevisiae*. *J. Biol. Chem.* **252**: 9010.

Hutchins, K. and Bussey, H. (1983) Cell wall receptor for yeast killer toxin: involvement of (1,6) β-D-glucan. *J. Bacteriol.* **154**: 161.

Julius, D., Blair, L., Brake, A., Sprague, G. and Thorner, J. (1983) Yeast alpha-factor is processed from a larger precursor polypeptide: the essential role of a membrane-bound dipeptidyl aminopeptidase. *Cell* **32**: 839.

Julius, D., Brake, A., Blair, L., Kunisawa, R. and Thorner, J. (1984) Isolation of the putative structural gene for the lysine-arginine cleaving endopeptidase required for processing of yeast prepro-α-factor. *Cell* **37**: 1075.

Kyte, J. and Doolittle, R.F. (1982) A simple method for displaying the hydropathic character of a protein. *J. Mol. Biol.* **157**: 105.

Lee, M., Pietras, D.F., Nemeroff, M.E., Corstanje, B.J., Field, L.J. and Bruenn, J.A. (1986) Conserved regions in defective interfering viral double-stranded RNAs from a yeast virus. *J. Virol.* **58**: 402.

Leibowitz, M.J., Hussain, I. and Williams, T.L. (1988) Transcription and translation of the yeast killer virus genome. In *Viruses of Fungi and Simple Eucaryotes*, eds. M.J. Leibowitz and Y. Koltin, Marcel Dekker, New York, 133.

Leibowitz, M.J. and Wickner, R.B. (1976) A chromosomal gene required for killer plasmid expression mating and sporulation in *Saccharomyces cerevisiae*. *Proc. Natl. Acad. Sci. USA* **73**: 2061.

Lolle, S.J. and Bussey, H. (1986) *In vivo* evidence for signal cleavage of the killer preprotoxin of *Saccharomyces cerevisiae*. *Mol. Cell. Biol.* **6**: 4274.

Lolle, S.J., Skipper, N., Bussey, H. and Thomas, D.Y. (1984) The expression of cDNA clones of yeast M_1 double-stranded RNA in yeast confers both killer and immunity phenotypes. *EMBO J.* **3**: 1383.

Makarow, M. (1985) Endocytosis in *Saccharomyces cerevisiae*: Internalization of enveloped viruses into spheroplasts. *EMBO J.* **4**: 1855.

Makower, M. and Bevan, E.A. (1963) The physiological basis of the killer character in yeast. *Proc. Int. Congr. Genet. XI.* **1**: 202.

Nakanashi, S., Inoue, A., Kita, T., Nakamura, M., Chang, A.C.Y., Cohen, S.N., and Numa, S. (1979) Nucleotide sequence of cloned cDNA for bovine corticotropin-β-lipotropin precursor. *Nature* **278**: 423.

Novick, P., Ferro, S. and Schekman, R. (1981) Order of events in the yeast secretory pathway. *Cell* **5**: 461.

Olsnes, S. and Phil, A. (1973) Isolation and properties of abrin, a toxic protein inhibiting protein synthesis. Evidence for differential functions of its two constituent peptide side chains. *Eur. J. Biochem.* **25**: 179.

Ouchi, K., Wickner, R.B., Toh-e, A. and Akiyama, H. (1979) Breeding of killer yeasts for sake brewing by cytoduction. *J. Ferment. Technol.* **57**: 483.

Perlman, D. and Halvorson, H.O. (1983) A putative signal peptidase recognition site and sequence in eukaryotic and prokaryotic signal peptides. *J. Mol. Biol.* **167**: 391.

Polonelli, L., Archibusacci, C., Sestito, M. and Morace, G. (1983) Killer system: a simple method for differentiating *Candida albicans* strains. *J. Clin. Microbiol.* **17**: 774.

Ridley, S.P., Sommer, S.S. and Wickner, R.B. (1984) Super-killer mutations in *Saccharomyces cerevisiae* suppress exclusion of M_2 double-stranded RNA by L_{A-HN} and confer cold sensitivity in the presence of M and L_{A-HN}. *Mol. Cell. Biol.* **4**: 761.

Rothman, J.H. and Stevens, T.H. (1986) Protein sorting in yeast: mutants defective in vacuole biogenesis mislocalize vacuolar proteins into the late secretory pathway. *Cell* **47**: 1041.

Skipper, N. and Bussey, H. (1977) Mode of action of yeast toxins: energy requirement for *Saccharomyces cerevisiae* killer toxin. *J. Bacteriol.* **129**: 668.

Skipper, N., Sutherland, M., Davies, R.W., Kilburn, D., Miller, R.C. *et al.* (1985) Secretion of a bacterial cellulase by yeast. *Science* **230**: 958.

Skipper, N., Thomas, D.Y. and Lau, P.C.K. (1984) Cloning and sequencing of the preprotoxin-coding region of the yeast M_1-dsRNA. *EMBO J.* **3**: 107.

Sommer, S.S. and Wickner, R.B. (1982) Yeast L dsRNA consists of a least three distinct RNAs; evidence that the non-Mendelian genes [HOK], [NEX] and [EXL] are on one of these dsRNAs. *Cell* **31**: 429.

Stanway, C.A. and Buck, K.W. (1984) Infection of protoplasts of the wheat take-all fungus, *Gaeumannomyces graminis* var. *triitici*, with double-stranded RNA viruses. *J. Gen. Virol.* **65**: 2061.

Sturley, S.L., Elliott, Q.E., LeVitre, J., Tipper, D.J., and Bostian, K.A. (1986*a*) Mapping of functional domians within the *Saccharomyces cerevisiae* type 1 killer preprotoxin. *EMBO J* **5**: 3381.

Sturley, S.L., El-Sherbeini, M., Kho, S.H., LeVitre, J., and Bostian, K.A. (1988*a*) In *Viruses of Fungi and Simple Eucaryotes*, eds. M.J. Leibowitz and I. Koltin, Marcel Dekker, New York, 179.

Sturley, S.L., Hanes, S.D., Burn, V.E. and Bostian, K.A. (1986*b*) Maturation and secretion of the M_1-dsRNA encoded killer toxin in *S. cerevisiae*. In *Yeast Cell Biology*, ed. J. Hicks, Alan Liss, New York, 537.

Sturley, S.L., LeVitre, J., Tipper, D.J., and Bostian, K.A. (1988*b*) Maturation of type 1 killer preprotoxin of *Saccharomyces cerevisiae* and the expression of the immunity phenotype (in press).

Sturley, S.L., Palmer, J.D., Douglas, C.M. and Bostian, K.A. (1988*c*) Endocytosis and vacuolar protein localization are required for the expression of immunity to type 1 killer toxin in *Saccharomyces cerevisiae* (in press).

Thrash, C., Voelkel, K., Dinardo, S., and Sternglanz, R. (1984) Identification of *Saccharomyces cerevisiae* mutants deficient in DNA topoisomerase I activity. *J. Biol. Chem.,* **259**: 1375.

Tipper, D.J. and Bostian, K.A. (1984) Double-stranded ribonucleic acid killer systems in yeasts. *Microbiol. Rev.* **48**: 125.

Tyagi, A.K., Wickner, R.B., Tabor, C.W. and Tabor, H. (1984) Specificity of polyamine requirements for the replication and maintenance of different double-stranded RNA plasmids in *Saccharomyces cerevisiae. Proc. Natl. Acad. Sci. USA* **81**: 1149.

Wang, J., Holden, D.W., and Leong, S.A. (1988) Gene transfer system for the phytopathogenic fungus *Ustilago maydis. Proc. Natl. Acad. Sci. USA* **85**: 865.

Wickner, R.B. (1976) Mutants of killer plasmid of *Saccharomyces cerevisiae* depending on chromosomal diploidy for expression and maintenance. *Genetics* **82**: 273.

Wickner, R.B. (1986) Double-stranded RNA replication in yeast: the killer system. *Ann. Rev. Biochem.* **55**: 373.

Wickner, R.B. and Leibowitz, M.J. (1976) Two chromosomal genes required for killer expression in killer strains of *Saccharomyces cerevisiae. Genetics* **82**: 429.

Wickner, R.B. and Leibowitz, M.J. (1979) *mak* mutants of yeast: mapping and characterization. *J. Bacteriol.* **140**: 154.

Wickner, R.B. Ridley, S.P., Fried, H., and Ball, S.G. (1982) Ribosomal protein L3 is involved in replications or maintenance of the killer double-stranded RNA genome of *Saccharomyces cerevisiae. Proc. Natl. Acad. Sci. USA* **79**: 4706.

Wickner, R.B. and Toh-e, A. (1982) [HOK] a new yeast non-Mendelian trait, enables a replication-defective killer plasmid to be maintained. *Genetics* **100**: 159.

Williams, T.L. and Leibowitz, M.J. (1987) Conservative mechanisms of the *in vitro* transcription of killer virus of yeast. *Virology* **158**: 231.

Woolford, C.A., Daniels, L.B., Park, F.J., Jones, E.W., Van Arsdell, J.N. and Innis, M.A. (1986) The *PEP4* gene encodes an aspartyl protease implicated in the posttranslational regulation of *Saccharomyces cerevisiae* vacuolar hydrolases. *Mol. Cell. Biol.* **6**: 2500.

Young, T.W. (1981) The genetic manipulation of killer character into brewing yeast. *J. Inst. Brew.* **87**: 292.

Young, T.W. and Yagiu, M. (1978) A comparison of the killer character in different yeasts and its classification. *Antonie van Leeuwenhoek. Microbiol. Serol.* **44**: 59.

Zhu, H., Bussey, H., Thomas, D.Y., Gagnon, J., and Bell A.W. (1987) Determination of the carboxyl termini of the α and β subunits of yeasts K1 killer toxin. *J. Biol. Chem.* **262**: 10 728.

9 Genetic manipulation of brewing yeasts

EDWARD HINCHLIFFE and DINA VAKERIA

9.1 Introduction

Yeasts have been used for the production of beer and other alcoholic beverages for thousands of years. The process of fermentation to produce alcohol, carbon dioxide and a variety of desirable flavour and aroma compounds has undergone little fundamental change over the centuries. In contrast, the brewing industry has been subject to major changes in recent years. Small uneconomic breweries have been closed in favour of larger production units capable of satisfying the demand for large volumes of beer. New plant and processes have enabled the improvement of production techniques with a concomitant demand for the more efficient utilization of raw materials and process plant. Consequently, there has arisen a need to develop new strains of brewing yeast which are better suited to the requirements of a more advanced, technically demanding industry. Progress in yeast molecular genetics affords the possibility of satisfying these requirements. The purpose of this chapter is to review recent advances in brewing yeast genetics and to highlight the progress toward the construction of new strains of yeast for commercial application.

Genetics has been suggested as having potential utility in reducing process costs, enhancing beer quality, improving aspects of process efficiency, developing new products and increasing the value of spent yeast (Panchal et al., 1984; Tubb, 1984; Tubb and Hammond, 1987). This potential has in most instances yet to be realized, since to our knowledge no strain of genetically manipulated yeast is currently being used in the commercial production of beer. Consequently, here we will concentrate upon the factors influencing the realization of this potential by discussing the commercial and technical constraints influencing the genetic manipulation of brewing yeast. This will be exemplified by describing strain development strategies and the progress and problems in four major areas of research activity: amylolysis, glucanolysis, diacetyl control and the utilization of waste yeast biomass.

The manufacture of beer consists of a large number of unit operations, commencing with the malting of barley and ending in the packaging of beer (Hough, 1985). The fermentation of hopped brewers' wort by yeast is crucial to the viability of this process. However, fermentation and product maturation account for only 14% and 27% respectively of the overall operating costs of the brewing process (Molzahn, 1985). This means that the economic advantage accruing from the application of yeast genetics is limited to a relatively small window of opportunity. Consequently when embarking on a strain improvement programme it is important to clearly identify the commercial target to which the genetic modification is to be directed and to derive the maximum benefit by widespread implementation. These requirements tend to limit the geneticist to yeast strains used in the manufacture of existing products which enjoy a large market share based upon high-volume sales. As a consequence, a major objective of any strain improvement strategy is not only the introduction of a new or modified characteristic, but also the retention of existing characteristics. In this respect genetic modification must have no detrimental effect upon either the fermentation performance of the yeast or the characteristics it imparts to the beer. For reasons which will become apparent this can prove technically difficult and may account for the minimal historical impact of genetics on the brewing industry.

9.2 The properties of brewing yeast

Brewing yeasts comprise a heterogeneous group of yeasts which have been classified into two taxonomic groups: *Saccharomyces uvarum* (*carlsbergensis*), capable of producing the extracellular enzyme α-galactosidase and therefore able to ferment melibiose, and *Saccharomyces cerevisiae* which includes strains negative in this regard.

Traditionally *S. uvarum* was used in the production of lager beer by bottom fermentation; whereas *S. cerevisiae* was used in the production of ale, frequently but not necessarily always by top fermentation, at higher temperatures. However, the difference between these two groups has recently been considered insufficient to justify their separation into two species, and it has been recommended that they consolidate within the species *S. cerevisiae* (Barnett *et al.*, 1983; Kregervan-Rij, 1984). Importantly, not all yeast belonging to this species are capable of producing palatable beer. Rainbow (1970) described brewing yeast as 'those yeasts capable of producing in subtly balanced proportions quantitatively minor metabolic products such as acids, esters, ketones and higher alcohols'. He went on to say that 'a yeast may be unsuitable for brewing because one or more of these minor metabolic products is produced in excessive amounts either in absolute terms or relative to one another'. This eloquent description embodies the essential physiological properties of yeast which have resulted in their empirical selection by the brewer. In addition to these physiological attributes, the genetic make-up of brewing yeast is important to their role as industrial micro-organisms. Brewing yeast are invariably polyploid or aneuploid (Molzahn, 1977; Aigle *et al.*, 1984). Unlike 'laboratory' haploid strains of yeast which have been subjected to constant selection for regular sexual reproduction and high spore viability, brewing yeast lack a mating type, they are reticent to mate and exhibit low spore viability (Johnston, 1965; Fowell, 1969; Anderson and

Martin, 1975). These properties confer a measure of phenotypic stability on brewing yeast which makes them less susceptible to the de-stabilizing forces of mutation and recombination.

9.3 Genetic techniques

9.3.1 The 'classic' genetics approach

Despite the obvious difficulties inherent in the genetic analysis of non-mating, non-sporulating polyploid yeast, some successes have been reported from the application of so-called 'conventional' genetic techniques. Mutations have been induced in production yeast by treatment with N-methyl-N-nitrosoguanidine (NTG) and ultraviolet light (Molzahn 1977) or by selection (Hockney and Freeman, 1981; Delgado and Conde-Zurita, 1983; Bilinski et al., 1984; Stewart et al., 1985; Galván et al., 1987). Hybrid brewing yeast have been obtained by rare-mating (Gunge and Nakatomi 1972) involving crosses between auxotrophic haploid and diploid 'laboratory' strains and respiratory-deficient (Spencer and Spencer, 1977; Tubb et al., 1981), or antibiotic-resistant mitochondrial mutants (Spencer and Spencer, 1980) or spore-clones (Keilland-Brandt et al., 1981) of brewing yeast. Cytoductants have been obtained in which the linear double-stranded RNA of S. cerevisiae, encoding killer toxin (zymocin), has been transferred to brewing yeast recipients by a specialized form of rare-mating involving kar1 (karyogamy defective) donor strains, incapable of transferring their own nuclear DNA (Young, 1983; Russell and Stewart, 1985; Hammond and Eckersley, 1984). Protoplast fusion has been used to produce amylolytic hybrids between brewing yeast and S. diastaticus (Freeman, 1981; Gillis-van-Maele and van Landschoot, 1985). Unfortunately, in addition to the acquisition of the desired genetic characteristics, such as the ability to secrete glucoamylase, many undesirable genetic traits can be transmitted to hybrids which are deleterious to their brewing performance (Barney et al., 1980). The best-characterized of such genetic traits is that associated with the gene POF1, which is present in laboratory haploid strains of both S. cerevisiae and S. diastaticus (Goodey and Tubb, 1982).

POF1 encodes an enzyme capable of decarboxylating ferulic acid, present in wort, to produce the herbal phenolic flavour compound 4-vinylguiacol. This results in beers with an unpleasant flavour or aroma of phenol. Although it is possible to avoid the transfer of the POF1 gene by the judicious choice of donor strain (Tubb et al., 1981), one cannot rule out the possibility of transferring additional, less well-characterized genetic traits with equally undesirable consequences for the brewing performance of hybrids.

One interesting and elegant approach to overcoming the possibility of hybrid yeast acquiring undesirable genetic markers is afforded by a low-temperature sporulation technique, which has been effective in obtaining haploid or near-haploid spore clones from a production lager yeast, previously believed to be non-sporulating (Gjermansen and Sigsgaard, 1981). Spore clones were isolated with a- and α-mating types, and subsequently re-hybridized with one another to produce hybrids with a brewing performance similar, and in some cases identical, to the parental strain (Gjermansen and Sigsgaard, 1981). This approach, although limiting in scope, facilitates the analysis of the genetic basis of 'brewing performance' and will undoubtedly yield useful information for future strain development strategies.

The problem with all the genetic techniques briefly described thus far is that they have a limited potential to create new strains with a markedly different genetic character. This is because these procedures are primarily concerned with either the modification and abolition of existing genetic traits or the reassortment of a limited pool of genetic variation by intraspecific hybridiz-ation. In contrast, recombinant DNA technology, coupled with transform-ation, facilitates both the construction of novel strains by the controlled introduction of a wide variety of new genes, and enables the directed modification of existing genetic traits.

9.3.2 Brewing yeast transformation

The most common vector system employed for yeast transformation is that based upon the endogenous 2-μm plasmid of S. cerevisiae (Beggs, 1978); this plasmid is present in all strains of brewing yeast that have been investigated, indicating that these yeasts are capable of replicating and maintaining plasmid DNA, and by implication recombinant 2-μm-based vectors (Tubb, 1980; Aigle et al., 1984; Hinchliffe and Daubney, 1986). This was first confirmed by the transformation of a leucine-requiring meiotic segregant of a production lager yeast using a complementary prototrophic gene (LEU2) carried on a 2-μm-based yeast/E. coli shuttle vector (Gjermansen, 1983). However, although transformant selection by complementation is used extensively for transform-ation of genetically marked haploid and diploid yeast strains, the routine transformation of prototrophic brewing yeast requires the use of dominant selectable genetic markers.

A list of dominant selectable markers effective in yeast transformation is presented in Table 9.1. These genes have been divided into two groups based upon their species of origin. If one considers the heterologous genetic markers, it is apparent that they all encode antibiotic resistance determinants which mediate their effect by the synthesis of antibiotic modifying enzymes. To facilitate the use of these genes for transformant selection, it is necessary to obtain efficient gene expression in yeast. Consequently, heterologous genes invariably require modification to incorporate 5' and 3' yeast gene expression signals, such as a yeast promoter and transcription termination signal, in order to produce sufficient of the appropriate antibiotic modifying enzyme. For example, Hadfield et al., (1986) fused the chloramphenicol resistance marker (Cm^R) of E. coli to the alcohol dehydrogenase gene promoter (ADC1) and the cytochrome C-1 transcription terminator (CYC1) of yeast to derive a high-

Table 9.1 Dominant selectable markers for yeast transformation.

	Yeast transformation	Brewing yeast transformation
Homologous		
CUP1—copper resistance (copper chelating protein)	Fogel and Welch (1982)	Henderson et al. (1985)
DEX1—growth on dextrin (glucoamylase)	Meaden et al. (1985)	
ILV2—sulphometuron methyl resistance	Falco and Dumas (1985)	Fleming and Hinchliffe (unpublished results)
Comp^R—(acetohydroxy acid synthase) CoA compactin resistance (HMG CoA)	Rine et al. (1983)	
MEL1—growth on melibiose (α-galactosidase)	Tubb et al. (1986)	
KIL-k1—killer toxin resistance (killer toxin and immunity factor)	Bussey and Meaden (1985)	
POF1—cinnamic acid resistance (ferulic acid decarboxylase)	Tubb (1987)	
Tun^R—tunicamycin resistance (UDP-N-acetylglucosomine-1-P-transferase)	Rine et al. (1983)	
Heterologous		
G418^R—G418 resistance (kanamycin phosphotransferase)	Webster and Dickson (1983)	
Hm^R—hygromycin B resistance (hygromycin B phosphotransferase)	Gritz and Davies (1983) Kaster et al. (1984)	
Cm^R—chloramphenicol resistance (chloramphenicol acetyltransferase)	Cohen et al. (1980)	Hadfield et al. (1986)
Pm^R—phleomycin resistance	Gatignol et al. (1987)	
MTX^R—methotrexate resistance (dihydrofolate reductase)	Zhu et al. (1985)	
TK—amethopterin and sulphanilamide resistance (thymidine kinase)	Zhu et al. (1984)	Goodey et al. (personal communication)

frequency transformation system for industrial yeast, which could be used for introducing both 2-μm-based vectors and integrating vectors into the yeast genome. In contrast, the unmodified Cm^R, without yeast gene expression signals, is incapable of producing sufficient chloramphenicol acetyl transferance in yeast to mediate effective resistance to chloramphenicol (Hollenberg 1979; Cohen et al., 1980).

It is not usually necessary to modify homologous genes, unlike heterologous genes, for expression in yeast, and these genetic markers can be used directly for transformant selection. One system which is receiving increasing attention for the transformation of brewing yeast is based upon the CUP1 gene

(Henderson et al., 1985). *CUP1* encodes a cysteine-rich copper-chelating protein whose synthesis is inducible by copper at the transcriptional level (Fogel et al., 1983; Karin et al., 1984). In copper-resistant yeast the *CUP1* gene is amplified some 10 to 15 times (Fogel and Welch, 1982); this results in the production of high levels of the copper-chelating protein enabling growth in the presence of high concentrations of copper. The utility of this gene for brewing yeast transformation stems from the innate copper sensitivity of these yeast, enabling the selection of copper-resistant transformants harbouring the *CUP1* gene on a 2-μm-based vector (Henderson et al., 1985). Although the frequency of protoplast-mediated transformation varies for different strains of brewing yeast (Henderson et al., 1985), experience shows that the alkali cation method of Ito et al. (1983) can be used effectively for transformation of a large number of different production brewing yeasts (Hinchliffe et al., 1987a; Fleming and Hinchliffe, unpublished results).

The choice of dominant selectable marker for introducing recombinant DNA into brewing yeast can be extremely important for subsequent strain evaluation. Although most of the selectable markers listed in Table 9.1 have been indicated for brewing yeast transformation (Knowles and Tubb, 1987) only a limited number have a reported utility (Table 9.1). The use of a particular genetic marker is influenced by its availability, the sensitivity of the recipient strain and the inherent efficiency of transformation. It is particularly important when introducing recombinant DNA into brewing yeast that the genetic marker itself is not deleterious to the properties of the yeast. In this respect, strains transformed with 2-μm-based vectors carrying the *CUP1* gene have been shown to posses identical fermentation properties to their untransformed parental counterparts (Meaden and Tubb, 1985; Hinchliffe and Daubney, 1986). Moreover, these yeasts also produce beer, in experimental brewery trials, which is indistinguishable from beer produced by the untransformed parental yeast strain (Hinchliffe and Daubney, 1986). Unfortunately, the same is not true of other selectable genes; for example the use of *POF1*, in the absence of subsequent genetic modification, would result in yeast with a propensity to produce phenolic off-flavours. The *ILV2* gene can be used at high copy number to select yeast resistant to the herbicide sulphometuron methyl (Falco and Dumas, 1985), and is very effective for brewing yeast transformation (Fleming and Hinchliffe, unpublished results). The product of this gene is the enzyme acetohydroxy acid synthase (AHAS) which catalyses both the conversion of pyruvate to α-acetolactate and α-ketobutyrate to α-aceto-hydroxybutyrate, intermediates in the biosynthesis of valine and isoleucine respectively. However, these two biosynthetic intermediates are non-enzymically decarboxylated during beer fermentation to the vicinal diketones, diacetyl and pentanedione. Therefore, high levels of α-acetolactate caused by the overexpression of *ILV2* result in high levels of diacetyl, which imparts an undesirable flavour or aroma of 'butterscotch' to beer.

In recent years, concern has been expressed over the use of heterologous genes, in particular antibiotic-resistance determinants, as dominant selectable markers. Although there is no evidence to suggest that such genes will have an adverse effect upon the flavour properties imparted by the yeast, concerns have been raised over the production of beer by organisms containing these DNA sequences. Accordingly, considerable effort has been directed toward the development of genetic strategies which circumvent these difficulties.

9.3.3 Recombinant DNA strategies for strain development

The inheritable stability of recombinant genes is an important requirement for the development of new strains of yeast. Whenever yeast is transformed with a 2-μm-based recombinant plasmid, the resultant transformed phenotype is unstable. This is usually a result of either intra- or intermolecular recombination with the endogenous 2-μm plasmid, or segregational instability through plasmid loss at cell division. This latter aspect can usually be controlled by growth on selective media, thereby exerting direct selection for either the gene of interest or the plasmid vector. However, such an approach is inappropriate for a brewing fermentation in which the growth medium is hopped brewers' wort. Therefore, when considering the use of 2-μm-based recombinant vectors it is necessary to closely evaluate the inherent stability of the plasmid when the yeast is grown on a medium non-selective for plasmid maintenance. In this respect, 2-μm-based plasmids appear to possess a high degree of stability in brewing yeast (< 0.2% plasmid loss per cell doubling), exhibiting greater stability in lager yeast over ale yeast (Meaden and Tubb, 1985; Hinchliffe and Daubney, 1986). This high level of stability may be due to the polyploid nature of most brewing yeast, since stability is greater in strains of higher ploidy (Mead et al., 1986). Whatever the explanation, it has been suggested that under certain circumstances unstable plasmid vectors may be sufficiently stable to facilitate their use in large-scale industrial fermentations (Hinchliffe and Daubney, 1986).

A novel approach to both transformation and plasmid stability has been described by Bussey and Meaden (1985), who used a 2-μm-based vector carrying an expressed cDNA copy of the *S. cerevisiae* killer toxin-immunity gene, to select for brewing yeast transformants. Yeasts expressing active toxin and therefore possessing the toxin-immunity gene, were seen to be at a selective advantage over cells which had lost the plasmid; these cells were subsequently killed. In addition, since the toxin is most active in the pH range 4–5, the pH at which beer fermentation normally proceeds, the expressed toxin-immunity gene offers a useful means of engineering plasmid stability in brewing yeast. Unfortunately, the authors did not indicate whether yeast modified in this manner could be used to produce acceptable beer. One might conjecture that the level of cell death which will inevitably arise from the effects of plasmid loss may adversely affect beer quality.

More recently, we have demonstrated exceedingly high plasmid stability in brewing yeast without recall to selective or autoselective procedures (Hinchliffe et al., 1987a). By incorporating the CUP1 gene between two directly oriented homologous repeats of 2-μm plasmid DNA on a 2-μm-based vector, and transforming brewing yeast, it was possible to direct the integration of the CUP1 gene into the endogenous 2-μm plasmid of the host. One outcome of this integration was the excision, and subsequent segregational loss, of the transforming vector which carried unwanted extraneous DNA sequences of bacterial origin. When the stability of the copper-resistant 2-μm integrant was tested by growth on non-selective medium, substantially no phenotypic plasmid loss was observed over more than 100 cell doublings. Furthermore, the properties of the recipient lager strain were unchanged by the genetic procedure employed (Hinchliffe et al., 1987a). Thus 2-μm-based plasmids can be used to stably maintain recombinant genes in brewing yeast. Furthermore methods are available to remove unwanted DNA sequences, such as those providing selection and replication functions for plasmid maintenance in E. coli.

Of course, not all recombinant DNA strategies are limited to the use of 2-μm plasmids. Recombinant DNA can be introduced into yeast using a wide variety of different vectors or transformation systems; these include autonomously replicating plasmids (ARS) (Williamson, 1985); centromeric (CEN) and telomeric vectors (Blackburn and Szostak, 1984); and integrative transformation (Hinnen et al., 1978). The latter is increasingly preferred for the development of new strains of brewing yeast, since chromosomally integrated genes should be stably inherited and integration itself can be restricted to the gene(s) of interest, thereby avoiding the insertion of unwanted DNA sequences. Several different procedures have now been developed for the integrative transformation of brewing yeast (Knowles and Tubb, 1987). In all cases, these methods require the provision of DNA homology between the incoming DNA and the yeast chromosome, wherein the choice of homology determines the site of gene integration. Moreover, the efficiency and specificity of integration can be enhanced by transformation with linear DNA (Orr-Weaver et al., 1983). Clearly, when directing the integration of a gene to a particular locus it is important to choose a site which will not interrupt the integrity of an essential gene, unless this is an objective of the integrative procedure. In this respect one approach has been described which makes use of the HO locus as a site for gene insertion (Yocum, 1986); using a two-step recombination protocol, the amyloglucosidase (AMG) gene of Aspergillus niger was integrated into the chromosome of brewing yeast. Subsequently, exogenous vector DNA sequences were excised by intrachromosomal recombination. In this case the transformation step was effected by selection for the integration of a bacterial G418 resistance gene co-linear with the AMG gene, whereas the excision event was monitored by screening for the loss of β-galactosidase activity. Additional methods have been described whereby linear DNA sequences carrying the gene(s) of interest are integrated by homologous recombination following co-transformation with a selectable

2-μm-based plasmid vector. Co-transformation necessitates a relatively high frequency of transformation, since co-transformants arise at less than 1–5% of all transformation events. Using this approach, the α-amylase gene of *A. niger* (G. Loison, personal communication, cited by Knowles and Tubb, 1987), the glucoamylase gene, *DEX1*, of *S. diastaticus* (Vakeria and Hinchliffe, unpublished results) and the endoglucanase gene, *egl1*, of *T. reesei* (Penttilä *et al.*, 1987*b*) have been inserted into the brewing yeast genome. Subsequently, stable transformants harbouring an integrated gene and no additional unwanted DNA sequences can be isolated which have spontaneously lost the 2-μm-based plasmid through non-disjunction at cell division.

9.4 Changing the genetic and phenotypic characteristics of brewing yeast

9.4.1 *Amylolytic brewing yeast*
Amylolytic brewing yeasts have been a long-term goal of the brewing and related fermentation industries. They have a suggested utility for the production of both low-carbohydrate (light) beer and premium beers with high alcohol content. These products are normally produced with the aid of exogenous commercial enzymes added at the beginning of fermentation or during wort production (Aschengreen, 1987); a yeast capable of producing its own starch degradative enzyme offers a significant reduction in process costs by removing the requirement for exogenous enzyme.

Examination of the carbohydrate composition of a 'typical' brewers' wort indicates that approximately 20% of the sugar is inaccessible to yeast as non-fermentable dextrin (Table 9.2). This dextrin is composed of starch and related malto-oligosaccharides which can be converted to fermentable mono-, di- and trisaccharides by hydrolysis of the component α-1, 4- and α-1,6-glycosidic linkages with the enzyme glucoamylase (EC 3.2.1.3). A variety of microorganisms produce glucoamylases which hydrolyse starch progressively from its non-reducing end to yield glucose (Tubb, 1986). Glucoamylase genes have been cloned from the filamentous fungi *Aspergillus awamori* (Nunberg *et al.*, 1984), *Aspergillus niger* (Boel *et al.*, 1984) and *Rhizopus* sp. (Ashikari *et al.*,

Table 9.2 Carbohydrate composition of a 'typical' brewers' wort.

Carbohydrate	% Total	Fermentable by *S. cerevisiae*
Sucrose	1	+
Fructose	2	+
Glucose	10	+
Maltotriose	13	+
*Dextrin (DP 10)	20	−
Maltose	54	+

*DP 10: degree of polymerization in excess of 10 glucose residues.

1985; Ashikari *et al.*, 1986) and expressed in the yeast *S. cerevisiae* (Innis *et al.*, 1985; Yocum, 1986; Ashikari *et al.*, 1985, respectively). In all cases efficient gene expression was obtained by using the structural glucoamylase gene, lacking introns, or cDNA fused to yeast 5' and 3' expression signals, although the 5' non-coding region of the *Rhizopus* glucoamylase was transcriptionally active in yeast (Ashikari *et al.*, 1986). Interestingly, the signal sequences of these fungal genes were effective in mediating the correct N-terminal processing and secretion of the glucoamylase into the extracellular culture medium, a feature of other fungal hydrolases expressed in yeast (Yoshizumi and Ashikari, 1987). Yeasts expressing the fungal glucoamylase were able to grow on starch as the sole carbon source, giving credence to the belief that brewing yeast secreting this enzyme would be capable of the complete hydrolysis of wort dextrin, enabling fermentation to proceed to the lowest specific gravity it can reach, the attenuation limit. Such yeast could then be used for the production of low-carbohydrate beer. Although the glucoamylase gene of *A. niger* under the control of a yeast gene promoter has been integrated into the genome of brewing yeast (Yocum, 1986; G. Loison, cited by Knowles and Tubb, 1987), no information is available to indicate how such yeasts perform under beer production conditions, nor is it known whether these yeasts are capable of producing a satisfactory product.

The yeast *Saccharomyces diastaticus* produces an extracellular glucoamylase, which is capable of hydrolysing starch. This enzyme is encoded by a family of genes variously referred to as *STA* or *DEX*. The allelic relationship between these two designations has been established such that *STA1*, *STA2* and *STA3* are allelic to *DEX2*, *DEX1* and *DEX3* genes respectively (Erratt and Nasim, 1986). The genes coding for *STA1* and *STA3* have been cloned (Yamashita *et al.*, 1985*a,b*), as have those encoding *STA2* (Pretorius *et al.*, 1986*a*) and *DEX1* (Meaden *et al.*, 1985). DNA sequence analysis of *STA1* indicates that the gene encodes a 778-amino-acid protein containing 13 potential N-glycosylation sites. The enzyme is believed to be synthethized as a large precursor protein containing a signal peptide for secretion, a threonine-rich region and a functional domain (Yamashita *et al.*, 1985*a*). Similar studies on *DEX1* indicate that it also encodes a precursor protein of 806 amino acid residues with a signal sequence of 21 amino acids (Tubb, 1987). The enzyme encoded by *DEX1* (STA2) is highly glycosylated; in its native form it exists as a dimer of approximately 300-kD, whereas the completely deglycosylated protein corresponds to 56-kD (Modena *et al.*, 1986). Interestingly, the three unlinked polymorphic *DEX* genes of *S. diastaticus* (*DEX1*, *DEX2* and *DEX3*) show strong DNA sequence homology, suggesting that they were probably all derived from a common ancestral gene (Yamashita *et al.*, 1985*b*; Pretorius *et al.*, 1986*a*). Whilst *S. cerevisiae* is known to produce a sporulation-specific glucoamylase, which is not secreted, DNA hybridization studies of the *S. cerevisiae* genome indicate three regions of homology with the *STA1* gene, designated S1, S2 and SGA. The S1 and S2 regions are homologous to the 5' end of the *STA1* gene, whereas SGA shows homology with the 3' structural gene sequence (Yamashita *et al.*, 1985*b*, 1987). The extent of these DNA homologies is sufficient to suggest that the extracellular glucoamylase produced by *S. diastaticus* may have resulted from gene fusions between the S1, S2 and SGA regions of *S. cerevisiae* genomic DNA (Yamashita *et al.*, 1985*b*, 1987).

Early attempts to create amylolytic brewing yeast concentrated on the fusion of *S. diastaticus* with brewing yeast; although technically feasible, the resultant hybrid yeast were unsuitable for final application for the reasons indicated earlier. Consequently, attention has focused upon the use of the cloned glucoamylase genes, and in particular *DEX1*, for the construction of amylolytic yeast. Unlike the fungal glucoamylases, the *DEX1* specific enzyme does not have an α-1,6-glycosidase activity and is therefore incapable of the complete hydrolysis of wort and beer dextrins. However, the enzyme has potential in the production of certain beers and can be used to substitute for the addition of exogenous enzyme preparations.

The *DEX1* gene was first introduced into a brewing lager yeast on a 2-μm-based vector carrying the *CUP1* selectable marker (Meaden and Tubb, 1985). Transformants were selected on the basis of their resistance to copper rather than their ability to grow on starch, although transformants were capable of growth on a medium containing starch as the sole carbon source. Fermentations with these amylolytic brewing yeasts indicated that they were capable of superattenuating wort by the partial utilization of dextrin, whereas the control yeast lacking the *DEX1* gene could not superattenuate (Meaden and Tubb, 1985).

More recently we have introduced the *DEX1* gene into several different production lager yeasts using the *CUP1–DEX1* plasmid pLHCD6 described by Meaden and Tubb (1985). Successive laboratory fermentations and experimental brewery trials have clearly demonstrated that yeast producing this extracellular enzyme can ferment brewers' wort in a manner comparable to the parental untransformed yeast, in which the fermentation was supplemented with an appropriate enzyme. More importantly, beers produced in experimental brewery trials have been tested by a combination of both analytical and organoleptic analyses and shown to be indistinguishable from those produced by the parental yeast-enzyme supplemented control (Hinchliffe, 1986). Unfortunately, whilst these experiments have provided encouraging information on the suitability of amylolytic yeast for beer production, the *DEX1* gene incorporated in a 2-μm-based recombinant vector is inheritably unstable (Figure 9.1). Although this instability does not manifest itself in an immediate depreciation in fermentation performance, it is envisaged that this would be unacceptable on a large commercial scale. Furthermore, pLHCD6 also harbours *E. coli* plasmid DNA sequences which are considered undesirable in a yeast strain intended for commercial application. Consequently, the integration of the *DEX1* gene into the chromosome of an appropriate host is a desirable objective.

Stability   of   Plasmid   Mediated

Glucoamylase   Production

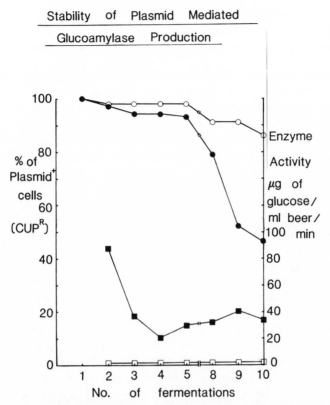

Figure 9.1 Plasmid stability and glucoamylase production during wort fermentations by dextrin-utilizing brewing yeast. Stability of BB9 (pET13:1), (O—O); BB9 (pLHCD6), (●—●); enzyme activity in wort supernatant of BB9 (pET13:1), (□—□); BB9 (pLHCD6), (■—■).

It is interesting to speculate upon the problems likely to be encountered with the chromosomal integration of *DEX1* and its subsequent expression in brewing yeast. It has previously been indicated that *S. cerevisiae* harbours DNA sequences with extensive homology to the *DEX1* gene; in this respect brewing yeasts are no exception (Meaden, 1986; Vakeria and Hinchliffe, unpublished results). One might consider using the *DEX1* DNA sequence homology to design and implement gene integration protocols for brewing yeasts; however, this may cause recombinational rearrangement in polyploid yeast, resulting in *DEX1* instability. Furthermore, it is not yet clear whether a single- or low-copy integrated *DEX1* gene will be capable of supplying sufficient enzyme to effect a suitable fermentation. Several genes have been described which appear to repress *DEX1* gene expression in *S. cerevisiae*, including *STA10* (Polaina and Wiggs, 1983; Pardo *et al.*, 1986); *INH1* (Yamashita and Fukui, 1984); *MAT* (Yamashita and Fukui, 1983; Yamashita *et al.*, 1985c; Pretorius *et al.*, 1986b); and *CDX* (Searle 1982; Tubb, 1983).

Although the expression of the cloned *DEX1* gene on a multicopy plasmid does not appear to be repressed in *S. cerevisiae* (Meaden *et al.*, 1985), this does not preclude the possibility of repression when the gene is present at a reduced copy number. Moreover, it should be noted that the *DEX1* gene described by Meaden *et al.* (1985) lacks a complete 5' non-translated region and may not be subject to the same mechanisms of gene regulation in *S. cerevisiae* as its chromosomal homologue in *S. diastaticus*. Thus a further requirement with regard to the construction of a stable *DEX1* strain by chromosomal gene integration may be the provision of appropriate gene expression signals, in particular a *S. cerevisiae* gene promoter.

9.4.2 Glucanolytic brewing yeast

Another area of major interest to the brewer is the development of β-glucanolytic yeast. β-glucan is the major structural polysaccharide, comprising the cell walls surrounding the starchy endosperm of barley. This glucose polymer consists primarily of $1 \rightarrow 3$ and $1 \rightarrow 4$ β-glycosidic linkages and is broken down during malting by the combined action of β-glucan solubilase and an endo-1, 3-1, 4-β-glucanase (barley β-glucanase) (Bamforth, 1982). The latter enzyme is heat-labile and is destroyed during malt kilning; consequently, significant quantities of high-molecular-weight β-glucan can survive into wort and beer to impede the filtration rate of brewery-conditioned beer and cause the formation of gels and hazes in the finished product (Bamforth, 1982). Some of these problems can be alleviated by the addition of exogenous β-glucanases during fermentation. The provision of a yeast capable of secreting 1, 3-1, 4-β-glucanase during fermentation, would circumvent the requirement for enzyme addition and result in beers which are more efficiently filtered.

An endo-1, 3-1, 4-β-glucanase derived from the bacterium *Bacillus subtilis* has been shown to have a marked improvement upon beer filtration rate (Hinchliffe, 1985; Hinchliffe and Box, 1985). The gene encoding the synthesis of this extracellular enzyme in *B. subtilis* has been cloned in *E. coli* (Cantwell and McConnell, 1983; Hinchliffe, 1984) and fully characterized by DNA sequencing (Murphy *et al.*, 1984). The structural gene encodes a protein which is approximately 25.5-kD in molecular mass and capable of the endolytic hydrolysis of $1 \rightarrow 4$ β-glycosidic linkages adjacent to $1 \rightarrow 3$. This gene is fortuitously expressed at an extremely low level in laboratory haploid strains of both *S. cerevisiae* (Hinchliffe and Box, 1984) and brewing yeast (Hinchliffe and Box, 1985); in both cases active enzyme was detected in intracellular cell free extracts. High-level expression of the β-glucanase gene in yeast has been achieved by fusion to the alcohol dehydrogenase (*ADC1*) promoter; such constructs, when introduced into both brewing ale and lager yeast on a 2-μm *CUP1* vector, produce significant levels of biologically active intracellular enzyme (Cantwell *et al.*, 1985). When these yeasts were grown in brewers' wort, β-glucan degradation was observed, suggesting that the enzyme was either secreted from the yeast or liberated as a result of autolysis (Cantwell *et al.*,

1987). The inclusion of the full length β-glucanase signal sequence, in gene fusions to the *ADC1* promoter, further increased the level of extracellular enzyme in cultures of a laboratory haploid yeast, suggesting that the *Bacillus* signal sequence is effective at mediating secretion of the enzyme in yeast (Cantwell *et al.*, 1987). The structural gene for β-glucanase has also been fused to the α-factor pheromone (MF-α-1) promoter and signal sequence, resulting in the successful secretion of the enzyme from both a haploid laboratory strain, DBY746 (Cantwell *et al.*, 1987) and an α-mating type maltose utilizing strain of yeast, BRG6003 (Lancashire and Wilde, 1987). Both of these strains are capable of hydrolysing barley β-glucan; the latter can also reduce beer viscosity and enhance filtration rate (Lancashire and Wilde, 1987). However, it is not known whether any of these gene fusion constructs, when incorporated in a commercial brewing yeast host, are capable of producing sufficient enzyme to efficiently degrade the β-glucan present in wort and beer. It has been suggested that this β-glucanase may not be ideal for synthesis by yeast for β-glucan degradation during fermentation (Hinchliffe, 1985). This is because the enzyme has maximum activity at pH 6.2, whereas at wort pH (typically 5.2) it is only 20% efficient, and at beer pH (typically 4.0) there is essentially no activity at all (Hinchliffe, 1985). This pH-dependent activity does not affect the ability of the *B. subtilis* enzyme to improve beer filtration, when added at the beginning of fermentation (Hinchliffe, 1985); however, the effect of declining pH on the activity of an enzyme released gradually into wort, concomitant with yeast growth during fermentation, may be critical in determining the efficiency of β-glucanolysis (Hinchliffe, 1985).

Alternative β-glucanases have been investigated which are capable of hydrolysing barley β-glucan in the acid milieu of a beer fermentation. The most promising is the major endoglucanase (EGI) produced by the filamentous fungus *Trichoderma reesei*. This β-glucanase has a pH optimum between 4 and 5. The gene, *egl1*, encoding the enzyme has been cloned and characterized (Penttilä *et al.*, 1986; Arsdell *et al.*, 1987). The full-length *egl1* cDNA has been expressed from the glycolytic phosphoglycerate kinase (*PGK*) gene promoter, and it has been shown that the natural *egl1* signal sequence is capable of mediating secretion of active enzyme in a laboratory strain of *S. cerevisiae* (Knowles *et al.*, 1985; Penttilä *et al.*, 1987a). The PGK–*egl1* expression constructs have also been introduced into brewing yeast strains (Penttilä *et al.*, 1987b). The gene is successfully expressed and secreted in brewing yeast when introduced on either a 2-μm plasmid carrying the *CUP1* selectable marker or integrated into the chromosome. The strains harbouring the integrated copy of the gene produced approximately 10-fold less β-glucanase than those harbouring a mulitcopy plasmid; however, both strains were capable of hydrolysing the β-glucan present in wort. More importantly, both strains produced beers with a reduced β-glucan content and an improved filtration rate; these beers were indistinguishable by organoleptic analysis from the control beers (Enari *et al.*, 1987). It may yet be desirable to reconstruct the integrants described by Pentillä *et al.* (1987*a*) so that they do not possess exogenous bacterial DNA sequences. Such strains would then be appropriate for use in a commercial process for improving the filtration performance of beers.

A further genetic route to β-glucanolytic yeast has been described by Jackson *et al.* (1986). These authors have expressed a full-length cDNA clone of the barley endo-1, 3-1, 4-β-glucanase in a laboratory strain of yeast, and succeeded in secreting active enzyme by fusion to the signal peptide coding region of the mouse salivary α-amylase. The secretion of active enzyme has also been reported in brewing yeast, although no information is available concerning the ability of these yeasts to degrade β-glucan during fermentation (von Wettstein, 1987*a*). Interestingly, the availability of the barley β-glucanase gene now facilitates attempts to elevate the thermotolerance of the enzyme by *in-vitro* mutagenesis. Ultimately it would be desirable to reintroduce such a gene into barley so that malt with a more robust β-glucanase activity might be produced. At present, this can not be achieved due to the lack of a successful transformation system for barley (von Wettstein, 1987*a*).

9.4.3 *Enhanced beer quality—diacetyl control*

Although brewing yeasts have been selected for their ability to produce beers with a desirable flavour and aroma, it is still necessary to ensure flavour quality by process manipulations. In this respect the prevention of yeast-derived off-flavours, such as diacetyl, pentane-2, 3-dione, esters and sulphur compounds, is essential. The vicinal diketones (VDKs) diacetyl and pentane-2, 3-dione impart a characteristic taste and aroma to beer, variously described as 'toffee-like' and butterscotch. This flavour characteristic is most easily detected in lager beers, where it is considered undesirable at levels in excess of $0.1-0.14$-μg ml^{-1}.

Diacetyl and pentane-2, 3-dione are formed in wort and beer from the non-enzymic oxidative decarboxylation of yeast-derived α-acetolactate and α-acetohydroxy butyrate, respectively. These products are intermediates of the valine and isoleucine biosynthetic pathways, respectively, and develop in beer as a result of leakage from yeast cells during fermentation. Diacetyl and pentane-2, 3-dione once formed in the fermenting wort are converted to acetoin and pentane-2, 3-diol respectively, by the action of yeast reductases. Thus the amount of VDK in beer is influenced by a combination of factors which include: (i) the quantity of α-aceto hydroxyacids in the fermenting wort, (ii) oxidative decarboxylation of α-aceto hydroxyacids to VDKs, and (iii) enzymic reduction of VDKs (Stewart and Russell, 1985). These reactions are themselves affected by raw materials, conditions of wort production, fermentation conditions, and the choice of yeast strain (Hough, 1985). The most common practice to minimize the adverse effect of VDKs on flavour is to store beer for extended periods of time at low temperature in the presence of small quantities of yeast. This process, traditionally referred to as 'lagering', is both

PRODUCTION OF DIACETYL AND PENTANEDIONE IN BEER

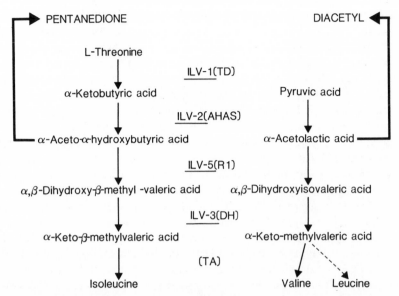

Figure 9.2 Formation of diacetyl and pentanedione. The biosynthetic pathway to isoleucine and valine. Abbreviations for enzymes: TD, threonine deaminase; AHAS, acetohydroxyacid synthetase; RI, reductoisomerase; DH, dehydrase; TA, transaminase.

time-consuming and expensive in terms of capital equipment utilization and running costs. A better control over VDK production or its removal will undoubtedly afford significant quality and process cost advantages. Consequently, several laboratories have explored genetic strategies, either to reduce the propensity of yeast to produce diacetyl, or to enhance the ability of yeast to remove it.

The biosynthetic pathway to isoleucine and valine synthesis and the formation of VDKs is indicated in Figure 9.2. One would predict from this pathway that the construction of mutants in the *ILV2* gene would result in yeast unable to synthesize α-acetohydroxy acids and therefore incapable of producing diacetyl and pentane-2, 3-dione. Ramos-Jeunehomme and Masschelein (1977) demonstrated that this was indeed the case; a recessive *ilv2* mutant of a laboratory haploid yeast produced neither diacetyl nor pentane-2, 3-dione in trial fermentations. Initial clones produced excessive amounts of propanol which was subsequently avoided by the use of an *ilv2, ilv1* double mutant (Ramos-Jeunehomme and Masschelein, 1977). Unfortunately, however, the acquisition of a double recessive mutation in a polyploid brewing yeast is essentially impossible. Kielland-Brandt *et al.* (1979) were able to obtain mutants of lager yeast with dominant alleles of the *ILV1* gene, by selecting for resistance to the amino acid analogue thiaisoleucine; however, the brewing properties of these yeast were not reported. More recently, dominant mutations of the *ILV2* gene have been obtained in laboratory haploid yeast (Falco and Dumas, 1985) and production brewing yeast (Gjermansen and Sigsgaard, 1987; Galván *et al.*, 1987) by selecting for spontaneous resistance to the herbicide sulphometuron methyl. Galván *et al.* (1987) showed that the α-acetohydroxy acid synthases associated with their sulphometuron methyl resistant yeast had an altered specific activity, but this did not result in a reduction of diacetyl levels in trial fermentations. In contrast, a group of sulphometuron methyl-resistant lager yeasts have been shown to produce reduced levels of diacetyl, although this reduction may not be sufficient to warrant their commercial implementation (Gjermansen and Sigsgaard, 1987).

The genes *ILV1*, *ILV2*, *ILV3* and *ILV5*, encoding enzymes involved in the biosynthesis of isoleucine and valine (Figure 9.2), have been isolated and DNA sequenced (Kielland-Brandt *et al.*, 1984; Polaina, 1984; Falco *et al.*, 1985; Casey *et al.*, 1986; Petersen *et al.*, 1986). The synthesis of these enzymes is believed to be regulated by the presence of isoleucine, valine and leucine in the growth medium, and also by general amino acid control (von Wettstein, 1987*b*). Clearly the availability of the cloned *ILV1* (threonine deaminase) and *ILV2* (α-acetohydroxy acid synthase) genes affords the potential for the directed mutagenesis of the corresponding chromosomal homologues in brewing yeast. However, the practical construction of an *ilv1*, *ilv2* mutant of a polyploid strain of brewing yeast may not be straightforward. It has been shown that at least two different alleles of both *ILV1* and *ILV2* exist in the Carlsberg lager yeast (Petersen *et al.*, 1983; von Wettstein, 1987*b*). Thus it may be necessary to clone all the relevant biosynthetic genes encoding the different isozymes, and use integrative transformation to disrupt their respective coding regions (Petersen, 1985).

An alternative to preventing the formation of the α-acetohydroxy acids, precursors of VDK production, is to enhance their removal. In this respect two very different approaches have been developed. The first of these involves increasing the levels of intracellular acetohydroxy acid reductoisomerase and acetohydroxy acid dehydratase, the products of the *ILV5* and *ILV3* genes respectively (Figure 9.2), with a view to increasing the anabolic flux of acetohydroxy acid intermediates into isoleucine and valine. Amplification of the *ILV5* gene by transformation on a multicopy 2-μm plasmid resulted in a 50–60% reduction in maximal VDK levels compared with the control yeast (Dillemans *et al.*, 1987). Overexpression of *ILV3* using a similar approach resulted in only a 10% reduction in maximal VDK levels (Goossens *et al.*, 1987). In both cases, amplification experiments were performed in a laboratory yeast strain, so it is not known whether a similar strategy would be effective in a production brewing yeast.

The second approach to enhancing the removal of VDK precursors has recently been described by Sone *et al.* (1987). The enzyme α-acetolactate decarboxylase (ALDC) converts α-acetolactate to acetoin directly, thereby preventing the formation of diacetyl. Exogenous ALDC can be added to fermenting beer to significantly reduce the level of VDK formation (Godtfredsen and Ottesen, 1982; Godtfredsen et al., 1983a; Godtfredsen et al., 1983b). The gene encoding ALDC has been cloned from *Enterobacter aerogenes* (Sone et al., 1987a) and more recently expressed in a production brewing yeast strain by fusion to the alcohol dehydrogenase gene promoter (ADC1) of *S. cerevisiae* (Sone et al., 1987b). Trial fermentations with a brewing yeast strain harbouring the ALDC producing plasmid showed that the total diacetyl concentration at the end of fermentation was significantly reduced relative to fermentations performed with the untransformed parental yeast. Moreover, there was no change in either fermentation rate or beer flavour compounds between both yeast strains. However, the 2-μm plasmid used in the transformation harbours a bacterial G418 resistance gene for transformant selection. In addition, the plasmid was unstable over successive fermentations, indicating that the large-scale use of the ALDC producing yeast would require a more stable genetic construct.

9.4.4 *Improving the value of spent yeast*

Normal brewing practice results in the production of large quantities of yeast which is surplus to the brewing process. Traditionally, this yeast is sold as a cheap source of biomass to be used in animal or microbial feeds or as yeast extracts for human consumption. The brewer obtains an extremely low price for this potentially valuable commodity. Interest has therefore been stimulated in developing alternative uses for waste yeast which would enhance its intrinsic value. One approach is to genetically manipulate the yeast to produce high-value proteins.

We have recently described a process in which heterologous gene expression can be controlled to regulate the synthesis of a protein of interest in brewing yeast following a beer fermentation (Hinchliffe et al., 1987b). In this respect, a cDNA encoding the human plasma protein, serum albumin (HSA), was fused to a hybrid promoter formed between the upstream activation sequence of the *GAL1–10* genes and the transcription initiation site of the *CYC1* gene. When introduced into brewing yeast, the resultant *CUP1*-bearing plasmid was shown to induce HSA synthesis in the presence of galactose, whereas in the absence of galactose no HSA was produced. This gene regulation system was found to be particularly useful for controlling gene expression in brewing yeast, since induction is very tightly regulated by the presence of galactose in the growth medium. Brewers' wort does not contain galactose; consequently, it was observed that yeasts harbouring the galactose-regulated HSA gene were capable of producing palatable beer, without any detectable synthesis of HSA. Beers produced were analysed and found to be indistinguishable from those produced by the genetically unmodified parental strain. Moreover, following several successive rounds of beer production, yeast could be induced with galactose to synthesize substantial quantities of HSA (Hinchliffe et al., 1987b).

Thus the technology is available to produce beer with a commercial strain of brewing yeast and to subsequently enhance the value of that yeast by the synthesis of a high-value protein.

9.5 Conclusions

Techniques for the genetic manipulation of brewing yeast are now established and are being applied to commercial targets, some of which have been outlined in detail here. It is becoming increasingly apparent that brewing yeasts are more amenable to recombinant DNA technology than 'classical' genetic techniques; consequently, it is likely that we will see an increase in the use of these methods in the future. However, the extent to which the brewing industry embraces new varieties of yeast will depend to a great extent upon the perceived acceptability of the strains derived. In this respect there are two important aspects to consider. Firstly, the yeast must satisfy the commercial constraints imposed—that is, it must be capable of producing the desired product. Secondly, and perhaps more importantly, the yeast must be compatible with the manufacture of a food product. Consequently, customer concerns over the nature of the genetic manipulations undertaken must be taken into account.

Acknowledgements

We wish to thank the Directors of Delta Biotechnology Ltd and Bass PLC for permission to publish this paper.

References

Aigle, M., Erbs, D. and Moll, M. (1984) Some molecular structures in the genome of lager brewing yeasts. *J. Amer. Soc. Brew. Chem.* **47**: 1.

Anderson, E. and Martin, P.A. (1975) The sporulation and mating of brewing yeast. *J. Inst. Brew.* **81**: 242.

Arsdell, J. Nv., Kwok, S., Schweickart, V.L., Ladner, M.B., Gelfand, D.H. and Innis, M.A. (1987) Cloning, characterization and expression in *Saccharomyces cerevisiae* of endoglucanase I from *Trichoderma reesei. Bio/Technology* **5**: 60.

Aschengreen, N.H. (1987) Enzyme technology. *Proc. Eur. Brew. Conv. 21st Congr. Madrid*: 221.

Ashikari, T., Nakamura, N., Tanaka, Y., Kiuchi, N., Shibano, Y., Tanaka, T., Amachi, T. and Yoshizumi, H. (1985) Cloning and expression of the *Rhizopus* glucoamylase gene in yeast. *Agric. Biol. Chem.* **49**: 2521.

Ashikari, T., Nakamura, N., Tanaka, Y., Kiuchi, N., Shibano, Y., Tanaka, T., Amachi, T. and Yoshizumi, H. (1986) *Rhizopus* raw-starch-degrading glucoamylase: its cloning and expression in yeast. *Agric. Biol. Chem.* **50**: 957.

Bamforth, C.W. (1982) Barley β-glucans: their role in malting and brewing. *Brewers' Digest*, June: 22.

Barnett, J.A., Payne, R.W. and Yarrow, D. (1983) *Yeasts—Characteristics and Identification.* Cambridge University Press, London.

Barney, M.C., Jansen, G.P. and Helbert, J.R. (1980) Use of spheroplast fusion and genetic transformation to introduce dextrin utilization into *Saccharomyces uvarum. J. Amer. Soc. Brew. Chem.* **38**: 1.

Beggs, J.D. (1978) Transformation of yeast by a replicating hybrid plasmid. *Nature* **275**: 104.

Bilinski, C.A., Sills, A.M. and Stewart, G.G. (1984) Morphological and genetic effects of benomyl on polyploid brewing yeasts: isolation of auxotrophic mutants. *Appl. Envir. Microbiol.* **48**: 813.

Blackburn, E.H. and Szostak, J.W. (1984) The molecular structure of centromeres and telomeres. *Ann. Rev. Biochem.* **53**: 163.

Boel, E., Hjort, I., Svensson, B., Norris, F., Norris, K.E. and Fiil, N.P. (1984) Glucoamylases G1 and G2 from *Aspergillus niger* are synthesized from two different but closely related mRNAs. *EMBO J.* **3**: 1097.

Bussey, H. and Meaden, P. (1986) Selection and stability of yeast transformants expressing cDNA of an M1 killer toxin-immunity gene. *Curr. Genet.* **9**: 285.

Cantwell, B.A. and McConnell, D.J. (1983) Molecular cloning and expression of a *Bacillus subtilis* β-glucanase gene in *Escherichia coli*. *Gene* **23**: 211.

Cantwell, B.A., Brazil, G., Hurley, J. and McConnell, D.J. (1985) Expression of the gene for endo-β-1,3-1,4-glucanase from *Bacillus subtilis* in *Saccharomyces cerevisiae*. *Proc. Eur. Brew. Conv. 20th Congr. Helsinki*: 259.

Cantwell, B.A., Ryan, T., Hurley, J.C., Doherty, M. and McConnell, D.J. (1987) Degradation of barley β-glucan by brewers' yeast. *Eur. Brew. Conv., Brewers' Yeast Symp., Vuoranta (Helsinki), Monograph XII*: 186.

Casey, G.C. (1986) Cloning and analysis of two alleles of the *ILV3* gene from *Saccharomyces cerevisiae*. *Carlsberg Res. Commun.* **51**: 327.

Cohen, J.D., Eccleshall, T.R., Needleman, R.B., Federhoff, H., Buchferer, B. and Marmur, J. (1980) Functional expression in yeast of the *Escherichia coli* plasmid gene coding for chloramphenicol acetyltransferase. *Proc. Natl. Acad. Sci. USA* **77**: 1078.

Delgado, M. and Conde-Zurita, J.C. (1983) Benomyl as a tool for the parasexual genetic analysis of brewing yeast. *Proc. Eur. Brew. Conv. 19th Congr. London*: 465.

Dillemans, M., Goossens, E., Goffin, O. and Masschelein, C.A. (1987) The amplification effect of the *ILV5* gene on the production of vicinal diketones in *Saccharomyces cerevisiae*. *J. Amer. Soc. Brew. Chem.* **45**:81.

Enari, T.-M., Knowles, J., Lehtinen, U., Nikkola, M., Penttilä, M., Suikko, M.-L., Home, S. and Vilpola, A. (1987) Glucanolytic brewers' yeast. *Proc. Eur. Brew. Conv., 21st Congr., Madrid*: 529.

Erratt, J.A. and Nasim, A. (1986) Allelism within the DEX and STA gene families in *Saccharomyces diastaticus*. *Mol. Gen. Genet.* **202**: 255.

Falco, S.C. and Dumas, K.S. (1985) Genetic analysis of mutants of *Saccharomyces cerevisiae* resistant to the herbicide sulphometuron methyl. *Genetics* **109**: 21.

Falco, S.C., Dumas, K.S. and Livak, K.L. (1985) Nucleotide sequence of the yeast *ILV2* gene which encodes acetolacetate synthase. *Nucl. Acids. Res.* **13**: 4011.

Fogel, S. and Welch, J. (1982) Tandem gene amplification mediates copper resistance in yeast. *Proc. Natl. Acad. Sci. USA* **79**: 5342.

Fogel, S., Welch, J.W., Cathala, G. and Karin, M. (1983) Gene amplification in yeast: *CUP1* copy number regulates copper resistance. *Curr. Genet.* **7**: 1.

Fowell, R.R. (1969) Life cycles in yeast. In *The Yeasts*, ed. Rose, A.H. and Harrison, J.S., Vol. 1, Academic Press, London and New York, 461.

Freeman, R.F. (1981) Construction of brewing yeasts for production of low carbohydrate beers. *Proc. Eur. Brew. Conv. 18th Congr. Copenhagen*: 497.

Galván, L., Peréz, A., Delgado, M. and Conde, J. (1987) Diacetyl production by sulfometuron resistant mutants of brewing yeast. *Proc. Eur. Brew. Conv. 21st Congr. Madrid*: 385.

Gatignol, A., Baron, M. and Tiraby, G. (1987) Phleomycin resistance encoded by the *ble* gene from transposon Tn5 as a dominant selectable marker in *Saccharomyces cerevisiae*. *Mol. Gen. Genet.* **207**: 342.

Gillis-van-Maele, A. and van Landschoot, A. (1985) Fluorochrome monitoring of yeast protoplast fusion. *Proc. Eur. Brew. Conv. 20th Congr. Helsinki*: 227.

Gjermansen, C. (1983) Mutagenesis and genetic transformation of meiotic segregants of lager yeast. *Carlsberg Res. Commun.* **48**: 557.

Gjermansen, C. and Sigsgaard, P. (1981) Construction of a hybrid brewing strain of *Saccharomyces carlsbergensis* by mating of meiotic segregants. *Carlsberg Res. Commun.* **46**: 1.

Gjermansen, C. and Sigsgaard, P. (1987) Cross-breeding with brewers' yeasts and evaluation of strains. *Eur. Brew. Conv. Symp. on Brewers' Yeast, Helsinki, Finland, Monograph XII*: 156.

Godtfredsen, S.E. and Ottesen, M. (1982) Maturation of beer with α-acetolactate decarboxylase. *Carlsberg Res. Commun.* **47**: 93.

Godtfredsen, S.E., Lorck, H. and Sigsgaard, P. (1983a) On the occurence of α-acetolactate decarboxylases among microorganisms. *Carlsberg Res. Commun.* **48**: 239.

Godtfredsen, S.E., Rasmussen, A.N. and Ottesen, M. (1983b) Application of the acetolactate decarboxylase from *Lactobacillus casei* for accelerated maturation of beer. *Carlsberg Res. Commun.* **49**: 69.

Goossens, E., Dillemans, M., Debourg, A. and Masschelein, C.A. (1987) Control of diacetyl formation by the intensification of the anabolic flux of acetohydroxyacid intermediates. *Proc. Eur. Brew. Conv. 21st Congr. Madrid*: 553.

Gritz, L. and Davies, J. (1983) Plasmid-encoded hygromycin B resistance: the sequence of hygromycin B phosphotransferase gene and its expression in *Escherichia coli* and *Saccharomyces cerevisiae*. *Gene* **25**: 179.

Gunge, N. and Nakatomi, Y. (1972) Genetic mechanisms of rare-matings of the yeast *Saccharomyces cerevisiae* heterozygous for mating type. *Genetics* **70**: 41.

Hadfield, C., Cashmore, A.M. and Meacock, P.A. (1986) An efficient chloramphenicol-resistance marker for *Saccharomyces cerevisiae* and *Escherichia coli*. *Gene* **45**: 149.

Hammond, J.R.M. and Eckersley, K.W. (1984) Fermentation properties of brewing yeast with killer character. *J. Inst. Brew.* **90**: 167.

Henderson, R.C.A., Cox, B.S. and Tubb, R.S. (1985) The transformation of brewing yeast with a plasmid containing the gene for copper resistance. *Curr. Genet.* **9**: 133.

Hinchliffe, E. (1984) Cloning and expression of a *Bacillus subtilis* endo-1,3-1,4-β-glucanase gene in *Escherichia coli* K12. *J. Gen. Microbiol.* **130**: 1285.

Hinchliffe, E. (1985) β-Glucanase: the successful application of genetic engineering. *J. Inst. Brew.* **91**: 384.

Hinchliffe, E. (1986) Enzymes by genetic manipulation. *The Brewer* July: 256.

Hinchliffe, E. and Box, W.G. (1984) Expression of the cloned endo-1,3-1,4-β-glucanase gene of *Bacillus subtilis* in *Saccharomyces cerevisiae*. *Curr. Genet.* **81**: 471.

Hinchliffe, E. and Box, W.G. (1985) Beer enzymes and genes: the application of a concerted approach to β-glucan degradation. *Proc. Eur. Brew. Conv. 20th Congr. Helsinki*: 267.

Hinchliffe, E. and Daubney, C.J. (1986) The genetic modification of brewing yeast with recombinant DNA. *J. Amer. Soc. Brew. Chem.* **44**: 98.

Hinchliffe, E., Fleming, C.J. and Vakeria, D. (1987a) A novel stable gene maintenance system for brewing yeast. *Proc. Eur. Brew. Conv. 21st Congr. Madrid*: 505.

Hinchliffe, E., Kenny, E. and Leaker, A.J. (1987b) Novel products from surplus yeast via recombinant DNA technology. *Eur. Brew. Conv. Brewers' Yeast Symp. Vuoranta (Helsinki). Monograph XII*: 139.

Hinnen, A., Hicks, J.B. and Fink, G.R. (1978) Transformation of yeast. *Proc. Natl. Acad. Sci. USA* **75**: 1929.

Hockney, R.C. and Freeman, R.F. (1981) Gratuitous catabolite repression by glucosamine of maltose utilization in *Saccharomyces cerevisiae*. *J. Gen. Microbiol.* **121**: 479.

Hollenberg, C.P. (1979) The expression in *Saccharomyces cerevisiae* of bacterial β-lactamase and other antibiotic resistance integrated in a 2 μm DNA vector. In *Extrachromosomal DNA*, eds. Cummings, D.J., Borst, P., David, I., Weissman, S. and Fox, C.F. *ICN-UCLA Symp. Vol. 15*, Academic Press, New York, 325.

Hough, J.S. (1985) *The Biotechnology of Malting and Brewing*. Cambridge University Press, London and New York.

Innis, M., Holland, M.J., McCabe, P.C., Cole, G.E., Wittman, V.P., Tal, R., Watt, K.W.K., Gelfand, D.H., Holland, J.P. and Meade, J.H. (1985) Expression, glycosylation, and secretion of an *Aspergillus* glucoamylase by *Saccharomyces cerevisiae*. *Science* **228**: 21.

Ito, H., Fukuda, Y., Murata, K. and Kimura, A. (1983) Transformation of intact yeast cells treated with alkali cations. *J. Bacteriol.* **153**: 163.

Jackson, E.A., Ballance, G.M. and Thomsen, K.K. (1986) Construction of a yeast vector directing the synthesis and release of barley (1 → 3, 1 → 4)-β-glucanase. *Carlsberg Res. Commun.* **51**: 445.

Johnston, J.R. (1965) Breeding yeasts for brewing. *J. Inst. Brew.* **71**: 130.

Karin, M., Najaran, R., Haslinger, A., Valenzuela, P., Welch, J.W. and Fogel, S. (1984) Primary structure and transcription of an amplified genetic locus: the *CUP1* locus of yeast. *Proc. Nat. Acad. Sci. USA* **81**: 337.

Kaster, K.R., Burgett, S.G. and Ingolia, T.D. (1984) Hygromycin B resistance as dominant selectable marker in yeast. *Curr. Genet.* **8**: 353.

Kielland-Brandt, M.C., Gjermansen, C., Nilsson-Tilgren, T., Petersen, J.G.L., Holmberg, S. and

Sigsgaard, P. (1981) Genetics of a lager production yeast. *MBAA Techn. Quar.* **18**: 185.

Kielland-Brandt, M.C., Holmberg, S. and Petersen, J.G.L. (1984) Nucleotide sequence of the gene for threonine deaminase (*ILV1*) of *Saccharomyces cerevisiae*. *Carlsberg Res. Commun.* **49**: 567.

Knowles, J.K.C., Penttilä, M., Teeri, T.T., Andre, L., Salvuori, I. and Lehtovaara, P. (1985) Transfer of genes coding for fungal glucanases into yeast. *Proc. Eur. Brew. Conv. 20th Congr. Helsinki*: 251.

Knowles, J.K.C. and Tubb, R.S. (1987) Recombinant DNA: gene transfer and expression techniques with industrial yeast strains. *Eur. Brew. Conv., Brewers' Yeast Symp., Vuoranta (Helsinki), Monograph XII*: 169.

Kreger-van-Rij, N.J.W. (1984) *The yeasts, a Taxonomic Study*. 3rd edn., Elsevier, Amsterdam.

Lancashire, W.E. and Wilde, R.J. (1987) Secretion of foreign proteins by brewing yeasts. *Proc. Eur. Brew. Conv. 21st Congr. Madrid*: 513.

Mead, D.J., Gardner, D.C.J. and Oliver, S.G. (1986) Enhanced stability of a 2μ-based recombinant plasmid in diploid yeast. *Biotech. Lett.* **8**: 391.

Meaden, P.G. (1986) Cloning and characterisation of a gene (*DEX1*) conferring extracellular amyloglucosidase production on brewing and other strains of *Saccharomyces cerevisiae*. PhD thesis, University of London, London.

Meaden, P.G., Ogden, K., Bussey, H. and Tubb, R. (1985) A *DEX* gene conferring production of an extracellular amyloglucosidase on yeast. *Gene* **34**: 325.

Meaden, P.G. and Tubb, R.S. (1985) A plasmid vector system for the genetic manipulation of brewing strains. *Proc. Eur. Brew. Conv. 20th Congr. Helsinki*: 219.

Modena, D., Vanoni, M., Englard, S. and Marmur, J. (1986) Biochemical and immunological characterization of the *STA2*—encoded extracellular glucoamylase from *Saccharomyces diastaticus*. *Arch. Biochem. Biophys.* **248**: 138.

Molzahn, S.W. (1977) A new approach to the application of genetics to brewing yeast. *J. Amer. Soc. Brew. Chem.* **35**: 54.

Molzahn, S.W. (1985) Alternative techniques of fermentation and maturation—their relative merits. *Proc. Eur. Brew. Conv. 20th Congr. Helsinki*: 159.

Murphy, N., McConnell, D.J. and Cantwell, B.A. (1984) The DNA sequence of the gene and genetic control sites for the excreted *B. subtilis* enzyme β-glucanase. *Nucleic Acids Res.* **12**: 5355.

Orr-Weaver, T.L., Szostak, J.W. and Rothstein, R.J. (1983) Genetic applications of yeast transformation with linear and gapped plasmids. In *Methods in Enzymology*, eds. Wu, R., Grossman, L. and Moldave, K. **101**, Academic Press, New York and London, 228.

Nunberg, J.H., Meade, J.H., Cole, G., Lawyer, F.C., McCabe, P., Schweickart, V., Tal, R., Wittman, V.P., Flatgaard, J.E. and Innis, M.A. (1984) Molecular cloning and characterization of the glucoamylase gene of *Aspergillus awamori*. *Mol. Cell. Biol.* **4**: 2306.

Panchal, C.J., Russell, I., Sills, A.M. and Stewart, G.G. (1984) Genetic manipulation of brewing and related yeast strains, *Food Technol.*, February: 99.

Pardo, J.M., Polaina, J. and Jimenez, J. (1986) Cloning of the *STA2* and *SGA* genes encoding glucoamylases in yeasts and regulation of their expression by the *STA10* gene in *Saccharomyces cerevisiae*. *Nucleic Acids Res.* **14**: 4701.

Penttilä, M., Lehtovaara, P., Nevalainen, H., Bhikhabhai, R. and Knowles, J. (1986) Homology between cellulase genes of *Trichoderma reesei*: complete nucleotide sequence of the endoglucanase I gene. *Gene* **45**: 253.

Penttilä, M., Andre, L., Saloheimo, M., Lehtovaara, P. and Knowles, J.K.C. (1987a) Expression of two *Trichoderma reesei* endoglucanases in the yeast *Saccharomyces cerevisiae*. *Yeast* **3**: 175.

Penttilä, M., Suikko, M.-L., Lehtinen, U., Nikkola, M. and Knowles, J.K.C. (1987b) Construction of brewers' yeasts secreting fungal endo-β-glucanase. *Curr. Genet.* **12**: 413.

Petersen, J.G.L., Holmberg, S., Nilsson-Tillgren, T. and Kielland-Brandt, M.C. (1983) Molecular cloning and characterization of the threonine deaminase (*ILV1*) gene of *Saccharomyces cerevisiae*. *Carlsberg Res. Commun.* **48**: 149.

Petersen, J.G.L. (1985) Molecular genetics of diacetyl formation in brewers' yeast. *Proc. Eur. Brew. Conv. 20th Congr. Helsinki*: 275.

Petersen, J.G.L. and Holmberg, S. (1986) The *ILV5* gene of *Saccharomyces cerevisiae* is highly expressed. *Nucleic Acids Res.* **14**: 9631.

Polaina, J. and Wiggs, M.Y. (1983) *STA10*: a gene involved in the control of starch utilization by *Saccharomyces*. *Curr. Genet.* **7**: 109.

Polaina, J. (1984) Cloning of the *ILV2*, *ILV3* and *ILV5* genes of *Saccharomyces cerevisiae*. *Carlsberg Res. Commun.* **49**: 577.

Pretorius, I.S., Chow, T., Modena, D. and Marmur, J. (1986a) Molecular cloning and characterization of the *STA2* glucoamylase gene of *Saccharomyces diastaticus*. *Mol. Gen. Genet.* **203**: 29.

Pretorius, I.S., Modena, D., Vanoni, M., England, S. and Marmur. J. (1986b) Transcriptional control of glucoamylase synthesis in vegetatively growing and sporulating *Saccharomyces* species. *Mol. Cell. Biol.* **6**: 3034.

Rainbow, C. (1970) Brewers' yeast. In *The Yeasts*, eds. Rose, A.H. and Harrison, J.S., Vol. 3, Academic Press, London and New York, 147.

Ramos-Jeunehomme, C.L. and Masschelein, C.A. (1977) Controle génétique de la formation des dicetones vicinales chez *Saccharomyces cerevisiae*. *Proc. Eur. Brew. Conv. 16th Congr. Amsterdam*: 267.

Rine, J., Hansen, W., Hardeman, E. and Davies, R.W. (1983) Targeted selection of recombinant clones through dosage effects. *Proc. Natl. Acad. Sci. USA* **80**: 6750.

Russell, I. and Stewart, G.G. (1985) Valuable techniques in the genetic manipulation of industrial yeast strains. *J. Amer. Soc. Brew. Chem.* **43**: 84.

Searle, B.A. (1982) Utilization of carbohydrates by brewing and amylolytic strains of *Saccharomyces*. PhD.thesis, University of London, London.

Sone, H., Fujui, T., Kondo, K., Tanaka, J. (1987a) Molecular cloning of the gene encoding α-acetolactate decarboxylase from *Enterobacter aerogenes*. *J. Biotechnol.* **5**: 87.

Sone, H., Kondo, K., Fujui, T., Shimuzu, F., Tanaka, J. and Inoue, T. (1987b) Fermentation properties of brewers' yeast having α-acetolactate decarboxylase gene. *Proc. Eur. Brew. Conv. 21st Congr. Madrid*: 545.

Spencer, J.F.T. and Spencer, D.M. (1977) Hybridization of non-sporulating and weakly sporulating strains of brewers' and distillers' yeasts. *J. Inst. Brew.* **83**: 287.

Spencer, J.F.T. and Spencer, D.M. (1980) The use of mitochondrial mutants in the isolation of hybrids involving industrial yeast strains. *Mol. Gen. Genet.* **177**: 355.

Stewart, G.G., Jones, R. and Russell, I. (1985) The use of derepressed yeast mutants in the fermentation of brewery wort. *Proc. Eur. Brew. Conv. 20th Congr. Helsinki*: 243.

Stewart, G.G. and Russell, I. (1985) Modern brewing technology. In *Comprehensive Biotechnology*, eds. Blanch, H.W., Drew, S. and Wang, D.I.C., Vol. 3, Pergamon, Oxford, 375.

Tubb, R.S. (1980) 2 μm DNA plasmid in brewery yeasts. *J. Inst. Brewing* **86**: 78.

Tubb, R.S., Searle, B.A., Goodey, A.R. and Brown, A.J.P. (1981) Rare mating and transformation for construction of novel brewing yeasts. *Proc. Eur. Brew. Conv. 18th Congr. Copenhagen*: 487.

Tubb, R.S. (1983) Genetic pathways to superattenuating (amylolytic) yeasts. In *Current Developments in Malting, Brewing and Distilling*, eds. Priest, F.G. and Campbell, I., Institute of Brewing, London, 67.

Tubb, R.S. (1984) Genetic development of yeast strains, *Brew. Guard.*, September: 34.

Tubb, R.S. (1986) Amylolytic yeasts for commercial applications. *Trends Biotechnol.* **4**: 98.

Tubb, R.S. (1987) Gene technology for industrial yeasts. *J. Inst. Brew.* **93**: 91.

Tubb, R.S., Liljestrom, P.L., Torkkeli, T. and Korhola, M. (1986) Melibiase (*MEL*) genes in brewing and distilling yeasts. In *Proc. Institute of Brewing, Aviemore Symposium*. eds. Priest, F.G. and Campbell, I. 298.

Tubb, R.S. and Hammond, J.R.M. (1987) Yeast genetics. In *Brewing Microbiology*, eds. Priest, F.G. and Campbell, I., Elsevier Applied Science, London and New York, 47.

Webster, T.D. and Dickson R.C. (1983) Direct selection of *Saccharomyces cerevisiae* resistant to the antibiotic G418 following transformation with a DNA vector carrying the kanamycin-resistance gene of Tn903. *Gene.* **26**: 243.

Welch, J.W., Fogel, S., Cathala, G. and Karin, M. (1983) Industrial yeasts display tandem gene iteration at the *CUP1* region. *Mol. Cell. Biol.* **3**: 1353.

von Wettstein, D. (1987a) Genetic engineering: barley. *Proc. Eur. Brew. Conv. 21st Congr. Madrid*: 189.

von Wettstein, D. (1987b) Molecular genetics in the improvements of brewers' and distillers' yeast. *Antonie van Leeuwenhoek J. Microbiol.* **53**: 299.

Williamson, D.H. (1985) The yeast ARS element, six years on; a progress report. *Yeasts* **1**: 1.

Yamashita, I. and Fukui, S. (1983) Mating signals control expression of both starch fermentation genes and a novel flocculation gene *FLO8* in the yeast *Saccharomyces*. *Agric. Biol. Chem.* **47**: 2889.

Yamashita, I. and Fukui, S. (1984) Genetic background of glucoamylase production in the yeast *Saccharomyces*. *Agric. Biol. Chem.* **48**: 137.

Yamashita, I., Suzuki, K. and Fukui, S. (1985a) Nucleotide sequence of the extracellular glucoamylase gene *STA1* in the yeast *Saccharomyces diastaticus*. *J. Bacteriol.* **161**: 567.

Yamashita, I., Maemura, T., Hatano, T. and Fukui, S. (1985b) Polymorphic extracellular glucoamylase genes and their evolutionary origin in the yeast *Saccharomyces diastaticus*. *J. Bacteriol.* **161**: 574.

Yamashita, I., Takano, Y. and Fukui, S. (1985c) Control of *STA1* gene expression by the mating-type locus in yeasts. *J. Bacteriol.* **164**: 769.

Yamashita, I., Nakamura, M. and Fukui, S. (1987) Gene fusion is a possible mechanism underlying the evolution of *STA1*. *J. Bacteriol.* **169**: 2142.

Yocum, R.R. (1986) Genetic engineering of industrial yeasts. *Proc. Bio. Expo. 86*, Butterworth, Stoneham, MA, 171.

Yoshizumi, H. and Ashikari, T. (1987) Expression, glycosylation and secretion of fungal hydrolases in yeast. *Trends Biotechnol.* **5**: 277.

Young, T.W. (1983) Brewing yeasts with anti-contaminant properties. *Proc. Eur. Brew. Conv. 19th Congr. London*: 129.

Zhu, J., Contreras, R., Gheysen, D., Ernst, J. and Fiers, W. (1985) A system for dominant transformation and plasmid amplification in *Saccharomyces cerevisiae*. *Bio/Technol.* **3**: 451.

Zhu, X.-L., Ward, C. and Weissbach, A. (1984) Control of *Herpes simplex* thymidine kinase gene expression in *Saccharomyces cerevisiae* by a yeast promoter sequence. *Mol. Gen. Genet.* **194**: 31.

10 Genetic manipulation of methylotrophic yeasts

P.E. SUDBERY and M.A.G. GLEESON

10.1 Introduction

Methylotrophic yeasts have the ability to utilize methanol as sole carbon source and methylamine as sole nitrogen source. They were first reported by Ogata et al. (1969). A relatively small number of yeasts has been identified with this ability. Taxonomically, they are closely related, and this is reflected in the consistent uniformity of the pathway between various yeasts. These pathways and their physiological regulation have been intensively studied and are now reasonably well understood. A key enzyme of the pathway, alcohol oxidase (AOX), can represent 30% of the total cell protein under optimal conditions. This enzyme is sequestered in microbodies called peroxisomes and is subject to strict regulation, being repressed by glucose and ethanol, and induced by methanol.

Interest in the industrial applications of these yeasts began with their use as as a source of single-cell protein (SCP) using methanol as a cheap carbon source. More recently, the potential of the AOX promoter to drive heterologous gene expression has led to its use as a second-generation yeast cloning and expression system. The development of genetic analysis has been stimulated by biotechnological applications, as well as by fundamental interest in the mechanisms operating to regulate synthesis of the enzymes of methanol metabolism. Considerable progress has been made in mutant isolation, complementation and recombination, genetic dissection of the control systems, and in the development of molecular transformation and associated technology. The purpose of this review is to chart progress in these areas. The biochemistry and physiology of methanol utilization have been fully reviewed by Veenhuis (1983). We have also reviewed more general aspects of the biology of methylotrophic yeast (Gleeson and Sudbery, 1988). and the reader is also referred to Sibirny et al. (1988) for a detailed review of genetic studies concerned with methanol utilization and its regulation.

10.2 Methanol metabolism

In methylotrophic yeasts, methanol is oxidized to carbon dioxide via a linear pathway (Figure 10.1). The primary oxidation step occurs within crystalline, membrane-bound, methanol-induced microbodies (peroxisomes) via alcohol oxidase (AOX), also known as methanol oxidase (MOX), to produce hydrogen peroxide and formaldehyde (Fujii and Tonomura, 1975; Kato *et al.*, 1976; Sahm and Wagner, 1973*a*; Tani *et al.*, 1972*a, b*; Van Dijken, 1975; Waites and Quayle, 1983). The hydrogen peroxide is subsequently reduced to water and molecular oxygen by catalase, which is found diffused within the peroxisomal matrix. Formaldehyde produced by methanol oxidation stands at the branch-point between the dissimilation and assimilation pathways of methanol metabolism. Dissimilation to carbon dioxide via formate proceeds within the cytoplasm by the action of formaldehyde dehydrogenase and formate dehydrogenase to generate energy in the form of $NADH_2$ (Fujii and Tonomura, 1975; Kato *et al.*, 1972; Kato *et al.*, 1980; Neben *et al.*, 1980; Sahm and Wagner, 1973*b*; Schutte *et al.*, 1976; Van Dijken *et al.*, 1976).

To generate biomass, formaldehyde enters the assimilatory pathway via the xylulose 5-phosphate cycle. The first critical step in this pathway is the reaction between formaldehyde (one carbon) and the xylulose 5-phosphate (five carbons), to give two three-carbon compounds, glyceraldehyde 3-phosphate and dihydroxyacetone. This reaction occurs within the peroxisome via a methanol-induced transketolase, dihydroxyacetone synthase (DHAS) (Douma *et al.*, 1985; Goodman, 1985; Waites and Quayle, 1983). By a series of rearrangement reactions, xylulose 5-phosphate is subsequently regenerated to form a cyclic pathway; one-third of the glyceraldehyde 3-phosphate molecules generated by the action of DHAS enter the central metabolism to become biomass (Babel and Hofmann, 1982).

Alcohol oxidase (AOX) oxidizes methanol catalytically to form formaldehyde, using oxygen as the electron acceptor and producing hydrogen peroxide as a by-product. No gain in energy is made to the cell by this reaction. AOX was first-purified and described in *Kloeckera* sp. no. 2201 (Tani *et al.*, 1982*a, b*) and later in *Hansenula polymorpha* (Van Dijken, 1976), *Candida boidinii* (Sahm and Wagner, 1973*a*), and *Candida* 25A (Yamada *et al.*, 1979). It is induced by growth on methanol and has a molecular mass of approximately 600 000. It is made up of eight subunits which form an octomer functional unit within a crystalline lattice (Kato *et al.*, 1976). Associated with the octomer aggregate are eight non-covalently bound FAD prosthetic groups.

In addition to methanol, AOX can use short-chain primary alcohols as a substrate, with reduced efficiency (Tani *et al.*, 1972*b*; Sahm and Wagner, 1973*b*; Van Dijken *et al.*, 1975; Van Dijken, 1976). Fujii and Tonomura (1975) showed that AOX is also capable of oxidizing formaldehyde. However, this reaction is thought to be of little consequence, since it was found to be strongly inhibited by hydrogen peroxide.

10.2.1 *Regulation of alcohol oxidase (AOX)*
The enzymes of the methanol pathway are regulated so that the organism may respond to the environment. They are subject to repression by glucose and ethanol and induction by methanol, but on alternative carbon sources, such as sorbitol, they are derepressed. Peroxisomal enzymes are also regulated by a

L

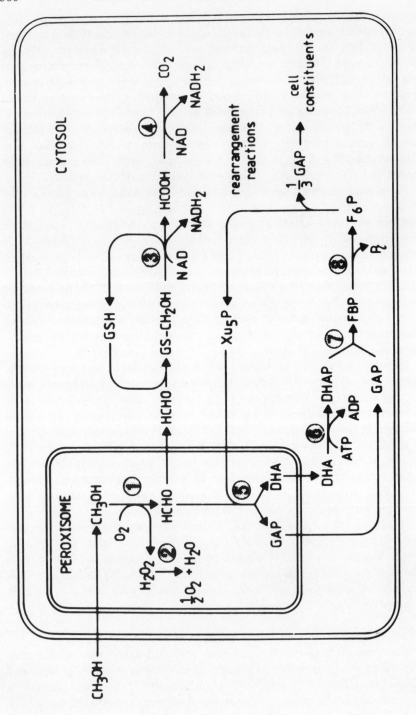

process known as carbon catabolite inactivation (Holtzer, 1976); that is, upon addition of ethanol or glucose to methanol-grown cells peroxisomal enzymes are very rapidly inactivated. The peroxisomes of such cells are subsequently degraded by a process of sequestration and proteolysis (Borman and Sahm, 1978; Veenhuis, 1983).

In *H. polymorpha* and *Candida boidinii*, maximum activity of AOX and catalase is observed during growth on methanol (Eggeling and Sahm, 1978, 1980*a*). Induction of AOX and catalase also occurs in the presence of formaldehyde and formate, although these compounds cannot support growth. In *H. polymorpha*, but not *C. boidinii*, significant amounts of AOX and catalase are observed during growth on non-C_1 compounds, such as glycerol, ribose, xylose or sorbitol, while in *Pichia pastoris* the AOX promoter is not active during growth on glycerol (Tschopp *et al.*, 1987*a*). In *H. polymorpha*, AOX activity was observed during growth in a glucose-limited chemostat (Egli *et al.*, 1980; Egli *et al.*, 1982). The maximum levels obtained were 10–20% of the levels formed in cells grown on a methanol/glucose mixture (C_1:$C_6 = 62.2\%$:38.8% w/w). Parallel investigations with *Kloeckera sp.* no. 2201 (*C. boidinii*) showed that AOX formation could also occur in the absence of methanol, but only at lower dilution rates. Estimates of the concentration of residual glucose showed that at the same dilution rate the residual glucose concentration is higher for *Kloeckera sp.* no. 2201 than for *H. polymorpha*, which presumably explains the requirement for a lower dilution rate. Thus, in *H. polymorpha* and, to a lesser extent, *Kloeckera sp.* no. 2201, a dual control over AOX activity is operative. High concentrations of glucose repress activity, and methanol induces activity. The appearance of AOX activity during growth on methanol is paralleled by the capacity of a cell-free extract to direct the synthesis of AOX by *in vitro* translation (Roggenkamp *et al.*, 1984), indicating that AOX mRNA is accumulating. Giuseppin *et al.* (1988) have determined the rate-limiting step in AOX activity. Measurements of AOX mRNA were made using the cloned gene as a probe, along with

Figure 10.1 The methanol pathway and its compartmentalization in *Hansenula polymorpha*. Methanol is oxidized by alcohol oxidase (1) within the peroxisome to produce formaldehyde and hydrogen peroxide. Hydrogen peroxide is reduced to water and molecular oxygen by catalase (2). Formaldehyde may diffuse into the cytosol, where it spontaneously reacts with reduced glutathione. This complex is subsequently oxidized to formate and then carbon dioxide by the action of formaldehyde dehydrogenase (3) and formate dehydrogenase (4), respectively. Alternatively, formaldehyde may react with xylulose 5-phosphate (Xu5P) by the action of dihydroxyacetone synthase (5) within the peroxisome to produce glyceraldehyde 3-phosphate (GAP) and dihydroxyacetone (DHA). Phosphorylation of DHA to dihydroxyacetone phosphate by the action of dihydroxyacetone kinase (DHAK) in the cystol facilitates the production of fructose 1,6-bisphosphate (FBP) by fructose 1,6-bisphosphate aldolase (7) and its dephosphorylation by fructose 1,6-bisphosphatase (8). Subsequent rearrangement reactions regenerate Xu5P, which is recycled to react with formaldehyde. The net product of this pathway is the generation of one-third of the GAP molecules made available for biomass. (From Douma *et al.*, 1985.)

protein both in the monomeric and octameric form, the ratio of FAD/AOX and the actual AOX activity. Cells were grown in a chemostat using glucose/methanol mixtures. They concluded that at low dilution rates, the level of AOX mRNA was limiting, but that at high dilution rates, other factors were important, such as the availability of FAD. This conclusion may be important in evaluating genetic studies on the regulation of the AOX gene in which enzyme measurements are used to measure gene activity which implicitly assumes that gene transcription is rate-limiting.

The regulation of AOX and DAS (dihydroxyacetone synthase) expression has been studied in *P. pastoris* using gene fusions in which β-galactosidase was used as a reporter enzyme fused to either AOX or DAS upstream sequences (Tschopp, 1987a). Integration of the constructs was directed to the *HIS4* gene allowing simultaneous measurements of AOX and β-galactosidase from separate copies of the methanol-regulated promoters. When fused to the AOX promoter, β-galactosidase expression rose from 1 to 120 units during growth in medium lacking glucose. After 25 h, addition of 0.5% methanol caused a further increase in activity to a maximum of 5 to 13×10^3 units. The activity of AOX increased coordinately with β-galactosidase but measurements of protein levels showed that there was between 5 and 10 times more AOX than β-galactosidase. A decreased expression of heterologous gene products compared to homologous products is commonly observed in *Saccharomyces cerevisiae* (Mellor *et al.*, 1985).

As with *Hansenula* and *Kloeckera* sp. no. 2201, the *P. pastoris* AOX promoter is active in the absence of methanol under conditions of carbon starvation. However, under these conditions the level is only 1–2% of the maximum observed in the presence of methanol, and no AOX activity is observed during growth on glycerol. It thus seems likely that while the *Hansenula* AOX gene can be derepressed to quite high levels in the absence of methanol induction, expression of AOX in *P. pastoris* is more tightly coupled to the presence of an inducer. The differences between the two organisms are quantitative rather than qualitative. In *P. pastoris*, AOX activities do increase 120-fold by carbon starvation in the absence of an inducer.

The addition of glucose or ethanol to methanol-grown cells of *C. biodinii* or *H. polymorpha* causes the rapid inactivation of AOX and catalase activity, as well as the eventual degradation of the enzymes themselves and the peroxisomes in which were located (Veenhuis, 1983). This process is apparently similar to that studied by Holtzer (1976) in *S. cerevisiae*. Catabolite inactivation is clearly separate from catabolite repression. First, mutants which relieve glucose catabolite repression in *C. boidinii* (Sakai *et al.*, 1987); and either glucose (*gcr*) or ethanol (*ecr*) catabolite repression in *Pichia pinus* (Sibirny *et al.*, 1986a) do not affect catabolite inactivation. Second, 2-deoxyglucose, which does induce catabolite repression, does not cause catabolite inactivation (Veenhuis, 1983).

10.3 Genetic analysis

10.3.1 *The life cycle of methylotrophic yeasts*

Methylotrophic yeasts are found in the genera *Candida*, *Torulopsis*, *Pichia* and *Hansenula* (Lodder, 1970; Lee and Komagata, 1980*a*, *b*). The early taxonomic studies (Lodder, 1970) reported that while *Pichia* and *Hansenula* are homothallic, ascosporogenic yeast, no sexual phase is evident for *Candida* and *Torulopsis*. Genetic studies have therefore been confined to the *Pichia* and *Hansenula* genera, in particular *P. pinus* and *H. polymorpha*. The life cycle of these two yeasts is quite similar (Figure 10.2). Germination of asci produces four haploid ascospores, displaying a bipolar mating type with two α and two *a* cells per ascus (Tolstorukov and Benevolenskii, 1978; Gleeson, 1986; Gleeson and Sudbery, 1988). Cells of opposite mating type mate to form diploids, which can be stably maintained on a rich medium. If ascospores are separated after germination, they give rise to haploid clones which are also stable on rich medium. When these haploid cells are subject to nutritional deprivation, mating-type switching occurs, followed by the formation of conjugation tubes and mating of cells of opposite mating type to produce diploids. The physiological control of mating-type switching has been analysed in detail for *P. pinus* (Tolstorukov and Benevolenskii, 1980). Mating-type switching only occurs at high frequency during the last few rounds of cell division, in response to nitrogen starvation. A pedigree study of cells going through *a*⇌α transitions revealed that a cell must go through at least one round of DNA replication and cell division in starvation medium in order for mating-type switching to occur. Switching of mating type is random; it can occur in either the mother or daughter cell and is not influenced by the mating type of the pre-transition cell

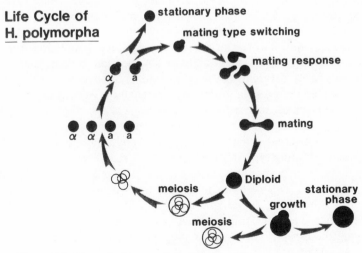

Figure 10.2 Schematic diagram of the life cycle of *Hansenula polymorpha*.

or the relationship of the cells to one another, i.e. there is an equal likelihood that an $a \rightleftharpoons \alpha$ transition will occur in either the mother or daughter cell. Formation of a conjugation tube occurs only in cells which have undergone a mating type transition irrespective of the actual mating type of the cell (Tolstorukov *et al.*, 1977; Benevolenskii and Tolstorukov, 1980; Gleeson, 1986). Sporulation is also induced during conditions of nutritional deprivation. In *H. polymorpha*, colonies of cells which have sporulated develop a pink colour, while in *P. pinus* they develop a brown colour (Gleeson, 1986; Benevolenskii and Tolstorukov, 1980). This means that colonies initiated by a single haploid cell growing on the appropriate medium will eventually contain cells at all stages of the life-cycle, including haploid cells, cells both with and without mating processes, diploid cells and asci. Nevertheless, diploid cells can readily be distinguished from haploid cells because they produce colonies which form asci more quickly and achieve a higher percentage of asci. This can conveniently be monitored in *H. polymorpha* by the colour of the colony. Diploid cells give rise to pink colonies after three days of growth on malt extract, while haploid cells give rise to colonies which have a pink tinge only after 8 days. Haploid and diploid colonies of *P. pinus* can similarly be distinguished (Benevolenskii and Tolstrorukov, 1980); spore formation occurs in diploid colonies within two days of growth on restrictive growth medium and reaches 73% after five days, while haploid cells show no spore formation within two days and only reach 39% after five days.

Genetic control of mating-type switching has been investigated in *P. pinus* by Benevolenskii and Tolstorukov (1980). Twenty-three mutants that were unable to complete the life-cycle when plated as haploid cells on restrictive growth medium were isolated. These mutants formed four phenotypes, representing defects in different stages of the life-cycle. These were: (i) mutants capable of mating with other clones, but which did not form copulation processes within clones; (ii) mutants incapable of mating with any cells; (iii) mutants that fail to mate but do produce copulation processes; (iv) mutants that mate to form diploids but then fail to sporulate.

Mutants in the first class are heterothallic, and the mating type switching mechanism does not function, so that cells remain of one mating type. These cells mated normally with heterothallic mutants of opposite mating type and with homothallic strains. A backcross between a heterothallic mutant and the parental strain revealed that homothallism/heterothallism segregated independently of mating type. This mutant therefore defines a defect in a gene, *HTH1*, that controls homothallism, and this gene is unlinked to the mating-type locus. It is possible that this gene is analogous to the *HO* locus of *S. cerevisiae*. Similarly in *H. polymorpha*, analysis of a cross between a strain of low fertility and a fertile strain resulted in a 2:2 segregation for homothallism, implying the existence of a gene analogous to the *HTH1* gene of *P. pinus* (Gleeson, 1986).

10.3.2 Techniques for genetic analysis

The life-cycle of methylotrophic yeast facilitated the development of genetic analysis. As haploids are stable on rich medium, conditional mutations could be induced by mutagenesis. Mating-type switching, induced by nutritional deprivation, allows complementary haploids to be mated and diploids selected on a suitable selective medium. Any pair of haploid strains can be mated in this way, regardless of their initial mating type. Complementation studies between pairs of mutant alleles in diploids allow the definition of genes. After sporulation, recombination covering the progeny allows the construction of linkage maps. In addition, natural and radiation-induced loss of chromosomes from diploid cells provides a parasexual cycle for genetic mapping (Tolstorukov et al., 1983). The first study of the genetics of a methylotrophic yeast has been carried out by Tolstorukov, Benevolenskii and colleagues, using P. pinus. More recently we have carried out genetic analysis on H. polymorpha (Gleeson et al., 1984; Gleeson, 1986; Gleeson and Sudbery, 1988).

Tolstorukov et al. (1977) isolated 84 auxotrophic mutants of P. pinus after NTG and UV mutagenesis. These mutants showed a bias with respect to different phenotypes, with an excess of adenine, methionine, lysine and cysteine requirers recovered. A similar situation occurred in H. polymorpha (Gleeson et al., 1984; Gleeson, 1986; Gleeson and Sudbery, 1988; an excess of adenine- and methionine-requirers was also found. The reason for the bias is not clear; it could be due to the relative sizes of the genetic targets (pathways with more enzyme-catalysed steps involve more genes that may be hit to give the same mutant phenotype); to the intrinsic mutability of certain loci; or to the clearness of certain phenotypes, which initially must be recognized by failure to grow on minimal medium. A specific search for leucine auxotrophs (Gleeson et al., 1986; Gleeson, 1986) provided some evidence for the latter possibility.

Complementation studies of the P. pinus mutants assigned the 84 mutants to 35 different loci. Sporulation of the diploids followed by tetrad dissection showed markers segregated in a regular Mendelian fashion with most markers segregating 2:2 in the tetrads and dihybrid crosses giving the tetrad types and ratio of types to be expected. Two alleles, ade5 and lys4 exhibited linkage showing the possibility for the construction of genetic maps.

P. pinus cells which are heterozygous for an auxotrophic marker give rise to segregants expressing this marker at a frequency of 2×10^{-3} per cell, per division (Tolstorukov et al., 1979). The bulk of these segregants are to be found in 'mature' colonies, that is colonies which appear only four days after inoculation on to YEPD plates instead of the two days which 'normal' colonies take. Mature colonies contain a higher percentage than normal of inviable cells, and are unstable with respect to the expression of heterozygous markers contained in the original diploid. About 50% of mature colonies are diploid on the basis of sporulational capability and 50% haploid on the basis of conjugation tube formation. It was concluded that mature colonies are

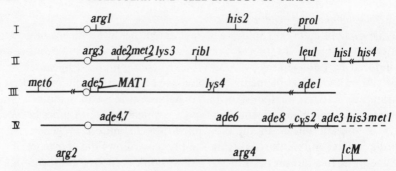

Figure 10.3 Genetic map of *Pichia pinus* reproduced from Tolstorukov *et al.* (1984).

aneuploid due to the loss of one or more chromosomes, and that chromosome loss will continue until the stable haploid state is reached. The expression of recessive heterozygous alleles is due to loss of the homologous chromosome carrying the dominant marker. This deduction is confirmed by the observation that, in a diploid carrying multiple pairs of heterozygous alleles, every possible combination of alleles was observed in mature colonies except nonparental combinations of alleles that are linked.

The incidence of mature colonies and thus of haploidization can be increased to 10% in survivors of gamma-irradiated diploid cells (Tolstorukov, 1982). This method of induced mitotic haploidization provides a method of assigning alleles to linkage groups (Tolstorukov *et al.*, 1983). Seventeen markers were assigned to four linkage groups in this way. A comprehensive study based on tetrad analysis of 164 separate crosses allowed 25 loci to be located on the four linkage groups revealed by haploidization and two additional fragments (Figure 10.3). Centromere-linked markers were identified for each of the linkage groups, which allowed the position of each centromere to be mapped (Tolstorukov *et al.*, 1984).

Techniques of genetic analysis have been developed for *H. polymorpha* (Gleeson *et al.*, 1984; Gleeson, 1986; Gleeson and Sudbery, 1988). A collection of 218 auxotrophic and temperature-sensitive mutants has been produced by NTG mutagenesis of strain NCYC495. As mentioned above, a bias in the type of auxotrophs recovered was observed with an excess of adenine- and methione-requirers. Complementary mutants crossed readily after pregrowth on malt extract and suitable selection for diploids. This allowed complementation studies to be carried out which assigned the mutants to 57 complementation groups. A particularly comprehensive study was possible with adenine-requirers. The analysis of 106 adenine-requiring mutants revealed 12 different complementation groups. In *S. cerevisiae* there are 13 adenine genes, so it is likely that all but one adenine gene has been identified in *H. polymorpha*. Of particular interest was the observation of intragenic complementation in the

ADE2 locus with two subgroups *ADE2.1* and *ADE2.2* containing 16 and 8 alleles respectively. This closely parallels the *ADE5 ADE7* complex of *S. cerevisiae* (Roman, 1956; Costello and Bevan, 1964).

Heteroallelic diploids can be readily sporulated on malt extract. Tetrad dissection is possible, but is made difficult by the presence of a thread-like structure which binds the ascospores together. Ascospores have a high viability and markers segregate in a regular fashion giving the expected tetrad types. One example of linkage has been discovered: in a cross between a methanol-requiring strain, *mut2-1 met6-1* and a second marked strain, *ade3-3* no recombination was observed between *mut2-1* marker and *met6-1*. Tests for recombination can be more conveniently carried out by random spore analysis using diethyl ether treatment. This allows a larger number of progeny to be tested, but is prone to problems of differential viability between mutant and wild-type alleles (Gleeson, 1986; Gleeson and Sudbery, 1988). The initial observation of linkage between *MUT2* and *MET6* has been confirmed and extended to include linkage to two further markers, *ADE4* and *ARG1*, by random spore analysis.

NCYC495 is a strain that is amenable to genetic analysis. Marked haploid strains can be crossed readily, complementation relationships tested in the stable diploid form, and recombination measured in the meiotic progeny. Not all strains lend themselves so readily to genetic analysis. Initial attempts to develop techniques of genetic analysis in *H. polymorpha* using a sample of the strain CBS4732 were frustrated by the low mating frequency of haploids (Gleeson, 1986). CBS4732 is the strain which has been used for biochemical and physiological studies. The sample used had been received from a microbiology laboratory and stored routinely on malt extract. This was probably responsible for the semisterility of the sample used.

An analysis of *H. polymorpha* strain VKM-U-1397 found that inheritance of auxotrophic markers was highly irregular. It was deduced that the strains were aneuploid, a feature associated with their origin from auxotrophic mitotic segregants of a heterozygous hybrid obtained by protoplast fusion (Bodunova *et al.*, 1986*a*). Other factors which were found to play a role in the non-Mendelian segregation of markers through meiosis included a non-random segregation of chromosomes in mitosis and a dependent separation of heterologous chromosomes in meiosis, due to heterologous centromere pairing (Bodunova *et al.*, 1986*b*). Genetic analysis of *P. pastoris* has also encountered aberrant meiotic segregations of auxotrophic markers, with a preponderance of wild-type ascospores (Digan and Lair, 1987). In this case, the stock strain was initially developed for high-level growth in fermenters prior to the initiation of genetic studies. This is consistent with the observation that aneuploidy, polyploidy, and a reduction in spore viability and classical gene segregation do occur with prolonged maintenance of strains by repeated subculturing (Fowell, 1969; Lindegren, 1949; Gleeson, 1986; Bodunova *et al.*, 1986*a, b*).

10.3.3 *Mutants of methanol utilization*

Mutant studies of the methanol pathway were first initiated to lend additional evidence to the biochemistry and physiology of the methanol pathway. Methanol utilization (*mut*) defective mutants have been isolated in strains of *H. polymorpha, C. biodinii, P. pinus* and *P. pastoris*. In the majority of such mutants, a clear enzyme defect was not identified. Sahm and Wagner (1973*a*) first reported the isolation of Mut⁻ mutants of *C. boidinii*. One, impaired for AOX activity, had a functional alcohol dehydrogenase, indicating that AOX played the initial key role in methanol oxidation. Induction of enzyme levels in a selection of these mutants in *C. boidinii* points to the role of methanol as the primary inducer of AOX, dihydroxyacetone synthase, and fructose 1, 6-bisphosphatase, and methanol or formaldehyde for the induction of formaldehyde dehydrogenses and formate dehydrogenase, while dihydroxyacetone kinase is induced by dihydroxyacetone. A more extensive analysis of Mut⁻ mutants has been reported for *P. pinus* and *H. polymorpha*. In an analysis of 17 methanol-defective mutants in *P. pinus*, Popova and Pugovtina (1982) identified nine which were defective for growth on methanol alone. Complementation analysis revealed five groups corresponding to five nuclear genes, which segregate 2⁺:2⁻ in tetrad analysis. Linkage was detected for two sets of these mutants.

An analysis of Mut⁻ mutants of *H. polymorpha* (Gleeson 1986; Gleeson and Sudbery, 1988) resulted in the allocation of 18 mutants to five complementation groups. The mutant alleles of these genes, termed *MUT1* to *MUT5*, were specifically defective in methanol utilization. Surprisingly, an enzymatic analysis of these mutants revealed only one class which was totally defective in a methanol-pathway enzyme (dihydroxyacetone synthase). The remaining four had appreciable levels of enzymes for methanol utilization. This result may reflect the necessity of having a fine balance of enzyme levels for efficient methanol utilization and/or the involvement of other genes for the co-ordinated functioning of peroxisomes.

A further group of methanol non-utilizing mutants was found also to be defective in growth on a number of carbon sources other than glucose (*acu* mutants). These are thought to be defective in the lifting of catabolite repression and are discussed further below. O'Connor and Quayle (1979) experienced similar difficulties in demonstrating an enzyme defect in a collection of mutants. Only one defect was found (in dihydroxyacetone kinase). Cregg and Madden (1988) generated AOX⁻ mutants in *P. pastoris* by gene disruption and demonstrated that there are two structural genes for AOX in this organism, both of which must be disrupted to generate a methanol⁻ phenotype. Molecular and physiological analysis of these genes, *AOX1* and *AOX2*, revealed that they are co-regulated; however, *AOX1* is expressed at a level tenfold higher than *AOX2*. No physiological significance can be attributed to this observation, since replacement of one gene by the other in all configurations has no effect on the growth of the cells on methanol (J. Cregg,

personal communication). Similarly, two copies of the gene for dihydroxyacetone synthase (*DAS*) are present in *P. pastoris* (G.P. Thill, personal
communication). In this case, the genes are expressed at approximately the
same level.

Mutants of *H. polymorpha* specifically defective in the AOX gene have been
isolated by selecting for cells resistant to allyl alcohol (Sibirny and Vitvitskaya,
1986; Sibirny *et al.*, 1986c). This compound is oxidized by AOX to acrolein,
which is toxic. Forty-nine mutants were isolated with either low or no AOX
activity. From one of these mutants a temperature-sensitive revertant was
isolated. It contained AOX protein which was temperature sensitive *in vitro*,
demonstrating that the structural gene itself had been mutated. This study
yielded evidence that the two *AOX* genes produced enzymes with different
thermal stabilities. However, Southern hybridization using the *AOX* gene
cloned by Ledeboer *et al.* (1985) shows that there is probably only one copy of
the gene in *H. polymorpha* strain CBS4732 (R.A. Veale, personal communication). Other enzyme defects generated by mutant programmes include
catalase (Eggeling and Sahm, 1980b) and formaldehyde reductase (Sibirny and
Ksheminskaya, 1987; Sibirny *et al.*, 1988) and dihydroxyacetone kinase (de
Koning *et al.*, 1987). Although not directly relevant to the present discussion it
may be noted that mutations affecting enzymes of the TCA cycle and
glyoxalate pathway have been isolated by Titorenko *et al.* (1987) and Sibirny
and Gonchar (1987). It was reported that a defective citrate synthase prevented
growth on ethanol, glycerol and methanol, while growth on glucose required
glutamate supplementation. As expected, other defects of the TCA cycle and
glyoxalate pathway prevented growth on ethanol and acetate.

10.3.4 *Regulatory mutants of methanol metabolism*

Since AOX production is under the dual control of glucose and ethanol
repression and methanol induction, it might be expected that mutants
affecting each facet of the control mechanism could be isolated. The glucose
repression of AOX is similar to carbon catabolite repression, which has been
extensively investigated in *S. cerevisiae* by mutant isolation. Such mutants may
either prevent the expression of repressible genes even in the absence of
glucose, or cause constitutive derepression of repressible genes in the presence
of glucose (Zimmermann *et al.*, 1977; Ciriacy, 1977; Zimmermann and Scheel,
1977; Entian and Zimmermann, 1982; Carlson *et al.*, 1981; Neigeborn and
Carlson, 1984; Entian, 1981; Entian and Zimmermann, 1980). Mutants of the
former category are pleiotropically defective for growth on different non-
optimal carbon sources. Mutants of the latter category were selected for the
ability to grow on a poor carbon source in the presence of glucose analogue.

Mutant studies of methylotrophic yeasts readily yield Mut⁻ strains which
display a phenotype typical of that expected in mutants of catabolite
repression. Gleeson (1986) found that 15 of 33 Mut⁻ mutants of *H. polymorpha*
are also unable to grow on ethanol, glycerol, maltose, and dihydroxyacetone.

These mutants, named alternative carbon utilization (*acu*) mutants, were not defective in respiration (*H. polymorpha* is a petite negative yeast). Since the hydrolysis of maltose by maltase yields glucose, upon which the cells can grow, it seems unlikely that the *acu* mutants reflect a defect in the central metabolism. A preliminary analysis of such mutants established a minimum of three complementation groups.

Revertants of the *Acu* phenotype were isolated by selection for growth on maltose. One class of revertants was found to be resistant to 2-deoxyglucose irrespective of the carbon source available. This mutation was dominant in crosses with the wild type. Mutants derepressed for glucose repressible genes were also isolated by selecting for growth in the presence of 2-deoxyglucose using either maltose or methanol as a carbon source. These mutants were relieved of repression by 2-deoxyglucose and were not specific to a particular carbon source, implying that a common control element exists for carbon metabolism. However, the extent of growth in the presence of 2-deoxyglucose varies for each carbon source, suggesting that additional regulatory mechanisms exist which are not common and may be particular to each carbon source.

The effect of mutations conferring resistance to 2-deoxyglucose on the activity of the *AOX* promoter has been investigated by measurements of AOX activity itself, and also by following the activity of an α-galatosidase gene fused to *AOX* 5′ sequences (R.A. Veale and M. Hodgkins, personal communication). Expression of AOX in such mutants is not repressed by glucose but is dependent on the presence of methanol. This result suggests that AOX expression is subject to dual control, repression by glucose and induction by methanol; the former is relieved in 2-deoxyglucose-resistant mutations, but a requirement for the latter remains. Similar findings were reported by Sakai *et al.* (1987), who isolated 2-deoxyglucose-resistant mutants of *C. boidinii*. Such dual control is consistent with the results of the physiological studies discussed above.

A detailed genetic study of catabolite repression has been carried out in *P. pinus* by Sibirny and his colleagues. Two mutants were isolated: *gcr1*, which alleviates glucose repression of AOX, and *ecr1*, which alleviates ethanol repression (Sibirny *et al.*, 1986*a*). *gcr1* mutants were isolated by selecting for growth on methanol plus 2-deoxyglucose. They were shown to be recessive and monogenic by crossing to a wild-type strain. *gcr1* mutants express AOX when grown on a mixture of glucose plus methanol, but not on glucose alone. There is thus alleviation of glucose repression, but a requirement for methanol induction remains. Glucose *inactivation* of AOX is not affected.

gcr1 mutants were isolated through the use of a mutant which is defective for isocitrate lyase, *icl1*. This strain is unable to grow on ethanol because the glyoxalate cycle cannot function. However, ethanol still acts to repress AOX; thus *icl1* mutants are also unable to grow on a mixture of methanol and ethanol. This provides a selection for mutants that are not repressed by ethanol and so can use methanol as a carbon source. A representative isolate was shown to carry a recessive, monogenic mutation. The *ecr1* mutation allows the expression of AOX in the presence of ethanol. This response is independent of the *icl1* mutation but a requirement for methanol induction remains. Although inactivation of AOX by ethanol is unaffected, the *icl1* mutation does alleviate ethanol inactivation of AOX (Sibirny *et al.*, 1986*b*). An interesting further effect of the *ecr1* mutation is that it renders the ethanol induction of enzymes of the glyoxalate cycle (isocitrate lyase and malate synthase) sensitive to the presence of methanol, whereas the wild type is insensitive (Sibirny and Titorenko, 1987). There is thus a reciprocal mode of action of the *ECR1* locus which depends on its allelic state. In *ECR1* strains, ethanol represses AOX, but methanol has no effect on the synthesis of enzymes of the glyoxalate cycle. In *ecr1* strains, ethanol has no effect on AOX; however, methanol acts to repress the synthesis of the glyoxalate cycle enzymes. These enzymes are found in microbodies (glyoxysomes). It has been suggested that microbodies for the utilization of C_1 (peroxisomes) and C_2 (glyoxysomes) substrates cannot simultaneously be present in the same cell. The lack of isocitrate lyase and malate synthase activities in *ecr1* strains in the presence of ethanol could be the result of an impaired translocation of the enzymes into glyoxysomes rather than a lack of gene expression (Sibirny *et al.*, 1988).

Sibirny *et al.* (1987) have tested the capability of a wide range of carbon substrates to repress the induction of AOX by methanol in wild-type, *gcr1* and *ecr1* strains. The compounds could be divided into five classes. Some compounds, including maltose and glycerol, did not repress, or only slightly repressed, the induction. The second group of compounds, exemplified by glucose, but including a range of hexose sugars, maltose and xylose, blocked the induction in the wild-type and *ecr1* strains but not in *gcr1* strains. The third group, exemplified by ethanol and including acetate and 2-oxoglutarate, repressed the induction in wild-type and *gcr1* strains but not in an *ecr1* strain. The fourth group, comprising malate and dihydroxyacetone, strongly repressed AOX in all three strains. Lastly, a group of compounds including formate and xylitol repressed the synthesis of AOX in the wild-type strain but not in either *gcr1* or *ecr1* strains. It is important to note that ethanol and glucose respectively are only one of a group of compounds that behave in a similar fashion in each case. When the activities of central metabolism were assayed for each strain, *ecr1* was found to have only 25% of wild-type 2-oxoglutarate dehydlogenase activity, while *gcr1* had only 16% of wild-type phosphofructokinase activity. Other enzyme activities were normal, including hexokinase activity in *gcr1* strains.

Clearly the pathways of glucose and ethanol catabolite repression can be genetically differentiated. The former is mediated through the *GCR1* gene product and the latter through the *ECR1* product. However, the role of the *ECR1* and *GCR1* genes remains uncertain, and the immediate effectors of catabolite repression unidentified. One possibility, although an unlikely one, is that *ECR1* and *GCR1* are structural genes respectively for 2-oxoglutarate dehydrogenase and phosphofructokinase, which are involved in the metabolism of ethanol and glucose into the true effectors of repression, (Sibirny *et al.*, 1987). In the case of *GCR1*, the repressing sugars 2-deoxyglucose and glucosomine are substrates for hexokinases yielding the 6-phosphate derivative, but this is not a substrate for phosphofructokinase and the diphosphate is not formed. Fructose-1, 6-diphosphate cannot, therefore, be the immediate effector of repression. Moreover, mutants of *S. cerevisiae* are known which are defective in phosphofructokinase activity but which nevertheless exert catabolite repression. In the case of *ecr1*, 2-oxoglutarate or a subsequent metabolite cannot be the true effector since ethanol and malate still exert repression in strains carrying mutations in genes coding for enzymes required for the production of 2-oxoglutarate from ethanol. The repressing potential of both ethanol and 2-oxoglutarate could be explained by the *ECR1* protein having an affinity for both molecules (Sibirny *et al.*, 1987). However, it seems more likely that *ECR1* and *GCR1* are sensitive to the general metabolic state and energy charge of the cell, possibly through interaction of low-molecular-mass indicators such as NADH, ATP and cAMP.

10.4 Applications of methylotrophic yeasts in biotechnology and the development of molecular genetics

10.4.1 *The advantages of methylotropic yeasts*
Methanol is a major by-product of the oil refining process. The industrial potential of methylotrophic yeasts initially arose out of their ability to utilize methanol as a carbon source. The production of single-cell protein (SCP) from methanol has successfully been commercialized as a feedstock additive, 'Provesta', by Philips Petroleum Company.

The potential benefits of methanol as a fermentation feedstock have been analysed recently by Linton and Niekus (1987). The price has remained stable over the fourteen-year period from 1970 to 1984, and there are good reasons to expect a low stable price will continue for some years, due to a surplus of production over consumption. Over the same period, the price of sugar has been extremely variable, ranging between $100 and $650 per tonne. Methylotrophic yeasts also grow well on glycerol, either as a sole carbon source or in mixed substrate fermentations with methanol or glucose. Like methanol, glycerol is also cheap and plentiful. The low cost of suitable fermentation substrates makes methylotrophic yeast attractive for fermentations where the substrate costs are significant. The ability of methylotrophic yeasts to grow on other carbon sources besides methanol, such as sugar and whey, prevents a potentially damaging reliance on a single carbon source.

The development of fermentation conditions for high-cell-density production on a simple, defined medium consisting of salts and a carbon source was accompanied by a greater understanding of the physiology of the methanol pathway enzymes. The fact that, under optimal conditions of derepression/induction, AOX can comprise up to 30% of the total cell protein clearly indicated that the gene could be highly expressed and was also strictly regulated. These features make the AOX promoter an ideal system to develop as an expression system for the production of heterologous protein.

In this area, methylotrophic yeasts must be directly compared with *S. cerevisiae*, in which heterologous protein production is well developed. Clearly, the price advantage of growth substrate is not as relevant for the production of high-value substances such as tissue plasminogen-activating factor or hepatitis B surface antigen. However, for higher-volume protein products, such as enzymes for household detergents or enzymes used in the food processing industry, the price of the growth substrate may become an important factor. Aside from these cost considerations, glycerol and methanol are both clean fermentation substrates which allow easy purification of secreted products.

Expression systems have been developed in both *P. pastoris* and *H. polymorpha*. For both of these systems, a heterologous product can therefore be expressed at a high level, from a single integrated copy and under strict regulation. In *S. cerevisiae*, high-level expression of a heterologous product usually requires the use of a multicopy plasmid, entailing a penalty in the form of decreased stability and an additional load to the cell in terms of plasmid maintenance. Stability may be achieved by integration, but this results in a lower level of expression from the single copy of the gene. The AOX promoter of methylotrophic yeasts allows the stability of a single integrated copy combined with a high level of expression. The tight regulation available allows separation of the growth from the production phases of the fermentation, thus avoiding the deleterious effects which may arise from any cellular toxicity of the heterologous protein. Studies of the molecular genetics of the methylotrophic yeasts were initiated in the early 1980s and have been developed into a highly efficient system for the commercial production of heterologous protein.

10.4.2 *Transformation of methylotrophic yeasts*

Transformation in methylotrophic yeasts has been reported in *P. pastoris* (Cregg *et al.*, 1985) and *H. polymorpha* (Gleeson *et al.*, 1986; Roggenkamp *et al.*, 1986; Tikhomirova *et al.*, 1986). In each case, a broadly similar strategy was employed. Mutants were isolated in the methylotrophic yeast which corresponded to the auxotrophic mutants in *S. cerevisiae*, which can be complemented either by existing vectors carrying the *Saccharomyces* gene or by the homologous sequence. Plasmid replication can be sustained either by fortuitous ARS activity in *S. cerevisiae* sequences or by the isolation of homologous ARS sequences following the same strategy originally employed by Stinchcomb *et al.* (1980) to isolate *Saccharomyces* ARS sequences.

Table 10.1 Plasmids used by Cregg *et al.*, (1985) to transform GS115 (his4) strain of *Pichia pastoris*. S.c = *Saccharomyces cerevisiae*; P.p = *Pichia pastoris*.

Plasmid	Construction	Transform-ation frequency (1)	*In vivo* fate of plasmid	ARS
pYA4	YEp13 + P.p HIS4	9.7×10^4	Unstable autonomous replication for 10 generations then integration	YEp13 LEU2 (2)
pYA2	pBR325 + S.c HIS4	1.6×10^4	Unstable autonomous replication for over 50 generations	S.c HIS4 (3)
pJ28	pYA2 + P.p HIS4	1.2×10^4	Integration after 10 generations	S.c HIS4 (4)
pYJ8	P.p HIS4 + pBR325	5.0×10^1	Integrates–	–
pYA63	pYJ8 + *TaqI* chromosomal fragments	1×10^5	Unstable autonomous replication for over 50 generations	PARSI
pYA90	"	1.5×10^5	"	"

(1) Number of transformants μg DNA^{-1}.
(2) The ARS activity was not found in the 2 micron origin of replication but was thought to reside in the *LEU2* sequence.
(3) The 9.4 kb *PstI* fragment from *S. cerevisiae* contained two ARS: a weak ARS in the *HIS4* gene itself and a strong ARS in the flanking sequences.
(4) The activity of the *HIS4* ARS is deduced from the increased transformation frequency of pYJ28 over pYJ8 and the observation that replication is autonomous during the first 10 generations.

Cregg *et al.* (1985) developed the system for *P. pastoris* based upon the isolation of mutants defective in histidinol dehydrogenase—the enzyme defective in *his4* mutants of *S. cerevisiae*. The *Pichia* gene was isolated by complementation of *S. cerevisiae his4* mutants with a *Pichia* gene library made in YEp13 (a 2 μm-based *LEU2* vector). Transformation of *Pichia* cells was achieved using the protoplasting procedure, and it was found that complementation would occur using either the homologous or heterologous sequences. The various plasmids and their properties are described in Table 10.1. DNA fragments originating from *Saccharomyces* DNA found in the *LEU2* gene of YEp13 (pYA4) or in the fragment carrying the *HIS4* sequence (pYA2) were found to have ARS activity in *Pichia* (they do not in *Saccharomyces*). *Pichia* ARS were isolated from a library of *Pichia* DNA made in a pYJ8 derivative. Two such sequences, PARS1 (pYA63) and PARS2, were analysed in detail. They allowed autonomous replication for over 50 generations of plasmids carrying both the homologous and heterologous *HIS4* genes. Although their structure is very similar to *Saccharomyces cerevisiae* ARS sequences, they do not function in that organism. An interesting conclusion is that the homologous *HIS4* gene with a weak ARS (pJ28) will eventually integrate, but can be maintained indefinitely from a strong ARS (pYA63, pYA90). A *heterologous HIS4* gene is maintained indefinitely in the autonomous form even from a weak ARS (pYA4, pYA2). It is presumed that the lack of sequence homology between plasmids composed entirely of heterologous DNA prevents recombination and integration into the *Pichia* genome.

Transformation systems developed for *H. polymorpha* are described in

Table 10.2 Transformation systems used in *Hansenula polymorpha* (H.p.); S.c = *Saccharomyces cerevisiae*; P.p = *Pichia pastoris*.

Host mutation	Plasmid	Construction	Transformation frequency (1)	In vivo fate of plasmid	ARS	Reference
Ura3	YRp17	S.c URA3 + S.c ARSI (2)	2.2×10^2	Unstable replication	S.c ARSI	Roggenkamp et al. (1986)
"	YIp5	S.c URA3 + pBR322	3×10^0	Integration	–	"
"	pHARI	YIP5 + SalI chromosomal fragments	1.5×10^3	Unstable replication	HARSI	"
		"				
leu2 (leu1-1) (3)	pHARSII YEp13 (4)	S.c 2 micron origin	4.6×10^2 1.5×10^2	" "	HARSII YEp13 sequence	Gleeson et al. (1986)
"	pSG12 (5)	YIp5 + H.p LEU2	5×10^0	Integration	–	Sudbery (unpublished)
leu2	PL2	S.c LEU2 + mt DNA fragments from *Candida utilis*	$1\text{-}3 \times 10^3$	Unstable replication which stabilizes upon concatemerization	*Candida utilis*	Tikhomirova et al. (1986)
"	pH13 and pH21	PL2 + H.p. chromosomal fragments	$1\text{-}5 \times 10^1$	Integration		"
"	pMH series	PL2 + Hansenula mt + DNA fragments	$2 \times 10^2 - 1 \times 10^3$	Unstable replication	Hansenula mt DNA	"

(1) Transformations μg DNA^{-1}.
(2) Stinchcomb et al. (1980).
(3) The *Hansenula polymorpha* mutant lacking β-isopropylmalate dehydrogenase which is the defect in *Saccharomyces cerevisiae leu2* strain is designated leu1-1 (Gleeson et al., 1986).
(4) Broach et al. (1979).
(5) Isolated by constructing a chromosomal library in YIP5 using partial digestion with Sau3a. The required fragment was isolated by complementation of JA221 (Beggs, 1978) a leuB6 strain of E. coli.

Table 10.2. Again, both homologous and heterologous genes were found to be able to complement the *Hansenula* mutation. Roggenkamp *et al.* (1986) isolated two ARS (HARSI and HARSII) elements from *Hansenula* chromosomal fragments using a method similar to that described above for *Pichia* PARS sequences. Again, it was found that *Saccharomyces* sequences (in YRp17) had weak ARS activity. The HARS elements were stronger, giving 2–5 times the transformation frequency of YRp17. Following a similar approach, Tikhomirova *et al.* did not obtain ARS sequences, since the plasmids (pH13 and pH21) did not have a high transformation frequency and did not replicate autonomously. Presumably, the *Hansenula* chromosomal sequences acted to provide homologous sequences for recombination with the chromosome, leading to integration. An interesting observation is that YEp13 plasmids linearised by *BamHI*, which cuts outside the *LEU2* gene, slightly increased transformation frequencies and resulted in recircularization *in vivo* and maintenance in the autonomous state. Digestion with *XhoI*, which does cut inside the *LEU2* sequence, failed to produce transformants (Gleeson *et al.*, 1986).

A general conclusion from both the *Pichia* and *Hansenula* studies is that auxotrophic mutations in the methylotrophic yeast can be complemented by either the homologous or heterologous gene. In the absence of ARS activity, the homologous gene transforms with a low frequency and quickly integrates. *Saccharomyces* DNA frequently contains sequences which fortuitously have ARS activity in the methylotrophic yeast, increasing transformation frequencies and allowing autonomous replication.

10.4.3 *Expression of heterologous genes*

In order to exploit the attractiveness of the *AOX* promoter, the structural gene and attendant regulatory sequences must first be cloned. This has successfully been achieved for both the *AOX* gene (*AOX* or *MOX*) itself and the dihydroxyacetone synthase gene (*DAS*) for *H. polymorpha* (Janowicz *et al.*, 1985; Ledeboer *et al.*, 1985) and *P. pastoris* (Ellis *et al.*, 1985). The *MOX* or *DAS* promoter can then be used to drive the high-level expression of a heterologous gene product with the attractive regulation already discussed.

The feasibility of this approach has been demonstrated both in *Hansenula* and *Pichia*. Tschopp *et al.* (1987a) constructed fusions of the *lacz* gene to both the AOX and DAS genes of *P. pastoris*. The hybrid constructs were integrated into the chromosomal *HIS4* gene, thus removing plasmid stability as a factor affecting the level of *lacz* expression. When driven from the *AOX* promoter, β-galactosidase activity increased from 0 to 17 U/mg protein upon glucose derepression, and methanol induction caused a further increase to $5–13 \times 10^3$ U/mg protein. Simultaneous AOX measurements showed increases which were perfectly coordinated. Expression from the *DAS* promoter paralleled physiological findings that expression required both derepression and methanol induction. A most interesting observation was that when the *AOX* or

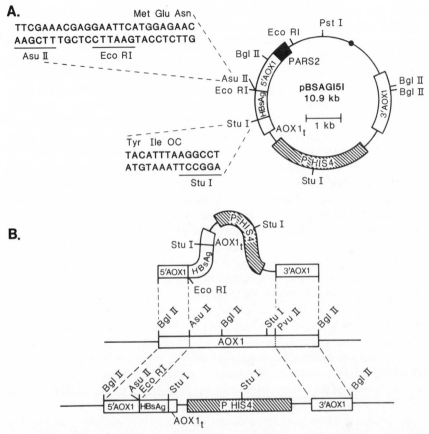

Figure 10.4 A *Pichia pastoris* vector for hepatitis B surface antigen expression. (*A*) The expression vector which can replicate autonomously or be used to integrate into the genome. Reading in a clockwise direction, beginning at the *EcoRI* site corresponding to the 11 o'clock position, the plasmid consists of: a 3.4-kb fragment of pBR322 (for proliferation in *E. coli*); a 1.5-kb fragment of the *AOX1* 3' sequence (open box); a 0.6-kb fragment of pBR322; a 2.7-kb fragment encoding the *P. pastoris* histidinol dehydrogenase gene (*PHIS4*, cross-hatched box); a 0.35-kb fragment of pBR322; a 0.2-kb fragment containing the *AOX1* transcriptional terminator; a 0.6-kb fragment encoding the hepatitis B surface antigen (HBsAg, open box); a 1.1-kb fragment containing the *AOX1* promoter; and a 0.4-kb fragment containing a *P. pastoris* autonomous replication sequence (PARS2, solid box). The DNA sequences at the *AOX1* promoter-HBsAg gene junction and the HBsAg gene *AOX1* terminator junction are shown to the left of the plasmid map. (*B*) A description of the recombinant events which lead to replacement of the *AOX1* structural gene with the hepatitis B surface antigen coding sequence (Cregg *et al.*, 1987).

DAS promoter was cloned on to a 2-μm vector and transformed into *S. cerevisiae*, it was subject to catabolite repression in the heterologous environment and was as powerful as the *Saccharomyces GAL2* or *CYC1* promoters.

As an example of the commercial application of this system, the SIBIA group has expressed hepatitis B surface antigen (HBsAg) (Cregg *et al.*, 1987). A cassette was constructed (Figure 10.4) which consisted of the HBsAg structural gene sandwiched between the AOX promoter and terminator genes, along with the *HIS4* gene. The linear cassette was targeted into the *AOX1* chromosomal locus by gene replacement in such a way that the *AOX1* gene was deleted, thus ensuring stability. *P. pastoris* encodes a second *AOX* gene (*AOX2*) which provides a much less efficient source of AOX. However, enough AOX is synthesized from the *AOX2* transcripts to allow slow growth of *Pichia* on methanol. Expression of HBsAg was undetectable during the growth phase on glycerol and induced to 3–4% of soluble cell protein upon addition of methanol. The non-secreted, unglycosylated product was apparently identical to the unglycosylated product recovered from human serum. Of particular interest was the high proportion of monomers that spontaneously assembled into 22-nm particles, which are necessary for immunogenicity. The process was successfully scaled up to 240-litre, two-stage batch cultures, which yielded 90 g of product—enough for an estimated nine million vaccine doses! The success of integration in achieving stability was demonstrated by the fact that 25 colonies recovered from the culture at the end of the 240-litre fermentation run all still produced high levels of product. This system achieves high-level expression from a single integrated copy, thus avoiding the penalty of instability which results from high-level expression from multiple copies of autonomously replicating plasmids.

Very often a heterologous gene product is required to be secreted and glycosylated. The capability of *P. pastoris* in this respect was demonstrated by the expression of the *S. cerevisiae* invertase gene driven by the AOX promoter (Tschopp *et al.*, 1987b). Invertase is normally secreted into the periplasmic space where it hydrolyses extracellular sucrose to glucose and fructose. The glycoprotein has a heterogeneous molecular mass of 110–125-kD consisting of a 58-kD polypeptide and variable amounts of N-limited mannose-rich oligosaccharide. The oligosaccharide is arranged into inner and outer layers. *sec18* mutants of *S. cerevisiae* which fail to add the outer layer produce a core glycosylated product of molecular mass 85–90-kD. *P. pastoris* normally lacks invertase, and thus has a suc⁻ phenotype. Engineered cells containing the *AOX*-invertase cassette integrated at either the *HIS4* or *AOX* locus are suc⁺. Expression is tightly regulated and requires methanol induction. Invertase protein is secreted into the medium, producing a yield of up to 2.5 g L^{-1} of culture medium. This is an exceptionally high yield of secreted product being at least an order of magnitude higher than is normally the case for *S. cerevisiae* or *Aspergillus nidulans*. The product is glycosylated, but it resembles the product from *sec18* mutants of *S. cerevisiae* more closely than that from wild-type cells. Another noteworthy difference between the two yeasts is the fate of the secreted product. In *S. cerevisiae* it is trapped in the periplasmic space, whereas in *P. pastoris* it is secreted into the medium.

Besides β-galactosidese, HBsAg, and invertase, the cloning and expression of human tumour necrosis factor at a level of $6-10\,g\,L^{-1}$ has been demonstrated (Sreekrishna et al., 1988).

Heterologous gene expression has also been reported in H. polymorpha. The structural gene for neomycin phosphotransferase from Tn5 has been cloned between the AOX promoter and terminator, and integrated disruptively at the AOX locus (Ledeboer et al., 1987). The integrated NEO gene conferred high-level resistance ($> 15\,mg\,mL^{-1}$) to the neomycin analogue G418. The integrated gene retains the strict physiological control of the AOX promoter. Another gene which has been expressed to a high level is α-galactosidase from Cyamposis tetragnobla (guar). This enzyme is active in the processing of guar gum. A cassette consisting of the AOX promoter-S. cerevisiae invertase signal sequence-α-galactosidase gene-AOX terminator has been introduced into cells, either on the autonomously replicating plasmid YEp13 or after integration by what was probably a duplicative process, so that an intact copy of the AOX gene has been retained. This enzyme is normally secreted and glycosylated in guar plants and a high level of secretion is observed in the engineered yeast cells. Expression is strictly regulated as expected, indeed the convenient colorimetric assays available to detect and quantify α-galactosidase activity make this an ideal reporter enzyme to study the regulation of the AOX gene, and its use in this respect has been alluded to above.

10.5 Concluding remarks

It is evident that progress has advanced in methylotrophic yeast to the point where genetic analysis is fully fledged and complex studies of regulation are possible. The potential for heterologous gene expression has been fully realized in P. pastoris and has been demonstrated in H. polymorpha. Compared to S. cerevisiae, methylotrophic yeast can produce a higher yield of heterologous products using a cheaper fermentation substrate while maintaining the advantages of secretion and glycosylation.

Acknowledgements

We are grateful to J.M. Cregg, K. Sreekrishna, A.A. Sibirny and I.I. Tolstorukov for making available material prior to publication and for permission to reproduce diagrams.

References

Babel, W. and Hofmann, K.H. (1982) The relationship between the assimilation of methanol and glycerol in yeasts. Arch. Microbiol. 132: 179.

Benevolenskii, S.V. and Tolstorukov, I.I. (1980) Study of the mechanisms of mating and self-diploidization in haploid yeasts Pichia pinus. III. Study of heterothallic mutants. Genetika 16: 1342.

Bodunova, E.N., Donich, V.N., Nesterova, G.F. and Soom, Y.A. (1986a) Genetic lines of Hansenula polymorpha yeast. Communication I. Preparation and characterisation of genetic lines. Genetika 22: 741.

Bodunova, E.N., Donich, V.N. and Nesterova, G.F. (1986b) Genetic lines of *Hansenula polymorpha* yeast. II Inheritance of abnormalities in meiotic segregation. *Genetika* **22**: 939.

Borman, C. and Sahm, H. (1978) Degradation of microbodies in relation to activities of alcohol oxidase and catalase in *Candida boidinii*. *Arch. Microbiol.* **117**: 67.

Carlson, M., Osmond, B. and Botstein, D. (1981) Mutants of yeast defective in sucrose utilisation. *Genetics* **98**: 25.

Ciriacy, M. (1977) Isolation and characterisation of yeasts defective in intermediary carbon metabolism and in carbon catabolite repression. *Mol. Gen. Genet.* **154**: 213.

Costello, W.P. and Bevan, E.A. (1964) Complementation between *ad5* and *ad7* alleles in yeast. *Genetics* **50**: 1219.

Cregg, J.M. and Madden, K.R. (1988) Development of yeast transformation systems and construction of methanol-utilization-defective mutants of *Pichia pastoris* by gene disruption. In *Biological Research on Industrial Yeasts*, eds. Stewart, G.G., Russell, I., Klein, R.D. and Hiebsch, R.R. CRC Press, Baton Rauge (in press).

Cregg, J.M., Barringer, K.J., Hessler, A.Y. and Madden, K.R. (1985) *Pichia pastoris* as a host system for transformation. *Mol. Cell. Biol.* **5**: 3376.

Cregg, J.M., Tschopp, J.F., Stillman, C., Siegel, R., Akong, M., Craig, W.S., Buckholz, R.G., Madden, K.R., Kellaris, P.A., Davies, G.R., Smiley, B.L., Cruze, J., Torregrossa, R., Velicelebi, G. and Thill, G.P. (1987) High-level expression and efficient assembly of hepatitis B surface antigen in the methylotrophic yeast *Pichia pastoris*. *Bio/Technology* **5**: 479.

de Koning, W., Gleeson, M.A., Harder, W. and Dijkhuizen, L. (1987) Regulation of methanol metabolism in the yeast *Hansenula polymorpha*. Isolation and characterization of mutants blocked in methanol assimilatory enzymes. *Arch. Microbiol.* **147**: 375.

Digan, M.E. and Lair, S.V. (1987) Genetics of the methylotrophic yeast *Pichia pastoris*. *Abstr., XIIth Int. Spec. Symp. on Genetics of Non-conventional Yeasts*, 90.

Douma, A.C., Veenhuis, M., de Koning, W., Evers, M. and Harder, W. (1985) Dihydroxyacetone synthase is localised in the peroxisomal matrix of methanol-grown *Hansenula polymorpha*. *Arch. Microbiol.* **143**: 237.

Eggeling, L. and Sahm, H. (1978) Derepression and partial insensitivity to carbon catabolite repression of the methanol dissimilatory enzymes in *Hansenula polymorpha*. *Eur. J. Appl. Microbiol. Biotechnol.* **5**: 197.

Eggeling, L. and Sahm, H. (1980a) Regulation of Alcohol Oxidase synthesis in *Hansenula polymorpha*: oversynthesis during growth on mixed substrates and induction by methanol. *Arch. Microbiol.* **127**: 119.

Eggeling, L. and Sahm, H. (1980b) Direct enzymatic assay for alcohol oxidase, alcohol dehydrogenase and formaldehyde dehydrogenase in colonies of *Hansenula polymorpha*. *Appl. Environ. Microbiol.* **42**: 268.

Egli, T., Van Dijken, J.P., Veenhuis, M., Harder, W. and Fiechter, A. (1980) Methanol metabolism in yeasts: Regulation of the synthesis of catabolite enzymes. *Arch. Microbiol.* **124**: 115.

Egli, T., Kappeli, O. and Fiechter, A. (1982) Regulatory flexibility of methylotrophic yeast in chemostat cultures. Simultaneous assimilation of glucose and methanol at a fixed dilution rate. *Arch. Microbiol.* **131**: 1.

Ellis, S.B., Brust, P.F., Koutz, P.J., Waters, A.F., Harpold, M.M. and Gingeras, T.R. (1985) Isolation of alcohol oxidase and two other methanol regulatable genes from the yeast *Pichia pastoris*. *Mol. Cell. Biol.* **5**: 1111.

Entian, K-D. (1981) A carbon catabolite repression mutant of *Saccharomyces cerevisiae* with elevated hexokinase activity: Evidence for regulatory control of hexokinase PII synthesis. *Mol. Gen. Genet.* **184**: 278.

Entian, K-D. and Zimmermann, F.K. (1980) Glycolytic enzymes and intermediates in carbon catabolite repression mutants of *Saccharomyces cerevisiae*. *Mol. Gen. Genet.* **177**: 345.

Entian, K-D. and Mecke, D. (1982) Genetic evidence for a role of hexokinase isozyme PII in carbon catabolite repression in *Saccharomyces cerevisiae*. *J. Biol. Chem.* **257**: 870.

Fujii, T. and Tonomura, K. (1975) Oxidation of methanol and formaldehyde by a system containing alcohol oxidase and catalase purified from *Candida sp. N16*. *Agric. Biol. Chem.* **39**: 2325.

Giuseppin, M.L.F., van Eijk, H.M.J. and Bes, B.C.M. (1988) Molecular regulation of methanol oxidase activity in continuous cultures of *Hansenula polymorpha*. *Biotechnol. Bioeng.* (in press).

Gleeson, M.A. (1986) Genetic analysis of the methylotrophic yeast *Hansenula polymorpha*. PhD Thesis, University of Sheffield, UK.

Gleeson, M.A.G. and Sudbery, P.E. (1988) Genetic analysis in the methylotrophic yeast *Hansenula polymorpha*. *Yeast* (in press).

Gleeson, M.A. and Sudbery, P.E. (1988) The methylotrophic yeasts. *Yeast* **4**: 1.

Gleeson, M.A., Waites, M.J. and Sudbery, P.E. (1984) Development of techniques for genetic analysis in the methylotrophic yeast *Hansenula polymorpha*. In *Microbial Growth on C1 Compounds*, eds. Crawford, R.L. and Hanson, R.S. American Society for Microbiology, Washington DC, 228.

Gleeson, M.A., Ortori, G.S. and Sudbery, P.E. (1986) Transformation of the methylotrophic yeast *Hansenula polymorpha*. *J. Gen. Microbiol.* **132**: 3459.

Goodman, J.M. (1985) Dihydroxyacetone synthase is an abundant constituent of the methanol-induced peroxisome of *Candida boidinii*. *J. Biol. Chem.* **260**: 7108.

Holtzer, H. (1976) Catabolite inactivation in yeast. *Trends Biochem. Sci.* **1**: 178.

Janowicz, Z.A., Eckart, M.R., Drewke, C., Roggenkamp, R., Hollenberg, C.P., Maat, J., Ledeboer, A.M., Visser, C. and Virrips, C.T. (1985) Cloning and characterization of the DAS gene encoding the major methanol assimilatory enzyme from the methylotrophic yeast *Hansenula polymorpha*. *Nucleic Acids Res.* **13**: 3043.

Kato, N., Tamaoki, H., Tani, Y. and Ogata, K. (1972) Purification and characterization of formaldehyde dehydrogenase in a methanol utilizing yeast, *Kloeckera sp.* no. 2201. *Agric. Biol. Chem.* **36**: 2411.

Kato, N., Omori, Y., Tani, Y. and Ogata, K. (1976) Alcohol oxidase of *Kloeckera sp.* no. 2201 and *Hansenula polymorpha*. Catalytic properties and subunit structure. *Eur. J. Biochem.* **64**: 341.

Kato, N., Sakazawa, C., Nishizawa, T., Tani, Y. and Yamada, H. (1980) Purification and characterization of S formylglutathione hydrolase from a methanol utilizing yeast *Kloeckera sp.* no. 2201. *Biochim. Biophys. Acta* **611**: 323.

Ledeboer, A.M., Maat, J., Visser, C., Bos, J.W., Verrips, C.T., Janowicz, Z., Eckart, M., Roggenkamp, R. and Hollenberg, C.P. (1985) Molecular cloning and characterization of a gene coding for methanol oxidase in *Hansenula polymorpha*. *Nucleic Acids Res.* **13**: 3063.

Ledeboer, A.M. Zoetmulder, M.C.M., Veale, R. and Sudbery, P.E. (1987) Development of the methylotrophic yeast *Hansenula polymorpha* into a host for the efficient expression of heterologous genes. *Proc. Fourth Eur. Biotechnol. Conf. Vol. 1*, 510.

Lee, J.D. and Komagata, K. (1980a) *Pichia cellobiosa, Candida cariosilignicola* and *Candida succiphilia*, new species of methanol utilising yeast. *Int. J. Syst. Bacteriol.* **30**: 514.

Lee, J.D. and Komagata, K. (1980b) Taxonomic study of methanol assimilating yeast. *J. Gen. Appl. Microbiol.* **26**: 133.

Linton, J.D. and Niekus, H.G.D. (1987) The potential of one-carbon compounds as fermentation feedstocks. *Antonie van Leewenhoek* **53**: 55.

Lodder, J. (1970) *The yeasts, a Taxonomic Study*. North Holland, Amsterdam.

Mellor, J., Dobson, M.J., Roberts, N.A., Kingsman, A.J. and Kingsman, S.M. (1985) Factors affecting heterologous gene expression in *Saccharomyces cerevisiae*. *Gene* **33**: 215.

Neben, I., Sahm, H. and Kula, M. (1980) Studies on an enzyme S formylglutathione hydrolase of the dissimilatory pathway of methanol in *Candida boidinii*. *Biochem. Biophys. Acta* **614**: 81.

Neigeborn, L. and Carson, M. (1984) Genes affecting the regulation of *SUC2* gene expression by glucose repression in *Saccharomyces cerevisiae*. *Proc. Natl. Acad. Sci. USA* **81**: 1172.

O'Conner, M.L. and Quayle, J.R. (1979) Mutants of *Hansenula polymorpha* and *Candida boidinii* impaired in their ability to grow on methanol. *J. Gen. Microbiol.* **113**: 203.

Ogata, K., Nishikawa, H. and Ohsugi, M. (1969) A yeast capable of utilising methanol. *Agr. Biol. Chem.* **33**: 1519.

Popova, I.A. and Pugovkina, N.A. (1982) Genetic control of the assimilation of methanol in the yeast *Pichia pinus*. Communication I. Production and genetic characterization of mutants not assimilating methanol. *Genetika* **18**: 916.

Roggenkamp, R., Janowicz, Z., Stanikowski, B. and Hollenberg, C.P. (1984) Biosynthesis and regulation of the peroxisomal methanol oxidase from the methylotrophic yeast *Hansenula polymorpha*. *Mol. Gen. Genet.* **194**: 489.

Roggenkamp, R., Hansen, H., Eckart, M. Janowicz, Z. and Hollenberg, C.P. (1986) Transformation of the methylotrophic yeast *Hansenula polymorpha* by autonomous replication and integration vectors. *Mol. Gen. Genet.* **202**: 302.

Roman, H. (1956) A system selective for mutations affecting synthesis of adenine in yeast. *Compt. Rend. Lab. Carlsberg Ser. Physiol.* **26**: 299.

Sahm, H. and Wagner, F. (1973a) Microbial assimilation of methanol. The ethanol- and methanol-oxidizing enzymes of the yeast *Candida boidinii*. *Eur. J. Biochem.* **36**: 250.

Sahm, H. and Wagner, F. (1973b) Microbial assimilation of methanol. Properties of formaldehyde dehydrogenase and formate dehydrogenase from *Candida boidinii*. *Arch Microbiol.* **90**: 263.

Sakai, Y., Sawai, T. and Tani, Y. (1987) Isolation and characterisation of a catabolite repression-insensitive mutant of a methanol yeast, *Candida boidinii* A5, producing alcohol oxidase in glucose-containing medium. *Appl. Env. Microbiol.* **53**: 1812.

Schutte, H., Flossdorf, J., Sahm, H. and Kula, M. (1976) Purification and properties of formaldehyde dehydrogenase and formate dehydrogenase from *Candida boidinii*. *Eur. J. Biochem.* **62**: 151.

Sibirny, A.A. and Ksheminskaya, G.P. (1987) Mutants of methylotrophic yeast *Hansenula polymorpha* deficient in formaldehyde reducing enzyme. *Abst., XIIth Int. Spec. Symp. on Yeast: Genetics of Non-conventional Yeast*, 59.

Sibirny, A.A. and Vitvitskaya, O.P. (1987) Mutants of methylotrophic yeast *Hansenula polymorpha* deficient in formaldehyde reducing enzyme. *Abst., XIIth Int. Spec. Symp. on Yeast: Genetics of Non-conventional Yeast*, 60.

Sibirny, A.A. and Gonchar, M.V. (1987) The mutants of methylotrophic yeast *Hansenula polymorpha* deficient in citrate synthase, the key enzyme of the tricarboxylic acid cycle. *Abstr., XIIth Int. Spec. Symp. on Yeast: Genetics of Non-conventional Yeast*, 58.

Sibirny, A.A. and Titorenko, V.I. (1987) The inhibition by methanol of isocitrate lyase and malate synthase induction in the mutant of the yeast *Pichia pinus ecrI* with affected repression of the enzymes of methylotrophic metabolism by ethanol. *Biotekhnologiya* **5** (in Russian).

Sibirny, A.A., Titorenko, V.I., Benevolenskii, S.V. and Tolstorukov, I.I. (1986a) Differences in the mechanism of ethanol and glucose catabolite repression of the enzymes of methanol metabolism in the yeast *Pichia pinus*. *Genetika* **22**: 439.

Sibirny, A.A., Titorenko, V.I., Benevolenskii, S.V. and Tolstorukov, I.I. (1986b) On regulation of methanol metabolism in the mutant of *Pichia pinus* yeast deficient in isocitrate lyase. *Biokhimya* **51**: 16.

Sibirny, A.A., Titorenko, V.I., Ubiyvovk, V.M., Ksheminskaya, G.P., Gonchar, M.V. and Vitvitskaya, O.P. (1986c) Regulation of synthesis and activity of enzymes involved in methylotrophic metabolism in yeasts. In *Abstr. 5th Int. Symp. on Microbial Growth on C1 Compounds*, eds. Divine, J.A. and van Kerseveld, H.W., Free University Press, Amsterdam, 101.

Sibirny, A.A., Titorenko, V.I., Efremov, B.D. and Tolstorukov, I.I. (1987) Multiplicity of mechanisms of carbon catabolite repression involved in the synthesis of alcohol oxidase in the methylotrophic yeast *Pichia pinus*. *Yeast* **3**: 233.

Sibirny, A.A., Titorenko, V.I., Gonchar, M.V., Ubiyvovk, V.A., Ksheminskaya, G.P. and Vitvitskaya, O.P. (1988) Genetic control of methanol utilization in yeasts. *J. Basic Microbiol.* **28** (in press).

Sreekrishna, K., McCombie, W.R., Potenz, R., Parker, K.A., Mazzaferro, P.K., Holden, K.A., Hubbard, C. and Fuke, M. (1987) Molecular basis for high level expression of heterologous proteins in *Pichia pastoris*. *Abstr., XIIth Int. Spec. Symp. Yeast: Genetics of Non-conventional Yeast*, 34.

Sreekrishna, K., Potenz, R.H.B., Cruze, J.A., McCombie, W.R., Parker, K.A., Paul, L.N., Mazzaferro, K., Holden, K.A., Harrison, R.G., Wood, P.J., Phelps, D.A., Hubbard, C.E. and Fuke, M. (1988) High level expression of heterologous proteins in methylotrophic yeast *Pichia pastoris*. *J. Basic Microbiol.* **28** (in press).

Stinchcomb, D.T., Thomas, M., Kelly, J., Silker, E. and Davis, R.W. (1980) Eukaryotic DNA segments capable of autonomous replication in yeast. *Proc. Natl. Acad. Sci. USA* **77**: 4559.

Tani, Y., Miya, T., Nishikawa, H. and Ogata, K. (1972a) Part I. Formation and crystallization of methanol oxidizing enzyme in a methanol utilizing yeast *Kloeckera sp.* no. 2201. *Agric. Biol. Chem.* **36**: 68.

Tani, Y., Miya, T. and Ogata, K. (1972b) Part II. Properties of crystalline alcohol oxidase from *Kloeckera sp.* no. 2201 *Agric. Biol. Chem.* **36**: 76.

Tikhomirova, L.P., Ikonomova, R.N. and Kuznetsova, E.N. (1986) Evidence for autonomous replication and stabilization of recombination plasmids in the transformants of yeast *Hansenula polymorpha*. *Curr. Genet.* **10**: 741.

Titorenko, V.I., Efremov, B.D., Tolstorukov, I.I. and Sibirny, A.A. (1987) Genetic and biochemical analysis of ethanol non-utilising mutants of the yeast *Pichia pinus*. *Abstr., XIIth Int. Spec. Symp. Yeasts: Genetics of Non-conventional Yeast*, 61.

Tolstorukov, I.I. and Benevolenskii, S.V. (1978) Study of the mechanism of mating and self-diploidization in haploid yeasts Pichia pinus. Communication I. Bipolarity of mating. Genetika 14: 519.

Tolstorukov, I.I. and Benevolenskii, S.V. (1980) Study of the mechanisms of mating and self-diploidization in haploid yeasts Pichia pinus. II. Mutations in the mating type locus. Genetika 16: 1335.

Tolstorukov, I.I. and Efremov, B.D. (1984) Genetic mapping of yeast Pichia pinus. II. Mapping by tetrad analysis. Genetika 20: 1099.

Tolstorukov, I.I., Dutova, T.A., Benevolenskii, S.V. and Soom, Y.O. (1977) Hybridization and genetic analysis of the methanol yeasts Pichia pinus. Genetika 13: 322.

Tolstorukov, I.I., Bliznik, K.M. and Korogodin, V.I. (1979) Mitotic instabilities of diploid cells of the yeast Pichia pinus. I. Spontaneous segregation. Genetika 15: 2140.

Tolstorukov, I.I., Bliznik, K.M. and Korogodin, V. (1982) Mitotic instability of diploid Pichia pinus cells. II. Segregation induced by gamma irradiation. Genetika 18: 1276.

Tolstorukov, I.I., Efremov, B.D. and Bliznik, K.M. (1983) Construction of a genetic map of the yeast Pichia pinus. I. Determination of linkage groups using induced mitotic haploidization. Genetika 19: 897.

Tschopp, J.F., Brust, P.F., Cregg, J.M., Stillman, C.A. and Gingeras, T.R. (1987a) Expression of the lacZ gene from two methanol-regulated promoters in Pichia pastoris. Nucleic Acids Res. 15: 3859.

Tschopp, J.F., Sverlow, G., Kosson, R., Craig, W. and Grinna, L. (1987b) High-level secretion of glycosylated invertase in the methylotrophic yeast Pichia pinus. Biotechnology 5: 1305.

Van Dijken, J.P., Veenhuis, M., Kreger-van-rij, N.J.W. and Harder, W. (1975) Microbodies in methanol assimilating yeast Arch. Microbiol. 102: 41.

Van Dijken, J.P., Oostra-Demkes, G.J., Otto, R. and Harder, W. (1976) S formylglutathione: The substrate for formate dehydrogenase in methanol utilizing yeasts. Arch. Microbiol. 111: 77.

Veenhuis, M., Van Dijken, J.P. and Harder, W. (1983) The significance of peroxisomes in the metabolism of one carbon compounds in yeast. Adv. Microb. Physiol. 24: 2.

Waites, M.J. and Quayle, J.R. (1983) Dihydroxyacetone synthase: A special transketolase for formaldehyde fixation from the methylotrophic yeast Candida boidinii CBS5777. J. Gen. Microbiol. 129: 935.

Yamada, H., Shin K., Kato, N., Shimizu, S. and Tani, Y. (1979) Purification and characterization of alcohol oxidase from Candida 25A. Agric. Biol. Chem. 43: 877.

Zimmermann, F.K. and Scheel, I. (1977) Mutants of Saccharomyces cerevisiae resistant to carbon catabolite repression. Molec. Gen. Genet. 154: 75.

Zimmermann, F.K., Kaufman, Rasenberger, H. and Haubmann, P. (1977) Genetics of carbon catabolite repression in Saccharomyces cerevisiae. Genes involved in the depression process. Molec. Gen. Genet. 151: 95.

11 The two-micron circle: model replicon and yeast vector

P.A. MEACOCK, K.W. BRIEDEN and A.M. CASHMORE

11.1 Introduction and overview

The 2-μm circle is an autonomously replicating 6.3-kb double-stranded DNA plasmid found in most strains of *Saccharomyces cerevisiae*. There are 50–100 copies per haploid genome, comprising 1–2% of the total DNA. Attempts to identify a role for the plasmid in the general biology of the yeast cell have been unsuccessful, and no phenotype has been consistently associated with its presence. A claim that the circles specified oligomycin resistance (Guerineau *et al.*, 1974) has failed to be substantiated.

Early biochemical studies suggested that 2-μm circles may be located in the cytoplasm (Clark-Walker and Miklos, 1974). However, Livingston (1977) concluded from cytoduction experiments that the 2-μm circle is sequestered in the nucleus but inherited like a cytoplasmic element, i.e. demonstrating 4:0 segregation at meiosis. There are now several lines of evidence consistent with a nuclear association. The recombination of 2-μm-based recombinant plasmids carrying nuclear genes with homologous chromosomal DNA, indicates that the 2-μm circle can come within close proximity to the chromosomes (Kielland-Brandt *et al.*, 1980). Furthermore the plasmid is packaged into nucleosomes, like chromosomal DNA, containing the normal composition of core histones (Livingston and Hahne, 1979; Nelson and Fangman, 1975) and also cosediments with a nuclear DNA complex (Taketo *et al.*, 1980).

Plasmid replication is, to a certain extent, under the same controls as that of chromosomal DNA. Wild-type *CDC* functions which are necessary for either initiation of the DNA synthesis phase (*CDC* 28, 4 and 7) or execution of chromosomal DNA replication (*CDC* 8) are also required for 2-μm replication (Petes and Williamson, 1975; Livingston and Kupfer, 1977). Density transfer experiments indicate that replication of the circles occurs near the start of S-phase, and proceeds in a semi-conservative manner, with all the molecules in the cell normally replicating once per cell cycle (Zakian *et al.*, 1979). In addition, the 2-μm circle is extensively transcribed by RNA polymerase II, to produce polyadenylated mRNA species (Broach *et al.*, 1979).

Electron microscopy studies of reannealed plasmid DNA revealed that the molecule is composed of two unique regions separated by two inverted repeat sequences of approximately 600-bp (Guerineau *et al.*, 1976, Royer and Hollenberg, 1976). Restriction enzyme digestions further demonstrated that the population of molecules was composed of two isomeric forms, differing in the relative orientation of one unique region to the other (Beggs *et al.*, 1976; Guerineau *et al.*, 1976; Hollenberg *et al.*, 1976*a*; Gubbins *et al.*, 1977).

The plasmid has now been fully sequenced, revealing the presence of four major open reading frames (ORFs), which have been termed *A*, *B*, *C* and *D* in decreasing order of size (Hartley and Donelson, 1980). The two inverted repeats comprise a duplication of a 599-bp sequence, and separate the plasmid molecule into two unique sequence regions of 2774-bp and 2346-bp. Reading frames *B* and *D* are located within the larger unique region, and ORFs *A* and *C* within the smaller region. Figure 11.1 shows the two molecular forms of the molecule and the locations of key genetic loci.

Reading frame *A* has been shown to encode a *trans*-acting product (*FLP*) which catalyses the inversion of the molecule via a site-specific recombination between the inverted repeat sequences (Broach and Hicks, 1980). This intramolecular recombination process is the underlying mechanism of a copy-amplification system (Futcher, 1986; Volkert and Broach, 1986).

In addition to the origin of replication (*ORI*), the other elements that have a major role in plasmid maintenance are *trans*-acting functions encoded by ORFs *B* (*REP1*) and *C* (*REP2*), and a *cis*-acting locus *STB* (*REP3*) (Kikuchi, 1983; Jayaram *et al.*, 1983*a*). It is now clear that the principal role of these functions is in plasmid partitioning (Kikuchi, 1983; Cashmore *et al.*, 1986). Recent evidence also implicates *D* in plasmid maintenance (Murray *et al.*, 1987, Cashmore *et al.*, 1988).

From this brief consideration of the roles of 2-μm-encoded functions it is clear that the plasmid is dedicated solely to its own maintenance. Indeed, Futcher and Cox (1983) propose the 2-μm circle to be an intracellular parasite which efficiently ensures its own propagation while causing minimal damage to its host. Circular DNA plasmids which have a strikingly similar genetic organization to the 2-μm circle have been detected in other yeast species (Utatsu *et al.*, 1987; Murray *et al.*, 1988).

In addition to being a good model system for the study of eukaryotic cell DNA replication and maintenance, the 2-μm circle has provided the basis for many yeast cloning vectors. We have therefore included in this review a discussion of the salient features of 2-μm biology, their relevance and how they might be best utilized in future developments.

We have attempted to standardize 2-μm genetic terminology. Historically, functions encoded by ORFs *B* and *C* and the *cis*-acting stability locus have become termed *REP1*, *REP2* and *REP3* respectively (Jayaram *et al.*, 1983*a*). This arose from the original suggestion that their role was in replication and copy-amplification. Since then it has become apparent that their role is in plasmid partioning. In order to eliminate confusion by the '*REP*' terminology we have preferred to designate these loci *B*, *C* and *STB*. Also, we have used [cir⁺] to denote yeast strains with native 2-μm circles and [cir⁰] to identify plasmid-free strains.

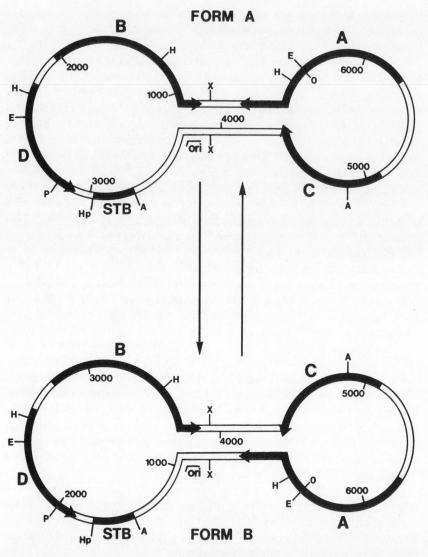

Figure 11.1 Structure of the 2-μm circle. The figure shows the two isomeric molecular forms (A and B) of the 6.3-kb 2-μm circle, annotated with the four open reading frames *A* (*FLP*), *B* (*REP1*), *C* (*REP2*) and *D*, and the two *cis*-acting loci *ORI* and *STB* (*REP3*). The molecules are drawn as a dumbbells with the bars representing the two copies of the inverted repeat sequences. The locations of cleavage sites for restriction endonucleases *Ava*I (A), *Eco*RI (E), *Hind*III (H), *Hpa*I (Hp), *Pst*I (P) and *Xba*I (X) are as indicated. Data from Hartley and Donelson (1980).

11.2 Expression of the 2-μm circle genes

The four long open reading frames identified by sequence analysis (Hartley and Donelson, 1980) are arranged as divergently expressed pairs in each of the unique regions (Figure 11.1). All four of the open reading frames are transcribed into mRNAs which are apparently not derived by extensive processing of precursor transcripts. The mRNAs are protected by a 5′-cap and 3′-polyadenylation. S1 nuclease protection experiments and primer extension analysis have been used to identify the position of the initiation and polyadenylation sites. ORFs B, C and D appear to utilize multiple transcription initiation sites, and more than one polyadenylation site has been mapped in the case of the ORF D mRNA. The three largest ORFs are preceded by recognizable TATA-like sequences (Sutton and Broach, 1985), consistent with their being transcribed by the nuclear RNA polymerase II.

Besides the four transcripts encompassing each of the four ORFs, two more large RNA species, '1950' and '1620', mapped by Northern analysis (Broach *et al.*, 1979; Sutton and Broach, 1985), are transcribed from the 2-μm circle in the large unique region of the plasmid. The '1950' transcript coterminates with the '1325' ORF B transcript, and the position of its 5′ terminus is situated somewhere in the region of STB-distal. The '1950' transcript therefore carries the messages of ORF D (antisense) and ORF B. By use of single-stranded DNA hybridization probes, Murray *et al.* (1987) have recently detected a 600-base RNA that is also antisense to ORF D and is colinear with the 5′ portion of the '1950' RNA. The relationship between the '600' and '1950' RNAs is not known. If the smaller is not derived from the larger by endonucleolytic processing, then there must be some sort of signal, perhaps within the 5′ region of ORF D, that terminates the '600' species but not the '1950'. Presumably the '1620' transcript coterminates with the '700' ORF D mRNA (Sutton and Broach, 1985). Although its 5′ terminus is not yet accurately mapped, it can be deduced that transcription of '1620' starts from somewhere in the region between map positions 1200 and 1300. It is not clear whether the '1950' and '1620' transcripts are translated into protein.

No transcripts have been found which traverse the inverted repeats. In fact, they contain terminators for RNA transcripts that originate from within the unique sequence regions. Additionally, the region of the large unique sequence containing the two *cis*-acting loci ORI and STB is also transcriptionally silent. A transcription map of the 2-μm circle plasmid (A-form isomer) is shown in Figure 11.2, and an overview of the precise locations of the ORF transcripts is presented in Table 11.1.

Hybridization analysis of the RNAs showed that the '1325' (ORF B) and '1275' (ORF C) species were easily detected in [cir⁺], cells whereas the '1425' (ORF A) was not, indicating differences in their relative abundances (Broach *et al.*, 1979; Sutton and Broach, 1985). By comparing the level of mRNA expressed from integrated single copies of individual 2-μm genes in [cir⁺] and

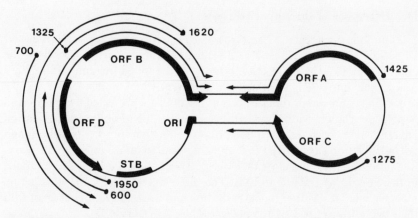

Figure 11.2 Transcription map of the yeast 2-μm plasmid. The map positions of the major transcripts are indicated on a diagram of the A-form of the 2-μm genome. Each transcript is identified by its length in bases as determined by Northern analysis, and the directions of transcription are indicated by the arrowheads. Data from Broach *et al.* (1979*a*), Sutton and Broach (1985) and Murray *et al.* (1987).

Table 11.1 Two-micron open reading frames and RNA transcripts. Positions of ORFs are identified by the coordinate system of Hartley and Donelson (1980) for the A form of the 2-μm circle. Positions of the major 5′-cap and 3′-polyadenylation sites and transcript sizes are from Sutton and Broach (1985).

Gene	ORF position (size)	5′ cap site of mRNA (map position)	3′ poly-adenylation site of mRNA (map position)	Transcript size (from Northern analyses)	Transcript size (from map data
A (FLP)	5570 ⇒ 520 (1269 bp)	5549	545	1425 b	1305 b
B (REP 1)	2008 ⇒ 890 (1119 bp)	2008, 2035	833	1325 b	1200 b
C (REP 2)	5198 ⇒ 2814 (888 bp)	5201–5198 5186–5184	4107	1275 b	1090 b
D	2271 ⇒ 2814 (544 bp)	2251, 2257	2841, 2860	700 b	600 b

[cir⁰] cells, Som *et al.* (1988) found that the abundances of the '1950', '1425', '1325' and '700' species were reduced by the presence of other 2-μm gene products. In contrast, the level of the '1275' RNA remained unaffected. Thus it seems likely that synthesis of several of the 2-μm RNAs is subject to negative regulation by other 2-μm products. In an elegant series of experiments using strand-specific probes and overexpression of various combinations of 2-μm genes, Murray *et al.* (1987) were able to analyse this in more detail. Production of the '1950', '1425' and '700' RNAs was reduced when ORFs *B* and *C* were co-ordinately over-expressed. Furthermore, overexpression of ORF *D* enhanced the abundances of the '1950', '1425' and '1325' species. The implications of these transcriptional regulatory circuits for the co-ordination of 2-μm functions is discussed elsewhere in this chapter.

The four ORFs could potentially encode proteins of molecular masses 48 546 (A), 43 164 (B), 33 140 (C) and 21 247 (D). Introduction of recombinant 2-μm plasmids into the E. coli minicell system resulted in production of a number of 2-μm specified polypeptides in the size range 15–48 kD (Hollenberg, 1978). Since this is a heterologous system in which expression is based on fortuitous recognition of expression signals on the plasmid DNA, it is difficult to draw meaningful conclusions from the data. However, high-level expression in E. coli has facilitated purification of the FLP (A) protein (Meyer-Leon et al., 1987) which has an apparent molecular weight of 45 kD on SDS-PAGE and an aminoterminal sequence identical to that predicted from the DNA sequence. As this protein has yet to be purified from yeast, it is not clear whether the 45-kD size of the product from E. coli represents an aberrant SDS-PAGE migration rate or post-translational processing.

Fusions to the E. coli lacZ gene have shown that all four reading frames are translated in yeast (Reynolds et al., 1987; Brieden and Meacock, unpublished observations). However, only the product of reading frame B has been directly detected. Antiserum raised to a B-β-galactosidase fusion protein has been shown to cross-react to a 48-kd protein in [cir$^+$] yeast cells. This protein, which approximates to the expected size for the ORF B product, is not found in [cir^0] cells. In addition, it has the correct aminoterminal sequence for B and is overproduced in cells carrying plasmids that express ORF B under the control of the GAL10 promoter (Broach et al., 1983a; Wu et al., 1987). Interestingly, the B protein was found to copurify with the nuclear matrix fraction, suggesting that it may be a karyoskeletal protein. Consistent with this view is the fact that the protein shows amino acid sequence similarities to three fibrous proteins, myosin, vimentin and tubulin (Wu et al., 1987). The implications of these observations for the mechanism of plasmid partitioning are discussed below.

11.3 The plasmid-specific intramolecular recombination system

Initial insertional mutagenesis studies identified the FLP (A) ORF as encoding a trans-acting function necessary for the inversion of the 2-μm circle (Broach and Hicks, 1980; Broach et al., 1982). These authors also demonstrated that the recombination event was site-specific and required sequences within a 65-bp region inside the 599-bp inverted repeat sequences (Broach et al., 1982). It is now known that FLP-mediated recombination can occur not only between inverted copies of the 2-μm repeats, but also between copies of the repeats arranged in a direct orientation, and between repeats on separate molecules. This intermolecular recombination results in dimer formation (Jayaram et al., 1983b). The site-specific recombination process has now been extensively studied using purified components to reconstruct the system *in vitro*. These studies of the molecular mechanism of the recombination process are highlighted in this section.

The *FLP* protein is encoded by the largest ORF on the 2-μm circle, and has a predicted molecular mass of 48 546. There are several lines of evidence that indicate that this protein is the only one required for the site-specific recombination process. Extensive mutagenesis has failed to reveal any essential host genes (McLeod *et al.*, 1984; Volkert and Broach, 1987), and studies of the operation of the system in a heterologous host, *E. coli*, suggest that either *FLP* acts alone, or that any other yeast functions required can be complemented by *E. coli* functions. Studies *in vitro* also support these conclusions (Vetter *et al.*, 1983; Sadowski *et al.*, 1984; Meyer-Leon, 1984; Babineau *et al.*, 1985).

The *FLP* protein has been purified using various protocols (Meyer-Leon *et al.*, 1984; Sadowski *et al.*, 1984; Babineau *et al.*, 1985; Meyer-Leon *et al.*, 1987). Using the method as described by Babineau *et al.* (1985), Gronostajski and Sadowski (1985c) demonstrated a concentration effect on the functioning of the *FLP* protein. At low concentrations, efficient intramolecular recombination was promoted, whereas at high levels, intermolecular recombination was favoured. At higher levels still, all types of *FLP*-mediated recombination were inhibited. However, Meyer-Leon *et al.* (1987) found that this concentration effect was not observed using a more highly purified *FLP* protein. They used a purification procedure yielding milligram quantities of protein which was 85% pure. This procedure involved affinity chromatography with an immobilized DNA ligand, whose design was based on knowledge of the recombination site.

Purified *FLP* protein has also enabled investigation of the topological features of the recombination mechanism. Beatty *et al.* (1986) studied the topological changes generated in supercoiled substrates after exposure to *FLP* protein *in vitro*. When the substrate had two *FLP* target sites in inverse orientation, the molecules recovered following exposure to *FLP* protein were multiply-knotted structures. Similarly, a substrate with target sites in direct orientation yielded multiply-interlocked catenates. This study also demonstrated that the *FLP* protein acts as a site-specific topoisomerase during the recombination reaction.

Early studies *in vivo* indicated that the minimal sequence required for *FLP* recombination is considerably less than the 599 bp of the repeat sequence (Broach *et al.*, 1982), and DNase protection studies carried out with purified *FLP* protein *in vitro* (Andrews *et al.*, 1985) demonstrated that the protected region was about 50 bp in length. This region contains an 8-bp core sequence (5'-TCTAGAAA-3') and three 13-bp repeat elements (5'-GAAGTTCCTATTCC-3'). The first two repeats are in direct orientation, and the third is inverted with respect to them, and also separated from them by the 8-bp core.

Gronostajski and Sadowski (1985a) used exonuclease digestion in an attempt to determine the minimal sequence required for recombination. They used exonucleases that degrade a single strand of the duplex DNA from a unique terminus, to generate recombination substrates consisting of a duplex DNA molecule with a single-stranded tail. These studies indicated that only part of two of the 13-bp repeats immediately adjacent to the 8-bp core, and the core itself, were needed for efficient recombination. The other 13-bp outside repeat was not required, an observation supported by other studies *in vitro* (Andrews *et al.*, 1985; Senecoff *et al.*, 1985) and *in vivo* (Jayaram, 1985). Proteau *et al.* (1986) have confirmed and more precisely defined the minimal recombination site, using fully duplex DNA as recombination targets. Their results define the smallest duplex sequence that is able to undergo recombination *in vitro* as the 8-bp core sequence surrounded by two 7-bp inverted sequences.

Andrews *et al.* (1985) demonstrated that *FLP* protein introduces single- and double-strand breaks into the substrate DNA. This cleavage occurs at the margins of the core spacer region, generating 8-bp 5' protruding ends with 5'-OH and 3' protein-bound termini. The outside 13-bp symmetry element was not important for this transient binding and subsequent cleavage. It was subsequently found (Gronostajski and Sadowski, 1985b) that the *FLP* protein was covalently bound to the 3' side of the cleavage site by a phosphotyrosyl linkage. The tyrosine residue which forms the transient linkage of the *FLP* protein with its substrate DNA is located near the carboxyl terminus of the *FLP* protein. Prasad *et al.* (1987) mutated tyrosine 343 of *FLP* to a phenylalanine or serine and, in studies with the mutant proteins, demonstrated that their binding to the substrate was unaffected. However, both mutant proteins were incapable of catalysing strand cleavage and recombination.

In order to analyse in more detail the binding of *FLP* to its target site, Andrews *et al.* (1987) carried out gel retardation assays to detect intermediates in the assembly of a functional protein-DNA complex. On a wild-type substrate, three specific complexes were found. Complexes I and II represented intermediates, while complex III was the stable end product of the reaction. The increments in size between the complexes were regular, corresponding to increases in molecular mass consistent with sequential binding of one, two or three molecules of *FLP* protein. The complexes were sensitive to heating in the presence of ionic detergent and were insensitive to competition with heparin, indicating that they were the result of specific non-covalent *FLP*–DNA interactions. Studies with truncated target sequences demonstrated that all three identical 13-bp sequences provided binding domains. This indicated that although, as discussed above, the outside 13-bp repeat is not apparently important for recombination, it is involved in protein binding. Analysis of defined mutations in both the *FLP* target site and the protein coding sequence should help to further elucidate the nature of the protein–DNA interaction.

M

This approach has already been initiated by Andrews et al. (1986) and Govind and Jayaram (1987).

This account demonstrates that a fairly extensive investigation of the molecular mechanisms of the recombination process has been carried out. But what is the biological significance of the process? Recent genetic studies have provided an answer to this question. It is now apparent that the recombination plays a role in replication control, allowing amplification of plasmid copy number (Volkert and Broach, 1986). This is discussed in a later section.

11.4 Plasmid replication

Despite behaving as an extrachromosomal genetic element, the 2-μm circle exhibits many features of its replication that are typical of nuclear DNA. Studies using synchronized cell cultures have shown that, in contrast to mitochondrial DNA (the major extrachromosomal genetic element of yeast cells), 2-μm DNA is replicated only during the S phase of the cell cycle (Zakian et al., 1979). Indeed, plasmid replication appears to occur during the early part of S phase. It is inhibited in $MATa$ cells that have been arrested in their cell cycle progression by treatment with α-factor mating hormone and, like chromosomal DNA replication, is dependent upon the products of a number of CDC genes (Livingston and Kupfer, 1977; Petes and Williamson, 1975).

Most strikingly, analysis of the density distributions of 2-μm molecules following the transfer of a culture from 'heavy' (^{13}C, ^{15}N) to 'light' (^{12}C, ^{14}N) density-labelled medium has led to the conclusion that the plasmid follows a strictly democratic and uniform pattern of replication (Zakian et al., 1979)— that is, every copy of the plasmid is replicated once, and once only during S phase. This is in marked contrast to other multicopy replicon systems such as bacterial plasmids or mammalian mitochondrial genomes, which demonstrate a 'random choice' relaxed pattern of replication (Rownd, 1969).

Although early electron micrographs of replicating 2-μm molecules led to ambiguous conclusions about the number and location of replication origins (Newlon et al., 1981), all recent data are consistent with the existence of a single origin at a fixed site on the 2-μm molecule. Molecular cloning techniques have demonstrated that the only 2-μm sequences capable of supporting autonomous replication of recombinant plasmids derive from a single region of the plasmid located at the junction of one of the inverted repeats and the large unique region (Broach and Hicks, 1980; Storms et al., 1979).

Deletion and subcloning analysis reveals that the fully functional activity is located between co-ordinates 3596 and 3817 of the Hartley and Donelson sequence, although fragments as short as the 74 bp between coordinates 3660 and 3734 also seem to have some replication ability (Broach et al., 1983b). Plasmids containing only this 2-μm region display extreme mitotic instability despite their multicopy nature. This is characteristic of chromosomal ARS-based plasmids (Murray and Szostak, 1983). Inspection of the DNA sequence in this region reveals the presence of a near-consensus ARS element (map position 3702), and various features of symmetry often found associated with replication origins (Broach, 1981).

More recently, two groups have devised analytical methods, based on two-dimensional gel electrophoresis of replicating molecules, which permit the direct assignment of origin (*ORI*) location and replication direction (Brewer and Fangman, 1987; Huberman et al., 1987). Both rely upon an initial size separation of DNA molecules by electrophoresis through neutral agarose. In one case (Brewer and Fangman, 1987) the second dimension uses high-voltage/high-concentration agarose electrophoresis to fractionate molecules according to their shape and conformation; non-linear and linear molecules of the same size migrate at different rates. The other system (Huberman et al., 1987) uses second-dimension electrophoresis through alkaline agarose to separate the individual DNA strands and so allows analysis of the lengths of nascent strands produced at replication forks. By transferring the fractionated molecules to membrane filters and hybridizing with suitable probes, it has been possible to determine the precise location of the replication origin and terminus, and the direction of replication.

These analyses have led to the following conclusions:

(i) Replication of the 2-μm circle proceeds bidirectionally from a single origin (*ORI*) located at map co-ordinate 3700 ± 100-bp (Form A map)

(ii) The two replication forks proceed at slightly different rates, the one traversing the large unique sequence moving more slowly, to a terminus located at map position 740 ± 200-bp

(iii) There is no evidence for secondary origins of replication, and both isomeric forms of the 2-μm replicate with equal efficiency.

Thus direct mapping of the replication origin (*ORI*) corresponds almost exactly with the ARS element located by genetic and DNA sequence analysis. This lends considerable support to the long-held assumption that ARS elements isolated from chromosomal DNA represent replication origins (Williamson, 1985).

From the preceding account it becomes clear that, in terms of its basic replication properties, the 2-μm circle is remarkably similar to chromosomal replicons. However, as well as obeying the 'eukaryotic' replication rule (one replication per origin per cell cycle), the plasmid is capable of circumventing normal cellular controls in order to maintain itself at high copy number. Cytoduction matings between a strain carrying two physically distinguishable species of 2-μm circles and a [cir^0] strain showed that molecules could be transmitted to the [cir^0] nucleus. In most cases, only one species was recovered, suggesting that only a single molecule was transferred. However, cells of the progeny clone contained that 2-μm species at the usual high copy level (Sigurdson et al., 1981). This finding suggests that multiple rounds of replication must have occurred at some stage of the colony's development.

More recently, it has become clear from molecular biological analysis that copy amplification does occur and is dependent on the 2-μm ORF *A* (*FLP*) gene product (Volkert and Broach, 1986).

11.5 Plasmid copy number amplification

Amplification of the plasmid copy number is mediated by the product of the *FLP* gene (Broach and Hicks, 1980; Broach *et al.*, 1982) which corresponds to ORF *A* ranging from position 5570 to position 520 on the A-form isomer. It encodes a site-specific recombinase (molecular mass 48 546) which catalyses highly efficient recombinational inversion of the unique segments about the 599-bp inverted repeats of the plasmid (Broach and Hicks, 1980; Cox, 1983; Vetter *et al.*, 1983). This results from recombination between short defined sites in the repeats, the *FLP* recombination target (*FRT*) sites (Broach *et al.*, 1982; Andrews *et al.*, 1985; Volkert and Broach, 1986).

Futcher (1986) has proposed a model for copy amplification which obviates forbidden multiple initiations during individual S-phases (see Figure 11.3). This model is based on the fact that the 2-μm circle replicates bidirectionally from its single origin (*ORI*), which is located adjacent to one of the plasmid's inverted repeat sequences. In Futcher's model, FLP-mediated recombination, between one replicated inverted repeat and the unreplicated inverted repeat, leads to amplification by reversing the direction of one replication fork so that the two forks travel around the 2-μm circle in the same direction, rather than converging. Continuing replication in this mode yields a multimeric replication intermediate. Subsequent FLP-recombination events resolve the multimer to monomers, and a reversal of the original event can restore the converging mode of replication fork movement. Indeed, the molecular architecture of the 2-μm circle, and similar plasmids found in other yeasts, appears to have evolved in keeping with this method of copy amplification. Thus, the location of the replication origin at the junction of the large unique sequence region and one of the inverted repeats maximizes the possibility of recombination between the distant unreplicated inverted repeat and a replicated copy.

This model predicts that, in order to be amplifiable, a plasmid molecule must have a pair of intact inverted repeat sequences and that recombination must occur during replication. Volkert and Broach (1986) devised a system to test these predictions. They integrated a *GAL10-FLP* fusion gene into a chromosomal location in a haploid strain. This was then mated with a second strain containing a single copy of a *flp* mutant 2-μm circle sequence, flanked by two copies of the *FLP* recombination site (*FRT*) arranged in a direct repeat configuration, integrated at a different chromosomal location. Induction of the *FLP* gene in the diploid, by growth on galactose, caused the 2-μm sequence to excise from the chromosome via recombination between the two *FRT* direct repeats. The excised 2-μm sequence was able to undergo intramolecular

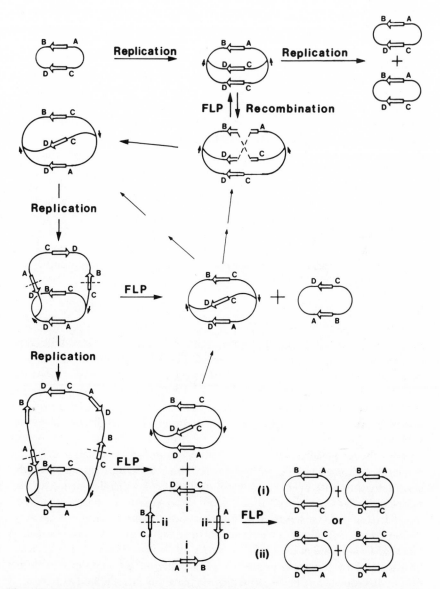

Figure 11.3 Two-micron copy-amplification by intramolecular recombination. The figure shows the *FLP*-mediated copy amplification model proposed by Futcher (1986), starting from the A-form of the plasmid. Open arrows within the plasmid molecules indicate inverted repeat sequences. Small closed arrows indicate direction of replication fork movement. Sites of recombination events are indicated by dashed lines. Recombinations between adjacent inverted repeats, that lead to inversion of the intervening sequence, are not shown.

inversion, also mediated by the induced *FLP* product, and was observed to amplify to a high copy level. However, mutation of one of the inverted repeat *FRT* sites, within the excised 2-μm sequence, abolished both inversion and copy amplification. Moreover, copy amplification occurred only if the *FLP* gene was continuously induced; pulse-induction gave excision and inversion of the 2-μm sequence, but the plasmid DNA remained at a low copy level. Other experiments by Murray *et al.* (1987) and ourselves (Praekelt and Meacock, unpublished observations), in which the *FLP* gene was inducibly expressed from a *GAL* promoter, have also shown that the copy level of the 2-μm circle is directly related to the production of this protein.

Since it has been demonstrated that under steady-state conditions the 2-μm circle exhibits a democratic mode of replication (during a cell cycle each molecule replicates once and once only), we must conclude that this process becomes operational only when the plasmid encounters an unbalanced low copy situation. It is important to note that a 2-μm circle that is only mutant in the *FLP* gene displays high mitotic stability even in a [cir⁰] host (Jayaram *et al.*, 1983a; Veit and Fangman, 1985). The *FLP*-mediated amplification is therefore not the primary mechanism of ensuring plasmid inheritance.

11.6 The plasmid partitioning system

Central to the stability of the plasmid is its ability to ensure transmission to both progeny cells during cell division. This partition process is promoted by the products of the two plasmid coding regions ORF *B* (1008–890) and ORF *C* (5198–4311) which code for the *B* (*REP1*) protein (molecular mass 43 164) and the *C* (*REP2*) protein (molecular mass 33 140). In addition, partitioning requires the *cis*-active stability locus *STB* (*REP3*). Murray and Cesareni (1986) divide this locus into two domains, *STB*-proximal and *STB*-distal, lying either side of the *Hpa* I site at position 2976. *STB*-proximal lies proximal to *ORI* between the *Hpa* I site and the *Ava* I site at position 3258. It is characterized by five tandem direct repeats of 62 bp (Hartley and Donelson, 1980) and has little chromatin structure (Veit and Fangman, 1985). *STB*-distal is distal to *ORI* between the *Pst* I site at position 2656 and the *Hpa* I site. It includes the 3' end of ORF *D* and a transcription terminator (Sutton and Broach, 1985).

The *B* and *C* products have been shown to act in concert at the *cis*-acting locus *STB* (Jayaram *et al.*, 1983a; Kikuchi, 1983; Jayaram *et al.*, 1985). In the absence of *B* or *C* proteins, or when *STB* is disrupted, 2-μm derivatives are rapidly lost due to the failure to segregate plasmids into the bud at mitosis, a phenotype normal for plasmids based on a chromosomal autonomously replicating sequence (Murray and Szostak, 1983). Such *ARS*-plasmids can be stabilized in a [cir⁺] strain by the inclusion of a *STB* sequence (Kikuchi, 1983). Although *STB*-proximal is sufficient for proper partitioning (Jayaram *et al.*, 1985) this function is severely disrupted by an active transcription from neighbouring sequences. Murray and Cesareni (1986) suggest that *STB*-distal is important to protect STB-proximal and ORI from such transcription by a transcription terminator and an additional 'silencer' element which represses expression from any upstream promoter.

The efficiency of 2-μm partitioning is further dependent upon the dosage of ORF B. Whereas one copy of ORF C integrated into a chromosomal location is sufficient to complement a multicopy C^- plasmid, this situation does not hold true for the B gene; one copy of ORF B does not produce enough product to stabilize multiple copies of a B^- plasmid (Cashmore et al., 1986). Therefore, although both B and C are necessary for the partitioning process, the B product appears to be required in large (stoichiometric?) amounts for efficient plasmid distribution, possibly because it is a structural component that interacts directly with the STB locus. As the B protein copurifies with the yeast nuclear matrix (Wu et al., 1987) and exhibits a high degree of homology with vimentin (an intermediate filament component) and with myosin heavy chain, Volkert et al. (1986) propose that B protein may also be incorporated into a yeast nuclear lamina structure. It may therefore act by providing dispersed binding sites for plasmid molecules in a nuclear structure that is equally shared between mother and daughter cells at mitosis.

From the investigations made to date it is not possible to distinguish whether the partioning process actively promotes equidistribution of plasmid molecules, or merely acts to neutralize the maternal bias in inheritance, so allowing the plasmid molecules to segregate stochastically. If the latter were the case, then the efficiency of the partitioning process, and in particular the fraction of cells failing to inherit plasmids, would be directly dependent on the plasmid copy level. Moreover, random partitioning should give rise to a broad distribution of plasmid copy numbers in daughter cells. In this context, it is interesting to note that Futcher and Cox (1983) have reported clonal variation in 2-μm copy number. A definitive answer to this problem will be possible only with the advent of methods which allow quantitation of plasmid copy numbers in the individual cells of a population. Methods based on both indirect (enzyme production levels) and direct (in-situ DNA hybridization) measurements are now under development (Albury, Cashmore and Wheals, unpublished observations).

11.7 Coordination of copy amplification and partitioning

The 2-μm plasmid is efficiently maintained in populations of yeast cells. Cells that have completely lost the 2-μm circle are generated at a rate as low as 1 in 10^6 per [cir$^+$] cell per generation, and the copy number remains remarkably constant (Futcher and Cox, 1983; Cashmore, 1984). There must therefore be some sort of coordination between the maintenance systems to ensure the steady state achieved. In a simple model, the B and C proteins could effect near-random segregation of the 2-μm circle between mother and daughter cells by interacting with the plasmid's domain STB (partition system).

Subsequently, the FLP-mediated copy-amplification process could correct the decreased plasmid copy number that arises in a subset of cells from partitioning error. The recent investigations support the interpretation that the 'partitioning system' and 'amplification system' are interdependent and that the stability of the 2-μm plasmid is indeed caused by an interplay of the two systems. It appears that these may be linked by the concerted regulation of B and C proteins.

Veit and Fangman (1985) examined the contributions that 2-μm-encoded factors make to the chromatin organization of the plasmid by comparing micrococcal nuclease sensitivity of the native 2-μm plasmid to that of 2-μm derivatives in which one or more of the major ORFs had been disrupted. They detected changes in the nuclease sensitivity of STB that resulted from disruption of ORF B indicating an interaction between STB and the B protein. Since disruption of ORF C did not affect the nuclease cleavage pattern of STB, although C appears to act at STB to promote stable maintenance (Jayaram et $al.$, 1983a; Kikuchi, 1983), they propose that the C gene product interacts indirectly with STB; perhaps by association with the B gene product to form a complex required for segregation. Moreover, they found that chromatin structure at the 5′ end region of ORF A was altered after interruption of either ORF B or ORF C. In the case of ORF C disruption, the chromatin alteration in ORF A is associated with an increase in copy number, due to increased FLP activity, suggesting that the B and C proteins act in their second role interdependently as repressors of FLP transcription.

Reynolds et $al.$ (1987) have shown that the total level of FLP activity decreases with increasing 2-μm copy number. A FLP-$lacZ$ fusion allowed them to identify the negative regulation of FLP gene expression and to confirm that expression of both B and C, acting in concert, is required to repress FLP expression. Additionally, expression of B or C from their own promoters at low copy numbers failed to repress FLP. As the FLP-$lacZ$ fusion is a protein fusion they could not distinguish between transcriptional and translational regulation of the FLP gene. However, the alteration in the chromatin structure upstream of the FLP gene in a B^- and C^- mutant suggests that at least some FLP regulation occurs at the level of mRNA abundance. Data from measurement of RNA abundances are also consistent with transcriptional regulation of ORF A by the B and C products (see section 11.2).

A model accounting for the maintenance of the 2-μm circle was proposed by Reynolds et $al.$ (1987), wherein the level of B and C proteins is proportional to 2-μm copy number. In cells with high copy levels, the B and C products are sufficient to form a complex which can bind to the cis-acting STB locus. This promotes near-random segregation at cell division by displacing the 2-μm plasmid from a nuclear structure that is normally retained by the mother cell. The same BC-complex binds also to the 5′ end of FLP repressing mRNA synthesis, and thereby inhibiting amplification. However, in cells acquiring a reduced copy number of the 2-μm plasmid due to segregation inaccuracy, the concentration of the BC-complex is diminished by gene dosage of B and C.

This causes derepression of *FLP* activity and induces amplification of copy number, which in turn allows cellular concentrations of *B* and *C* to increase again and to re-establish repression of *FLP*.

Som *et al.* (1988) also used gene fusion technology to analyse the interaction of 2-μm circle genes and gene products, and confirmed that expression of *FLP* is regulated by other 2-μm gene products. Consequently there is now a growing body of evidence for the existence of dual roles for certain of the gene products.

The unravelling of this complex network of interactions which appears to coordinate partitioning and copy number control will necessitate the facility to analyse the role of distinct protein domains. Veit (personal communication) has generated missense mutations in several of the 2-μm ORFs using hydroxylamine-induced *in vitro* mutagenesis of plasmid DNA. Such mutants will undoubtedly prove to be invaluable in these studies. Already, the chromatin organization analysis of missense mutations of *B* and *C* has provided some interesting results. Two such mutants, although mapping to *C*, gave a B⁻-like nuclease cleavage pattern in the *STB* region. This indicates that the altered *C* gene products may interfere with *B* protein's interaction with *STB* sequences. However, the altered ORF *C* products still retain their ability to interact with the ORF *A* region, since a wild-type nuclease cleavage pattern was observed throughout the *FLP* gene. A third missense mutation, mapping to *B*, also retained an apparent wild-type interaction with *ORF A*, but displayed an altered chromatin pattern at the *STB* region, indicating that interaction with this site was affected.

More recent investigations also implicate ORF *D* in the coordination process. Cashmore *et al.* (1988) carried out studies on 2-μm derivatives with both insertional and frameshift mutations in *D*. The results indicate that there is a protein product encoded by ORF *D* which is involved in plasmid maintenance. A single, chromosomally integrated copy of ORF *C* could complement a $C^- B^+ D^+$ plasmid but not a $C^- B^+ D^-$ plasmid. Striking differences were observed in the partitioning of D^+ and D^- derivatives. Absence of *D* function could be compensated by an increase in dosage of the *C* gene, indicating that the *D* product may act to regulate the expression of *C*. Since the *C* product has been implicated in copy number control as well as partitioning these data suggest that the *D* product may be involved in coordination of both processes.

The system used by Cashmore *et al.* (1988) excluded the copy number amplification mechanism mediated by *FLP* and the inverted repeats. However, Murray *et al.* (1987) have carried out investigations supporting the hypothesis that *D* is important in coordination of maintenance functions. These workers used inducible overexpression of plasmid gene products from a chromosomal locus to either disrupt the normal balance of plasmid proteins or to complement mutant plasmids. Their data confirmed that *FLP* expression regulates plasmid copy number, and indicated that its control involves not only the products of ORFs *B* and *C* but also *D*. Their data also show that

N

overexpression of D causes 2-μm copy amplification by stimulating transcription of ORF A (FLP). They propose that the D protein acts to antagonize the B–C repression of FLP expression. Whether or not D turns out to affect FLP expression directly, or to function via an interaction with C, requires further investigation. However, it is certainly clear that the hitherto neglected D plays a central role in regulation.

From the preceding account it is obvious that transcriptional regulation is important in regulating maintenance functions. Emphasis has been placed on the regulation of mRNAs corresponding to the ORFs. However, there also exist other transcripts (particularly those of 1950 bases and 1620 bases) which are apparently not involved in protein production (see section 11.2). Transcription from the promoter of the '1950' transcript lying in the STB region has been examined (Jayaram et al., 1985). An investigation of the functional domains of the cis-acting STB ($REP3$) locus was made, which indicated that there was a control site at the 5' end of the sequence from which the transcript is made, i.e. within STB-distal, as defined by Murray and Cesarini (1986). They proposed that regulation of the '1950' transcript was brought about by differential amounts of B protein binding to the STB locus, and that this element of regulation was important in some way in copy number control. One possibility could be that this transcript acts by modulating the level of D expression via an RNA–RNA interaction. D protein in turn would interact with C and/or B to modulate FLP expression and thus influence copy number.

Such hypotheses are alluring, but nonetheless purely speculative. A substantial amount of investigation is still required to elucidate the nature of the interactions. However, we can conclude that the system will provide not only a good model for copy number regulation, but also for the regulation of gene expression.

11.8 Other factors influencing plasmid maintenance

To date, most of the studies of 2-μm maintenance have concentrated on the role played by the plasmid itself. It is now clear that plasmid molecules have considerable interaction with the host cell, both at the levels of replication and partitioning. Consequently, one would expect there to be host-encoded functions which affect these processes. In addition, the effect of the environment and growth conditions on maintenance is an important consideration, particularly when using 2-μm plasmids as vector systems for the maintenance of other DNA sequences in yeast. Despite the relative sparsity of information available in the literature, these aspects warrant some discussion.

11.8.1 *Host-encoded functions*
We have already indicated that certain of the CDC gene products play a role in 2-μm replication. More recently, other classes of mutants affecting plasmid

replication have been isolated. Maine et al. (1984) used centromere-based ARS plasmids and isolated mutations, in sixteen complementation groups. One of the groups of mutants also affected the 2-μm circle. Kikuchi and Toh-e (1986) used plasmid pSLel, a derivative of 2-μm carrying only *ORI* and the yeast *LEU2* gene, to screen for replication mutants. Using this approach they identified a nuclear gene *MAP1* that appears to play a role in initiation of replication. Again this mutant was not specific to 2-μm, but also affected *ARS* and centromere-based plasmids.

Holm (1982) described a mutation *nib1*, in a nuclear gene which resulted in lethal sectoring of colonies when 2-μm plasmids were present in the cells. The lethal sectoring was apparently due to mortality of cells, which also correlated with enhanced copy numbers of 2-μm. The role of the wild-type gene *NIB1* in 2-μm maintenance has not been further elucidated at the molecular level, but it is tantalizing to speculate that it may be important in copy-number regulation.

The partitioning mechanism of 2-μm, although now being intensively investigated at the level of plasmid involvement, is still a 'black box' with respect to the role of host proteins. Evidence to date, particularly the observation that the 2-μm *B* protein is a karyoskeletal component (Wu *et al.*, 1987), indicates that the plasmid molecules utilize some form of anchorage to a cellular structure to ensure their inheritance at cell division. Host mutants defective in 2-μm partitioning may therefore provide invaluable tools for the study of the structure of the yeast nucleus.

We have initiated work in this area and have identified a number of strains that are good candidates for partition mutants. The first of these is a spontaneous [cir^0] derivative which arose during long-term batch culturing of strain MC16. This isolate cannot be transformed with 2-μm-based plasmids, although ARS and CEN-based plasmids behave normally. When the isolate is crossed with a [cir$^+$] strain and the resultant diploid sporulated, plasmids are inherited by all four segregants, but there is a 2:2 co-segregation of plasmid instability and inability to be transformed (Cashmore and Williamson, unpublished observations).

The other class of strains which appears defective in maintenance has been isolated from strains JRY188 and S150-2B. During chasing out of the native 2-μm from these strains using selection for a hybrid plasmid, as originally described by Dobson et al. (1980), isolates of these strains were identified that had a low copy number of 2-μm (approximately five per cell). On subculturing of the strains, the copy number in the population remained low, and colony hybridization experiments demonstrated that in fact the distribution of plasmids was affected, most of the population (80–90%) at any one time being [cir^0], and the remaining 10–20% containing a wild-type level of plasmids (Cashmore and Coxon, unpublished observations). Of course, such strains may be intermediates in the 'chasing out' process. However, the fact that copy number does not revert to wild-type on subsequent culturing would argue against this. Also, we have been unable to transform the isolates with 2-μm-based plasmids, a further indication that they may be good candidates for host-encoded maintenance mutants.

11.8.2 Growth conditions

The literature available on this aspect of yeast plasmid biology is not only scant but also confusing in its conclusions, depending to a great extent on the particular strains used. Spontaneous [cir°] isolates have been observed for a variety of strains (Walmsley et al., 1983; Futcher and Cox, 1983; Cashmore, 1984) and at widely differing frequencies. Nonetheless, their appearance does indicate that the molecular maintenance mechanisms do not operate with 100% efficiency, and that there must be other environmental factors which affect maintenance of the plasmids in the population. Several investigations have been made of the effect of the 2-μm circle on growth-rate of cells. It is clear from studies of plasmids in both chemostat (Walmsley et al., 1983) and serial batch cultures (Futcher and Cox, 1983) that spontaneous [cir°] cells in a population have a slight growth advantage. This is not the case for induced [cir°] isolates which have been generated by 'chasing out' with a hybrid plasmid according to the method of Dobson et al. (1980). Mead et al. (1987) determined maximum specific growth rates and found that, unlike the spontaneous [cir°] isolates, these induced strains did not display a growth advantage over plasmid-containing isolates.

If there is any growth advantage at all to loss of the 2-μm circle, as observed for certain spontaneous isolates, one would expect the plasmid to be lost rapidly from the population. This is not the case and we are left with the question why this is not so. Mead et al. (1986) observed that the rate of plasmid loss was reduced by allowing haploid populations to enter stationary phase periodically, and that plasmids were more stable in diploids. They propose that yeast cells in nature are usually homothallic, and that the 2-μm circle exploits the diploid and stationary phases of this lifecycle for its maintenance in the population. The effect of stationary phase does not appear to be the same for 2-μm-based hybrid plasmids which are in fact destabilized by periods in stationary phase and other culture conditions, such as growth on glycerol and increased temperature, which generally reduce growth rate (Cashmore, 1984; Kleinman et al., 1987; Cashmore, unpublished observations).

The implications of these observations for yeast transformation and the use of yeast vectors will probably be widespread. This is certainly an area where much work of a systematic nature is required.

11.9 Two-micron-like plasmids in other yeasts

Plasmids with structural similarities to the 2-μm circle have now been isolated and characterized in other yeast species. Most of those studied to date have been identified in Zygosaccharomyces and Saccharomyces species. Examples are pSR1 and pSR2 from Z. rouxii (Toh-e et al., 1982); pSB1 and pSB2 from S.

bailii, and pSB3 and pSB4 from *S. bisporus* (Toh-e *et al.*, 1984); pSM1 from *Z. fermentati* (Utatsu *et al.*, 1987). Another plasmid, pKD1, was isolated from *Kluyveromyces drosophilarum* (Falcone *et al.*, 1986). Other yeast strains have been surveyed, but plasmids appear to be a rarity. For example, Toh-e *et al.* (1982) screened 100 different strains and found only two with plasmids; both were strains of *Z. rouxii*, each harbouring a single type of plasmid pSR1 or pSR2. All of these plasmids have a similar molecular organization to 2-μm, with inverted repeats separating two unique sequence regions, and exist as two isomeric forms arising from intramolecular recombination between the repeats. To date, five of the 2-μm-like plasmids have been sequenced (pSR1, Araki *et al.*, 1985; pSB3, Toh-e and Utatsu, 1985; pSB2 and pSM1, Utatsu *et al.*, 1987 and pKD1, Chen *et al.*, 1986), allowing molecular as well as functional analyses of plasmid coding regions to be carried out.

The most intensively studied is pSR1 from *Z. rouxii*. This is a 6251-bp plasmid with a pair of 959-bp inverted repeats separating unique regions of 2654 and 1679 bp. It has two ARS elements, one in each inverted repeat, which can promote replication not only in the natural host *Z. rouxii* but also in *S. cerevisiae*. Three open reading frames (*P*, *S*, and *R*) have been found, and insertional inactivation has indicated that *R* is involved in the intramolecular recombination process (Araki *et al.*, 1985) and *P* and *S* in maintenance (Jearnpipatkul *et al.*, 1987a). A *cis*-acting locus *Z*, also involved in maintenance, has been identified (Jearnpipatkul *et al.*, 1987b). This region has been delimited to a sequence of approximately 380 bp in the small unique region of the plasmid, a region which is devoid of open reading frames. The similarities with 2-μm are remarkable. It is now evident that the other plasmids that have been analysed also share those features. Each has three open reading frames, one of which is involved in intramolecular recombination between the inverted repeats, and two important for maintenance (cf. *FLP, B* and *C* of 2-μm). Plasmid pSM1, isolated from *Z. fermentati*, also has a fourth open reading frame which could potentially encode a 200-amino-acid protein; perhaps a candidate for a 2-μm *D*-like function?

These functional similarities are not necessarily reflected in nucleotide sequence homologies. Partial similarities have been found between the ORF-encoded *FLP*-like functions at the protein level when comparing the deduced amino acid sequences (Utatsu *et al.*, 1987). Other than homology between *FLP* proteins, these workers reported homology between only one other pSB2 gene product (*B*, 357 amino acids) and a pSR1 protein (*P*, 410 amino acids). Murray *et al.* (1988) reported a limited similarity between the products of the second largest open reading frames of various plasmids (including *B* of 2-μm), and no consistent homologies have been found in the case of the third largest ORFs (including *C* of 2-μm).

The maintenance systems, although organizationally similar to each other and to the *B, C, STB* system of 2-μm, are operationally exclusive. In general, mutant functions cannot be complemented by the equivalent protein from a heterologous plasmid (Toh-e and Utatsu, 1985; Araki *et al.*, 1985; Chen *et al.*, 1986). Similarly, *FLP*-like proteins recognize only the inverted repeat of their own plasmids. Despite this, the host specificity of the plasmids is somewhat relaxed. All plasmids can replicate when artificially introduced in *S. cerevisiae* (Toh-e *et al.*, 1984; Toh-e and Utatsu, 1985; Araki *et al.*, 1985; Chen *et al.*, 1986). However, stability is poor because the maintenance systems do not appear to operate in the heterologous host. There are exceptions to this. For example, pSB3 is stably inherited in its natural host *Z. bisporus* and in the heterologous *Z. rouxii* (Toh-e and Utatsu, 1985). The converse is also true, because pSR1 (from *Z. rouxii*) appears identical to an as yet unsequenced plasmid pSB4 found in *Z. bisporus* (Toh-e *et al.*, 1984; Araki *et al.*, 1985). As these yeasts are taxonomically close, this equivalence of the host environment is perhaps not surprising (Murray *et al.*, 1988). Also, pKD1 isolated from *K. drosophilarum* is stable in the closely related *K. lactis* (Chen *et al.*, 1986).

The discovery and investigation of these 2-μm-like plasmids will extend the range of yeasts for which transformation systems can be developed. In fact, the *Kluyveromyces* plasmid pKD1 has already been used to construct recombinant plasmids, enabling stable transformation and maintenance in a *K. lactis* host (Bianchi *et al.*, 1987). This system has been successfully used for the cloning of a *K. lactis* chromosomal gene (Wesolowski-Louvel *et al.*, 1988).

11.10 Use of the 2-μm circle as a yeast cloning vector

The properties of autonomous multicopy replication and efficient mitotic segregation have made the 2-μm circle an attractive molecule from which to develop plasmid vectors for gene cloning in yeast. However, this has not been without attendant difficulties because, as discussed above, the sole purpose of 2-μm gene products, genetic loci and molecular structure is to ensure plasmid propagation. Thus even the mere insertion of a piece of 'non-functional' extra DNA, notwithstanding any gene disruption that might arise, will alter the plasmid's molecular architecture, and so can be expected to affect the pattern of replication fork movement and duration of plasmid replication. It is therefore not surprising that no 2-μm recombinant plasmid constructed to date exhibits exactly the same maintenance characteristics as the parental plasmid. All are segregated less efficiently and often have reduced copy level (Futcher and Cox, 1984). This generally arises because the most convenient restriction enzyme sites for vector construction lie within the four plasmid genes, and insertion of DNA inactivates a required maintenance function. To the best of our knowledge, there has been no systematic study of the maintenance behaviour of 2-μm derivatives containing variously-sized pieces of foreign DNA at different sites in the plasmid molecule.

11.10.1 *Types of 2-μm vector*
Two basic strategies have been adopted in 2-μm vector development. On the one hand segments of the 2-μm circle that contain only the *cis*-acting loci that

are absolutely required for maintenance (*ORI* and *STB*) have been combined with other DNA fragments for selection in yeast and propagation in *E. coli.* Such 'shuttle-vectors' are generally termed YEP (yeast episomal plasmids) vectors. On the other hand, the complete 2-μm circle has been combined with similar fragments (whole-2-μm vectors); see Broach (1983) for a description of several vectors of each type. The *E. coli* sequences, although not necessary for yeast propagation, are included because they allow usage of gene cloning and manipulation techniques that are more highly developed in the 'shuttle' bacterial host. However, in some instances their incorporation in vectors may not be acceptable (e.g. when the yeast is to be used as a food source).

For construction of YEP vectors, the 2-μm *cis*-acting loci are most conveniently recovered from the *B*-form of the molecule on either a 2240-bp *Eco*RI fragment or a 2212-bp *Hind*III fragment (see Figure 11.1). In both cases the fragment also contains other restriction sites that can be used for cloning foreign DNA; most notably the *Pst*I site within the 3' remnant of ORF *D*. A major advantage of this type of vector is that it can be kept to a reasonably small size, thus easing DNA manipulations. However, since YEP vectors contain only the *cis*-acting maintenance functions, they are dependent upon the presence of the endogenous 2-μm circle for provision of the *trans*-acting segregation products *B* (REP1) and *C* (REP2). Thus in a [cir^0] host they exhibit maternal bias of inheritance, like *ARS* plasmids with no centromere sequences, and a high loss rate of over 10% loss per generation (Murray and Szostak, 1983), whereas in a [cir$^+$] host their loss rate is much smaller, around 1–3% per generation. YEP vectors are maintained in a multicopy state at 25–60 copies per cell (Futcher and Cox, 1984). Since they do not contain the two inverted repeat sequences necessary for intramolecular recombination-based copy amplification, we presume that they achieve this from asymmetric plasmid segregation during establishment of the initial transformant colony. YEP vectors based upon the *Hind*III or *Eco*RI fragments discussed above do contain one copy of the 2-μm inserted repeat. They can therefore undergo *FLP*-mediated intermolecular recombination with the resident 2-μm circle, resulting in the formation of a complex chimaeric molecule. Such recombination products have been reported (Dobson *et al.*, 1980) and have been observed to undergo further rearrangements, resulting in the loss of non-2-μm sequences. Thus the potential exists for elimination of foreign DNA cloned within 2-μm sequences.

Whole-2-μm vectors have the advantage that they can encode all necessary maintenance functions particularly *B* (*REP1*) and *C* (*REP2*), and hence are not necessarily dependent on the presence of the native molecule. Their major disadvantage is their large molecular size, which makes their DNA more difficult to isolate and manipulate. Vectors of this type normally have the non-2-μm sequences inserted at restriction sites in the ORF *D* and/or ORF *A* regions of the plasmid. Insertions into ORF *A* result in inactivation of the *FLP* recombination system, and the plasmid is therefore propagated in one molecular form only in a [cir^0] host. Recombinant plasmids which are *A*$^+$ can undergo intramolecular recombination and hence copy amplification. In this context, it is of interest to note that the vector pJDB248 (Beggs, 1981), which has foreign DNAs inserted only into the ORF D region, is extremely stable and even during growth under non-selective conditions is lost only at a very low rate, less than 1% per generation (Cashmore, 1984; Futcher and Cox, 1984).

A variation of the whole-2-μm vector has recently been described by Hinchliffe et al., (1987). This is based upon a previous observation that DNA molecules containing a single copy of the 2-μm inverted repeat sequence will undergo recombination with the native 2-μm circle when transformed into [cir⁺] yeast cells. Thus if the gene of interest is physically linked to, or inserted within, transforming 2-μm inverted repeat DNA, it too will become inserted into the 2-μm circle. The authors report that 2-μm-recombinants produced in this way are remarkably stable and rarely lost even during growth under non-selective conditions.

11.10.2 Selection regimes and their effect on vector behaviour

As the 2-μm circle confers no overt phenotype on its host cell, it is necessary for yeast vectors to carry a selectable marker in order that plasmid-transformed cells can be identified and recovered. Additionally, presence of a selectable marker on the vector enables the imposition of a suitable regime that will maximize maintenance of the plasmid molecule during growth of the culture. Since cells containing 2-μm plasmid exhibit a reduced growth rate compared to plasmid-free cells (Futcher and Cox, 1983; Mead et al., 1986) they would be quickly eliminated by plasmid-free segregants during growth in a non-selective medium. The most commonly used markers are yeast chromosomal genes, such as URA3, LEU2, TRP1 etc., so that the vector confers prototrophy upon a suitable auxotrophic host strain. In order for the selection to be effective, host mutations must have very low reversion frequencies, and alleles of most commonly used genes are now available that contain either small deletions or multiple point mutations. Although suitable for use with genetically well-characterized laboratory strains, these markers are not suitable for use with industrial yeasts which are often prototrophic and polyploid, thus making mutant isolation almost impossible. Therefore recent years have seen the development of a number of dominant direct selection markers, which have been incorporated into 2-μm vectors. Many of these confer resistance to antibiotics such as chloramphenicol (Hadfield et al., 1986), methotrexate (Zhu et al., 1985) hygromycin B (Kaster et al., 1984), and the aminoglycoside G418 (Jiminez and Davies, 1980). However, as these are generally developed from bacterial resistance genes, expressed from a yeast promoter, they are undesirable within food yeasts, and dominant markers based on endogenous yeast genes such as the copper resistance marker CUP1 (Fogel and Welch, 1982; Henderson et al., 1985; Meaden and Tubb, 1985) may be preferred.

As well as facilitating yeast transformation, the genetic marker provides a selection pressure for plasmid maintenance during culture growth. Indeed, the imposition of a powerful selective regime can often circumvent problems of plasmid instability and low copy. For example, the whole-2-μm vector pJDB219 (Beggs, 1978) and YEP-type vectors derived from it, such as pJDB207 (Beggs, 1981) are found to exhibit an extremely high copy level, *c.* 200 per cell (Erhart and Hollenberg, 1983), and to be very stably maintained even in [cir^0] hosts, during non-selective growth (Futcher and Cox, 1984). The explanation for this unusual behaviour lies with the particular *leu2* allele that constitutes the selective marker on the plasmids. This particular allele, isolated from randomly sheared DNA inserted at the *Pst*I site of the 2-μm ORF *D* by homopolymer tailing, lacks sequences of the natural LEU2 promoter and so is very poorly expressed. It is even possible that transcription of this allele originates from the *B-C*-repressed '1950' mRNA promoter of the 2-μm '*STB*-distal' locus (Jayaram *et al.*, 1985; Sutton and Broach, 1985). Thus, as the dosages of 2-μm genes *B* and *C*, also on the plasmid, are increased with plasmid copy level, expression of the *leu2*-d allele will become progressively repressed. The net result is that only cells that have accumulated very many copies of the plasmid can produce enough of the LEU2 gene product and thereby propagate under the selective regime. As the high plasmid copy level is found with both YEP and whole-2-μm vectors, it probably arises from asymmetric segregation rather than induction of the *FLP*-mediated amplification process. One consequence of the use of the *leu2*-d allele is difficulty in obtaining plasmid transformant clones (Futcher and Cox, 1984; Henderson *et al.*, 1985). In our experience, the spheroplasting method (Beggs, 1978) is more successful, probably because this can result in the simultaneous uptake of several plasmid molecules.

Similar high plasmid copy levels achieved in response to a stringent selection regime have been reported with dominant markers. Yeast cells containing very many copies of 2-μm-based vectors with either a dihydrofolate reductase gene, or the herpes simplex virus (HSV) thymidine kinase gene, can be obtained by use of the drug methotrexate (Zhu *et al.*, 1985; Zealey *et al.*, 1988). In both cases the plasmid copy level could be varied by altering the concentration of antibiotic supplied. In each case where elevated plasmid copy levels have been reported, there has been a concomitant improvement in mitotic stability of the vector.

The selection systems discussed above all require either the use of a defined synthetic or semi-synthetic growth medium (auxotrophic complementation selection), or addition of toxic chemicals to the yeast growth medium (drug resistance selection). Both are costly procedures, not readily welcomed in yeast fermentations designed to produce 'low-value high-volume' products. In this situation, it would be advantageous to have self-selection systems for vector maintenance that would function in growth media of any composition. This is most easily achieved by inclusion on the vector of a wild-type copy of an essential gene without which a mutant host cell would not survive. Such a gene might encode an enzyme of central cellular metabolism, such as triose-phosphate isomerase, as suggested by Kawasaki (1984). A variation on this theme has been published by Loison and co-workers (Loison *et al.*, 1985; Marquet *et al.*, 1987), who introduced a 2-μm vector carrying the URA3 gene into a host strain that was mutant in both the *URA3* and *FUR1* genes. Without the vector the double mutant host strain would die, since it could synthesize UMP neither *de novo* by decarboxylation of orotidine-5-monophosphate (*ura3*) nor by salvage of uracil (*fur 1*). These authors demonstrated that even in an undefined molasses-based industrial fermentation, the 2-μm plasmid vector was maintained in 100% cells over at least 30 generations of growth. Selection systems of this type are still limited to one particular host strain, namely that carrying the lethal mutation. A more general self-selection system for plasmid maintenance has been described by Lolle *et al.* (1984), Bussey and Meaden (1985), and Lee and Hassan (1988). This selection is imposed by utilization of a cDNA clone of the killer-immunity dsRNA M of *S. cerevisiae*, which is expressed from a plasmid-borne promoter. Thus presence of the vector confers not only host immunity but also the selective agent–killer toxin which is secreted from the cells into the culture medium. This system is entirely yeast-based, has the potential to operate in any strain (providing it is initially sensitive to killer-toxin) and in any growth medium (providing it is not inhibitory for toxin production or action), and prevents contamination by undesirable yeast strains.

11.10.3 *Dosage effects on gene expression levels*
Two-micron vectors are generally used to maximize expression of a cloned gene by increasing its gene dosage (Chevallier *et al.*, 1980). However, there have been few rigorous studies on the relationship between expression levels and 2-μm copy number. Our own work (Hadfield *et al.*, 1987) has shown that a foreign protein, chloramphenicol acetyl transferase (CAT), is expressed at proportionately higher levels when the gene is present on a multicopy 2-μm vector than when integrated as a single copy into a chromosomal locus. This may not always be the case, as required transcriptional activators may become limiting for yeast promoters cloned on multicopy vectors (see for example Baker *et al.*, 1984). Also, it should be noted that high-level expression of a cloned promoter can be detrimental to the 2-μm vector and result in a dramatic reduction of plasmid copy level. For instance, transcription from a powerful promoter placed adjacent to the *ORI-STB* locus will reduce plasmid copy level despite the presence of 2-μm *ORF D* terminator sequences (Jayaram *et al.*, 1985; C. Hadfield, personal communication). Since many of the DNA sequences inserted into yeast vectors are from heterologous sources, it is impossible to predict what transcriptional activities they will exhibit in the yeast cell and what effect this may have on the behaviour of the supporting 2-μm vector (see example Marczynski and Jaehning, 1985).

Notwithstanding these possible complications, the highest expression levels of foreign proteins (1–5%) total cell protein have generally been achieved with high-copy 2-μm vectors, especially those based on the *leu2*-d allele (see for example Tuite *et al.*, 1982). There is therefore considerable scope for 2-μm vectors which can be switched from a low to high copy level by an external stimulus. Such vectors have proved useful in *E. coli* for maximizing gene expression levels (Yarranton *et al.*, 1984; Wright *et al.*, 1987). By co-ordinately inducing gene expression and copy amplification, rapid synthesis of the required protein can be achieved at a predetermined stage of a fermentation. The observation that 2-μm copy amplification is dependent on the expression level of the *FLP* product suggests one route by which such vectors might be achievable in yeast.

Acknowledgements

We are grateful to the many yeast workers who sent us copies of their works. In particular we thank Drs W. Fangman, C. Hadfield, J. Huberman, M. Jayaram, J. Murray and B. Veit who allowed us to cite their results prior to publication.

References

Andrews, B.J., Proteau, G.A., Beatty, L.G. and Sadowski, P.D. (1985) The *FLP* recombinase of the 2-μm circle DNA of yeast: interaction with its target sequences. *Cell* **40**: 795.

Andrews, B.J., McLeod, M., Broach, J.R., and Sadowski, P.D. (1986). Interaction of the *FLP* recombinase of the *Saccharomyces cerevisiae* 2-μm plasmid with mutated target sequences. *Mol. Cell Biol.* **7**: 2482.

Andrews, B.J., Beatty, L.G. and Sadowski, P.D. (1987) Isolation of the intermediates in the binding of the *FLP* recombinase of the yeast plasmid 2-micron circle to its target sequence. *J. Mol. Biol.* **193**: 345.

Araki, H., Jearnpipatkul, A., Tatsumi, H., Sakura, T., Ushio, K., Mata, J. and Oshima, Y. (1985) Molecular and functional organisation of yeast plasmid pSR1. *J. Mol. Biol.* **182**: 191.

Babineau, D., Vetter, D., Andrews, B.J., Gronostajski, R.M., Proteau, G.A., Beatty, L.G. and Sadowski, P.D. (1985) The *FLP* protein of the 2-micron plasmid of yeast. Purification of the protein from *E. coli* cells expressing the cloned *FLP* gene. *J. Biol. Chem.* **260**: 12313.

Baker, S.M., Okkema, P.F. and Jaehning, J.A. (1984) Expression of the *Saccharomyces cerevisiae GAL7* gene on autonomous plasmids. *Mol. Cell Biol.* **4**: 2062.

Beatty, L.G., Babineau-Clary, D., Hogrefe, C. and Sadowski, P.D. (1986) *FLP* site-specific recombinase of the yeast 2-μm plasmid: topological features of the reaction. *J. Mol. Biol.* **188**: 529.

Beggs, J.D. (1978) Transformation of yeast by a replicating hybrid plasmid. *Nature* **275**: 104.

Beggs, J.D. (1981) Multiple-copy yeast vectors. In *Molecular Genetics in Yeast*, eds. D. von Wettstein, J. Friis, M. Kielland-Brandt and A. Stenderup, Alfred Benzon Symp. **16**: Munksgaard, Copenhagen, 383.

Beggs, J.D. Guerineau, M. and Atkins, J.F. (1976) A map of the restriction targets in yeast 2 micron plasmid DNA cloned on bacteriophage lambda. *Mol. Gen. Genet.* **148**: 287.

Bianchi, M.M., Falcone, C., Chen, X.J., Wesolowski-Louvel, M., Frontali, L. and Fukahara, H. (1987) Transformation of the yeast *K. lactis* by new vectors derived from the 1.6-μm circular plasmid. *Curr. Genet.* **12**: 185.

Brewer, B.J. and Fangman, W.L. (1987) The localisation of replication origins on *ARS* plasmids. *Cell* **51**: 463.

Broach, J.R. (1981) The yeast plasmid 2-μ circle. In *The Molecular Biology of the Yeast Saccharomyces cerevisiae: Life Cycle and Inheritance*, eds. J.N. Strathern, E.W. Jones and J.R. Broach, Cold Spring Harbor Laboratory, New York, 445.

Broach, J.R. (1983) Construction of high copy yeast vectors using 2-μm circle sequences. In *Methods in Enzymology*, Vol. 101, part C, eds. R. Wu, L. Grossman and K. Moldave, Academic Press, New York, 307.

Broach, J.R., Atkins, J.F., McGill, C. and Chow, L. (1979) Identification and mapping of the transcriptional and translational products of the yeast plasmid 2-μm circle. Cell 16: 827.

Broach, J.R. and Hicks, J.B. (1980) Replication and recombination functions associated with the yeast plasmid 2-μ circle. Cell 21: 501.

Broach, J.R., Guarascio, V.R. and Jayaram, M. (1982) Recombination with the yeast 2-μm circle is site-specific. Cell 29: 227.

Broach, J.R., Li Y-Y., Wu L-CC., Jayaram, M. (1983a) Vectors for high-level, inducible expression of cloned genes in yeast. In Experimental Manipulation of Gene Expression, ed. M. Inouye, Academic Press, New York, 83.

Broach, J.R., Li Y-Y., Feldman, J., Jayaram, M., Abraham, J., Nasmyth, K.A. and Hicks, J.B. (1983b) Localization and sequence analysis of yeast origins of DNA replication. Cold Spring Harbor Symp. Quant. Biol. 47: 1165.

Bussey, H. and Meaden, P. (1985) Selection and stability of yeast transformants expressing cDNA of an M1 killer toxin-immunity gene. Curr. Genet. 9: 285.

Cashmore, A.M. (1984) Factors affecting maintenance of plasmids in yeast. PhD Thesis, CNAA, London.

Cashmore, A.M., Albury, M.S., Hadfield, C. and Meacock, P.A. (1986) Genetic analysis of partitioning functions encoded by the 2-μm circle of Saccharomyces cerevisiae. Mol. Gen. Genet. 203: 154.

Cashmore, A.M., Albury, M.S., Hadfield, C. and Meacock, P.A. (1988) The 2-μm D region plays a role in yeast plasmid maintenance. Mol. Gen. Genet. 212: 426.

Chen, X.J., Saliola, M., Falcone, C., Bianchi, M.M. and Fukuhara, H. (1986) Sequence organisation of the circular plasmid pKD1 from the yeast K. drosophilarum. Nucleic Acids Res. 14: 4471.

Chevallier, M-R., Bloch, J-C. and Lacroute, F. (1980) Transcriptional and translational expression of a chimeric bacterial-yeast plasmid in yeasts. Gene 11: 11.

Clark-Walker, G.D. and Miklos, G.L.G. (1974) Localisation and quantitation of circular DNA in yeast. Eur. J. Biochem. 41: 359.

Cox, M. (1983) The FLP protein of the 2-μm plasmid: expression of a eukaryotic genetic recombination system in Escherichia coli. Proc. Natl. Acad. Sci. USA. 80: 4223.

Dobson, M.J., Futcher, A.B. and Cox, B.S. (1980) Loss of 2-μm DNA from S. cerevisiae transformed with the chimaeric plasmid pJDB219. Curr. Genet. 2: 201.

Erhart, E. and Hollenberg, C.P. (1983) The presence of a defective LEU2 gene on 2-μm DNA recombinant plasmids of Saccharomyces cerevisiae is responsible for curing and high copy number. J. Bacteriol. 156: 625.

Falcone, C., Saliola, M., Chen, X.J., Frontali, L. and Fukuhara, H. (1986) Analysis of a 1.6-μm circular plasmid from the yeast K. drosophilarum: structure and molecular dimorphism. Plasmid 15: 248.

Fogel, S. and Welch, J. (1982) Tandem gene amplification mediated copper resistance in yeast. Proc. Natl. Acad. Sci. USA 79: 5342.

Futcher, A.B. (1986) Copy number amplification of the 2-μm circle plasmid of Saccharomyces cerevisiae. J. Theoret. Biol. 119: 197.

Futcher, A.B. and Cox, B.S. (1983) Maintenance of the 2-μm circle plasmid in populations of Saccharomyces cerevisiae. J. Bacteriol. 154: 612.

Futcher, A.B. and Cox, B.S. (1984) Copy number and stability of 2-μm circle-based artificial plasmids of Saccharomyces cerevisiae. J. Bacteriol. 157: 283.

Govind, N.S. and Jayaram, M. (1987) Rapid localization and characterization of random mutations within the 2-μ circle site-specific recombinase: a general strategy for analysis of protein function. Gene 51: 31.

Gronostajski, R.M. and Sadowski, P.D. (1985a) Determination of the DNA sequences essential for FLP-mediated recombination by a novel method. J. Biol. Chem. 260: 12320.

Gronostajski, R.M. and Sadowski, P.D. (1985b) The FLP recombinase of Saccharomyces cerevisiae 2-μm plasmid attaches covalently to DNA via its phosphotyrosyl linkage. Mol. Cell. Biol. 5: 3274.

Gronostajski, R.M. and Sadowski, P.D. (1985c) FLP protein of the 2-micron plasmid of yeast: inter- and intramolecular reactions. J. Biol. Chem. 260: 12328.

Gubbins, E.J., Newlon, C.S., Kann, M.D. and Donelson, J.E. (1977) Sequence organisation and expression of a yeast plasmid DNA. Gene 1: 185.

Guerineau, M., Sloninski, P.P. and Avner, P.R. (1974) Yeast episome: oligomycin resistance associated with a small covalently closed non-mitochondrial circular DNA. *Biochem. Biophys. Res. Commun.* **61**: 462.

Guerineau, M., Grandchamp, C. and Sloninski, P.P. (1976) Circular DNA of a yeast episome with two inverted repeats: structural analysis by a restriction enzyme and electron microscopy. *Proc. Natl. Acad. Sci. USA* **73**: 3030.

Hadfield, C., Cashmore, A.M. and Meacock, P.A. (1986) An efficient chloramphenicol-resistance marker for *Saccharomyces cerevisiae* and *Escherichia coli*. *Gene* **45**: 149.

Hadfield, C., Cashmore, A.M. and Meacock, P.A. (1987) Sequence and expression characteristics of a shuttle chloramphenicol-resistance marker for *Saccharomyces cerevisiae* and *Escherichia coli*. *Gene* **52**: 59.

Hartley, J.L. and Donelson J.E. (1980) Nucleotide sequence of the yeast plasmid. *Nature* **286**: 860.

Henderson, R.C.A., Cox, B.S. and Tubb, R. (1985) The transformation of brewing yeasts with a plasmid containing the gene for copper resistance. *Curr. Genet.* **9**: 133.

Hinchliffe, E., Fleming, C.J. and Vakeria, D. (1987) A novel 'stable' gene maintenance system for brewing yeast. *Proc. Eur. Brew. Conv. 21st Congr. Madrid*: 505.

Hollenberg, C.P. (1978) Mapping of regions on cloned *Saccharomyces cerevisiae* 2-μm DNA coding for polypeptides synthesised in *E. coli* minicells. *Mol. Gen. Genet.* **162**: 23.

Hollenberg, C.P., Degelman, A., Kusterman-Kuhn, B. and Royer, H.D. (1976) Characterisation of 2-μm DNA of *Saccharomyces cerevisiae* by restriction fragment analysis and integration in an *Escherichia coli* plasmid. *Proc. Natl. Acad. Sci. USA* **73**: 2072.

Holm, C. (1982) Clonal lethality caused by the yeast plasmid 2-μm DNA. *Cell* **29**: 585.

Huberman, J.A., Spotila, L.D., Nawotka, K.A., El-Assouli, S.M. and Davies, L.R. (1987) The *in vivo* replication orgin of the yeast 2-μm plasmid. *Cell* **51**: 473.

Jayaram, M. (1985) Two-micrometer circle site-specific recombination: the minimal substrate and the possible role of flanking sequences. *Proc. Natl. Acad. Sci. USA* **82**: 5875.

Jayaram, M., Li, Y-Y., and Broach, J.R. (1983a) The yeast plasmid 2-μm circle encodes components required for its high copy propagation. *Cell* **34**: 95.

Jayaram, M., Li Y-Y., McLeod, M. and Broach, J.R. (1983b) Analysis of site-specific recombination associated with the yeast plasmid 2-micron circle. In *Mechanisms of DNA Replication and Recombination*, ed. N.R. Cozzarelli, Alan R. Liss, New York, 685.

Jayaram, M., Sutton, A. and Broach, J.R. (1985) Properties of *REP3*: a cis-acting locus required for stable propagation of the yeast plasmid 2-micron circle. *Mol. Cell. Biol.* **5**: 2466.

Jearnpipatkul, A., Araki, H. and Oshima, Y. (1987a) Factors encoded by and affecting the holding stability of yeast plasmid pSR1. *Mol. Gen. Genet.* **206**: 88.

Jearnpipatkul, A., Hutacharoen, R., Araki, H. and Oshima, Y. (1987b) A cis-acting locus for the stable propagation of yeast plasmid pSR1. *Mol. Gen. Genet.* **207**: 355.

Jiminez, A. and Davies, J. (1980) Expression of a transposable antibiotic resistance element in *Saccharomyces*. *Nature* **287**: 869.

Kaster, K., Burgett, S.G. and Ingolia, T.D. (1984) Hygromycin B resistance as a dominant selectable marker in yeast. *Curr. Genet.* **8**: 353.

Kawasaki, G. (1984) Stable maintenance of yeast expression vectors on complex media. *Abstr., 12th Int. Symp. on Yeast Genetics and Molecular Biology*, Edinburgh: 284.

Kielland-Brandt, M.C., Wilken, B., Holmberg, S., Petersen, J.G.L. and Nilsson-Tillgreñ, T. (1980) Genetic evidence for nuclear location of 2-micron DNA in yeast. *Carlsberg Res. Commun.* **45**: 119.

Kikuchi, Y. (1983) Yeast plasmid requires a cis-acting locus and two plasmid proteins for its stable maintenance. *Cell* **35**: 487.

Kikuchi, Y. and Toh-e, A. (1986) A nuclear gene of *Saccharomyces cerevisiae* needed for stable maintenance of plasmids. *Mol. Cell. Biol.* **6**: 4053.

Kleinman, M.J., Gingold, E.B. and Stanbury, P.F. (1987) The stability of the yeast plasmid pJDB248 depends on growth rate of the culture. *Biotechnol. Lett.* **8**: 225.

Lee, F.-J.S. and Hassan, H.-M. (1988) Stability and expression of a plasmid-containing killer toxin cDNA in batch and chemostat cultures of *Saccharomyces cerevisiae*. *Biotech. Bioengg* **31**: 783.

Livingston, D.M. (1977) Inheritance of 2-μm DNA from *Saccharomyces*. *Genetics* **86**: 73.

Livingston, D.M. and Hahne, S. (1979) Isolation of a condensed intracellular form of the 2 μm DNA plasmid of *Saccharomyces cerevisiae*. *Proc. Natl. Acad. Sci. USA* **76**: 3727.

Livingston, D.M. and Kupfer, D.M. (1977) Control of *Saccharomyces cerevisiae* 2-micron DNA replication by cell division cycle genes that control nuclear DNA replication. *J. Mol. Biol.* **116**: 249.

Loison, G., Nguyen-Juillert, M., Alouani, S. and Marquet, M. (1985) Plasmid-transformed *ura3 fur1* double mutants of *S. cerevisiae*: an autoselection system applicable to the production of foreign proteins. *Bio/Technology* **4**: 433.

Lolle, S., Skipper, N., Bussey, H. and Thomas, D.Y. (1984) The expression of cDNA clones of yeast M1 double-stranded RNA in yeast confers both killer and immunity phenotypes. *EMBO J.* **3**: 1383.

Maine, G.T., Sinka, P. and Tye B-K. (1984) Mutants of *S. cerevisiae* defective in the maintenance of minichromosomes. *Genetics* **106**: 365.

Marczynski, G.T. and Jaehning, J.A. (1985) A transcriptional map of a yeast centromere plasmid: unexpected transcripts and altered gene expression. *Nucleic Acids Res.* **13**: 8487.

Marquet, M., Alouani, S., Hass, M.L., Loison, G. and Brown, S.W. (1987) Double mutants of *Saccharomyces cerevisiae* harbour stable plasmids: stable expression of an eukaryotic gene and the influence of host cell physiology during continuous culture. *J. Biotechnol.* **6**: 135.

McLeod, M., Volkert, F. and Broach, J.R. (1984) Components of the site-specific recombination system encoded by the yeast plasmid 2-micron circle. *Cold Spring Harbor Symp. Quant. Biol.* **49**: 779.

Mead, D.J., Gardner, D.C.J. and Oliver, S.G. (1986) The yeast 2-μ plasmid: strategies for the survival of a selfish DNA. *Mol. Gen. Genet.* **205**: 415.

Mead, D.J., Gardner, D.C.J. and Oliver, S.G. (1987) Phenotypic differences between induced and spontaneous 2-μm-plasmid-free segregants of *Saccharomyces cerevisiae.*, *Curr. Genet.* **11**: 415.

Meaden, P.G. and Tubb, R.S. (1985) A plasmid vector for the genetic manipulation of brewing strains. *Proc. Eur. Brew. Conv. 20th Congr. Helsinki*: 219.

Meyer-Leon, L., Senecoff, J.F., Bruckner, R.C. and Cox, M.M. (1984) Site-specific genetic recombination promoted by the *FLP* protein of the yeast 2-micron plasmid *in vitro*. *Cold Spring Harbor Symp. Quant. Biol.* **49**: 797.

Meyer-Leon, L., Gates, C.A., Attwood, J.M., Wood, E.A. and Cox, M.M. (1987) Purification of the *FLP* site-specific recombinase by affinity chromatography and re-examination of the basic properties of the system. *Nucleic Acid. Res.* **15**: 6469.

Murray, A.W. and Szostak, J.W. (1983) Pedigree analysis of plasmid segregation in yeast. *Cell* **34**: 961.

Murray, J.A.H. and Cesareni, G. (1986) Functional analysis of the yeast plasmid partitioning locus *STB*. *EMBO J.* **5**: 3391.

Murray, J.A.H., Scarpa, M., Rossi, N. and Cesareni, G. (1987) Antagonistic controls regulate copy number of the yeast 2-μ plasmid. *EMBO J.* **6**: 4205.

Murray, J.A.H., Cesareni, G. and Argos, P. (1988) Unexpected divergence and molecular coevolution in yeast plasmids. *J. Mol. Biol.* **200**: 601.

Nelson, R.G. and Fangman, W.L. (1979) Nucleosome organisation of the yeast 2-μm plasmid: a eukaryotic minichromosome. *Proc. Natl. Acad. Sci. USA* **76**: 6515.

Newlon, C.S., Devenish, R.J., Suci, A. and Roffis, C.J. (1981) Replication origins used *in vivo* in yeast. *ICN-UCLA Symp. Mol. Cell. Biol.* **22**: 501.

Petes, T.D. and Williamson, D.H. (1975) Replicating circular DNA molecules in yeast. *Cell* **4**: 249.

Prasad, P.V., Young, L-G. and Jayaram, M. (1987) Mutations in the 2-μm circle site-specific recombinase that abolish recombination without affecting substrate recognition. *Proc. Natl. Acad. Sci. USA* **84**: 2189.

Proteau, G., Sidenberg, D. and Sadowski, P.D. (1986) The minimal duplex DNA sequence required for site-specific recombination promoted by the *FLP* protein of yeast *in vitro*. *Nucleic Acids Res.* **14**: 4787.

Reynolds, A.E., Murray, A.W. and Szostak, J.W. (1987) Roles of the 2-μm gene products in stable plasmid maintenance of the 2-μm plasmid of *Saccharomyces cerevisiae*. *Mol. Cell. Biol.* **7**: 3566.

Rownd, R. (1969) Replication of a bacterial episome under relaxed control. *J. Mol. Biol.* **44**: 387.

Royer, H.D. and Hollenberg, C.P. (1976) *Saccharomyces cerevisiae* 2-μm DNA. An analysis of the monomers and its multimers by electron microscopy. *Mol. Gen. Genet.* **150**: 271.

Sadowski, P.D., Lee, D.D., Andrews, B.J., Babineau, D., Beatty, L., Morse, M.J., Proteau, G. and Vetter, D. (1984) *In vitro* systems for genetic recombination of the DNAs of bacteriophage T7 and yeast 2-micron circle. *Cold Spring Harbor Symp. Quant. Biol.* **49**: 789.

Senecoff, J.F., Bruckner, R.C. and Cox, M.M. (1985) The *FLP* recombinase of the yeast 2-μm plasmid: characterisation of its recombination site. *Proc. Natl. Acad. Sci. USA* **82**: 7270.

Sigurdson, D.C., Gaarder, M.E. and Livingston, D.M. (1981) Characterisation of the transmission during cytoductant formation of the 2-μm DNA plasmid from *Saccharomyces*. *Mol. Gen. Genet.* **183**: 59.

Som, T., Armstrong, K.A., Volkert, F.C. and Broach, J.R. (1988) Autoregulation of 2-μm circle gene expression provides a model for maintenance of stable a plasmid copy levels. *Cell* **52**: 27.

Storms, R.K., McNeil, J.B., Khandekar, P.S., An, G., Parker, J. and Friesen, J.D. (1979) Chimeric plasmids for cloning of deoxyribonucleic acid sequences in *Saccharomyces cerevisiae*. *J. Bacteriol.* **140**: 73.

Sutton, A. and Broach, J.R. (1985) Signals for transcription initiation and termination in the *Saccharomyces cerevisiae* plasmid 2-μm circle. *Mol. Cell. Biol.* **5**: 2770.

Taketo, M., Jazwinski, M. and Edelman, G.M. (1980) Association of the 2-μm DNA plasmid with yeast folded chromosomes. *Proc. Natl. Acad. Sci. USA* **77**: 3144.

Toh-e, A. and Utatsu, I. (1985) Physical and functional structure of a yeast plasmid, pSB3, isolated from *Zygosaccharomyces bisporus*. *Nucleic Acids Res.* **13**: 4267.

Toh-e, A., Tada, S. and Oshima, Y. (1982) 2-μm DNA-like plasmids in the osmophilic haploid yeast *Zygosaccharomyces rouxii*. *J. Bacteriol.* **151**: 1380.

Toh-e, A., Araki, H., Utatsu, I. and Oshima, Y. (1984) Plasmids resembling 2-μm DNA in the osmotolerant yeasts *Saccharomyces bailii* and *Saccharomyces bisporus*. *J. Gen. Microbiol.* **130**: 2527.

Tuite, M.F., Dobson, M.J., Roberts, N.A., King, R.M., Burke, D.C., Kingsman, S.M. and Kingsman, A.J. (1982) Regulated high efficiency expression of human interferon-alpha in *Saccharomyces cerevisiae*. *EMBO J.* **1**: 603.

Utatsu, I., Sakamoto, S., Imura, T. and Toh-e, A. (1987) Yeast plasmids resembling 2-μm DNA: regional similarities and diversities at the molecular level. *J. Bacteriol.* **169**: 5537.

Veit, B.E. and Fangman, W.L. (1985) Chromatin organisation of the *Saccharomyces cerevisiae* 2-μm plasmid depends on plasmid-encoded products. *Mol. Cell. Biol.* **5**: 2190.

Vetter, D., Andrews, B.J., Roberts-Beatty, L. and Sadowski, P.D. (1983) Site-specific recombination of yeast 2-μm DNA *in vitro*. *Proc. Natl. Acad. Sci. USA* **80**: 7284.

Volkert, F.C. and Broach, J.R. (1986) Site-specific recombination promotes plasmid amplification in yeast. *Cell* **46**: 541.

Volkert, F.C. and Broach, J.R. (1987) The mechanism of propagation of the yeast 2-μm circle plasmid. In *Biology of Industrial Yeasts*, eds. G.G. Stewart, I. Russell, R.D. Klein and R.R. Hiebsch, CRC Press, Boca Raton: 145.

Volkert, F.C., Wu, L-C.C. Fisher, P.A. and Broach, J.R. (1986) Survival strategies of the yeast plasmid two-micron circle. In *Extrachromosomal Elements in Lower Eukaryotes*, eds. R.B. Wickner, A. Hinnebusch, A.M. Lambowitz, I.C. Gunsalus and A. Hollaender, Plenum, New York, 375.

Walmsley, R.M., Gardner, D.C.J. and Oliver, S.G. (1983) Stability of a cloned gene in yeast grown in chemostat culture. *Mol. Gen. Genet.* **192**: 361.

Wesolowski-Louvel, M., Tanguy, C. and Fukuhara, H. (1988) A nuclear gene required for expression of the linear DNA-associated killer system in the yeast *Kluyveromyces lactis*. *Yeast* **4**: 71.

Williamson, D.H. (1985) The yeast *ARS* element, six years on: a progress report. *Yeast* **1**: 1.

Wright, E.M., Humphreys, G.O. and Yarranton, G.T. (1987) Dual-origin plasmids containing an amplifiable ColEl *ori*; temperature-controlled expression of cloned genes. *Gene* **49**: 311.

Wu, L-C.C., Fisher, P.A. and Broach, J.R. (1987) A yeast plasmid partitioning protein as a karyoskeletal component. *J. Biol. Chem.* **267**: 883.

Yarranton, G.T., Wright, E., Robinson, M.K. and Humphreys, G.O. (1984) Dual-origin plasmid vectors whose origin of replication is controlled by the coliphage lambda promoter P_L. *Gene* **28**: 293.

Zakian, V.A., Brewer, B.J. and Fangman, W.L. (1979) Replication of each copy of the yeast 2-micron DNA plasmid occurs during S phase. *Cell* **17**: 923.

Zealey, G.R., Goodey, A.R., Piggott, J.R., Watson, M.K., Cafferkey, R.C., Doel, S.M., Carter, B.C.A. and Wheals, A.E. (1988) Amplification of plasmid copy number by thymidine kinase expression in *Saccharomyces cerevisiae*. *Mol. Gen. Genet.* **211**: 155.

Zhu, J., Contreras, R., Gheysen, D., Ernst, J. and Fiers, W. (1985) A system for dominant transformation and plasmid amplification in *Saccharomyces cerevisiae*. *Bio/Technology* **3**: 451.

Index

360